CHICAGO PUBLIC LIBRARY
BUSINESS / SCIENCE / TECHNOLOGY
400 S. STATE ST.   60605

HICAGO PUBLIC LIBRARY
USINESS / SCIENCE / TECHNOLOGY
00 S. STATE ST.   60605

Introduction to Control Systems

Dedicated to:

Alexander David who is the Future. (D.K.A.)

My father who led me into Engineering; my teachers who showed
me the way in Control Engineering; and to my children who, in
using this book, will lead us to the promised land. (R.B.Z.)

Introduction to Control Systems

Third edition

D. K. Anand
Department of Mechanical Engineering
University of Maryland, College Park
Maryland, USA

R. B. Zmood
Department of Electrical Engineering
Royal Melbourne Institute of Technology
Melbourne, Victoria, Australia

Butterworth-Heinemann Ltd
Linacre House, Jordan Hill, Oxford OX2 8DP

℞ A member of the Reed Elsevier plc group

OXFORD LONDON BOSTON

MUNICH NEW DELHI SINGAPORE SYDNEY

TOKYO TORONTO WELLINGTON

First published by Pergamon Press 1974
Second edition 1984
Third edition published by Butterworth-Heinemann Ltd 1995

© Butterworth-Heinemann Ltd 1995

All rights reserved. No part of this publication
may be reproduced in any material form (including
photocopying or storing in any medium by electronic
means and whether or not transiently or incidentally
to some other use of this publication) without the
written permission of the copyright holder except
in accordance with the provisions of the Copyright,
Designs and Patents Act 1988 or under the terms of a
licence issued by the Copyright Licensing Agency Ltd,
90 Tottenham Court Road, London, England W1P 9HE.
Applications for the copyright holder's written permission
to reproduce any part of this publication should be addressed
to the publishers

British Library Cataloguing in Publication Data

A catalogue record for this book is available from the British Library

ISBN 0 7506 2298 9

Library of Congress Cataloguing in Publication Data

A catalogue record for this book is available from the Library of Congress

Printed in Great Britain by Hartnolls Limited, Bodmin, Cornwall

R01228 59395

CHICAGO PUBLIC LIBRARY
BUSINESS / SCIENCE / TECHNOLOGY
400 S. STATE ST.   60605

Contents

11 Non Linear Control Systems   **565**
  11.1 Introduction 565
  11.2 The Phase Plane-Method 566
  11.3 Limit Cycles in Non-Linear Control Systems 592
  11.4 Describing Function Technique 599
  11.5 Stability Criteria 614
  11.6 Stability Region for Non-Linear Systems 627
  11.7 Summary 629
  11.8 References 630
  11.9 Problems 632

12 Systems with Stochastic Inputs   **643**
  12.1 Introduction 643
  12.2 Signal Properties 645
  12.3 Input-Output Relationships 650
  12.4 Linear Correlation 655
  12.5 Summary 656
  12.6 References 656
  12.7 Problems 657

13 Adaptive Control Systems   **659**
  13.1 Introduction 659
  13.2 Adaptive Control Methods 660
  13.3 Controller Design Methods 663
  13.4 System Parameter Estimation 669
  13.5 Adaptive Control Algorithms 673
  13.6 Stability of Adaptive Controllers 684
  13.7 An Application Example 685
  13.8 Summary 687
  13.9 References 687
  13.10 Problems 687

A Laplace and $\mathcal{Z}$-Transforms   **689**
  A.1 Laplace Transforms 689
  A.2 $\mathcal{Z}$-Transforms 692
  A.3 References 695

B Symbols, Units and Analogous Systems   **697**
  B.1 Systems of Units 697
  B.2 Symbols and Units 698
  B.3 Comparison of Variables in Analogous Systems 700
  B.4 References 700

About the Authors

Dr D. K. Anand is both a Professor and Chairman of the Department of Mechanical Engineering at the University of Maryland, College Park, Maryland, U.S.A. He is a registered Professional Engineer in Maryland and has consulted widely in Systems Analysis for the U.S. Government and Industry. He has served as Senior Staff at the Applied Physics Laboratory of the John Hopkins University and Director of Mechanical Systems at the National Science Foundation. Dr Anand has published over one hundred and fifty papers in technical journals and conference proceedings and has published two othe books on Introductory Engineering. As well he has a patent on Heat Pipe Control. He is a member of Tau Beta Pi, Pi Tau Sigma, Sigma Xi, and is a Fellow of ASME.

Dr R. B. Zmood is the Control Discipline Leader in the Department of Electrical Engineering at Royal Melbourne Institute of Technology, Melbourne, Victoria, Australia. He has consulted widely both in Australia and in the U.S. on the industrial and military applications of control systems. He has served as a staff member of the Telecom Research Laboratories (formerly A.P.O. Research Laboratories) and the Aeronautical Research Laboratories of the Australian Department of Defence, as well as having worked in industry for a considerable period. Dr Zmood joined RMIT in 1980 and since that time his research interests have centered on the control of magnetic bearings both from a theoretical and application viewpoint and he has published widely in this area. He is a member of IEEE.

Preface

Since the printing of the first two editions, the use of computer software by students has become an important adjunct to the teaching and learning of control systems analysis. With this in mind the entire text has been enlarged and strengthened in the third edition. In addition an attempt has been made to broaden the scope of the book so that it is suitable for mechanical and electrical engineering students as well as for other students of control systems. This revision has been largely carried out by the second author.

The advent of the desk-top computer based computer aided design (CAD) tools has removed the need for repeated hand computations previously required in control system design. While this has forced a fundamental review of the material taught in control courses, it is our contention that many of the analytical and graphical tools, developed during the early days of the discipline are still important for developing an intuitive understanding, or a "mind's eye model", of system design. The computer simply removes the drudgery of applying them.

In reviewing the content of the earlier editions we have sought to arrive at a balance between the material which has pedagogical value and that which has proved useful to the authors in research and industrial practice. This has led to the deletion of some material, and the inclusion of much new material. In addition the order of the material has been altered to assist in the assimilation of important concepts. Class room experience has shown, for example, that when the dominant pole concept is introduced at the same time as the root locus analysis method for feedback systems students identify this idea with the analysis method, rather than accepting it as a separate concept. By presenting it divorced from the root locus method it has been found they more readily accept the generality of the idea.

In the early chapters considerable attention is given to introducing the many methods of mathematical modeling physical systems. To this end the concept of the system S is emphasized and the mathematical models are viewed as approximate but useful descriptions of the system. Their relative utilities depend upon the application in question. While very little motivation for the adoption of these models is given at this time the rapid progress in later chapters to their use in design is felt to satisfy the question of the student. Why all these models? Consistent with our focus on the central role of the system S, the presentation of the various models is carefully developed so as to show their interrelationships.

Apart from discussing steady state and transient performance measures and the sensitivity function, we have introduced unstructured robustness concepts for investigating the effect upon system operation of large changes in its parameters. As the parameters of all practical control systems vary over some non-infinitesimal but defined range the robustness approach has been assuming an ever more important role in system design. Although there is a rich collection of research results on system robustness our treatment of this field is necessarily brief.

It has long been felt by the authors that, while most introductory control system texts dwell on various design techniques such as root locus and other methods at length, they gloss over two of the most important aspects of control system design. These being control strategies and component sizing. While in some instances these are only of minor concern, in many cases they are of utmost importance. Wrong decisions on these matters during the early stages of a project can lead to poor system operation or even failure. In both cases it can be very costly to correct the situation at a later stage after an expensive plant or machine has been constructed. This cost can be measured both in time and money.

The classical design techniques of the root locus and the frequency response methods involve sequentially adjusting the parameters of the assumed controller structures to determine if the performance specification is satisfied. These approaches involve a considerable amount of trial and error, as well as relying on designer inspiration for the selection of the appropriate controller structures. As an alternative approach we present here a state space pole placement design method where the performance specification leads systematically and directly to the controller design by a well defined numerical algorithm. State observers, which are needed to implement these designs, are also introduced, and it is shown how these designs are integrated to complete a total control system design.

The design methodologies discussed in earlier chapters of the book lead to controllers with fixed parameter settings. Adaptive control was developed for systems having large plant parameter changes where the controller settings are adjusted to accommodate these changes and so as to always give the desired performance. In the discussion only the basics of adaptive control are presented. Such important concepts as the certainty equivalence principle, model reference adaptive systems (MRAS), and self tuning regulators (STM) are introduced and applied to a number of examples of adaptive control systems.

The material in this book has been used in a variety of courses over the last twenty years by the authors, both at the University of Maryland, and the Royal Melbourne Institute of Technology (RMIT). At RMIT the material presented has been used as the basis for junior level and senior level courses in electrical engineering, each running over two semesters for $1\frac{1}{2}$ hours per week. At the University of Maryland both authors have covered the equivalent of Chapters 1 to 7 in a one semester course to mechanical engineering students taking their senior year. Other combinations of chapters could be easily be used as a basis for other courses. For example Chapters 1 to 4, 6, 7, 9 and 10 could be used as an introductory course on digital control systems. Apart from its use as an undergraduate text the book is well structured to be read by practicing engineers and applied scientists who need to utilize control techniques in their work.

A hallmark of earlier editions was the use of copious examples to illustrate the various concepts and techniques. This feature has been retained, with the range of problems in each chapter being greatly expanded, both in number and in spread of difficulty. To this end the teacher will find some problems are elementary exercises, some are challenging even to good students, some are open-ended, and some are design-oriented. These latter problems are intended to encourage the student to approach control design problems from a holistic or integrated point of view. As well they illustrate the power of computer analysis for control system design. Cautious selection of problems, suited to the audience who are using the book, will need to be exercised.

In carrying out a task of this magnitude many people, some of them unknowingly, have contributed to its success. First of all there are the many students who have suffered through our trying to get the presentation right. Then there are our colleagues with whom we have discussed the finer points of presentation. Dr. G. Feng of the University of New South Wales deserves special mention for it was he who wrote the first draft of Chapter 13. Also Dr. T. Vinayagalingam of RMIT critically read the complete manuscript and offered many suggestions for improvement of presentation. Mr. T. Bergin has read and critiqued some of the key chapters, while Daniel Zmood, the son of the second author, read many of the sections from a student perspective and made useful suggestions for clarifying the text. Ms R. Luxa painstakingly typed the entire manuscript from the handwritten notes and Mr R. Wang drew many of the figures. To all we express our thanks. Finally to our wives Asha and Devorah, and to our families, who at various times saw us disappear for long hours to write the manuscript we express our gratitude. Their forbearance is much appreciated.

D. K. ANAND
Bethesda, Maryland 1994

R. B. ZMOOD
Melbourne, Australia 1994

Chapter 1

Introduction

1.1 Historical Perspective

The desire to control the forces of nature has been with man since early civilizations. Although many examples of control systems existed in early times, it was not until the mid-eighteenth century that several steam operated control devices appeared. This was the time of the steam engine, and perhaps the most noteworthy invention was the speed control flyball governor invented by James Watt.

Around the beginning of the twentieth century much of the work in control systems was being done in the power generation and the chemical processing industry. Also by this time, the concept of the autopilot for airplanes was being developed.

The period beginning about twenty-five years before World War Two saw rapid advances in electronics and especially in circuit theory, aided by the now classical work of Nyquist in the area of stability theory. The requirements of sophisticated weapon systems, submarines, aircraft and the like gave new impetus to the work in control systems before and after the war. The advent of the analog computer coupled with advances in electronics saw the beginning of the establishment of control systems as a science. By the mid-fifties, the progress in digital computers had given engineers a new tool that greatly enhanced their capability to study large and complex systems. The availability of computers also opened the era of data-logging, computer control, and the state space or modern method of analysis.

The Russian sputnik ushered in the space race which led to large governmental expenditures on the U.S. space program as well as on the devel-

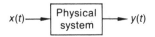

Figure 1.1: A physical system

opment of advanced military hardware. During this time, electronic circuits became miniaturized and large sophisticated systems could be put together very compactly thereby allowing a computational and control advantage coupled with systems of small physical dimensions. We were now capable of designing and flying minicomputers and landing men on the moon. The post sputnik age saw much effort in system optimization and adaptive systems.

Finally, the refinement of the micro chip and related computer developments has created an explosion in computational capability and computer-controlled devices. This has led to many innovative techniques in manufacturing methods, such as computer-aided design and manufacturing, and the possibility of unprecedented increases in industrial productivity via the use of computer-controlled machinery, manipulators and robotics.

Today control systems is a science; but with the art still playing an important role. Much mathematical sophistication has been achieved with considerable interest in the application of advanced mathematical methods to the solution of ever more demanding control system problems. The modern approach, having been established as a science, is being applied not only to traditional engineering control systems, but to newer fields like urban studies, economics, transportation, medicine, energy systems, and a host of fields which are generating similar problems that affect modern man.

1.2   Basic Concepts

Control system analysis is concerned with the study of the behavior of dynamic systems. The analysis relies upon the fundamentals of system theory where the governing differential equations assume a **cause-effect (causal)** relationship. A physical system may be represented as shown in Fig. 1-1, where the excitation or **input** is $x(t)$ and the response or **output** is $y(t)$. A simple control system is shown in Fig. 1-2. Here the output is compared to the input signal, and the difference of these two signals becomes the excitation to the physical system, and we speak of the control

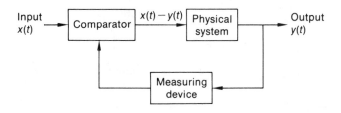

Figure 1.2: A simple control system

system as having **feedback**. The **analysis** of a control system, such as described in Fig. 1-2, involves the determination of $y(t)$ given the input and the characteristics of the system. On the other hand, if the input and output are specified and we wish to design the system characteristics, then this is known as **synthesis**.

A generalized control system is shown in Fig. 1-3. The **reference** or **input variables** $r_1, r_2, \ldots, r_m$ are applied to the **comparator** or **controller**. The **output variables** are $c_1, c_2, \ldots, c_n$. The signals $e_1, e_2, \ldots, e_p$ are **actuating** or **control variables** which are applied by the controller to the **system** or **plant**. The plant is also subjected to **disturbance inputs** $u_1, u_2, \ldots, u_q$. If the output variable is not measured and fed back to the controller, then the total system consisting of the controller and plant is an **open loop system**. If the output is fed back, then the system is a **closed loop system**.

1.3 Systems Description

Because control systems occur so frequently in our lives, their study is quite important. Generally, a control system is composed of several subsystems connected in such a way as to yield the proper cause-effect relationship. Since the various subsystems can be electrical, mechanical, pneumatic, biological, etc., the complete description of the entire system requires the understanding of fundamental relationships in many different disciplines. Fortunately, the similarity in the dynamic behavior of different physical systems makes this task easier and more interesting.

As an example of a control system consider the simplified version of the attitude control of a spacecraft illustrated in Fig. 1-4. We wish the satellite to have some specific attitude relative to an inertial coordinate system. The actual attitude is measured by an attitude sensor on board the satellite. If the desired and actual attitudes are not the same, then

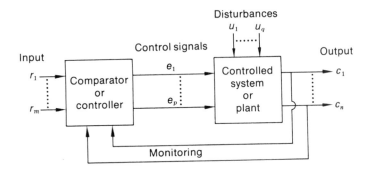

Figure 1.3: A general control system

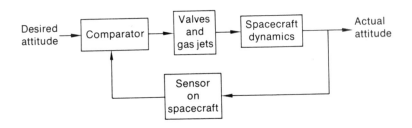

Figure 1.4: Control of satellite attitude

the comparator sends a signal to the valves which open and cause gas jet firings. These jet firings give the necessary corrective signal to the satellite dynamics thereby bringing it under control. A control system represented this way is said to be represented by **block diagrams**. Such a representation helps in the partitioning of a large system into subsystems. This allows each subsystem to be studied individually, and the interactions between the various subsystems to be studied at a later time.

If we have many inputs and outputs that are monitored and controlled, the block diagram appears as illustrated in Fig. 1-5. Systems where several variables are monitored and controlled are called **multivariable** systems. Examples of multivariable systems are found in chemical processing, guidance and control of space vehicles, the national economy, urban housing growth patterns, the postal service, and a host of other social and urban problems.

As another example consider the system shown in Fig. 1-6. The figure

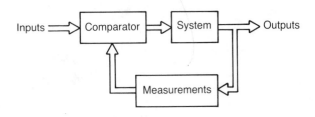

Figure 1.5: Representation of a multivariable system

shows an illustration of the conceptual design of a proposed Sun Tracker. Briefly, it consists of an astronomical telescope mount, two silicon solar cells, an amplifier, a motor, and gears. The solar cells are attached to the polar axis of the telescope so that if the pointing direction is in error, more of the sun's image falls on one cell than the other. This pair of cells, when connected in parallel opposition, appear as a current source and act as a positional error sensing device. A simple differential input transistor amplifier can provide sufficient gain so that the small error signal produces an amplifier output sufficient for running the motor. This motor sets the rotation rate of the polar axis of the telescope mount to match the apparent motion of the sun. This system is depicted in block diagram form in Fig. 1-7. The use of this device is not limited to an astronomical telescope, but can be used for any system where the Sun must be tracked. For example, the output of a photovoltaic array or solar collector can be maximized using a Sun Tracker.

In recent years, the concepts and techniques developed in control system theory have found increasing application in areas such as economic analysis, forecasting and management. An interesting example of a multivariable system applied to a corporation is shown in Fig. 1-8. The inputs of Finance, Engineering and Management when compared to the output which include products, services, profits, etc., yield the excitation variables of available capital, labor, raw materials and technology to the plant. There are two feedback paths, one provided by the company and the other by the marketplace.

The number of control systems that surround us is indeed very large. The essential feature of all these systems is the same. They all have input, control, output, and disturbance variables. They all describe a controller and a plant. They all have some type of a comparator. Finally, in all cases we want to drive the control system to follow a set of preconceived commands.

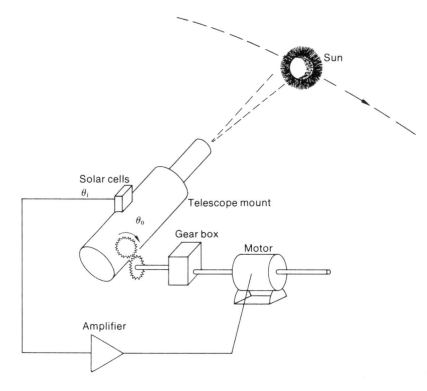

Figure 1.6: Schematic of a Sun Tracker

1.4   Design, Modeling, and Analysis

Prior to the building of a piece of hardware, a system must be designed, modeled and analyzed. Actually the analysis is an important and essential feature of the design process. In general, when we design a control system we do so conceptually. Then we generate a mathematical model which is analyzed. The results of this analysis are compared to the performance specifications that are desired for the proposed system. The accuracy of the results depends upon the quality of the original model of the proposed design. The Sun Tracker proposed in Fig. 1-6 is a conceptual design. We shall show, in Chapter 9, how it is analyzed and then modified so that its performance satisfies the system specifications. The objective then may be considered to be the prediction, prior to construction, of the dynamic behavior that a physical system exhibits, i.e. its natural motion when

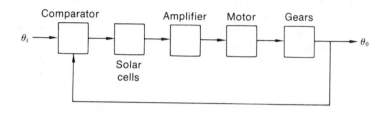

Figure 1.7: Block diagram of the Sun Tracker

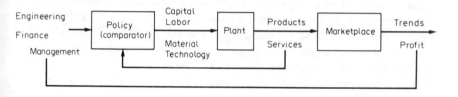

Figure 1.8: Block diagram of a corporation

disturbed from an equilibrium position and its response when excited by external stimuli. Specifically we are concerned with the speed of response or **transient response**, the accuracy or **steady state response**, and the **stability**. By stability we mean that the output remains within certain reasonable limiting values. The relative weight given to any special requirement is dependent upon the specific application. For example, the air conditioning of the interior of a building may be maintained to $\pm 1°C$ and satisfy the occupants. However, the temperature control in certain cryogenic systems requires that the temperature be controlled to within a fraction of a degree. The requirements of speed, accuracy and stability are quite often contradictory and some compromises must be made. For example, increasing the accuracy generally makes for poor transient response. If the damping is decreased, the system oscillations increase and it may take a long time to reach some steady state value.

It is important to remember that all real control systems are nonlinear; however, many can be approximated within a useful, though limited, range as linear systems. Generally, this is an acceptable first approximation. A very important benefit to be derived by assuming linearity is that the superposition theorem applies. If we obtain the response due to two different inputs, then the response due to the combined input is equal to the sum of the individual responses. Another benefit is that operational mathematics can be used in the analysis of linear systems. The operational method allows us to transform ordinary differential equations into algebraic equations which are much simpler to handle.

Traditionally, control systems were represented by higher-order linear differential equations and the techniques of operational mathematics were employed to study these equations. Such an approach is referred to as the **classical method** and is particularly useful for analyzing systems characterized by a single input and a single output. As systems began to become more complex, it became increasingly necessary to use a digital computer. The work on a computer can be advantageously carried out if the system under consideration is represented by a set of first-order differential equations and the analysis is carried out via matrix theory. This is in essence what is referred to as the **state space** or **state variable** approach. This method, although applicable to single input-output systems, finds important applications for multivariable systems. Another very attractive benefit is that it also enables the control system engineer to study variables inside a system.

It is perhaps interesting to note that much of the work in the classical theory of dynamics rests on the state variable viewpoint. In writing the equations of motion of a system using Lagrange's principle, it is necessary to use linearly independent variables or generalized coordinates. The number of these coordinates is equal to the number of degrees of freedom. Hamilton, however, showed that the use of generalized momentum coordinates lead to greatly simplified equations of motion. What this meant was that the state of a second-order system, for example, could be represented by the independent variable and its time derivative. Therefore, the system under consideration is represented by first-order differential equations.

Since this is an introductory course, it is our intention to expose you to both the classical and state space viewpoints. We must note, however, that although the easier route is to initially begin with the classical viewpoint, it is the state approach that is more natural for the more complex and interesting problems. At this level, a thorough study should necessarily include both viewpoints.

Regardless of the approach used in the design and analysis of a control system, we must at least follow the following steps:

1. Postulate a control system and state the system specification to be satisfied.

2. Generate a functional block diagram and obtain a mathematical representation of the system.

3. Analyze the system using any of the analytical or graphical methods applicable to the problem.

4. Check the performance (speed, accuracy, stability, or other criterion) to see if the specifications are met.

5. Finally, optimize the system parameters so that (1) is satisfied.

Whatever the physical system or specific arrangement, we shall see that there are only a few basic concepts and analytical tools that are pivotal to the prediction of system behavior. The fundamental concepts that are learned here and applied to a few examples have therefore a much wider range of applicability. The real range will only be clear when you start working with the ideas to be developed here.

1.5   Text Outline

With the assumption that the student is familiar with Laplace transforms, Chapter 2 introduces mathematical modeling of analogous physical systems. Various systems are represented in operational form as well as by a set of first-order equations. Representation of control systems by classical as well as state space techniques is introduced in Chapter 3. It is seen that in the classical approach a system is represented by its transfer function, whereas in the state approach it is represented by a vector-matrix differential equation. The interrelationship between these representations and the approximation of non-linear systems by linear systems are also examined.

Response in the time domain is discussed using classical methods in Chapter 4. This development relies on operational mathematics, with which prior familiarity is assumed. The state space method of analysis is discussed in Chapter 5. Some fundamentals of matrix theory to support this chapter are given in Appendix C and should be reviewed at this time. Performance and specifications of control systems in the time domain are discussed in Chapter 6. In addition to the discussion of the classical performance measures some of the rudimentary ideas of system robustness are also introduced.

Complementing the time domain analysis, several analytical and graphical procedures for studying system stability are presented in Chapter 7. It is stressed that the utility of these procedures is greatly enhanced if a digital computer is used. A number of computer packages for analyzing control systems are listed in Appendix D.

Up to this point a wide range of control system analysis tools have been introduced. Before we can proceed to the final system design and optimization it is necessary to examine the control strategies to be used and the plant component sizing for the system. These are both studied in Chapter 8. While in many instances the selection of a control system strategy and/or the sizing of the plant components is straightforward, there are many examples of costly mistakes to show that the answers to these questions are often not trivial.

Once the control strategy selection and plant component sizing has been completed and the system performance is obtained, methods for altering it are introduced next in Chapter 9. This chapter includes the Sun Tracker problem we spoke about earlier in this chapter. Here we also show how the state space method of state variable feedback can be used for system pole placement design and for performance optimization. The important concept of a state observer is also introduced.

Whereas the first nine chapters are introductory, the last four are more advanced. Chapters 10 dwells on discrete systems. Here the classical method of analysis is introduced. Attention is given to the digital forms of the conventional PD, PI, and PID controllers.

The effect on system behavior due to nonlinearities is discussed in Chapter 11. A number of techniques which have proved useful for the study of non-linear systems are discussed. These include a modification of the classical method by using describing functions as well as the construction of phase portraits. In this chapter we also introduce Lyapunov's stability criterion. This is a method of ascertaining system stability via energy considerations. Additionally the Popov stability criterion as well as its graphical interpretation is examined. Based upon the ideas of Lyapunov a method for estimating the region of stability of a non-linear system is presented.

The analysis of systems so far has assumed that they are subjected to inputs that are deterministic and Laplace transformable. In Chapter 12 we remove this constraint and consider stochastic inputs. A methodology is developed that allows us to describe system behavior using correlation coefficients.

The limitations of the classical fixed structure control techniques discussed in earlier chapters have long been recognized. In Chapter 13 we introduce the concepts of adaptive control. The two principal approaches of model reference adaptive systems (MRAS) and self tuning regulators (STR) are examined. Apart from presenting useful parameter estimation algorithms for these systems results on their stability of operation are stated.

Chapter 2

Modeling of Physical Systems

2.1 Introduction

Before analyzing a control system it is necessary that we have a **mathematical model** of the system. The analysis of the mathematical model gives us insight into the behavior of the physical system. Naturally, the accuracy of the information obtained depends upon how well the system has been mathematically modeled.

The behavior of a real system is **nonlinear** in nature and often quite difficult to analyze. As a first step we can, however, construct models that are **linear** over a limited but satisfactory range of operating conditions. When this is done, we gain two important advantages. The first is the property of superposition. This means that the system initially at rest responds independently to different inputs applied simultaneously. If $r_1(t)$ and $r_2(t)$ are two inputs applied separately to a system, then the outputs may be represented as

$$r_1(t) \rightarrow x_1(t), \quad \text{and } r_2(t) \rightarrow x_2(t)$$

Now if $r_1(t)$ and $r_2(t)$ are applied together, then the property of superposition allows us to represent the output as

$$[r_1(t) + r_2(t)] \rightarrow [x_1(t) + x_2(t)]$$

The second property of linearity is concerned with proportional response. This implies that if the input is multiplied by a factor K, then the output

11

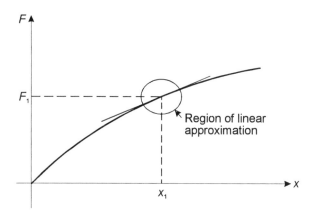

Figure 2.1: A nonlinear spring

is multiplied by the same factor, i.e.

$$[K_1 r_1(t) + K_2 r_2(t)] \rightarrow [K_1 x_1(t) + K_2 x_2(t)]$$

Although many mechanical and electrical systems do indeed behave in a linear fashion over fairly large useful ranges, fluid and thermal systems frequently do not exhibit this behavior. Also, active network elements such as diodes and transistors exhibit nonlinear characteristics. When the elements under investigation are non-linear they may often be linearized about a specified operating point provided the input and output signals of the system are replaced by their **incremental values**. In these circumstances the equations describing the system dynamics may be reduced to **linear differential equations**, and the **Laplace transform** can be applied for finding their solution.To illustrate the procedure of linearization consider the nonlinear spring whose behavior is shown in Fig. 2-1. The force F is given by

$$F = f(x)$$

where $f(x)$ is some nonlinear function of x. Let assume that we wish to use the spring about x_1 so that a Taylor series expansion about this point yields

$$F = f(x) = f(x_1) + (x - x_1)\left(\frac{df(x)}{dx}\right)_{x=x_1} + \cdots$$

If we assume that $(df(x)/dx)_{x=x_1}$ is a good estimate of the slope of the curve in the neighborhood of x_1 and denote this be k, then

$$F \approx f(x_1) + k(x - x_1)$$

or

$$F - F_1 \approx k(x - x_1)$$

which is the necessary linear approximation to the nonlinear element provided we restrict the application of this equation to a small neighborhood about the operating point x_1.

In addition to the property of linearity it sometimes simplifies work if we assume that the signals flowing in the control system are incremental values. This can be best illustrated by an example. Consider a mechanical system to which a torque $T(t)$ is applied,

$$I\frac{d}{dt}\omega(t) = T(t)$$

where I is the moment of inertia and $\omega(t)$ is the angular velocity. If we define $\Omega(s)$ and $T(s)$ as the Laplace transforms of the output signal $\omega(t)$ and the input signal $T(t)$ then the ratio $\Omega(s)/T(s)$ becomes

$$\frac{\Omega(s)}{T(s)} = \frac{1}{Is}$$

This is known as the **transfer function** of the system. Now let us vary the torque by $q(t)$ which causes a variation in the angular velocity by $\theta(t)$, then

$$I\frac{d}{dt}(\omega(t) + \theta(t)) = T(t) + q(t)$$

which may be simplified by taking the Laplace transform and using the previous result, to give

$$\frac{\Theta(s)}{Q(s)} = \frac{1}{Is}$$

We note that although the input signal is $Q(s)$ and the output is $\Theta(s)$, the right-hand side of the equation remains the same as previously. Comparison with the previous case shows that the input and output values have been replaced by their **incremental values**. It is customary in the analysis of control systems to employ incremental values, where there are variations about some nominal operating conditions. This operating condition is sometimes referred to as a **quiescent point** of operation. The approximation of nonlinear systems by linear systems and their representation by transfer functions will be discussed in more detail in Chapter 3.

Mathematical models can be found for a system by the application of one or more fundamental laws peculiar to the physical nature of the system or component. For example, mechanical translational and rotational problems use Newton's law and the d'Alembert principle; electrical circuit problems use Kirchhoff's and Ohm's laws; thermal systems employ the Fourier heat conduction equation and Newton's law; and finally, Darcy's law for flow and the continuity equation can be used to describe hydraulic systems. Regardless of the nature of the system however, the application of any of these laws yields differential equations that have the same basic form. The units and symbols used in these systems appear in Appendix B.

The following sections should be considered as a review since the equations we derive have in most cases been encountered before. Our objective in the remainder of this chapter will be to show how a system can be represented by an ordinary differential equation or set of first-order differential equations. Also, we shall restrict our considerations to systems that can be characterized by linear ordinary differential equations. As examples we shall include some very fundamental components, commonly used in control systems.

2.2   Mechanical Systems

Mechanical systems may be classified into two categories, viz. translational and rotational. Although the method of analysis is the same in both cases, the appearance of gears tends to make rotational systems somewhat more complex.

In the development which follows both free-body diagram and mechanical nodal network methods will be used to demonstrate the establishment of the equations of motion for mechanical systems. Both of these approaches rely on the application of **d'Alembert's principle**, which requires at the kth mechanical node that,

$$\sum_i f_{ki} + f_{mk} = 0 \qquad (2.1)$$

where $f_{mk} = -d(m_k \dot{x}_k)/dt$ is the inertial force on the kth mass as a function of its absolute velocity $\dot{x}_k$, and $\sum f_{ki}$ = sum of all external forces acting on the kth mass. In addition, all mechanical systems must satisfy the **geometric consistency condition**,

$$\left[\sum_k x_k \right]_{\text{loop } \ell} = 0 \qquad (2.2)$$

where the branch relative displacements x_k are summed around loop ℓ. The application of d'Alembert's principle and the geometric consistency condition to simple situations is illustrated in Fig. 2-2.

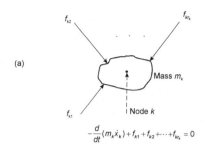

$$-\frac{d}{dt}(m_k\ddot{x}_k) + f_{k1} + f_{k2} + \cdots + f_{kr_k} = 0$$

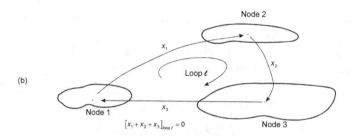

$$[x_1 + x_2 + x_3]_{loop\,\ell} = 0$$

Figure 2.2: Application of d'Alembert's principle

Translation Motion

The translational mode mentioned above refers to motion along a straight path. The physical elements employed to describe translation problems are masses, springs, and dampers. These are schematically shown in Fig. 2-3. Also shown is the relationship between the forces, displacements, and the physical properties of the elements. Mass is the element that stores kinetic energy. If a force $f(t)$ is acting on a body of mass m then $f(t) = m\,d^2x(t)/dt^2$. A spring is the element that stores potential energy. If a force $f(t)$ is applied to a **linear spring** (sometimes called a Hookean spring), then from Hooke's law, $f(t) = kx(t)$. A damper is the element that creates frictional force. In general, the frictional force experienced by a moving body consists of static friction (stiction), coulomb friction, and viscous or linear friction. We shall concern ourselves only with **linear friction**. When a force $f(t)$ is applied to a linear damper then $f(t) = B\,dx(t)/dt$.

Consider the mechanical system shown in Fig. 2-4(a) and its equivalent

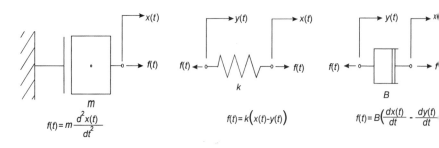

Figure 2.3: Mechanical elements

shown in Fig. 2-4(b). If initially the system is assumed to be under static equilibrium and the mass m is then perturbed, a **free body diagram** showing all the forces is drawn as shown in Fig. 2-4(c). If the mass is displaced by $x(t)$, then using d'Alembert's principle,

$$-(f_s(t) + f_d(t)) - m\frac{d^2x(t)}{dt^2} = 0$$

Substituting for the spring and damper force we have,

$$m\frac{d^2x(t)}{dt^2} + B\frac{dx(t)}{dt} + kx(t) = 0$$

Now if we also apply an input force $f(t)$, then

$$m\frac{d^2x(t)}{dt^2} + B\frac{dx(t)}{dt} + kx(t) = f(t) \qquad (2.3)$$

We now define

$$x_1(t) = x(t), x_2(t) = \frac{dx(t)}{dt}, u(t) = f(t)$$

Substituting these quantities into Eq. (2.3) yields two first-order differential equations,

$$\frac{dx_1(t)}{dt} = x_2(t)$$

$$\frac{dx_2(t)}{dt} = -\frac{B}{m}x_2(t) - \frac{k}{m}x_1(t) + \frac{1}{m}u(t)$$

(2.4)

which completely describes our system. These equations are known as the **state equations** for the spring-mass system shown in Fig. 2-4.

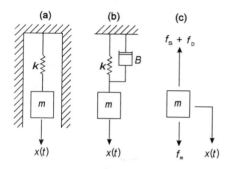

Figure 2.4: Schematic of a mechanical system

Defining the symbolic operators p and $1/p$ by

$$px = \frac{dx}{dt}, \quad p^2x = \frac{d^2x}{dt^2}, \cdots, \text{ etc}$$

and

$$\frac{1}{p}x = \int_{t_0}^{t} x(t)dt + x_0$$

enables us to write the above differential equations in simplified operator notation. Thus Eq.(2.3) becomes

$$m\,p^2x(t) + Bpx(t) + kx(t) = f(t) \tag{2.5}$$

The transfer function representation, which will be discussed in detail in Chapter 3, is obtained by taking the Laplace transform of Eq. (2.3), assuming zero initial conditions, and rearranging it to give

$$\frac{X(s)}{F(s)} = \frac{1}{ms^2 + Bs + k} \tag{2.6}$$

Notice that we obtained only one second-order differential equation for the system shown in Fig. 2-4. This is so because the system has only one **degree of freedom**. The number of degrees of freedom is equal to the number of masses in motion of this type. If there were two masses, the system would have two degrees of freedom and we would require two coordinates to describe the position of the masses.

Example 1 *A mass is vibrating as shown in Fig. 2-5. The motion of the mass is described by $x(t)$ whereas the other end of the spring is displaced by $y(t)$. Obtain an expression for $X(s)/Y(s)$.*

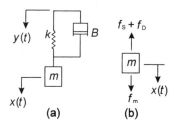

Figure 2.5: Spring mass system

The spring and damper are displaced by $y(t)$ at one end and by $x(t)$ at the other. From the free body diagram shown in Fig. 2-5 and assuming $x(t) > y(t)$ we obtain expressions for the spring force $f_s(t)$ and damping force $f_d(t)$ as

$$f_s(t) = k(x(t) - y(t)), \quad f_d(t) = Bp(x(t) - y(t))$$

The spring and damping force balance the inertia force,

$$-mp^2 x(t) - k(x(t) - y(t)) - Bp(x(t) - y(t)) = 0$$

which with some rearrangement becomes,

$$mp^2 x(t) + Bpx(t) + kx(t) = Bpy(t) + ky(t)$$

Laplace transforming and rearranging yields the transfer function

$$\frac{X(s)}{Y(s)} = \frac{Bs + k}{ms^2 + Bs + k}$$

Example 2 *Obtain the transfer functions for the double mass problem shown in Fig. 2-6.*

Assuming $x > y > z > 0$, $py(t) > 0$, and $pz(t) > 0$, we note that the two springs are in tension. Summing the forces on each mass,

$$\Sigma f_1 = m_1 p^2 y(t)$$

$$= k_1[x(t) - y(t)] - k_2[y(t) - z(t)] - B_1 py(t)$$

$$\Sigma f_2 = m_2 p^2 z(t) = k_2[y(t) - z(t)] - B_2 pz(t)$$

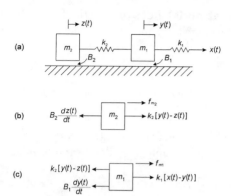

Figure 2.6: A two degree of freedom mechanical system

Rearranging these equations we obtain

$$m_1 p^2 y(t) + B_1 p y(t) + (k_1 + k_2) y(t) - k_2 z(t) \quad = \quad k_1 x(t)$$

$$m_2 p^2 z(t) + B_2 p z(t) + k_2 z(t) - k_2 y(t) \quad = \quad 0$$

The Laplace transform for zero initial conditions yields,

$$\left[m_1 s^2 + B_1 s + (k_1 + k_2) \right] Y(s) - k_2 Z(s) \quad = \quad k_1 X(s)$$

$$-k_2 Y(s) + \left[m_2 s^2 + B_2 s + k_2 \right] Z(s) \quad = \quad 0$$

Since there are two outputs, one for each mass, we expect to find two transfer functions. Solving the above equations simultaneously we obtain the two transfer functions,

$$\frac{Y(s)}{X(s)} = \frac{k_1 [m_2 s^2 + B_2 s + k_2]}{\Delta}$$

and

$$\frac{Z(s)}{X(s)} = \frac{k_1 k_2}{\Delta}$$

where

$$\Delta = \left[m_1 s^2 + B_1 s + (k_1 + k_2) \right] \left[m_2 s^2 + B_2 s + k_2 \right] - k_2^2$$

The **mechanical nodal network** method can also be used for analyzing mechanical systems. Symbolic representations of mass, spring, and damping elements are shown in Fig. 2-3. The procedure used is illustrated in Example 3 below.

Example 3 *Obtain the governing state equations for the double mass problem shown in Fig. 2-6.*

The mechanical nodal network for this system is shown in Fig. 2-7. Applying d'Alembert's principle at nodes Y and Z in turn gives,

$$-m_1 p^2 y(t) - B_1 p y(t) - k_1[y(t) - x(t)] - k_2[y(t) - z(t)] = 0$$

and

$$-m_2 p^2 z(t) - B_2 p z(t) - k_2[z(t) - y(t)] = 0$$

Rearranging these equations easily shows they are the same as those given in Example 2, and as a consequence the transfer functions are also the same.

We introduce two additional variables $v_y(t)$ and $v_z(t)$, defined by

$$\dot{y}(t) = v_y(t), \quad \text{and} \quad \dot{z}(t) = v_z(t)$$

Substituting for $\dot{y}(t)$ and $\dot{z}(t)$ in the above equations gives the four state equations for the double mass system shown in Fig. 2-7, as

$$\dot{y}(t) = v_y(t)$$

$$\dot{z}(t) = v_z(t)$$

$$\dot{v}_y(t) = -\frac{B_1}{m_1}v_y(t) - \frac{k_1}{m_1}[y(t) - x(t)] - \frac{k_2}{m_1}[y(t) - z(t)]$$

$$\dot{v}_z(t) = -\frac{B_2}{m_2}v_z(t) - \frac{k_2}{m_2}[z(t) - y(t)]$$

Rotational Motion

When the motion of the mechanical system is rotational, then Newton's law states that the sum of the torques is equal to the change of angular momentum. Since in most applications inertia is constant, this is identical to the statement that $T(t) = I\alpha(t)$ where $T(t)$ is torque, I is the moment of inertia and $\alpha(t)$ is the angular acceleration. The physical elements employed

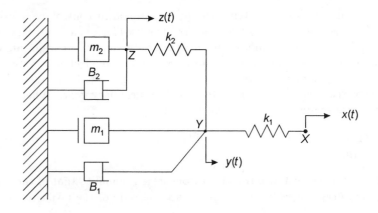

Figure 2.7: Nodal network for 2-degree of freedom mechanical system

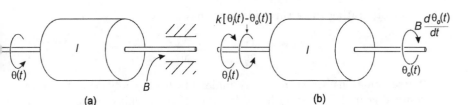

Figure 2.8: Single degree of freedom rotational system

to describe rotational problems are inertia, torsional spring, and torsional damper. The interpretation of these quantities is identical to those defined for translational motion.

Consider the single degree of freedom (one shaft position completely defines the position of the rotating element) rotational system shown in Fig. 2-8(a). The free body diagram is shown in Fig. 2-8(b) where $\theta_i > \theta_o$. The spring torque acts in the positive direction and the damping torque acts in the negative direction. Summing the torques we have

$$\Sigma T = Ip^2\theta_o(t) = k(\theta_i(t) - \theta_o(t)) - Bp\theta_o(t) \tag{2.7}$$

Rearranging, and taking the Laplace transform for zero initial conditions, we obtain the transfer function,

$$\frac{\Theta_o(s)}{\Theta_i(s)} = \frac{k}{Is^2 + Bs + k} \tag{2.8}$$

Rotational mechanical systems are quite common in control systems and are often used in conjunction with gears. The use of gears in control systems helps attain torque magnification and speed reduction. Gear trains are used as matching devices just like transformers are used in electrical systems. A gear train system is shown in Fig. 2-9. The gear with N_1 teeth is the primary gear and that with N_2 teeth is the secondary gear. When two or more gears are in contact we observe that:

1. The work done by one gear is equal to that of the other gear, i.e. $T_1\theta_1 = T_2\theta_2$.

2. The linear distance traveled by one gear is equal to that of the other, i.e. $\theta_1 r_1 = \theta_2 r_2$ where r_1 and r_2 are the radii of the respective gears.

3. The number of teeth on the surface of a gear is proportional to the radius of the gear, i.e. $r_1/r_2 = N_1/N_2$.

Incorporating these three ideas, we have

$$\frac{T_1}{T_2} = \frac{\theta_2}{\theta_1} = \frac{N_1}{N_2} \tag{2.9}$$

These equations are correct only under idealized conditions. In practice, coupled gears have backlash and friction. If the backlash is large, then the control system can become unstable as discussed in Example 18 of Chapter 11. For our purposes, we shall consider idealized conditions. Referring now to Fig. 2-9, the rotational equation for the secondary side, when the inertia of shaft and gears is neglected, becomes

$$T_2(t) = I_2 p^2 \theta_2(t) + B_2 p \theta_2(t)$$

where $T_2(t)$ is the torque developed by the secondary gear. We also know from Eq. (2-9) that

$$\frac{T_1(t)}{T_2(t)} = \frac{p^2\theta_2(t)}{p^2\theta_1(t)} = \frac{p\theta_2(t)}{p\theta_1(t)} = \frac{N_1}{N_2}$$

Substituting this result in the previous equation we obtain

$$T_1(t) = I_2 \left[\frac{N_1}{N_2}\right]^2 p^2\theta_1(t) + B_2 \left[\frac{N_1}{N_2}\right]^2 p\theta_1(t)$$

If we define

$$I = I_2 \left[\frac{N_1}{N_2}\right]^2, \text{ and } B = B_2 \left[\frac{N_1}{N_2}\right]^2$$

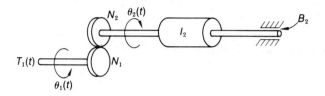

Figure 2.9: A simple gear train

then the governing equation becomes

$$T_1(t) = Ip^2\theta_1(t) + Bp\theta_1(t) \tag{2.10}$$

We have essentially reflected the inertia and damping from the secondary side to the primary side. This procedure may be extended to systems containing several gear trains as shown in the next example.

Example 4 *For the multiple-gear system shown in Fig. 2-10, obtain (a) the system transfer function and (b) set of first-order differential equations.*

(a) The inertia I_3 and damping B_3 may be reflected to shaft *2* and when added to I_2 and B_2 yields,

$$I_{2e} = I_2 + I_3 \left[\frac{N_3}{N_4}\right]^2$$

$$B_{2e} = B_2 + B_3 \left[\frac{N_3}{N_4}\right]^2$$

where I_{2e} and B_{2e} are the equivalent inertia and damping about shaft *2*. This inertia and damping are reflected in turn to shaft *1* and

$$I_{1e} = I_1 + I_{2e} \left[\frac{N_1}{N_2}\right]^2$$

$$B_{1e} = B_1 + B_{2e} \left[\frac{N_1}{N_2}\right]^2$$

Finally, the rotational equation for this system becomes

$$T_1(t) = I_{1e}\, p^2\theta_1(t) + B_{1e}\, p\theta_1(t) \tag{2.11}$$

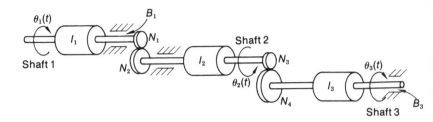

Figure 2.10: A multiple gear train

Taking the Laplace transform for zero initial conditions, the transfer function becomes

$$\frac{\Theta_1(s)}{T_1(s)} = \frac{1}{s(I_{1e}s + B_{1e})}$$

(b) In order to represent the system by a set of first-order differential equations we define $x_1(t) = \theta_1(t), x_2(t) = \dot{\theta}_1(t)$, and $u(t) = T_1(t)$, and substitute them into Eq. (2-15) to obtain

$$\dot{x}_1(t) = x_2(t)$$

$$\dot{x}_2(t) = -\frac{B_{1e}}{I_{1e}}x_2(t) + \frac{1}{I_{1e}}u(t)$$

which are the required equations governing the behavior of the rotational system.

2.3   Electrical Systems

Equations representing the behavior of electrical systems can be obtained by the application of Kirchoff's laws. The first states: the sum of all voltage drops around a closed loop in a circuit is zero,

$$\left[\sum_k e_k\right]_{\text{loop } \ell} = 0 \qquad (2.12)$$

where e_k is the kth voltage drop in loop ℓ. The second law states: the sum of the currents at the nth electrical node in a circuit is zero,

$$\left[\sum_m i_m\right]_{\text{node } n} = 0 \qquad (2.13)$$

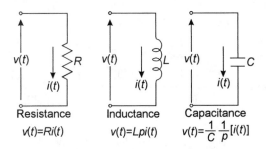

Figure 2.11: Electrical elements

where i_m is the mth current flowing into node n. The method employing the first law is referred to as the **loop method** while that employing the second is referred to as the **node method**. Either of the two methods are used, in conjunction with the **constitutive equations** that describe the physical nature of each component in a system, to derive the governing equations.

Elements used in electrical networks are shown in Fig. 2-11. If we apply a voltage $v(t)$ across a resistance, then from Ohm's law $v(t) = Ri(t)$. Similarly, if a voltage is applied across an inductor or capacitor, we have $v(t) = Lpi(t)$ and $i(t) = Cpv(t)$ from Faraday's and Coulomb's laws respectively. Resistance is an energy dissipating element, inductance stores kinetic energy, while capacitance stores potential energy.

Consider the network shown in Fig. 2-12(a). For deriving the necessary differential equation using the loop method, we observe that there are two loops. Summing the voltage drops across R, L, and C, we have the following two equations corresponding to the two loops,

$$R_1 i_1(t) + L \, p i_1(t) + \frac{1}{C}\frac{1}{p}(i_1(t) - i_2(t)) \;=\; 0$$

$$\frac{1}{C}\frac{1}{p}(i_2(t) - i_1(t)) + R_2 i_2(t) \;=\; v_2(t)$$

(2.14)

Taking the Laplace transform, assuming zero initial conditions, and re-

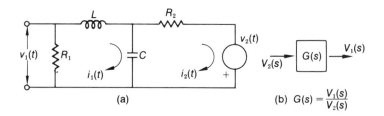

Figure 2.12: An electrical network

arranging, we obtain

$$I_1(s)\left[R_1 + Ls + \frac{1}{Cs}\right] - I_2(s)\left[\frac{1}{Cs}\right] = 0$$

$$-I_1(s)\left[\frac{1}{Cs}\right] + I_2(s)\left[\frac{1}{Cs} + R_2\right] = V_2(s)$$

Solving these simultaneous equations and noting that $V_1(s) = R_1 I_1(s)$, we obtain

$$\frac{V_1(s)}{V_2(s)} = \frac{R_1}{LCR_2s^2 + s(R_1R_2C + L) + (R_1 + R_2)} \tag{2.15}$$

where $V_1(s)/V_2(s)$ is the transfer function for the network. This is sometimes denoted by $G(s)$ as shown in Fig. 2-12(b).

As an alternate mathematical representation, we define

$$x_1(t) = i_1(t), \quad \text{and} \quad x_2(t) = \frac{1}{p}(i_1(t) - i_2(t))$$

Substituting these into Eqs. (2-14) gives

$$R_1x_1(t) + Lpx_1(t) + \frac{1}{C}x_2(t) = 0$$

$$-\frac{1}{C}x_2(t) + R_2i_2(t) = v_2(t) \tag{2.16}$$

The second equation can be rewritten as

$$-\frac{1}{C}x_2(t) + R_2(-px_2(t) + x_1(t)) = v_2(t)$$

Rearranging these equations we obtain two first-order differential equations describing the network,

$$
\begin{aligned}
\dot{x}_1(t) &= -\frac{R_1}{L}x_1(t) - \frac{x_2(t)}{LC} \\[2mm]
\dot{x}_2(t) &= x_1(t) - \frac{x_2(t)}{CR_2} - \frac{v_2(t)}{R_2}
\end{aligned}
\qquad (2.17)
$$

It is perhaps proper to note here that:

1. State variables can always be chosen to be currents through inductances and voltages across capacitances, provided they have independent connections. This is also true of energy storage variables in other systems.

2. The order of the system is equal to the order of the characteristic equation. This is the same as the minimum number of state variables needed for the systems that we shall consider.

We shall formalize the procedure of obtaining state equations from a transfer function in the next chapter. Here we are content to show how a system is represented by a set of first-order differential equations.

The nodal and loop methods of analysis may be generalized to include active networks. The study of active networks is very important since most feedback control systems include some active elements. Although such elements are not completely linear, they may be treated as such over a limited range of operation as mentioned earlier. Consider the transistor, used in the common-emitter connection, for a single stage amplifier shown in Fig. 2-13(a). The equivalent circuit is shown in Fig. 2-13(b). Employing the loop equations we obtain,

$$
(R_g + r_b + r_e)i_1(t) - r_e i_2(t) = v_1(t)
$$

$$
-r_e i_1(t) + [r_e + r_c(1 - \alpha) + R_L]i_2(t) = -r_m i_b(t)
$$

Here r_e is the emitter resistance, r_c the collector resistance, r_b the base resistance, and r_m the mutual resistance. Since the output voltage $v_0(t) = R_L i_2(t)$ and $i_b(t) = i_1(t)$ we have

$$
(R_g + r_b + r_e)i_1(t) - \left(\frac{r_e}{R_L}\right)v_0(t) = v_1(t)
$$

$$
-(r_e - r_m)i_1(t) + \frac{(r_e + r_c(1 - \alpha) + R_L)}{R_L}v_0(t) = 0
$$

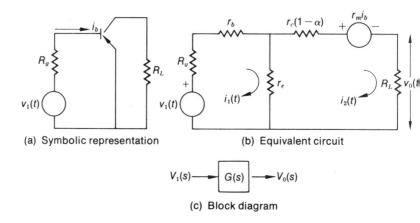

(a) Symbolic representation   (b) Equivalent circuit

$$V_1(s) \longrightarrow \boxed{G(s)} \longrightarrow V_0(s)$$

(c) Block diagram

Figure 2.13: Single stage amplifier

The transfer function $V_0(s)/V_1(s)$ may now be obtained. This is often referred to as the **amplifier gain** and is seen to be constant.

Transformers are often needed as coupling or matching devices in electrical networks. When a transformer is present in an electrical network, then any current change in the primary side causes an induced voltage in the secondary side and vice versa. This phenomenon is referred to as **mutual inductance**. Consider the circuit shown in Fig. 2-14(a), where the dots indicate that the mutual inductance is negative[1]. Using the loop method, we have

$$R_1 i_1(t) + L_1 p i_1(t) + \frac{1}{C}\frac{1}{p} i_1(t) - M p i_2(t) = v_1(t)$$

$$(2.18)$$

$$R_2 i_2(t) + (L_2 + L_3) p i_2(t) - M p i_1(t) = 0$$

The transfer function may now be obtained by Laplace transforming and solving the equations simultaneously as we did before. In order to represent the system by state equations, we can define $x_1(t) = \frac{1}{p} i_1(t), x_2(t) = i_1(t)$, and $x_3(t) = i_2(t)$ and obtain a set of first-order differential equations. We leave this as an exercise for the reader.

---

[1] If i_1 and i_2 both flow in or out of the dots then the fluxes aid and the mutual inductance treated as a voltage drop has the same sign as self inductance, *i.e.* positive. If i_1 flows into the dot whereas i_2 flows out of the dot, then the mutual inductance is negative.

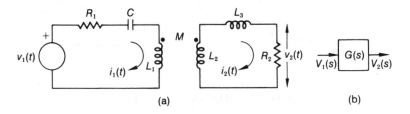

Figure 2.14: An electric circuit with a transformer

Example 5 *Obtain two first-order differential equations for the RLC electrical network shown in Fig. 2-15. Also find the transfer function $I(s)/E(s)$.*

Using Kirchoff's voltage law and noticing that the same current flows through each element, we have

$$Ri(t) + Lpi(t) + \frac{1}{C}\frac{1}{p}i(t) = e(t)$$

Next we define two variables, $x_1(t) = \frac{1}{p}i(t)$, and $x_2(t) = i(t)$ and write two equations as

$$\dot{x}_1(t) = x_2(t)$$

$$\dot{x}_2(t) = -\frac{1}{LC}x_1(t) - \frac{R}{L}x_2(t) + \frac{1}{L}e(t)$$

These two first-order equations then describe the electrical network. For obtaining the transfer function we return to the original equation and take its Laplace transform assuming zero initial conditions,

$$RI(s) + LsI(s) + \frac{1}{Cs}I(s) = E(s)$$

Collecting terms and solving gives the required transfer function,

$$\frac{I(s)}{E(s)} = \frac{Cs}{LCs^2 + RCs + 1}$$

While powerful computer based methods have been developed for finding state equations from circuit schematics a useful technique for hand computation is the method described by Kuh and Rohrer. This will be illustrated by an example.

30 CHAPTER 2. MODELING OF PHYSICAL SYSTEMS

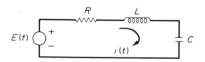

Figure 2.15: RLC electric circuit

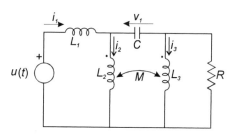

Figure 2.16: An electrical network with mutual inductance

Example 6 *For the circuit shown in Fig. 2-16 obtain the system state equations.*

As discussed above the first step is to identify the currents through inductors and voltages across capacitors as the state variables. Such a selection is shown in the above figure. The second step is to replace all inductors and all capacitors in the schematic diagram by current and voltage sources respectively. This modified schematic is shown in Fig. 2-17. The constitutive equations for the element are

$$v_L = L_1 p i_1$$

$$v_L = L_2 p i_2 + M p i_3$$

$$v_L = L_3 p i_3 + M p i_2$$

$$i_C = C p v_1$$

where M is the mutual inductance between coils L_1 and L_2. In addition we can write the mesh and nodal equations for the modified network using the Kirchoff voltage and current laws. The choice of meshes and nodes is not unique but the total number of loop and node equations must equal the

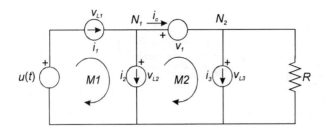

Figure 2.17: Modified schematic for circuit shown in Fig. 2.16

sum of the number of current and voltage sources replacing inductors and capacitors. For the example we have chosen nodes N_1 and N_2, and meshes M_1 and M_2. Thus the equations for nodes N_1 and N_2, and meshes M_1 and M_2 become

$$i_1 - i_2 - i_C = 0$$

$$i_C - i_3 - \frac{v_{L_3}}{R} = 0$$

$$u - v_{L_1} - v_{L_2} = 0$$

$$v_{L_2} - v_1 - v_{L_3} = 0$$

Substitution of the constitutive relationships in the above equations gives the state equations, after some algebraic manipulation,

$$pv_1(t) = \frac{1}{C}(i_1(t) - i_2(t))$$

$$pi_1(t) = \frac{1}{L}[u(t) - v_1(t) - R(i_1(t) - i_2(t) - i_3(t))]$$

$$pi_2(t) = \frac{L_3}{L_2L_3 - M^2}v_1(t) + \frac{L_3 - M}{L_2L_3 - M^2}R[i_1(t) - i_2(t) - i_3(t)]$$

$$pi_3(t) = \frac{L_2 - M}{L_2L_3 - M^2}R[i_1(t) - i_2(t) - i_3(t)] - \frac{M}{L_2L_3 - M^2}v_1(t)$$

2.4 Electromechanical Systems

Apart from the purely mechanical and purely electrical systems already discussed above there is another class where electromechanical coupling

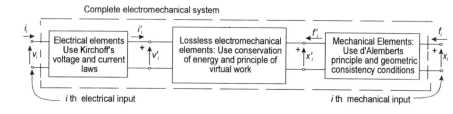

Figure 2.18: Partitioning of electromechanical system

occurs and these will now be examined. In these systems both electrical and mechanical energy flows both into and out of the system as well as interacting within the system. Two approaches may be used in analyzing these systems: one based on energy, and the other based upon the electromagnetic force laws. In the following we will restrict our consideration to magneto-mechanical interactions.

In analyzing these systems we follow the procedure of separating out all purely mechanical and purely electrical elements including dissipative elements so leaving only conservative electromechanical elements to be analyzed. This partitioning procedure is illustrated in Fig. 2-18. Since the analysis of purely mechanical and purely electrical systems has already been discussed in this section we will concentrate on the analysis of lossless electromechanical systems and then show how the results may be applied to model physical systems.

We start by examining the application of energy methods to determine the force in the electromagnetic actuator shown in Fig. 2-19. The magnetic flux linkage λ is defined by Faraday's law of induction

$$v = \frac{d\lambda}{dt} \tag{2.19}$$

It can be shown that the flux linkages are related to the current i and the physical position of the armature x, and is given by

$$\lambda = \lambda(i, x) \tag{2.20}$$

or

$$i = i(\lambda, x) \tag{2.21}$$

In general this relationship is non-linear, although often, if the magnetic components of the actuator are unsaturated then it can be approximated by a linear function. The constitutive relationship for the flux linkage λ is shown in Fig. 2-20.

Suppose the system is assembled to its final mechanical form, and we then increase the flux linkage from zero to λ. The instantaneous electrical power input P_m is

$$P_m(t) = i(t)v(t) = i(t)\frac{d\lambda}{dt}$$

Consequently the **stored energy** in the magnetic field W_m is

$$W_m(\lambda, x) = \int_0^t P_m(t)dt = \int_0^\lambda i(\lambda, x)d\lambda \tag{2.22}$$

It will be noted from Fig. 2-20 that this equation can interpreted graphically as the area above the curve for the armature position x_3. It is convenient to define the **co-energy** $W_m^c(i, x)$ as

$$W_m^c(i, x) = \int_0^i \lambda(i, x)di \tag{2.23}$$

and it will be noted from the figure that

$$W_m(\lambda, x) + W_m^c(i, x) = i\,\lambda(i) \tag{2.24}$$

From the conservation of energy and the principle of virtual work we have over an infinitesimal period dt,

Increase of stored magnetic energy = inflow of electric energy
   + inflow of mechanical energy

$$dW_m = i(t)v(t)dt + f_{ext}(t)\dot{x}(t)dt = i(\lambda, x)d\lambda + f_{ext}(x)dx \tag{2.25}$$

Assuming the coil current $i(t)$ is an independently controlled variable we note that

$$W_m = W_m(i, x)$$

Thus

$$dW_m = \frac{\partial W_m}{\partial i}di + \frac{\partial W_m}{\partial x}dx \tag{2.26}$$

Also from Eq. (2-20)

$$d\lambda = \frac{\partial \lambda}{\partial i}di + \frac{\partial \lambda}{\partial x}dx \tag{2.27}$$

Substituting Eqs. (2-26) and (2-27) into Eq. (2-25) gives

$$f_{ext}\,dx = \left[\frac{\partial W_m}{\partial i} - i\frac{\partial \lambda}{\partial i}\right]di + \left[\frac{\partial W_m}{\partial x} - i\frac{\partial \lambda}{\partial x}\right]dx \tag{2.28}$$

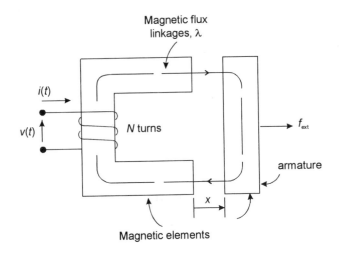

Figure 2.19: Electromagnetic actuator

From Eq. (2-24)

$$\frac{\partial W_m}{\partial i} = i\frac{\partial \lambda}{\partial i} + \lambda - \frac{\partial W_m^c}{\partial i} = i\frac{\partial \lambda}{\partial i}$$

and substituting for $\partial W_m/\partial i$ in Eq. (2-28) gives

$$f_{ext}\ dx = \left[\frac{\partial W_m}{\partial x} - i\frac{\partial \lambda}{\partial x}\right] dx$$

Since dx is an arbitrary differential we have

$$f_{ext} = \frac{\partial}{\partial x}(W_m - i\lambda)$$

and from Eq. (2-24)

$$f_{ext} = -\frac{\partial W_m^c(i, x)}{\partial x} \tag{2.29}$$

Example 7 *In Fig. 2-19 is shown the simplified construction of an elec-tromagnetic actuator for a MAGLEV railroad train. Find a relationship for the magnetic force as a function of the armature current i, and the armature airgap x.*

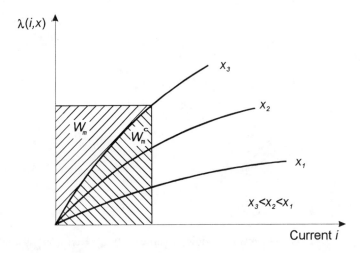

Figure 2.20: Flux linkages as a funtion of armature current and air gap

If we assume the magnetic elements are unsaturated then the curves shown in Fig. 2-20 can be approximated by the equation

$$\lambda = L(x)i$$

In this case

$$W_m^c(i, x) = \int_0^i L(x)i \; di = \frac{1}{2}L(x)i^2$$

Thus the magnetic force

$$f_{ext} = -\frac{\partial W_m^c}{\partial x} = -\frac{1}{2}i^2\frac{dL}{dx}$$

To complete the analysis we find the relationships for the electrical elements using Faraday's law, Ohm's law, and Kirchoff's voltage law. Thus

$$v(t) = i(t)R + p\left[L(x(t))i(t)\right] = i(t)R + L(x(t))pi(t) + i(t)\frac{dL}{dx}px(t)$$

It will be noted that both of the above equations are non-linear functions of the armature position and the coil current.

Direct Current Control Motor

The d.c. motor, which is an important element in many control systems, has a speed-torque characteristic which depends upon the field excitation. We apply energy methods for evaluating the performance equations.

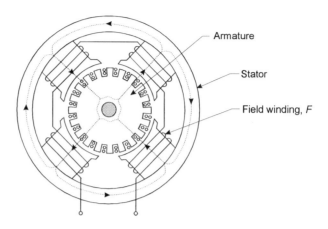

Figure 2.21: Cross sectional view of separately excited d.c. motor

Fig. 2-21 shows the cross-sectional view of a separately excited d.c. motor. The armature core is constructed of steel laminations stacked on a steel shaft, and slots are provided for accommodating the armature conductors which conduct the armature current. There is a constant magnetic field represented by the broken lines in the figure which shows the paths of the magnetic flux. This field is either generated by the field winding F, or by permanent magnets in the case of modern servo-motors. The field flux is shown passing across the airgap and into the armature core, and is thus cut by the armature conductors when the armature rotates.

Denoting the useful flux per pole by Φ webers, the number of pole-pairs by P, and the armature angular displacement by θ radian it can be shown that the flux linkages

$$\lambda(\theta) = \frac{PZ\Phi}{\pi c}\theta$$

In this equation Z is the total number of armature conductors, and c is the number of parallel paths between the positive and negative brushes. From Eq. (2-23)

$$W_m^c(i_a, \theta) = \int_0^{i_a} \frac{PZ\Phi}{\pi c}\theta di = \frac{PZ\Phi}{\pi c}\theta i_a$$

Hence from Eq. (2-29) the external torque applied to the armature is

$$T_{ext} = -\left[\frac{PZ\Phi}{\pi c}\right] i_a = -K_T i_a \qquad (2.30)$$

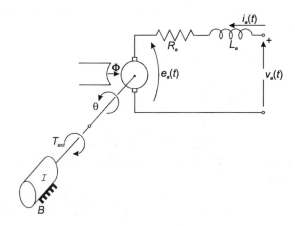

Figure 2.22: Equivalent circuit for separately excited d.c. motor

From Faraday's law the armature back e.m.f. e_a

$$e_a(t) = p \left[\frac{PZ\Phi}{\pi c} \theta \right] = \left[\frac{PZ\Phi}{\pi c} \right] \omega_r = K_B \omega_r \qquad (2.31)$$

where ω_r is the angular velocity of the armature. To complete the analysis we need to include the circuit elements due to the armature resistance R_a and inductance L_a, using Kirchoff's voltage law as shown in Fig. 2-22. The equation for the armature circuit is

$$v_a(t) = i_a(t)R_a + L_a p i_a(t) + e_a(t) \qquad (2.32)$$

The motor back e.m.f. constant K_B and torque constant K_T defined above can be seen to be equal if SI units are used. In practice the measured values of these constants will differ slightly because of frictional and brush to commutator volt-drop effects.

Example 8 *Find the transfer function for the separately excited armature controlled d.c. motor connected to an inertial load I and viscous load B shown in Fig. 2-22.*

From this figure and Eq. (2-30) we find

$$I p^2 \theta(t) + B p \theta(t) = -T_{ext}(t) = K_T i_a(t) \qquad (2.33)$$

Assuming zero initial conditions, we take the Laplace transform of Eqs. (2-31) to (2-33), and then after some algebraic manipulation obtain the transfer function

$$\frac{\Theta(s)}{V_a(s)} = \frac{K_T}{s[IL_a s^2 + (IR_a + L_a B)s + (BR_a + K_T K_B)]} \tag{2.34}$$

This equation can be simplified by introducing the electrical and mechanical time constants, τ_e and τ_m respectively, where we define

$$\tau_e = \frac{L_a}{R_a}, \quad \tau_m = \frac{IR_a}{R_a B + K_T K_B}$$

Assuming that $\tau_e \ll \tau_m$, which is often the case in practice, then it is easily shown that

$$\frac{\Theta(s)}{V_a(s)} \approx \frac{K_T/(K_T K_B + BR_a)}{s(\tau_m s + 1)(\tau_e s + 1)} \tag{2.35}$$

and if $\tau_e \to 0$ then

$$\frac{\Theta(s)}{V_a(s)} \approx \frac{K_T/(K_T K_B + BR_a)}{s(\tau_m s + 1)} \tag{2.36}$$

Example 9 *Find the transfer function for the field controlled d.c. motor shown schematically in Fig. 2-23.*

The armature current i_a is constant and the air gap flux Φ is given by

$$\Phi(t) = K_f i_f(t)$$

where K_f is constant. The developed torque T_{ext} is

$$T_{ext}(t) = -\left[\frac{PZ}{\pi c} i_a\right] \Phi(t) = -K_1 i_f(t)$$

The field voltage is

$$v_f(t) = R_f i_f(t) + L_f p i_f(t)$$

and the mechanical torque is

$$-T_{ext}(t) = Ip^2\theta(t) + Bp\theta(t)$$

Assuming zero initial conditions and taking the Laplace transform of the above equations,

$$T_{ext}(s) = -K_1 I_f(s)$$

$$V_f(s) = (L_f s + R_f)I_f(s)$$

$$-T_{ext}(s) = (Is^2 + Bs)\Theta(s)$$

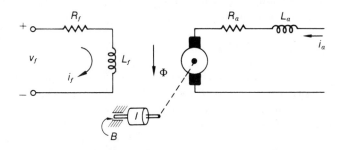

Figure 2.23: Field controlled d.c. motor

Substitution yields

$$\frac{\Theta(s)}{V_f(s)} = \frac{K}{s(\tau_m s + 1)(\tau_f s + 1)} \qquad (2.37)$$

where

$\tau_m = I/B$ = time constant of motor
$\tau_f = L_f/R_f$ = time constant of field
$K = K_1/BR_f$ = motor constant

2.5   Thermal Systems

Modeling of thermal systems by linear differential equations is not as common as the other systems since thermal systems generally tend to be nonlinear. However, in order to obtain a first approximation we shall linearize the systems about an appropriate operating point. Often this results in the assumption that the physical system under consideration be characterized by one uniform temperature.

The fundamental concept used for deriving the thermal system equation is that the difference of heat coming into and leaving a body is equal to the increase of the thermal energy of the system. The physical properties used are mass, specific heat, thermal capacitance, conductance, and resistance. Temperature is the driving potential and heat is the quantity which flows. Generally, thermal resistance is defined as

$$R = \frac{\Delta T}{q}$$

where ΔT is the temperature difference and q the heat flow rate. Depending upon the system, R may include the contributions of thermal conduction, convection, and radiation. The thermal capacitance C is the product of mass and specific heat and

$$q(t) = mc\,pT(t) = CpT(t)$$

Consider a mass m dropped into an oil bath at temperature T_h. We shall assume that the temperature is uniform inside the mass at any given time and also that the oil bath temperature is constant. The heat entering the mass from the oil at any time is

$$q_{in}(t) = \frac{(T_h - T_m(t))}{R} \tag{2.38}$$

where T_m is the temperature of the mass. Here $R = 1/hA$ where A is the surface area of mass in contact with the oil and h is the heat transfer coefficient due to convection. The heat entering the mass goes to increase the heat content (or internal energy) of the system, i.e.

$$q_{in}(t) = mc\,pT_m(t) \tag{2.39}$$

Equating Eqs. (2-38) and (2-39) we have

$$mc\,pT_m(t) = \frac{(T_h - T_m(t))}{R}$$

Defining $T(t) = T_m(t) - T_h$ we obtain

$$pT(t) = -\frac{1}{RC}T(t) \tag{2.40}$$

which is a first-order differential equation. For obtaining the transfer function we assume that the initial temperature of the mass is T_{0m}. Letting $T_0 = T_{0m} - T_h$, and Laplace transforming Eq. (2-40) yields the transfer function

$$\frac{T(s)}{T_0} = \frac{1}{s + \tau} \tag{2.41}$$

where

$$\tau = \frac{1}{RC}$$

Example 10 *A small sphere of mass m, at a temperature T_0, is dropped into a large oil bath at a constant temperature T_B. If the heat-transfer coefficient of convection is h, find an expression for the temperature of the sphere expressed in the Laplace domain.*

The heat gained by the sphere by convection is equal to the internal energy increase so that

$$c\rho V \frac{dT}{dt}(t) = Ah(T_B - T(t))$$

or

$$\frac{dT}{dt}(t) + \frac{Ah}{c\rho V}T(t) = \frac{Ah}{c\rho V}T_B$$

where T is the temperature of the sphere, V is the volume and A is the surface area. Defining $T_a(t) = T_B - T(t)$ and then taking the Laplace transform we have

$$T_a(s) = \frac{T_a(0)}{s + \alpha}$$

where $T_a(0) = T_B - T(0)$ and $\alpha = Ah/c\rho V$. Solving for $T(s)$ we obtain

$$T(s) = T_B - \frac{T_a(0)}{s + \alpha}$$

Example 11 *A satellite of mass m kg and surface area A m^2 is subjected to solar heating. Assuming that the satellite exhibits some average temperature T_m, obtain the equation governing the temperature of the satellite. Linearize the result about an equilibrium temperature T_0 of the satellite and obtain the system transfer function.*

The thermal input from the sun is

$$q_s = A\alpha E(t)$$

where A is the outer area of the satellite, α is the coefficient of absorptivity, and $E(t)$ is the incident energy of the sun at time t. The satellite is simultaneously losing heat[2] at the rate of

$$q_l = A\epsilon\sigma T_m^4$$

where ϵ is the emissivity of the body, and σ is the Stefan-Boltzmann constant. Since the difference between the incoming and outgoing energy increases the energy of the system, we have

$$q_s - q_l = q_{increase} = mc\, p T_m(t)$$

---

[2] This is the Stefan-Boltzmann law which states that a body at temperature T (in Kelvin) radiates energy at the rate $\epsilon\sigma T^4$. For the case of ideal radiation $\epsilon = 1$ and $\sigma = 5.67 \times 10^{-8}$ W/(m^{2}K^4). If T is in degree R, then $\sigma = 0.1714 \times 10^{-8}$ BTU/hr ft^{2}R^4.

Substituting for q_s and q_l,

$$mcpT_m(t) = A\alpha E(t) - A\epsilon\sigma T_m^4 \qquad (2.42)$$

which is a nonlinear differential equation. In order to linearize, we assume

$$T_m(t) = T_0 + T(t); \quad T_0 \text{ is constant}, \quad T(t) \text{ is small}$$

$$E(t) = E_0 + e(t); \quad E_0 \text{ is a constant}, \quad e(t) \text{ is small}$$

Substituting these into Eq. (2-42) and expanding T_m^4 we obtain

$$mcpT(t) = A\alpha E_0 - A\epsilon\sigma T_0^4 + A\alpha e(t) - 4A\epsilon\sigma T_0^3 T(t)$$

Under equilibrium conditions, $A\alpha E_0 = A\epsilon\sigma T_0^4$ so that

$$mcpT(t) = A\alpha e(t) - 4A\epsilon\sigma T_0^3 T(t)$$

Setting

$$C_1 = \frac{A\alpha}{mc}, \quad \text{and } C_2 = \frac{4A\epsilon\sigma T_0^3}{mc}$$

the governing equation becomes

$$pT(t) + C_2 T(t) = C_1 e(t)$$

which is a first order linear differential equation. The transfer function is

$$\frac{T(s)}{E(s)} = \frac{C_1}{s + C_2}$$

2.6   Hydraulic Systems

Process control in the chemical industry and the use of fluid power in many industrial and military applications requires the use of a variety of fluid or hydraulic systems. Like the electrical and mechanical systems, three basic elements exist for hydraulic systems. These are resistance, capacitance, and inertia (or inertance) elements as shown in Appendix B.

Consider the liquid level system shown in Fig. 2-24. The capacitance of the system is represented by the tank volume and the resistance by the inlet valve. If q_i is the flow rate into the tank and q_o is the flow rate out of the tank, then conservation of mass (assuming incompressible liquid) yields the following equation

$$q_i(t) - q_o(t) = Aph(t) \qquad (2.43)$$

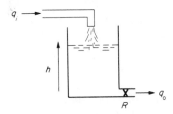

Figure 2.24: Single tank liquid level system

where A is the cross-sectional area of the tank. We can derive another equation by considering the flow through the valve,

$$q_o(t) = \frac{\rho g h(t)}{R} = \frac{P(t)}{R} \tag{2.44}$$

where R is the resistance, P is the fluid pressure and ρ is the density. Substituting for h we have

$$q_i(t) - q_o(t) = \frac{AR}{\rho g} p q_o(t) = CR\, p q_o(t) \tag{2.45}$$

where $C = A/\rho g$ is the fluid capacitance. The transfer function becomes

$$\frac{Q_o(s)}{Q_i(s)} = \frac{1}{1 + sCR}$$

If we represent $Q_i = P_i/R$ and $Q_o = P_o/R$, then the transfer function can also be written as

$$\frac{P_o(s)}{P_i(s)} = \frac{1}{1 + sCR}$$

The term **inertance** refers to the pressure difference between two points needed to cause a volumetric flow acceleration of unity. Consider, for example, two equal cross-sections of area A, a distance L apart at a pressure difference of ΔP. Then from Newton's second law,

$$A(\Delta P(t)) = m\, pv(t)$$

Since the flow rate is $q = Av$ and $m = A\rho L$ we have

$$\Delta P(t) = \frac{\rho L}{A} pq(t) = I\, pq(t)$$

or

$$I = \frac{\Delta P}{dq/dt} = \frac{\rho L}{A} \qquad (2.46)$$

where I is defined as the liquid flow inertance. In many fluid power applications fluid inertance is often neglected.

Example 12 *Obtain the transfer function $Q_o(s)/Q_i(s)$ for the fluid system shown in Figs. 2-25(a) and 2-25(b).*

Applying the conservation of mass to each tank shown in Fig. 2-25(a) yields

$$q_i(t) - q_2(t) = A_1 p h_1(t)$$

$$q_2(t) - q_o(t) = A_2 p h_2(t)$$

We also have two equations for the resistance,

$$q_o(t) = \frac{\rho g h_2(t)}{R} \text{ and } q_2 = \frac{\rho g(h_1(t) - h_2(t))}{R}$$

Substituting these in the previous equations, letting $C = A/\rho g$ and simplifying, yields the transfer function

$$\frac{Q_o(s)}{Q_i(s)} = \frac{1}{R^2 C_1 C_2 s^2 + (2C_1 + C_2)Rs + 1}$$

For the tanks shown in Fig. 2-25(b) we have the same two conservation of mass equations but the resistance equations are

$$q_o(t) = \frac{\rho g h_2(t)}{R} \text{ and } q_2(t) = \frac{\rho g h_1(t)}{R}$$

Substituting yields,

$$\frac{Q_o(s)}{Q_i(s)} = \frac{1}{(1 + sC_1 R)(1 + sC_2 R)}$$

Example 13 *In the hydraulic system shown in Fig. 2-26, the pistons are assumed massless and frictionless. The mass M is connected to the piston by a rigid, massless rod which slides in the frictionless bearing. Find the differential equation relating $f(t)$ to $y(t)$ and the transfer function $Y(s)/F(s)$ if the region between the pistons is filled with an incompressible fluid. The constriction in the line connecting the two cylinders has a flow resistance R.*

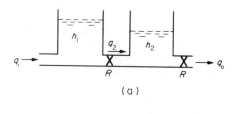

(a)

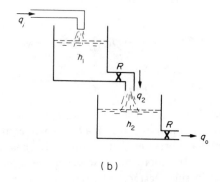

(b)

Figure 2.25: Double liquid level system

For the piston with area A_1 we write

$$m_1 a_1(t) = f(t) - P_1(t)A_1$$

where P_1 is the pressure on the piston face. Since the mass of the piston is assumed zero we have $f(t) = P_1(t)A_1$. The piston with area A_2 is attached to mass M, so that

$$M\, p^2 y(t) = P_2(t)A_2$$

and the flow through the orifice is

$$q(t) = \frac{P_1(t) - P_2(t)}{R} = A_2 p y(t)$$

Substituting for P_1 and P_2 we obtain

$$\left[\frac{A_1 M}{A_2}\right] p^2 y(t) + (RA_1 A_2)p y(t) = f(t)$$

Taking the Laplace transform and assuming zero initial conditions we obtain

$$s^2 Y(s) + \left[\frac{RA_2^2}{M}\right] sY(s) = F(s)\left[\frac{A_2}{A_1 M}\right]$$

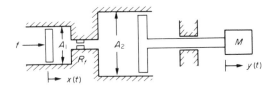

Figure 2.26: Hydraulic system

Setting $K = A_2/A_1M$ and $\tau = RA_2^2/M$ we have the required transfer function

$$\frac{Y(s)}{F(s)} = \frac{K}{s(s+\tau)}$$

In fluid power applications there are two types of hydraulic servomechanisms: displacement controlled systems and valve controlled systems. Although the former has better efficiency, the latter has high gain and quick response. Since it is the more popular of the two we consider the valve controlled systems in this section.

The fundamental relation in such a hydraulic system is

$$q(t) = C_d A v(t) \qquad (2.47)$$

where $q(t)$ is the volume flow rate, A is the cross sectional area, $v(t)$ is the fluid velocity, and C_d is the coefficient of discharge. In valve applications the flow of fluid is controlled by varying the area of the control orifice as shown in Fig. 2-27. The velocity through the orifice is given by

$$v(t) = \sqrt{2\Delta P/\rho}$$

where ΔP is the pressure drop across the orifice and ρ is the density of the working fluid. If the spool moves a distance x and the orifice is characterized as having width w (port width), then $A = wx(t)$ and substitution in Eq. (2-47) gives the flow rate as

$$q(t) = C_d w \sqrt{\frac{2\Delta P}{\rho}} x(t) \qquad (2.48)$$

Consider the valve shown in Fig. 2-28. If the spool moves by $x(t)$, then the flow into the main cylinder is given by Eq. (2-48). This flow causes the piston to move with a velocity $dy(t)/dt$. Substituting this into Eq. (2-48)

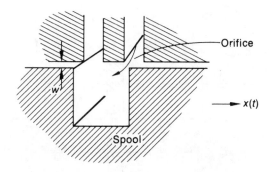

Figure 2.27: Configuration of port valve

we get

$$q(t) = C_d w \sqrt{\frac{2\Delta P}{\rho}} x(t) = A\, py(t)$$

where the coefficient of discharge for the main cylinder is unity. Assuming constant ΔP,

$$py(t) = K_v x(t)$$

where

$$K_v = \frac{C_d w}{A} \sqrt{\frac{2\Delta P}{\rho}}$$

The transfer function becomes

$$\frac{Y(s)}{X(s)} = \frac{K_v}{s} \qquad (2.49)$$

In this case, the amount of liquid that leaks has been neglected. The analysis of hydraulic systems is not complete unless this component is considered. The effect of fluid compressibility is another item that must be reckoned with.

Example 14 *For the hydraulic actuator shown in Fig. 2-28, derive the governing equation if compressibility effects are included. Obtain the system transfer function and show that it reduces to $1/As$ if compressibility and leakage are zero. The storage due to compressibility effects is $vpP(t)/\beta$ where β is the bulk modulus and v is the effective fluid volume under compression. Assume that the output shaft is connected to a mass of m kg.*

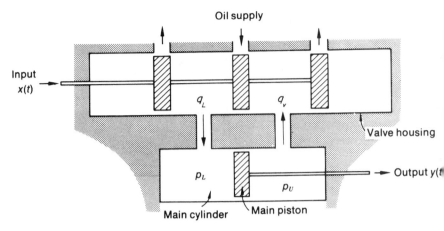

Figure 2.28: Hydraulic valve actuator

The flow into or out of the actuator is a linear combination of the following:

1. Flow resulting from the actuator motion. This is $Apy(t)$ where $y(t)$ is the actuator rod displacement.

2. Storage due to compressibility effects. This is given by $vpP(t)/\beta$ where β is the bulk modulus of the hydraulic fluid.

3. Finally, flow resulting from leakage across the actuator piston. This is $C_L(P_L - P_U)$ where C_L is the leakage coefficient and P_L, P_U represent pressures in the left and right hand chambers respectively.

The flow out of the right chamber is q_U and into the left is q_L where

$$q_U(t) = Apy(t) - \frac{v}{\beta}pP_U(t) + C_L(P_L(t) - P_U(t))$$

$$q_L(t) = Apy(t) + \frac{v}{\beta}pP_L(t) + C_L(P_L(t) - P_U(t))$$

Now if we consider the average flow $q(t)$ to be $(q_U(t) + q_L(t))/2$, then

$$q(t) = Apy(t) + \frac{v}{2\beta}p(P_L(t) - P_U(t)) + C_L(P_L(t) - P_U(t))$$

In addition to this equation, we have a force equation, i.e. the force generated by the actuator rod equals the inertia force,

$$A(P_L(t) - P_U(t)) = mp^2y(t)$$

Substituting this into the previous equation yields

$$q(t) = A\,py(t) + \frac{vm}{2\beta A}p^3 y(t) + \frac{C_L m}{A}p^2 y(t)$$

Taking the Laplace transform and assuming zero initial conditions we obtain the transfer function

$$\frac{Y(s)}{Q(s)} = \frac{1}{\dfrac{vm}{2\beta A}s^3 + \dfrac{C_L m}{A}s^2 + As}$$

For a simplified analysis it is often assumed that the fluid is incompressible $(\beta = \infty)$ and leakage is zero, so that the transfer function reduces to

$$\frac{Y(s)}{Q(s)} = \frac{1}{As}$$

If the flow rate through the valve can be controlled, then any desired output can be obtained.

2.7 System Components

Most control systems and components consist of a combination of the systems we have been discussing in the previous sections. For example, an electrical amplifier may be used to amplify an electrical signal that drives a motor which may be coupled to an inertia through a gear train. Clearly, we need to apply the method developed for analyzing electrical systems as well as Newton's law for mechanical systems in order to obtain the transfer function of the complete system.

In this section we consider systems that combine some of the concepts developed previously. We shall derive the differential equations and the transfer functions. Since you have been introduced to the representation of systems by a set of first-order differential equations, we shall not attempt to do so in each case here. Instead, we shall agree that such a representation can be obtained when necessary. Additionally, we need to note that the components considered here are very fundamental. Indeed, modern systems contain many other components that are far more complex. Here we are content with more common as well as simpler examples of control system components.

Amplifiers

The amplifier is a very important part of any control system. Basically, it is used to deliver an output signal which is larger, in a prescribed way,

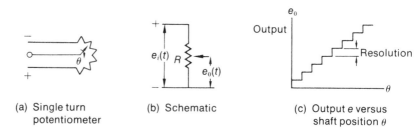

(a) Single turn   (b) Schematic   (c) Output e versus
   potentiometer   shaft position θ

Figure 2.29: A potentiometer

than the input signal. A good amplifier design generally requires that the input impedance be large, so that the source is not loaded, and that the output impedance be small so that the power element can be easily driven. Basically there are two types of amplifiers, viz. power amplifiers and voltage amplifiers.

In the power amplifier the output power is larger than that provided at the input. In the case of the voltage amplifier, the output voltage is larger than the input without regard to power. In electrical circuits, voltage amplification is derived by the use of an operational amplifier which forms the heart of analog computation.

Signal amplification can be obtained by valves, relays, gears or electronic amplifiers depending upon the particular application. For purposes of modeling an amplifier in a system, the bandwidth is normally quite large so that the response may be considered flat. In this case the

$$Output = Input \times Constant$$

Potentiometers

Potentiometers are devices that produce a signal directly proportional to a physical quantity. This signal may be mechanical, electrical, or whatever form is convenient. By far the most important type of potentiometer is the resistance potentiometer shown in Fig. 2-29. The transfer function is given by

$$\frac{E(s)}{\Theta(s)} = \frac{E}{\theta} = \text{constant} \qquad (2.50)$$

The actual transfer function deviates from this constant when the potentiometer is loaded. This departure from linearity is usually very small however. Since most potentiometers are wire-wound, the sliding brush contact touches the wires only discretely as the potentiometer shaft is rotated.

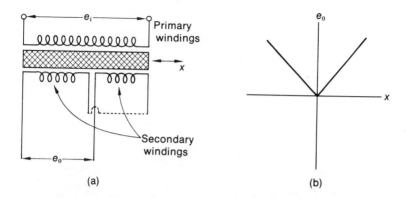

Figure 2.30: Linear variable differential transformer

This causes the output voltage, instead of being a continuous function of shaft position, to be discontinuous as shown in Fig. 2-29.

The minimum change in output voltage obtained by rotating the shaft, divided by total applied voltage is called the resolution of a potentiometer,

$$\text{Resolution} = \Delta e/E$$

If the potentiometer has n turns, then $\Delta e = E/n$ and the resolution is $1/n$. This resolution determines the accuracy of a control system. More modern potentiometers have helical resistance elements and also have many turns. This tends to smooth out the staircase effect of Fig. 2-29(c).

Linear Variable Differential Transformer (LVDT)

This device is used as a displacement transducer and is shown in Fig. 2-30(a). It consists of a primary winding, energized with a fixed a.c. voltage, and two secondary windings. A movable magnetic core provides the necessary coupling. The secondary windings are wired to oppose each other so that if the core is centered, the output voltage is zero. If it is moved, however, there is an increased induced voltage in one and a decrease in the other so that e_o is non-zero and has the same phase as the secondary with the increased voltage. This is shown graphically in Fig. 2-30(b).

The output of the LVDT therefore not only provides an output voltage proportional to the displacement, but yields a phase relationship dependent

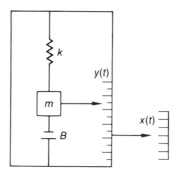

Figure 2.31: Schematic of an accelerometer

upon the direction of displacement. The transfer function is

$$\frac{E_o(s)}{X(s)} = \text{constant} \tag{2.51}$$

Accelerometer

As the name implies, this device measures the acceleration of a system. Basically this instrument measures the motion of a restrained mass when it is subjected to an acceleration. The instrument consists of a mass, spring, and damper as shown in Fig. 2-31.

The mass position is given by $y(t)$ whereas $x(t)$ is the displacement of the frame with respect to the body whose acceleration is to be measured. The transfer function may be obtained using Newton's law and is

$$\frac{Y(s)}{s^2 X(s)} = \frac{1}{s^2 + \dfrac{B}{m}s + \dfrac{k}{m}} \tag{2.52}$$

The output of the accelerometer is $y(t)$ and is generally measured with a potentiometer. If the values of m, k and B are selected correctly, a fairly linear device, under given operating conditions, is obtained.

Synchros

These devices are used as torque transmitters or position indicators. They are generally used in pairs. When they are used to transmit torques, one synchro is used to generate an electrical signal corresponding to a shaft

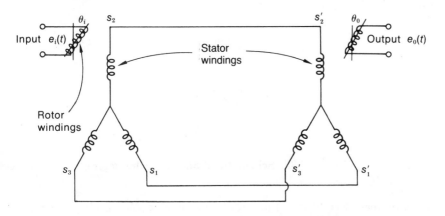

Figure 2.32: A synchro pair as an error detecting device

position. This signal is introduced into the receiver which assumes the same shaft position as the transmitter. Because of system losses the power at the receiver is less than that transmitted, therefore the angle at the receiver does depend upon the electrical efficiency. Synchros used in this mode are useful for presenting display information like dial readings.

When synchros are used as position indicators they become error detecting devices. The output is equal to the sine of the difference between the input and output angles. When the error is small it is approximately equal to the difference of the shaft position. This is shown in Fig. 2-32. The error equation is thus approximately given by

$$e_o(t) = K(\theta_i(t) - \theta_o(t)) \qquad (2.53)$$

Tachometer

Tachometers are used as velocity pickoff devices since they generate an output voltage proportional to the shaft angular velocity. A very important use of a tachometer is for damping position servomechanisms by placing them in the feedback loop as will be shown in a later chapter. Of the many types of velocity pickoff devices, the a.c. and d.c. tachometers are the most widely used.

An a.c. tachometer is shown schematically in Fig. 2-33. A known sinusoidal voltage $e_i(t)$ is applied, to the primary winding, setting up flux. Since the secondary winding is at right angles, the output voltage $e_o(t)$ is

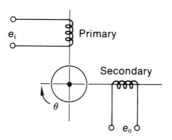

Figure 2.33: Schematic of an a.c. tachometer

zero, or very close to it, if the rotor is not moving. When the rotor moves the output voltage is proportional to $d\theta(t)/dt$,

$$e_o = K\frac{d\theta(t)}{dt}$$

and the transfer function becomes

$$\frac{E_o(s)}{\Theta(s)} = Ks \tag{2.54}$$

where K is the sensitivity of the tachometer. When the control system is a d.c. system, then it is usual to use a d.c. tachometer. Since these devices have brushes, they are generally less accurate than a.c. tachometers and also have drift. They are very similar to d.c. motors. The transfer function of a d.c. tachometer is the same as that shown in Eq. (2-54).

A.C. Control Motors

Control motors are employed for obtaining the necessary torque in control systems. Both a.c. and d.c. motors are used and the latter motor has been discussed above. A.c. motors generally have two phases and are similar to a.c. tachometers. The two phases are **excitation**, or fixed voltage, and **control**, or variable voltage. A reference voltage is applied to the fixed part and an error voltage to the variable phase. If the control or variable phase is zero, there is no output torque. This torque increases directly as a function of the error voltage magnitude. These motors produce a damping torque proportional to velocity and also one proportional to control voltage.

Consider an a.c. motor having the following characteristics

$$T(t) = b(t) - m\omega(t)$$

where $T(t)$ is the motor torque, $\omega(t)$ is the angular velocity, while m is a constant. The quantity $b(t)$ depends upon the control voltage $v(t)$, and

$$b(t) = Kv(t)$$

Therefore

$$T(t) = Kv(t) - m\omega(t)$$

Now if $T = 0$, then

$$m = \frac{Kv}{\omega}$$

and if we further assume that the torque speed curves are linearized as shown in Fig. 2-34, then

$$m = \frac{\text{stall torque (rated voltage)}}{\text{no-load speed (rated voltage)}}$$

If the motor has moment of inertia I, and viscous damping B, then the torque becomes

$$T(t) = I\,p^2\theta(t) + B\,p\theta(t)$$

Equating this torque to the previous expression yields

$$Kv(t) - m\,p\theta(t) = I\,p^2\theta(t) + B\,p\theta(t) \tag{2.55}$$

Rearranging and taking the Laplace transform for zero initial conditions we obtain the transfer function

$$\frac{\Theta(s)}{V(s)} = \frac{K}{s(Is + m + B)}$$

or

$$\frac{\Theta(s)}{V(s)} = \frac{K_m}{s(\tau s + 1)} \tag{2.56}$$

where

$$K_m = \frac{K}{m + B}, \quad \text{and } \tau = \frac{I}{m + B}$$

This is then the transfer function of a two-phase induction motor.

Gears

The use of gears is quite common in rotational mechanical systems for improving the transfer of power between drive and load or for matching

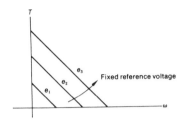

Figure 2.34: Linearized speed torque curves for 2-phase a.c. motor

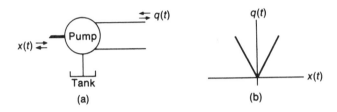

Figure 2.35: Schematic of hydraulic pump

speeds. Two gears are shown in Fig. 2-9 where one is considered the input and the other the output gear. We note that

$$\frac{N_2}{N_1} = \frac{\theta_1}{\theta_2} = \frac{p\theta_1}{p\theta_2} = \frac{p^2\theta_1}{p^2\theta_2} = \frac{T_2}{T_1} = \text{constant} = N$$

Generally if gears are manufactured carefully the constant N is maintained and there is negligible power loss, backlash, etc. If, however, these occur, different modeling techniques to handle such nonlinearities are discussed later in Chapter 11.

Hydraulic Pump

The source of all fluid power is the hydraulic pump. The pump that we consider here has positive but variable displacement. A schematic of a hydraulic pump is shown in Fig. 2-35. Such pumps are motor driven. The amount of fluid pumped is a function of the pump stroke $x(t)$ for a fixed motor speed. The direction of stroke dictates the flow direction. The flow rate is given by

$$q(t) = Kx(t) \qquad (2.57)$$

where K is a characteristic of the pump.

Hydraulic Actuator

A hydraulic actuator and valve arrangement is shown in Fig. 2-28. The transfer function for the idealized system is given by Eq. (2-49).

Error Detecting Devices (Subtractors)

Besides using a pair of synchros for obtaining positional error, there are several other devices commonly used as error sensing devices. Potentiometers and linear variable differential transformers when used in pairs are displacement error detecting devices. The hydraulic valve as well as a bevel and gear differential are commonly used error sensors in hydraulic and gear train systems. Some of these devices are shown in Fig. 2-36.

For a.c. control systems transformers may be used as error sensors. Another attractive method of differencing or adding signals is to use resistances in conjunction with a high gain amplifier. Consider the network of Fig. 2-37. Writing the nodal equations at node N,

$$\frac{e_i(t) - v(t)}{R_1} + \frac{e_f(t) - v(t)}{R_2} + \frac{e_o(t) - v(t)}{R_f} = 0$$

Since the amplifier has high gain, $v(t)$ is small so the above equation may be written as

$$e_o(t) = -\frac{R_f}{R_1}e_i(t) - \frac{R_f}{R_2}e_f(t) \tag{2.58}$$

Since $e_i(t)$ may be biased to be positive or negative, Eq. (2-58) may be used for summing or differencing. This technique is quite commonly used in analog computation.

Gyroscopes

The gyroscope (commonly called a **gyro**) is a basic element in many instruments for guidance and control of moving vehicles. A gyro is shown in Fig. 2-38. The inner gimbal supports the spinning wheel and is restrained by a spring or damper depending upon the application. The z-axis is the input axis and the x-axis is the output axis. Assume the outer gimbal is rigidly fixed to a moving vehicle. If the vehicle turns at a rate of $d\psi/dt$, then the rate of change of angular momentum is $I\omega(d\psi/dt)$. This gives rise to a moment about the output axis. If the outer gimbal has a spring of stiffness k, then the moment is $k\theta$ and

$$k\theta(t) = I\omega p\psi(t)$$

Item	Schematic	Error equation
Potentiometer	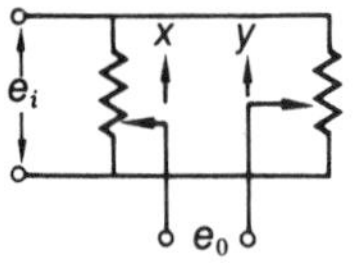	$e_0 = K(x-y)$
LVDT	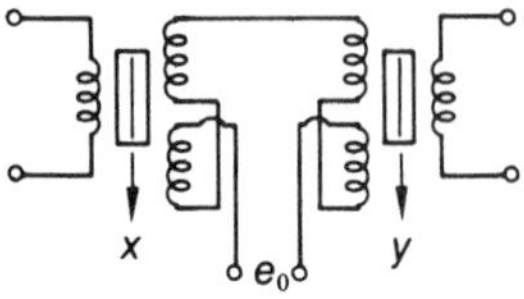	$e_0 = K(x-y)$
Hydraulic valve	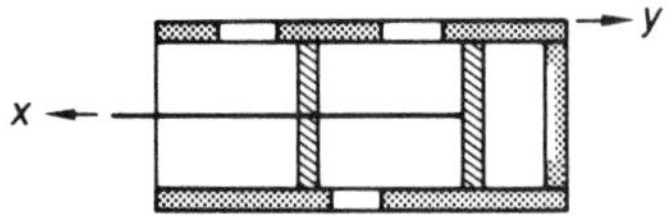	$\Delta = (x-y)$
Bevel and gear differential	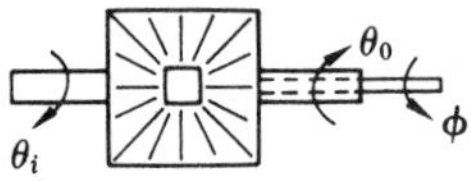	$\phi = \frac{1}{2}(\theta_i - \theta_0)$

Figure 2.36: Error sensors

or

$$\frac{\Theta(s)}{\Psi(s)} = \frac{I\omega s}{k} \tag{2.59}$$

When the output is proportional to the input rate $p\psi(t)$, the gyro is called a **rate gyro**.

If the torsional spring is replaced by a torsional damper having a damping coefficient B, then the moment about the output axis is $Bp\theta(t)$ and

$$Bp\theta(t) = I\omega p\psi(t)$$

or

$$\frac{\Theta(s)}{\Psi(s)} = \frac{I\omega}{B} \tag{2.60}$$

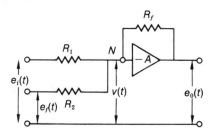

Figure 2.37: Operational amplifier circuit for computing sum or difference

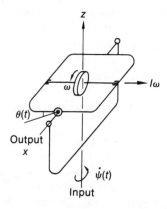

Figure 2.38: Single degree of freedom gyro

Here the output angle is proportional to the integral of the input angular rate, or the input angle. When this happens the gyro is called an **integrating gyro**. Whether the gyro is an integrating or rate gyro, we need to obtain the angle θ. This angle may be obtained by a potentiometer. A special potentiometer used for this is called an E-pickoff potentiometer.

Transducers

Transducers are a very important part of a control system since they provide a usable signal that measures a variable that must be either controlled or is useful as a control parameter. There is a large variety of transducers that are currently available. Their accuracy and cost is dependent upon the intended use. Most of them have linear characteristics around their point of operation. If the departure from linearity is significant a table or equation

Table 2-1 Transducers

Transducer	Use	Method of Operation
Capacitive probe	Liquid level	Capacitance change between electrodes probe due to level change which varies dielectric.
Diaphragm	Pressure	Deflection of a circular plate that is proportional to the pressure.
Geiger counter	Nuclear radiation	Ionization current produced by ion-pair in gas or solid subjected to incident radiation.
Gyroscope	Orientation and guidance	Change causes displacement relative to fixed axis of rotating wheel.
Photodetector	Light detection	Resistance change in semiconductor detection device junction due to light.
Photovoltaic cell	Light detection	Output voltage when a junction of two dissimilar metals is illuminated.
Piezoelectric crystal	Pressure	Mechanical distortion of crystal produces a voltage.
Potentiometer	Displacement	Change in voltage due to variable resistance or magnetic coupling.
Pyrometer	Solar radiation	Thermopile measures temperature of black and white surfaces to yield a temperature difference.
Resistance thermometer	Temperature	Temperature changes cause change in electrical resistance of material.
Solar cell	Orientation relative to sun	Provides a signal proportional to cosine of angle between cell and normal Sun vector.
Strain gauge	Strain	Electrical resistance change due to material deformation
Tachometer	Velocity	Voltage that is proportional to speed of the armature rotating in magnetic field.
Thermocouple	Temperature	EMF proportional to temperature.

for corrections is available. Detailed information on transducers is readily available from manufacturers' data. A list of some common transducers, their use and method of operation is given in Table 2-1.

2.8 Summary

We have shown how a linearized mathematical model of a physical system may be developed by using certain fundamental laws describing the behavior of the physical system. This mathematical model was seen to reduce to the form of an integro-differential equation.

From the integro-differential equation, the system transfer function was obtained. This is the ratio of the output to the input when the output and input are expressed as Laplace transforms and the initial conditions are zero. As an alternate representation, the system was described by a set of first-order differential equations.

It is important to note however that many complex components and real problems can often not be characterized by linear models and represented by linear equations. Nevertheless, linear models serve as very powerful tools for first approximations in studying system behaviors over limited operating ranges.

2.9 References

1. J. L. Shearer, *Introduction to System Dynamics*, Addison Wesley Publ. Co., Reading, Mass., 1967.

2. R. C. Dorf, *Modern Control Systems*, Addison Wesley Publ. Co., Reading, Mass, 1989.

3. R. H. Cannon, *Dynamics of Physical Systems*, Mc Graw Hill Book Co., New York, 1967.

4. J. J. D'Azzo, C. H. Houpis, *Linear Control System Analysis and Design — Conventional and Modern*, Mc Graw Hill Book Co., New York, 1988.

5. B. C. Kuo, *Automatic Control Systems*, Prentice Hall, Inc., Englewood Cliffs, N. J., 1987.

6. J. Meisel, *Principles of Electromechanical Energy Conversion*, Mc Graw Hill Book Co., New York, 1966.

7. D. C. White, H. H. Woodson, *Electromechanical Energy Conversion*, John Wiley and Sons, 1959.

8. K. Ogata, *System Dynamics*, Prentice Hall, Inc., Englewood Cliffs, N. J., 1978.

9. E. S. Kuh, R. A. Rohrer, "The State Variable Approach to Network Analysis", IEEE Proc., Vol. 53, No. 7 (July 1965), pp. 672-685.

10. R. J. Mayhan, *Discrete-Time and Continuous-Time Linear Systems*, Addison Wesley Publ. Co., Reading, Mass., 1983.

11. E. Hughes, *Electrical Technology*, Longman Scientific & Technical, Essex, England, 1987.

12. E. O. Doebelin, *Measurement Systems*, Mc Graw Hill Book Co., New York, 1983.

13. N. N. Norton, *Handbook of Transducers for Electronic Measuring Systems*, Prentice Hall, Inc., Englewood Cliffs, N. J., 1969.

2.10   Problems

2-1 Obtain the transfer functions for the circuits shown in Fig. P2-1.

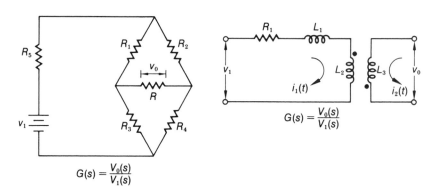

Figure P2-1

2-2 Obtain the transfer function for the single stage amplifier shown in Fig. 2-13.

2-3 Obtain $G(s)$ for the system shown in Fig. 2-14.

2-4 Obtain the transfer function of the lead and lag circuits shown in Fig. P2-4, which are often used as compensators in feedback control systems.

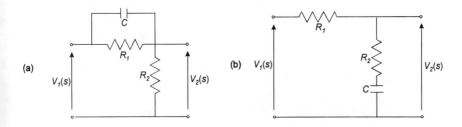

Figure P2-4

2-5 The parallel-T network shown in Fig. P2-5 can be used for compensating control systems having resonant modes. Obtain the transfer function for this network.

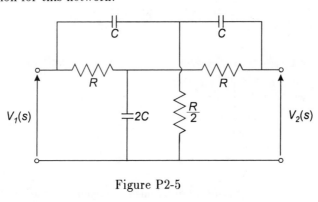

Figure P2-5

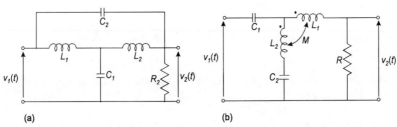

Figure P2-6

2-6 Obtain the transfer functions and state equations for the electrical networks shown in Fig. P2-6. In obtaining the state equations take the currents through the inductors and the voltages across the capacitors as the state variables.

2-7 An inverting summing amplifier using an operational amplifier is shown in Fig 2-37. Assuming the operational amplifier gain $A \gg 10^3$, prove that the output voltage $e_0(t)$ is given by Eq. (2-58).

2-8 A system using a d.c. generator driving a d.c. motor (Ward Leonard drive) is shown in Fig. P2-8. The d.c. generator which is driven at constant speed by a motor (not shown) has its output voltage controlled by the field current i_{fg}. For this system obtain the transfer function $\Omega_m(s)/V_f(s)$, and also obtain the state equations by assuming the state variables are i_{fg}, i_a, θ_m, and $d\theta_m/dt(= \omega_m)$. The motor and generator e.m.f.'s are given by Eq. 2-31.

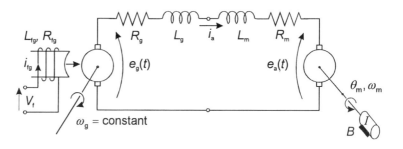

Figure P2-8

2-9 Obtain a set of first-order differential equations for each of the systems shown in Fig. P2-9.

2-10 Represent the system of Fig. 2-6 by a set of first-order differential equations.

2-11 Obtain the governing differential equation for the system shown in Fig. P2-11.

2-12 For the two fluid systems of Fig. P2-12, obtain $Q_o(s)/Q_1(s)$. Do these systems differ? How? Assume the three cylindrical tanks have cross-sectional area A in each case.

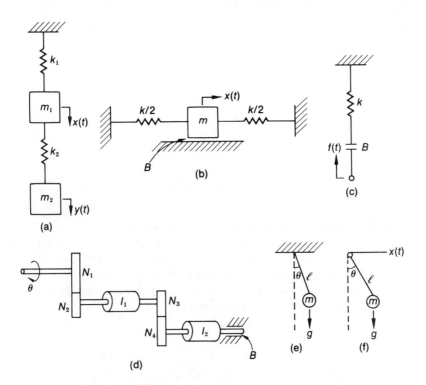

Figure P2-9

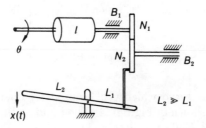

Figure P2-11

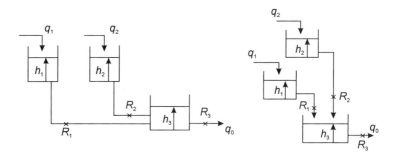

Figure P2-12

2-13 Consider the hydraulic servo actuator shown in Fig. P2-13 assuming the input to the servo is the displacement $x(t)$ of the spool valve, and the output piston is connected to the compliant load M_0 and M_1, where the piston mass is included in mass M_0. Derive the governing system transfer function $Y_1(s)/X(s)$, assuming fluid leakage and compressibility effects are present.

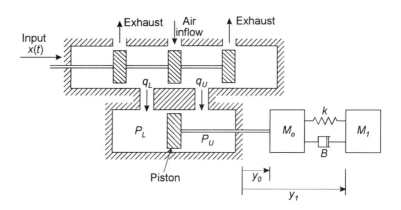

Figure P2-13

2-14 A hydraulic servo which can be used for operating aircraft control surfaces is shown in Fig P2-14. The input to the system is the control stick angular deflection $\theta(t)$, and the output is the elevator angle $\phi_e(t)$. In the analysis it should be assumed that angles $\theta(t)$ and $\phi_e(t)$ are small. Determine the state equations for this system, and find the transfer function $\Phi_e(s)/\Theta(s)$.

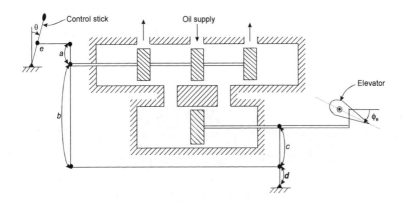

Figure P2-14

2-15 A shaker table is shown in Fig. P2-15. The potentiometer measures the difference between the position of the two masses. Neglecting gravity and initial conditions obtain $E_o(s)/Y(s)$.

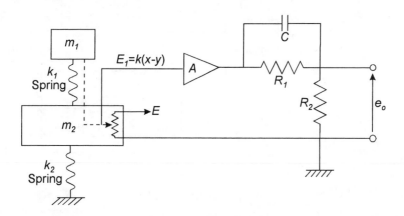

Figure P2-15

2-16 A simple integrating servomechanism is shown in Fig. P2-16. Derive the transfer function $V_o(s)/V_1(s)$ and discuss all your assumptions.

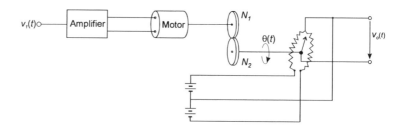

Figure P2-16

2-17 Obtain $X(s)/E(s)$ for the system of Fig. P2-17.

2-18 Represent the system of Fig. P2-17 by a set of first-order differential equations.

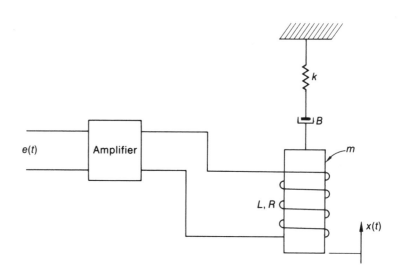

Figure P2-17

2-19 Obtain $\Theta(s)/E(s)$ for the system of Fig. P2-19.

2-20 Obtain a set of first-order differential equations to represent the system of Fig. P2-19.

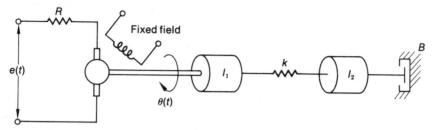

Figure P2-19

2-21 For the hydraulic actuator shown in Fig. 2-28, derive a set of first-order differential equations governing its behavior.

2-22 Obtain the equations of motion of the cart and the inverted pendulum shown in Fig. P2-22. Linearize the equations under the assumption that x and θ are small and $m_1 \gg m_2$. Neglect frictional forces.

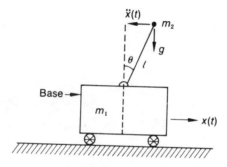

Figure P2-22

Chapter 3

Models for Control Systems

3.1 Introduction

In the previous chapter we saw how a system may be represented by a linear ordinary differential equation with constant coefficients. We also showed by using Laplace transform notation that a single-input single-output system may be represented by a **transfer function**. Additionally it was shown how instead of representing a system by a linear ordinary differential equation it could also be represented by **state equations** which are a set of first-order differential equations.

In this chapter we will explore the relationships between these various representations of a given system and show how they are inter-related. We start by examining the impulse and step responses of a system and their relationship to the system's transfer function. We will then show how the transfer functions of individual components can be combined in a systematic way so as to obtain the transfer function of an entire control system using **block diagram algebra** methods.

State equation representations of systems and their relationships to other models will be examined and it will be seen that state realizations are not unique. In the previous chapter it was seen that often the physical models of systems are non-linear. Building upon the discussion of the last chapter the linearization of non-linear systems about a given operating point using perturbation methods will be examined.

71

3.2   System Impulse and Step Responses

While the emphasis in the last chapter was on modeling of **components** we now approach the modeling from a **system viewpoint**. Here we view the system as being composed of many interacting components which in some instances have great complexity both in terms of numbers and in terms of interconnections. An alternative to the differential equation models developed in the last chapter which can aid the understanding of complex system operation is to view them from an **input-output** point of view. Here we assume that the system S, shown in Fig. 3-1 has some nodes which are identified as inputs u and others that are identified as outputs y. For the present we assume for simplicity the system S has only a single input and output, and it is linear so that superposition applies.

Suppose we apply an impulse $\delta(t - \tau)$ at time $t = \tau$ to the input of system S, where τ is arbitrary. In this case the output $y(t) = g(t, \tau)$ is called the **impulse response** of system S, and it is assumed the system response depends upon the time of application τ of the input disturbance. If the impulse response $g(t, \tau)$ satisfies the condition

$$g(t + a, \tau + a) = g(t, \tau) \tag{3.1}$$

where a can take any value, then we say the system is **time invariant**, because the impulse response only depends upon the difference between the observation time t and the disturbance time τ. This can be seen by taking $a = -\tau$ in Eq. (3-1) so that we find

$$g(t, \tau) = g(t - \tau, 0) \text{ (abbreviated as } g(t - \tau)) \tag{3.2}$$

In the following discussion it will be assumed the system S is time invariant.

The Superposition Integral

Considering the input signal $u(t)$ shown in Fig. 3-2(a), we partition the interval from zero to t into n sub-intervals of uniform length Δx. We define the signal $\tilde{u}(t)$ as shown in Fig. 3-2(b), which is a weighted sum of impulses, and is written as

$$\tilde{u}_n(t) = \sum_{i=0}^{n} [u(x_i)\Delta x] \delta(t - x_i) \tag{3.3}$$

If we let $\Delta x \to 0$ and the corresponding number of sub-intervals $n \to \infty$, then Eq. (3-3) becomes

$$\tilde{u}(t) = \lim_{n \to \infty} \tilde{u}_n(t) = \int_0^t \delta(t - x)u(x)dx \tag{3.4}$$

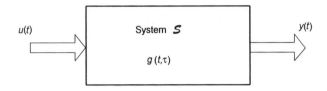

Figure 3.1: Symbolic representation of system $\mathcal{S}$

and from the **sifting property** of the impulse function $\delta(t)$ we have $\tilde{u}(t) = u(t)$. The response $\Delta y_i(t)$ due to the impulse $u(x_i)\Delta x \; \delta(t - x_i)$ is given by

$$\Delta y_i(t) = g(t - x_i)u(x_i)\Delta x$$

Since the system $\mathcal{S}$ is assumed linear we can apply the principle of super-position, and the system output becomes,

$$y(t) = \lim_{n \to \infty} \sum_{i=0}^{n} \Delta y_i(t) = \int_0^t g(t - x)u(x)dx \tag{3.5}$$

The expression given in Eq. (3-5), which is termed the **superposition** or **convolution integral**, gives the response of system $\mathcal{S}$ to a system input $u(t)$ and is sometimes denoted by

$$y(t) = g(t) * u(t)$$

The Du Hamel Integral

We can also represent the input signal $u(t)$ by a series of step functions as shown in Fig. 3-2(c), where the interval zero to t has been partitioned into n equal intervals of length Δx. Denoting the input unit step function at time t by $H(t)$, we can approximate $u(t)$ by the series

$$\tilde{u}_n(t) = \sum_{i=1}^{n} \left[\frac{du}{dt}(x_i)\Delta x \right] H(t - x_i) \tag{3.6}$$

It will be noted that if $\Delta x \to 0$, and correspondingly $n \to \infty$, then it can

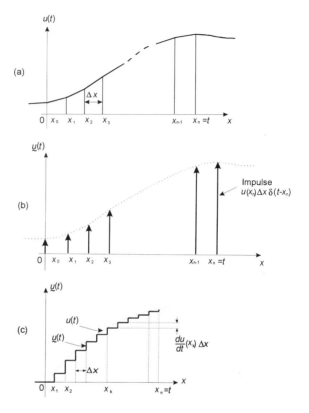

Figure 3.2: Decomposition of time function $u(t)$ into impulse and step function time series

be shown that

$$u(t) = \int_0^t H(t-x)\frac{du}{dt}(x)dx$$

We now find the output response of system $\mathcal{S}$ in terms of its step response $g_s(t)$, where the step is applied at time $t = 0$. Again, appealing to the principle of superposition, the response of the system to the input $\tilde{u}(t)$ given by Eq. (3-6) is

$$y(t) \approx \sum_{i=1}^{n} g_s(t-x_i)\frac{du}{dt}(x_i)\Delta x$$

As above, if $\Delta x \to 0$ then in the limit it follows that

$$y(t) = \int_0^t g_s(t-x)\frac{du}{dt}(x)dx \tag{3.7}$$

This expression which gives the response of the system to an input $u(t)$ in terms of the system step response is known as the **Du Hamel integral**.

Example 1 *Obtain the time response of the system shown in Fig. 3-3(a) to the input signal $u(t)$ shown in Fig. 3-3(b).*

It can be shown that the impulse and step responses, $g(t)$ and $g_s(t)$ respectively, are

$$g(t) = \begin{cases} 0.2\, e^{-0.2t} & t > 0 \\ 0 & t \leq 0 \end{cases} \tag{3.8}$$

$$g_s(t) = \begin{cases} 1 - e^{-0.2t} & t > 0 \\ 0 & t \leq 0 \end{cases} \tag{3.9}$$

Firstly we will apply the convolution integral to find the output $y(t)$. Substituting Eq. (3-8) into Eq. (3-5) gives

$$y(t) = \begin{cases} 0.2e^{-0.2t} \displaystyle\int_0^t e^{0.2x} x\, dx & t \leq 1 \\[2ex] 0.2e^{-0.2t} \left[\displaystyle\int_0^1 e^{0.2x} x\, dx + \int_1^t e^{0.2x} dx \right] & t > 1 \end{cases}$$

Evaluating the above integrals we find

$$y(t) = \begin{cases} 5(e^{-0.2t} - 1) + t & t \leq 1 \\ 5\, e^{-0.2t}(1 - e^{0.2}) + 1 & t > 1 \end{cases}$$

Let us now examine the application of the Du Hamel integral to finding the time response of the system for the input $u(t)$ given in Fig. 3-3(b). The derivative of $u(t)$ is shown in Fig. 3-3(c). Substituting Eq. (3-9) into Eq. (3-7) we find

$$y(t) = \begin{cases} \displaystyle\int_0^t \left(1 - e^{-0.2t(t-x)}\right) dx & t \leq 1 \\[2ex] \displaystyle\int_0^1 \left(1 - e^{-0.2(t-x)}\right) dx & t > 1 \end{cases}$$

Upon evaluating these integrals we have the same result as given above.

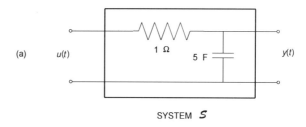

(a) $u(t)$ $y(t)$

SYSTEM S

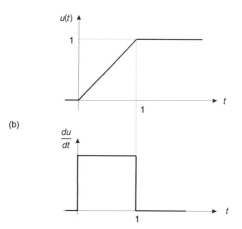

(b)

Figure 3.3: (a) Schematic diagram of electrical network. (b) Network input signal

3.3   The Transfer Function

The transfer function of the system represented by the block diagram shown in Fig. 3-1 will now be introduced.

It has been shown above that the time response $y(t)$ of the system for an input time function $u(t)$ is given by the convolution integral in Eq. (3-5) or by the Du Hamel integral in Eq. (3-7). Both of these integrals can be visualized as defining what are called integral operators. By this we mean that in each case an unambiguous procedure is defined whereby the input signal u is transformed into the output signal y. This operation can be presented in symbolic form as

$$y = G(u) \qquad (3.10)$$

where u and y represent the input and output time functions and G represents the integral operation of transforming u into y according to either Eq. (3-5) or Eq. (3-7).

In investigating the operation of complex systems it is difficult to work directly with these integral operators. Instead we utilize a result from Laplace transform theory. This shows that the Laplace transform $Y(s)$ of the time function $y(t)$ defined by the convolution operator given in Eq. (3-5) is

$$Y(s) = G(s)U(s) \tag{3.11}$$

where $G(s)$ and $U(s)$ are the Laplace transforms of the impulse response $g(t)$ and the input time function $u(t)$, respectively. Since $G(s)$ can be viewed as operating on the input signal $U(s)$ to yield the output $Y(s)$ the function $G(s)$ is termed the system **transfer function**, where Eq. (3-11) shows the signals $U(s)$ and $Y(s)$ are now algebraically related rather than by the integral expression given in Eq. (3-5).

3.4 Differential Equation Representation

While the concept of a transfer function has been defined for systems described by integral operators we can also define it for systems described by ordinary differential equations. We assume here that the system S is described by the differential equation with constant coefficients, $a_{n-1}, \cdots, a_0$, $b_{n-1}, \cdots, b_0$,

$$p^n y(t) + a_{n-1}p^{n-1}y(t) + \cdots + a_0 y(t) = b_{n-1}p^{n-1}u(t) + \cdots + b_0 u(t) \tag{3.12}$$

or more concisely as

$$P(p)y(t) = Q(p)u(t) \tag{3.13}$$

where $P(p)$ and $Q(p)$ are polynomials in the differential operator p, given by

$$P(p) = p^n + a_{n-1}p^{n-1} + \cdots + a_0 \tag{3.14}$$

$$Q(p) = b_{n-1}p^{n-1} + b_{n-2}p^{n-2} + \cdots + b_0 \tag{3.15}$$

To introduce the concept of the transfer function in this case we assume the initial conditions for the differential equation are all zero, and the input time function $u(t)$ has the Laplace transform $U(s)$. Taking the transform of both sides of Eq. (3-12) gives

$$P(s)Y(s) = Q(s)U(s)$$

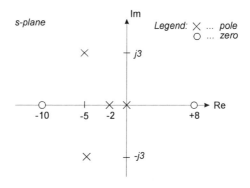

Figure 3.4: Pole-zero diagram for system $\mathcal{S}$

where $Y(s)$ is the transform of the time function $y(t)$. This equation can be written as

$$Y(s) = \frac{Q(s)}{P(s)}U(s) \qquad (3.16)$$

where the quotient $P(s)/Q(s)$ is called the **transfer function** of the system $\mathcal{S}$ described by Eq. (3-12). We have seen how these transfer functions may be computed for the examples given in Chapter 2.

Comparison of Eqs. (3-11) and (3-16) show that the transfer function $G(s)$ for system $\mathcal{S}$ described by an ordinary differential equation with zero initial conditions is given by

$$G(s) = \frac{Q(s)}{P(s)} \qquad (3.17)$$

This result can also be obtained by evaluating the impulse response $g(t)$ for Eq. (3-12) and then subsequently finding its Laplace transform $G(s)$. The roots of the numerator polynomial $Q(s)$ are called the **zeros** of the system, while those of the denominator polynomial $P(s)$ are referred to as the system **poles**. When the variable s assumes the value of one of the system poles it will be noted that the function $G(s)$ becomes infinite, while if it assumes the value of one of the system zeros then it becomes zero. Both the poles and zeros will be shown to play an important role in the

Classical solution of d. e.

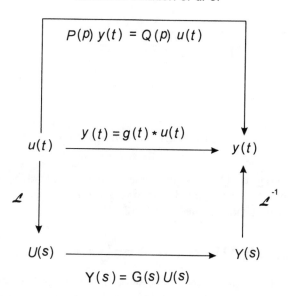

Figure 3.5: Commutative diagram showing the relationship between input and output time functions $u(t)$ and $y(t)$

understanding of the operation of a system and are often plotted as a **pole-zero diagram** on the complex plane. To illustrate this suppose a system S has the transfer function

$$G(s) = \frac{10(s+10)(s-8)}{s(s+2)(s+10s+34)}$$

In this case we note the system has zeros at $s = -10$ and $s = 8$, and poles at $s = 0$, $s = -2$, $s = -5 + j3$, and $s = -5 - j3$. Plotting the poles and zeros on the complex plane we obtain the pole-zero diagram shown in Fig. 3-4.

The relationship between the input and output time functions and their respective Laplace transforms is summarized in the commutative diagram shown in Fig. 3-5. This shows that the system output time function $y(t)$ can be directly evaluated using the convolution integral. Alternatively it can be obtained by the apparently convoluted procedure of finding the output time function $y(t)$ from its Laplace transform. This procedure offers both practical and analytical advantages over the direct evaluation of the

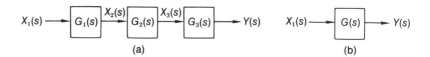

Figure 3.6: Cascaded elements

convolution integral. The most immediate advantage, and the one of great-est importance in the current discussions is the fact that $Y(s)$ is given by the algebraic operation defined in Eq. (3-11).

3.5   Block Diagram Analysis

We have seen how the transfer function of an individual component can be obtained in the previous section. Owing to the occurrence of many transfer functions in **series (or cascade)** and **parallel** connections it is necessary that we have rules to combine them systematically. Consider the block diagram of cascaded elements shown in Fig. 3-6(a). From the definition of a transfer function we have

$$X_2(s) \;=\; G_1(s)X_1(s)$$

$$X_3(s) \;=\; G_2(s)X_2(s)$$

$$Y(s) \;=\; G_3(s)X_3(s)$$

and substitution yields

$$Y(s) \;=\; G_3(s)X_3(s) = G_3(s)[G_2(s)X_2(s)]$$

$$=\; G_3(s)G_2(s)G_1(s)X_1(s)$$

which can be written as

$$G(s) = G_3(s)G_2(s)G_1(s) \tag{3.18}$$

The overall transfer function then is simply the product of individual trans-fer functions as shown in Fig. 3-6(b).

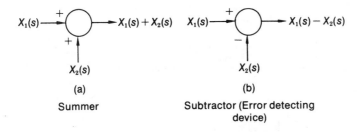

(a)   (b)
Summer   Subtractor (Error detecting
                                        device)

Figure 3.7: Addition and subtraction of signals

For applications where it is required to generate a signal which is the sum of two signals we define a **summer** or **summing junction** as shown in Fig. 3-7(a). If the difference is required, then we define a subtractor as shown in Fig. 3-7(b). Subtractors are often called **error detecting devices** since the output signal is the difference between two signals of which one is usually a **reference signal**. Examples of several components used for summing and subtracting signals were given in the previous chapter.

The combination of block diagrams in parallel is shown in Fig. 3-8(a). From the definition of the transfer function we have

$$Y_1(s) \;=\; G_1(s)X(s)$$

$$Y_2(s) \;=\; G_2(s)X(s)$$

$$Y_3(s) \;=\; G_3(s)X(s)$$

and the summer adds these signals

$$Y(s) = Y_1(s) + Y_2(s) + Y_3(s)$$

or

$$Y(s) = [G_1(s) + G_2(s) + G_3(s)]X(s)$$

The overall transfer function shown in Fig. 3-8(b) is thus

$$G(s) = G_1(s) + G_2(s) + G_3(s) \tag{3.19}$$

In summary, we observe that for cascaded elements the overall transfer function is equal to the product of the transfer function of each element,

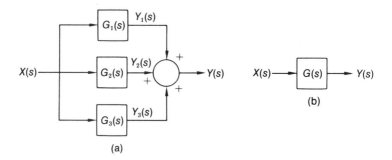

Figure 3.8: Parallel combination of elements

whereas the overall transfer function for parallel elements is equal to the sum of the individual transfer functions.

When a control system is represented in block diagram form we have not only **forward paths** but **feedback paths** as well. By giving due consideration to the direction of the signals we can derive a single transfer function to represent the entire system. This overall transfer function will be seen to be the ratio of the input and the output of the entire system.

Closed Loop Systems

Although control systems having a single input and output can be represented by a single block diagram and one transfer function, it is advantageous to show the individual elements since their transfer functions may be generally determined independently.

A single **closed loop** control system is shown in Fig. 3-9. The individual signals and blocks are identified as follows:

$$
\begin{aligned}
R(s) &= \text{Reference input} \\
B(s) &= \text{Feedback signal} \\
E(s) &= \text{Error or actuating signal} \\
C(s) &= \text{Output signal} \\
G_c(s) &= \text{Controller transfer function} \\
A(s) &= \text{Amplifier transfer function} \\
G_p(s) &= \text{Transfer function of system to be controlled} \\
&\quad \text{(commonly referred to as the plant)}
\end{aligned}
$$

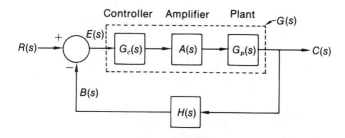

Figure 3.9: Block diagram of simple closed loop system

$H(s)$ $=$ Transfer function of feedback element
$G(s)$ $=$ $G_c(s)A(s)G_p(s) =$ Forward loop transfer function
$G_o(s)$ $=$ $G_c(s)G_p(s)A(s)H(s) =$ Open loop transfer function
$C(s)/R(s)$ $=$ Closed loop transfer function

If the system does not have any **feedback** $H(s) = 0$, and we have an **open loop system**. This system consists of three cascaded transfer functions. If however we do have feedback, then the closed loop transfer function may be obtained by first defining the error equation

$$E(s) = R(s) - B(s) \qquad (3.20)$$

In addition, the following transfer functions may be written,

$$C(s) = G(s)E(s)$$

$$B(s) = H(s)C(s)$$

$$G(s) = G_c(s)A(s)G_p(s)$$

Substituting for $E(s)$ and $B(s)$

$$\frac{C(s)}{G(s)} = R(s) - H(s)C(s)$$

$$C(s)\left[\frac{1}{G(s)} + H(s)\right] = R(s)$$

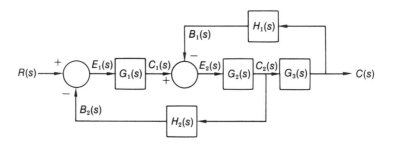

Figure 3.10: Block diagram of system with two feedback paths

and the **closed loop transfer function** becomes

$$\frac{C(s)}{R(s)} = \frac{G(s)}{1 + G(s)H(s)} \tag{3.21}$$

The **error-input transfer function** may be derived using Eq. (3-21),

$$\frac{E(s)}{R(s)} = \frac{1}{1 + G(s)H(s)} \tag{3.22}$$

When a control system consists of a feedback path where $H(s) = 1$, then it is referred to as a **unity feedback control system**. In general, $H(s)$ need not be unity and also the control system may consist of many feedback loops. The technique used to obtain the overall transfer function for multi-loop systems is identical to the previous method except that additional error equations are generally necessary.

Example 2 *Derive the overall transfer function for the control system shown in Fig. 3-10.*

From this figure we see

$$E_1(s) = R(s) - B_2(s)$$

$$E_2(s) = C_1(s) - B_1(s)$$

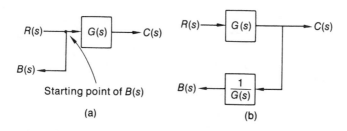

Figure 3.11: Moving the starting point of a signal

$$C_1(s) = G_1(s)E_1(s)$$

$$C_2(s) = G_2(s)E_2(s)$$

$$C(s) = G_3(s)C_2(s)$$

$$B_1(s) = H_1(s)C(s)$$

$$B_2(s) = H_2(s)C_2(s)$$

Substituting the transfer functions into the second error equation we obtain

$$E_1(s) = C(s)\frac{[1 + G_2(s)G_3(s)H_1(s)]}{G_1(s)G_2(s)G_3(s)}$$

Substituting the transfer functions into the first error equation,

$$E_1(s) = R(s) - [H_2(s)C(s)/G_3(s)]$$

Finally, equating these two equations we obtain the overall transfer function

$$\frac{C(s)}{R(s)} = \frac{G_1(s)G_2(s)G_3(s)}{1 + G_1(s)G_2(s)H_2(s) + G_2(s)G_3(s)H_1(s)}$$

Defining $G(s) = G_1(s)G_2(s)G_3(s)$ and $H(s) = H_1(s)/G_1(s) + H_2(s)/G_3(s)$, we obtain

$$\frac{C(s)}{R(s)} = \frac{G(s)}{1 + G(s)H(s)}$$

which has a form identical to Eq. (3-22).

The method used above is straightforward but very cumbersome. It can be greatly simplified if we employ some shortcuts which we consider next.

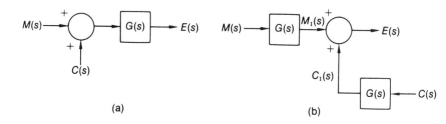

(a)   (b)

Figure 3.12: Moving the summing junction

Block Diagram Reduction

When the block diagram representation gets complicated, it is advisable to **reduce** the diagram to a simpler and more manageable form prior to obtaining the overall transfer function. We shall consider only a few rules for **block diagram reduction**. We already have two rules, viz. cascade and parallel connection.

Consider the problem of moving the starting point of a signal shown in Fig. 3-11(a) from behind to the front of $G(s)$. Since $B(s) = R(s)$ and $R(s) = C(s)/G(s)$, then $B(s) = C(s)/G(s)$. Therefore if the takeoff point is in front of $G(s)$, then the signal must go through a transfer function $1/G(s)$ to yield $B(s)$ as shown in Fig. 3-11(b).

Consider the problem of moving the summing point of Fig. 3-12(a). Since

$$E(s) \;=\; [M(s) + C(s)]G(s) = M(s)G(s) + C(s)G(s)$$

$$=\; M_1(s) + C_1(s)$$

where

$$M_1(s) = M(s)G(s) \text{ and } C_1(s) = C(s)G(s)$$

The generation of the signals $M_1(s)$ and $C_1(s)$ and adding them to yield $E(s)$ is shown in Fig. 3-12(b). A table of the most common reduction rules is given in Table 3-1.

Consider the transfer function of the system shown in Fig. 3-13(a). The final transfer function is shown in Fig. 3-13(d). Note that the first reduction involves a parallel combination; the second involves a cascade combination

Table 3-1 Some rules for block diagram reduction

Rule	Original block diagrams	Reduced block diagrams
Commutated elements		
Cascaded elements		
Addition or subtraction		
Moving feedforward element		
Moving summing element		
Moving summing point		
Moving starting point		
Converting non-unity system to unity feedback		
Closed loop system		

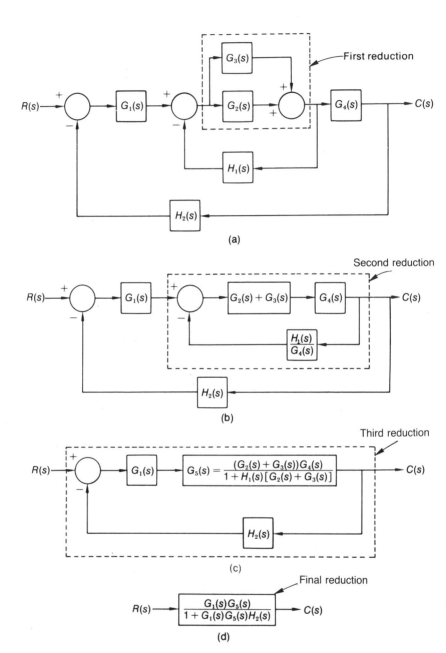

Figure 3.13: Obtaining transfer function by block diagram reduction

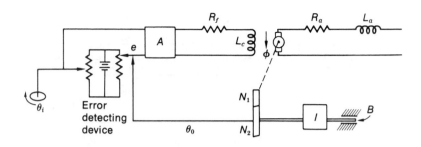

Figure 3.14: Schematic of position servomechanism

as well as the use of Eq. (3-22). The last reduction again involves the use of Eq. (3-22).

Before we continue, let us stop and consider a few examples of complete systems and derive their overall transfer functions.

Example 3 *An electromechanical servomechanism is shown in Fig. 3-14. This is a simple position servomechanism employing a d.c. motor. Find the transfer function between the input and output, $\Theta_o(s)/\Theta_i(s)$.*

The error detector measures the difference between the input θ_i and output θ_o

$$\theta_e(t) = \theta_i(t) - \theta_o(t)$$

and feeds this signal to the amplifier. The signal then goes to the d.c. motor whose transfer function is $G(s)$. The output, after being modified by the gear ratio $N = N_1/N_2$, is the feedback signal. The block diagram is shown in Fig. 3-15.

Since all the transfer functions in the forward loop are cascaded, the overall transfer function becomes

$$\frac{\Theta_o(s)}{\Theta_i(s)} = \frac{K_1 A G(s) N}{1 + K_1 A G(s) N}$$

Substituting for $G(s)$ from Eq. (2-37) we obtain

$$\frac{\Theta_o(s)}{\Theta_i(s)} = \frac{K_1 A N K}{s(1 + s\tau_m)(1 + s\tau_f) + K_1 A N K}$$

or

$$\frac{\Theta_o(s)}{\Theta_i(s)} = \frac{K_2}{s(1 + s\tau_m)(1 + s\tau_f) + K_2}$$

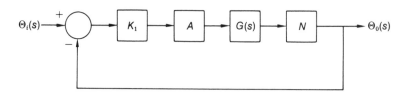

Figure 3.15: Block diagram of a position servomechanism

where $K_2 = K_1 A N K$ and τ_m, τ_f are the motor and field time constants.

Example 4 *We shall derive the governing equations and transfer functions of a gas turbine. It will be assumed that the turbine behavior is linear about some operating condition.*

The following variables are first defined:

$q(t)$ The variation of fuel input
$\omega_r(t)$ Speed setting of the turbine
$\omega(t)$ Actual speed of the turbine(must be controlled)
$T(t)$ Variation in driving torque
$T_1(t)$ Variation of torque disturbance

Assuming linear behavior we note that

$$q(t) \;=\; K_1[\omega_r(t) - \omega(t)]$$

$$T(t) \;=\; K_2 q(t) = K_1 K_2[\omega_r(t) - \omega(t)]$$

where K_1 is a gain constant of the governor, and K_2 is a constant determined by the turbine design. The rotational motion of the turbine is related to the torque by

$$T(t) - T_1(t) = I\frac{d\omega(t)}{dt} + B\omega(t)$$

where I is the load inertia and B is the damping. The block diagrams of the governor, turbine, and load are shown in Fig. 3-16. Writing these

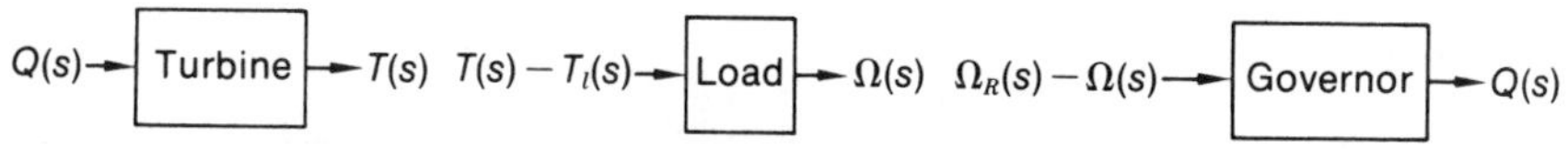

Q(s)→[Turbine]→T(s) T(s) − T_l(s)→[Load]→Ω(s) Ω_R(s) − Ω(s)→[Governor]→Q(s)

Figure 3.16: Transfer function of turbine, load and governor

equations in Laplace transform notation and letting $K = K_1 K_2$ we obtain

$$T(s) = K[\Omega_r(s) - \Omega(s)]$$

$$T(s) - T_1(s) = (Is + B)\Omega(s)$$

Substituting the first into the second we obtain

$$\Omega(s) = \frac{K}{Is + B + K}\Omega_r(s) - \frac{1}{Is + B + K}T_1(s)$$

The entire block diagram is shown in Fig. 3-17. We note that a single transfer function is not obtainable. This is due to the presence of two inputs. Since the system is linear we can also obtain the output by using superposition, i.e. obtaining the output when $\Omega_r(s) = 0$ and adding it to the output when $T_1(s) = 0$. The output $\Omega_1(s)$, when $T_1(s) = 0$, is

$$\Omega_1(s) = \frac{K_1 K_2 G(s)}{1 + K_1 K_2 G(s)}\Omega_r(s)$$

and the output $\Omega_2(s)$, when $\Omega_r(s) = 0$, is

$$\Omega_2(s) = -\frac{G(s)}{1 + K_1 K_2 G(s)}T_1(s)$$

where we have the negative sign to account for the input sign. The total output when both inputs are non-zero is

$$\Omega(s) = \Omega_1(s) + \Omega_2(s)$$

$$= \frac{K_1 K_2 G(s)}{1 + K_1 K_2 G(s)}\Omega_r(s) - \frac{G(s)}{1 + K_1 K_2 G(s)}T_1(s)$$

Substituting for $G(s) = 1/(Is + B)$ and letting $K = K_1 K_2$ we obtain the same equation as before.

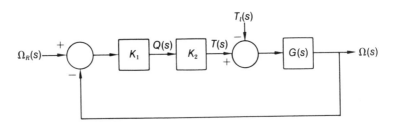

Figure 3.17: Block diagram of turbine

Example 5 *In this example we consider a stable platform. The function of a stable platform is to maintain a fixed angular reference using the property of a gyroscope, i.e. a torque about an input axis produces an angular velocity about the orthogonal axis. Naturally, the input axis must not be the spin axis of the gyro. (Why?) Consider a single axis platform shown in Fig. 3-18 where the y-axis is the input and the x-axis is the output axis. A disturbing torque $T_y(t)$ about the y-axis causes the spin axis direction to rotate through an angle θ. We wish to obtain $\Theta(s)/T_y(s)$ for the platform.*

Noting that the angular momentum of the wheel is h, we balance the torques,

$$T_y(t) - I_y \frac{d^2\phi_y(t)}{dt^2} = h\frac{d\theta(t)}{dt}$$

where I_y is the platform, gyro, and frame moment of inertia about the y-axis. The torque $T_y(t)$ gives rise to $\dot\theta$ (called precession) which causes a torque about the x-axis of

$$-T_x(t) = h\frac{d\phi_y(t)}{dt} - I_x \frac{d^2\theta(t)}{dt^2} = 0$$

We have set $T_x(t)$ to zero since there is actually no applied torque about the x-axis. Taking the Laplace transform of the above two equations for zero initial conditions and substituting yields

$$\frac{\Theta(s)}{T_y(s)} = \frac{h/I_x I_y}{s(s^2 + h^2/I_x I_y)}$$

The second term in the denominator is generally quite small and as a first

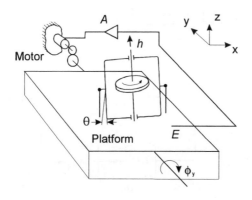

Figure 3.18: A single axis platform

approximation is often neglected. This term is referred to as the **nutation frequency**. Neglecting this yields

$$\frac{\Theta(s)}{T_y(s)} = \frac{1}{I_y s^2} \frac{h}{I_x s} = G_p(s)G_g(s)$$

The block diagram representation for this is shown in Fig. 3-19(a). Now since it is necessary to counteract the torque $T_y(t)$, we sense θ and use this signal to drive a motor and gears in order to apply a torque T_s about the y-axis. The difference between this torque T_s and T_y is the error that drives the platform as shown in Fig. 3-19(b). Denoting the transfer function of the motor and amplifier as $G_m(s)$ and A, we obtain

$$\frac{\Theta(s)}{T_y(s)} = \frac{G_p(s)G_g(s)}{1 + G_p(s)G_g(s)AG_m(s)}$$

If $G(s) = G_p(s)G_g(s)$ and $H(s) = AG_m(s)$, then the transfer function reduces to

$$\frac{\Theta(s)}{T_y(s)} = \frac{G(s)}{1 + G(s)H(s)}$$

Example 6 *As a final example we consider a d.c. position servomecha-*

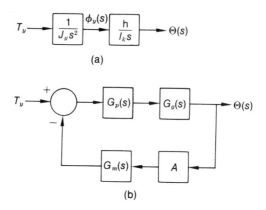

(a)

(b)

Figure 3.19: Stable platform

nism as shown in Fig. 3-20(a) where

$$G(s) = \frac{K_2}{s(s+25)}, \quad and \ H(s) = sK_3$$

Obtain the expression for $E(s)/R(s)$ and $C(s)/R(s)$.

We note that the control system has two feedback loops. The inner loop represents an armature-controlled d.c. motor. This part of the system can be represented by the following transfer function,

$$\frac{C(s)}{N(s)} = \frac{G(s)}{1 + G(s)H(s)} = F(s)$$

and the reduced block diagram is now shown in Fig. 3-20(b). We can now write the output transfer function as

$$\frac{C(s)}{R(s)} = \frac{K_1 F(s)}{1 + K_1 F(s)}$$

Substituting for $F(s)$ yields

$$\frac{C(s)}{R(s)} = \frac{K_1 G(s)}{1 + K_1 G(s) + G(s)H(s)}$$

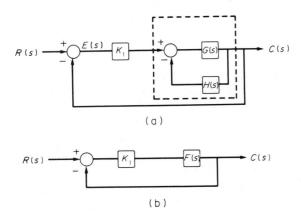

Figure 3.20: Block diagram for servomechanism

Finally substituting for $G(s)$ and $H(s)$ we have

$$\frac{C(s)}{R(s)} = \frac{K_1 K_2}{s^2 + (K_2 K_3 + 25)s + K_1 K_2}$$

The error transfer function is given by

$$\frac{E(s)}{R(s)} = \frac{1}{1 + K_1 F(s)}$$

Substituting for $F(s)$ yields

$$\frac{E(s)}{R(s)} = \frac{1 + G(s)H(s)}{1 + K_1 G(s) + G(s)H(s)}$$

Now substituting for $G(s)$ and $H(s)$ we have

$$\frac{E(s)}{R(s)} = \frac{s(s + K_2 K_3 + 25)}{s^2 + (K_2 K_3 + 25)s + K_1 K_2}$$

We note that the denominator is the same as that in the output transfer function.

3.6   State Equation Representation

When a system S is represented by its transfer function we are able to study the behavior of the system in terms of its input and output. The state

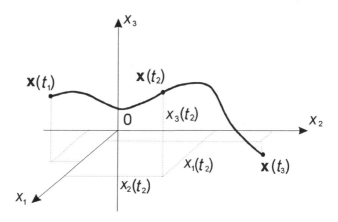

Figure 3.21: State space trajectory

variable method of system representation, however, allows us to investigate the behavior internal to the system $\mathcal{S}$ as well. In such a representation the state of a system is characterized by a collection of variables called **state variables**. Given the input they must be specified at any time in order to **uniquely predict** the behavior of the system, for $t \geq t_0$. The state of the system $\mathcal{S}$ is a function of the input for $t \geq t_0$ and the state at t_0, but is independent of the state and input before t_0. If there are n variables $x_1, x_2, \cdots, x_n$ to describe the state, then the vector $\mathbf{x}$ of n components is the **state vector**, and this vector is often denoted in matrix form as

$$\mathbf{x} = \begin{bmatrix} x_1 \\ x_2 \\ \vdots \\ x_n \end{bmatrix} \tag{3.23}$$

The n-dimensional space defined by the state variables as coordinates is called the **state space**, and as the state vector is a time function $\mathbf{x}(t)$, the path produced in the state space by $\mathbf{x}(t)$ as a function of time is called the **state trajectory**. A typical trajectory for a system having a 3-dimensional state space is shown in Fig. 3-21. This geometric interpretation of the state trajectory will prove useful in our later discussions.

The idea central to state representation is clarified if we consider the system shown in Fig. 3-22. We can ascertain the state of the system uniquely at any time if we know the angular position and velocity of the

Figure 3.22: A simple pendulum

pendulum, i.e. $\theta(t)$ and $\dot{\theta}(t)$. Let us define

$$x_1(t) = \theta(t), \quad x_2(t) = \dot{\theta}(t)$$

where $x_1(t)$ and $x_2(t)$ are the state variables of this system. If we consider θ to be small so that the governing equations are linear, then we can establish

$$\dot{x}_1(t) = x_2(t)$$

$$\dot{x}_2(t) = -\frac{g}{\ell}x_1(t)$$

which are the **state equations**. We have already shown how the mathematical model of a system may be written in this way in Chapter 2. It should be emphasized here that although the state of a dynamic system is uniquely determined from a specified set of state variables, the set itself is not unique. We can therefore select a variety of such sets. The particular one chosen is often dictated by convenience or the type of information desired.

In general, then, when we represent a system in terms of state variables we do so by a set of first-order linear differential equations of the form,

$$\dot{x}_1(t) = a_{11}x_1(t) + a_{12}x_2(t) + \cdots + a_{1n}x_n(t) + b_1u(t)$$

$$\dot{x}_2(t) = a_{21}x_1(t) + a_{22}x_2(t) + \cdots + a_{2n}x_n(t) + b_2u(t) \qquad (3.24)$$

$$\vdots$$

$$\dot{x}_n(t) = a_{n1}x_1(t) + a_{n2}x_2(t) + \cdots + a_{nn}x_n(t) + b_nu(t)$$

where $x_1(t), x_2(t), \cdots, x_n(t)$ are the state variables. The output of the system is expressed as a linear combination of the state variables,

$$y(t) = c_1x_1(t) + c_2x_2(t) + \cdots + c_nx_n(t) \qquad (3.25)$$

Eqs. (3-24) and (3-25) can be written compactly as

$$\dot{\mathbf{x}}(t) = \mathbf{A}\mathbf{x}(t) + \mathbf{b}u(t) \qquad (3.26)$$

$$y(t) = \mathbf{c}\mathbf{x}(t) \qquad (3.27)$$

where

$$\mathbf{x}(t) = \begin{bmatrix} x_1(t) \\ x_2(t) \\ \vdots \\ x_n(t) \end{bmatrix}, \quad \mathbf{A} = \begin{bmatrix} a_{11} & a_{12} & \cdots & a_{1n} \\ a_{21} & a_{22} & \cdots & a_{2n} \\ \vdots & \vdots & & \vdots \\ a_{n1} & a_{n2} & \cdots & a_{nn} \end{bmatrix}, \quad \mathbf{b} = \begin{bmatrix} b_1 \\ b_2 \\ \vdots \\ b_n \end{bmatrix}$$

and

$$\mathbf{c} = \begin{bmatrix} c_1 & c_2 & \cdots & c_n \end{bmatrix}$$

Here $\mathbf{x}(t)$ is the state vector, $\mathbf{A}$ is the **coefficient matrix** which is assumed constant, $\mathbf{b}$ is the **input vector** that essentially weights the effect of $u(t)$ which is the **system input** or forcing function, and $\mathbf{c}$ is the constant **output vector**. The vector-matrix differential equation shown in Eq. (3-26) is the **state equation** and that given by Eq. (3-27) is the **output equation**. When the input is zero in Eq. (3-26), the resulting equation becomes a **homogeneous equation.**

Before we continue, it is important that we establish the relationships between the state equation, system transfer function, and the governing differential equation of the system. This can be understood very easily as follows. We first assume zero initial conditions and take the Laplace transform of Eq. (3-26) and Eq. (3-27)

$$s\mathbf{X}(s) = \mathbf{A}\mathbf{X}(s) + \mathbf{b}U(s)$$

$$\mathbf{Y}(s) = \mathbf{c}\mathbf{X}(s)$$

Solving for $X(s)$ in the first equation and substituting into the second,

$$X(s) = [sI - A]^{-1}bU(s)$$

and

$$Y(s) = c[sI - A]^{-1}bU(s)$$

so that

$$\frac{Y(s)}{U(s)} = c[sI - A]^{-1}b \qquad (3.28)$$

Therefore $c[sI - A]^{-1}b$ is the **transfer function** of a system whose output is $Y(s)$ and input is $U(s)$.

An interesting form of Eq. (3-26) and Eq. (3-27) occurs when

$$A = \begin{bmatrix} 0 & 1 & 0 & \cdots & 0 \\ 0 & 0 & 0 & \cdots & 0 \\ \vdots & \vdots & \vdots & & \vdots \\ 0 & 0 & 0 & & 1 \\ -a_0 & -a_1 & -a_2 & \cdots & -a_{n-1} \end{bmatrix}, \quad b = \begin{bmatrix} 0 \\ 0 \\ \vdots \\ 0 \\ 1 \end{bmatrix}$$

$$c = \begin{bmatrix} c_0 & c_1 & \cdots & c_{n-1} \end{bmatrix}$$

so that

$$\begin{aligned} \dot{x}_1(t) &= x_2(t) \\[4pt] \dot{x}_2(t) &= x_3(t) \\ &\vdots \\ \dot{x}_{n-1}(t) &= x_n(t) \\[4pt] \dot{x}_n(t) &= -a_0 x_1(t) - a_1 x_2(t) - \cdots - a_{n-1} x_n(t) + u(t) \\[4pt] y(t) &= c_0 x_1(t) + c_1 x_2(t) + \cdots + c_{n-1} x_n(t) \end{aligned} \qquad (3.29)$$

Successive elimination of the variables $x_n, x_{n-1}, \ldots, x_2$ in Eq. (3-29) followed by the application of the Laplace transform to the resulting equation gives the transfer function

$$\frac{Y(s)}{U(s)} = \frac{c_{n-1}s^{n-1} + c_{n-2}s^{n-2} + \cdots + c_0}{s^n + a_{n-1}s^{n-1} + \cdots + a_0} \qquad (3.30)$$

This shows that the system whose transfer function is given by Eq. (3-30) has a state representation given by Eq. (3-29). We can additionally see that Eq. (3-30) corresponds to the differential equation

$$p^n y(t) + a_{n-1}p^{n-1}y(t) + \cdots + a_0 y(t) = c_{n-1}p^{n-1}u(t) + \cdots + c_0 u(t) \qquad (3.31)$$

We see therefore how the governing differential equation, transfer function, and state representation are related. They are, in fact, all equivalent but alternate ways of looking at the mathematical model of the same system.

Suppose a system is described by a linear ordinary differential equation. In the remainder of the section we will examine how state equations for this system can be obtained.

State Equations from Differential Equations

If the differential equation of a system is available, it may be directly represented in state variable form. Consider, for example, the equation of a harmonic oscillator,

$$p^2\theta(t) + k\theta(t) = f(t)$$

This equation may be rewritten as

$$p[p\theta(t)] = f(t) - k\theta(t)$$

Now let us define two state variables, $x_1(t) = \theta(t), x_2(t) = \dot{\theta}(t)$. These definitions combined with the previous equation yield

$$\dot{x}_1(t) \quad = \quad x_2(t)$$

$$\dot{x}_2(t) \quad = \quad -kx_1(t) + f(t)$$

which are two first-order linear differential equations replacing the original second-order differential equation. In matrix notation the above equations are written as,

$$\begin{bmatrix} \dot{x}_1(t) \\ \dot{x}_2(t) \end{bmatrix} = \begin{bmatrix} 0 & 1 \\ -k & 0 \end{bmatrix} \begin{bmatrix} x_1(t) \\ x_2(t) \end{bmatrix} + \begin{bmatrix} 0 \\ 1 \end{bmatrix} f(t)$$

or abbreviated as

$$\dot{\mathbf{x}}(t) = \mathbf{A}\mathbf{x}(t) + \mathbf{b}u(t)$$

where

$$\mathbf{x}(t) = \begin{bmatrix} x_1(t) \\ x_2(t) \end{bmatrix}, \quad \mathbf{A} = \begin{bmatrix} 0 & 1 \\ -k & 0 \end{bmatrix}, \quad \mathbf{b} = \begin{bmatrix} 0 \\ 1 \end{bmatrix}, \quad u(t) = f(t)$$

The coefficient matrix $\mathbf{A}$ is seen to be constant. If the previous differential equation had damping so that

$$p^2\theta(t) + Bp\theta(t) + k\theta(t) = f(t)$$

then the constant coefficient matrix becomes

$$A = \begin{bmatrix} 0 & 1 \\ -k & -B \end{bmatrix}$$

We can now easily generalize our observations and consider an nth-order differential equation,

$$p^n x(t) + a_{n-1}p^{n-1}x(t) + \cdots + a_0 x(t) = f(t) \qquad (3.32)$$

Again we employ the previous method of solving for the highest derivative,

$$p^n x(t) = -a_{n-1}p^{n-1}x(t) - \cdots - a_0 x(t) + f(t)$$

As before we define n state variables, $x_1 = x$, $x_2 = px$, $x_3 = p^2 x, \ldots$, $x_{n-1} = p^{n-2}x$, $x_n = p^{n-1}x$. Combining these definitions with the above equation shows

$$\dot{x}_1(t) = x_2(t)$$

$$\dot{x}_2(t) = x_3(t)$$

$$\dot{x}_3(t) = x_4(t) \qquad (3.33)$$

$$\vdots$$

$$\dot{x}_n(t) = -a_{n-1}x_n(t) - a_{n-2}x_{n-1}(t) - \cdots - a_0 x_1(t) + f(t)$$

$$x(t) = x_1(t)$$

In matrix notation the $n \times n$ **coefficient matrix**, the $n \times 1$ **input matrix** and the $1 \times n$ **output matrix**, become

$$A = \begin{bmatrix} 0 & 1 & 0 & 0 & \cdots & 0 \\ 0 & 0 & 1 & 0 & \cdots & 0 \\ 0 & 0 & 0 & 1 & \cdots & 0 \\ \vdots & \vdots & \vdots & \vdots & & \vdots \\ -a_0 & -a_1 & -a_2 & -a_3 & \cdots & -a_{n-1} \end{bmatrix}, \quad b = \begin{bmatrix} 0 \\ 0 \\ 0 \\ \vdots \\ 1 \end{bmatrix} \qquad (3.34)$$

$$c = \begin{bmatrix} 1 & 0 & 0 & \cdots & 0 \end{bmatrix}$$

where the matrix differential and output equations are defined by Eqs. (3-26) and (3-27).

The polynomial, $\lambda^n + a_{n-1}\lambda^{n-1} + \cdots + a_0$, which can be constructed, using the coefficients of the differential equation, is referred to as the **characteristic polynomial** of Eq. (3-32) as it plays an extremely important

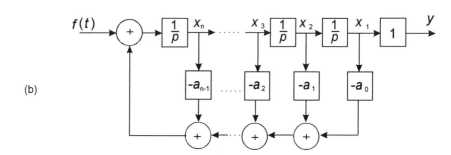

Figure 3.23: (a) Definition of integration block. (b) Block diagram for nth-order differential equation

part in the theory of differential equations. By comparing Eqs. (3-32) and (3-34) it can be seen that matrices $\mathbf{A}$, $\mathbf{b}$, and $\mathbf{c}$, can be written down by inspection. Since $\mathbf{A}$ is also easily constructed from the characteristic polynomial it is referred to as its **companion matrix**.

A graphical interpretation of Eq. (3-33) using block diagrams can be made by introducing the concept of an integrating block as shown in Fig. 3-23(a). Thus from Eq. (3-33) a block diagram can be constructed as shown in Fig. 3-23(b).

The examples considered so far have involved only one differential equation. In many situations it is necessary to study simultaneous equations that are coupled. The approach is the same as above and is illustrated in the next example.

Example 7 *Obtain state equations for a system described by*

$$p^2\phi(t) + k_1\phi(t) + B_1 p\psi(t) = f_1(t)$$

$$p^2\psi(t) + k_2\psi(t) + B_2 p\phi(t) = f_2(t)$$

We first write the governing equations as

$$p^2\phi(t) = -k_1\phi(t) - B_1 p\psi(t) + f_1(t)$$

$$p^2\psi(t) = -k_2\psi(t) - B_2 p\phi(t) + f_2(t)$$

As before, we define the following state variables,

$$x_1(t) = \phi(t)$$

$$x_2(t) = \dot{\phi}(t)$$

$$x_3(t) = \psi(t)$$

$$x_4(t) = \dot{\psi}(t)$$

so that the second derivatives of $\phi(t)$ and $\psi(t)$ are reduced to first derivatives in the newly introduced variables $x_2(t)$ and $x_4(t)$. From these definitions the state equations become

$$\dot{x}_1(t) = x_2(t)$$

$$\dot{x}_2(t) = -k_1 x_1(t) - B_1 x_4(t) + f_1(t)$$

$$\dot{x}_3(t) = x_4(t)$$

$$\dot{x}_4(t) = -k_2 x_3(t) - B_2 x_2(t) + f_2(t)$$

The coefficient matrix and the input matrix can now be written by inspection as

$$\mathbf{A} = \begin{bmatrix} 0 & 1 & 0 & 0 \\ -k_1 & 0 & 0 & -B_1 \\ 0 & 0 & 0 & 1 \\ 0 & -B_2 & -k_2 & 0 \end{bmatrix}, \quad \mathbf{b}u(t) = \begin{bmatrix} 0 \\ f_1(t) \\ 0 \\ f_2(t) \end{bmatrix}$$

where again the matrix differential equation is given by Eq. (3-26). We note that the procedure followed is identical to the case of the uncoupled problem. A block diagram for this system of state equations is shown in Fig. 3-24.

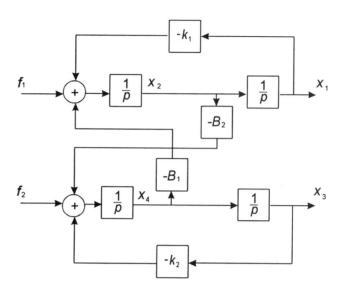

Figure 3.24: Block diagram for coupled differential equations

3.7   Relationship Between System Representations

We have seen in the previous section that if a system is modeled using an ordinary differential equation of the type shown in Eq. (3-32) then it may be transformed into an equivalent set of state equations as shown in Eq. (3-33). This transformation is **not unique**, and it is important to note that these equations are simply particular mathematical representations of the key element which is the **"system"**. The various representations however are all inter-related in the sense that each can be derived one from the other. In this section a number of canonical state equation realizations of a given system, which are derived from its transfer function, are developed.

Rather than examining the canonical realizations for a general nth-order transfer function we will examine the procedures used for obtaining these realizations for the 3rd order transfer function

$$\frac{Y(s)}{U(s)} = G(s) = \frac{c_2 s^2 + c_1 s + c_0}{s^3 + a_2 s^2 + a_1 s + a_0} \tag{3.35}$$

where there are no common factors in the numerator and denominator. The reader should be able to make the appropriate generalizations.

Jordan Canonical Form (Diagonal)

This technique is preferred when the transfer function appears in partial fraction form. Consider a transfer function that has been expanded in partial fraction form,

$$\frac{Y(s)}{U(s)} = \frac{1}{6s} - \frac{2}{3(s+3)} + \frac{3}{2(s+4)} \tag{3.36}$$

The block diagram is generated for each factor and then added as shown in Fig. 3-25. The state variables are defined using the convention of the previous section so

$$\dot{x}_1(t) = \tfrac{1}{6} u(t)$$

$$\dot{x}_2(t) = -\tfrac{2}{3} u(t) - 3x_2(t)$$

$$\dot{x}_3(t) = \tfrac{3}{2} u(t) - 4x_3(t)$$

$$y(t) = x_1(t) + x_2(t) + x_3(t)$$

The coefficient matrices and the state vector become

$$\mathbf{x}(t) = \begin{bmatrix} x_1(t) \\ x_2(t) \\ x_3(t) \end{bmatrix}, \quad \mathbf{A} = \begin{bmatrix} 0 & 0 & 0 \\ 0 & -3 & 0 \\ 0 & 0 & -4 \end{bmatrix}, \quad \mathbf{b} = \begin{bmatrix} 1/6 \\ -2/3 \\ 3/2 \end{bmatrix}$$

$$\mathbf{c} = \begin{bmatrix} 1 & 1 & 1 \end{bmatrix}$$

While no convention has been established for the names of the following canonical forms we shall follow the terminology adopted by Kailath.

Controller Canonical Form

Rearranging Eq. (3-35) we obtain

$$\frac{Y(s)}{c_2 s^2 + c_1 s + c_0} = \frac{U(s)}{s^3 + a_2 s^2 + a_1 s + a_0} = E(s) \tag{3.37}$$

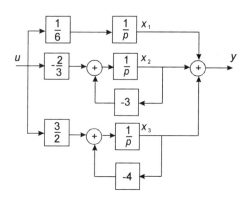

Figure 3.25: Block diagram for Jordan canonical form

With E(s) defined as in Eq. (3-37) we can write the first and second terms of this equation in differential equation form,

$$y(t) = (c_2 p^2 + c_1 p + c_0)e(t) \qquad (3.38)$$

$$(p^3 + a_2 p^2 + a_1 p + a_0)e(t) = u(t) \qquad (3.39)$$

Defining the state variables for the system as $x_1 = e, x_2 = pe, x_3 = p^2 e$, and combining them with Eq. (3-39) gives

$$\dot{x}_1(t) = x_2(t)$$

$$\dot{x}_2(t) = x_3(t) \qquad (3.40)$$

$$\dot{x}_3(t) = -a_0 x_1(t) - a_1 x_2(t) - a_2 x_3(t) + u(t)$$

It will be observed from these definitions and Eq. (3-38) that the output is

$$y(t) = c_2 x_3(t) + c_1 x_2(t) + c_0 x_1(t) \qquad (3.41)$$

If the state equations are written in vector form, then

$$\dot{\mathbf{x}}(t) = \mathbf{A}\mathbf{x}(t) + \mathbf{b}u(t) \qquad (3.42)$$

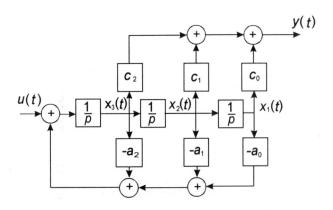

Figure 3.26: Controller canonical form

$$y(t) = \mathbf{cx}(t) \qquad (3.43)$$

where

$$\mathbf{x}(t) = \begin{bmatrix} x_1(t) \\ x_2(t) \\ x_3(t) \end{bmatrix}, \quad \mathbf{A} = \begin{bmatrix} 0 & 1 & 0 \\ 0 & 0 & 1 \\ -a_0 & -a_1 & -a_2 \end{bmatrix}, \quad \mathbf{b} = \begin{bmatrix} 0 \\ 0 \\ 1 \end{bmatrix}$$

$$\mathbf{c} = \begin{bmatrix} c_0 & c_1 & c_2 \end{bmatrix}$$

Paralleling the last section, where the transformation of differential equations to state equations was discussed, the polynomial $Q(s) = s^3 + a_2 s^2 + a_1 s + a_0$, derived from the denominator of the transfer function Eq. (3-35), will be referred to as its **characteristic polynomial**. As well the matrix $\mathbf{A}$ given above will also be referred to as the **companion matrix** of $Q(s)$.

The block diagram for Eqs. (3-40) and (3-41) is shown in Fig. 3-26.

Example 8 *Obtain the controller canonical form for the system transfer function*

$$\frac{Y(s)}{U(s)} = \frac{(s+1)(s+2)}{s(s+3)(s+4)}$$

Expanding the numerator and denominator polynomials and re-writing the transfer function in the differential equation form given in Eq. (3-38) gives

$$y(t) = (p^2 + 3p + 2)e(t)$$

and
$$(p^3 + 7p^2 + 12p)e(t) = u(t)$$

Defining the state variables $x_1 = e, x_2 = pe, x_3 = p^2 e$, we find the state equations are

$$\dot{x}_1(t) = x_2(t)$$

$$\dot{x}_2(t) = x_3(t)$$

$$\dot{x}_3(t) = -12x_2(t) - 7x_3(t) + u(t)$$

and
$$y(t) = 2x_1(t) + 3x_2(t) + x_3(t)$$

If the state equations are written in vector form then

$$A = \begin{bmatrix} 0 & 1 & 0 \\ 0 & 0 & 1 \\ 0 & -12 & -7 \end{bmatrix}, \quad b = \begin{bmatrix} 0 \\ 0 \\ 1 \end{bmatrix}$$

$$c = \begin{bmatrix} 2 & 3 & 1 \end{bmatrix}$$

Observability Canonical Form

We now approach the establishment of the state equations for Eq. (3-35) in a different manner. We can write Eq. (3-35) in differential equation form as

$$(p^3 + a_2 p^2 + a_1 p + a_0)x_1(t) = (c_2 p^2 + c_1 p + c_0)u(t) \tag{3.44}$$

where we define $x_1(t) = y(t)$. Transposing the first term from the right to the left side gives

$$p^2(px_1(t) - c_2 u(t)) + a_2 p^2 x_1(t) + a_1 p x_1(t) + a_0 x_1(t) = c_1 p u(t) + c_0 u(t)$$

The first state equation and the state variable x_2 are now defined by

$$x_2(t) = px_1(t) - \gamma_2 u(t) \tag{3.45}$$

where we define $\gamma_2 = c_2$. Substituting Eq. (3-45) into Eq. (3-44) gives

$$p^2 x_2(t) + a_2 p(x_2(t) + \gamma_2 u(t)) + a_1(x_2(t) + \gamma_2 u(t)) + a_0 x_1(t) = c_1 p u(t) + c_0 u(t)$$

and combining like terms in this equation gives

$$p^2 x_2(t) + a_2 p x_2(t) + a_1 x_2(t) + a_0 x_1(t) = \gamma_1 p u(t) + \delta_1 u(t) \tag{3.46}$$

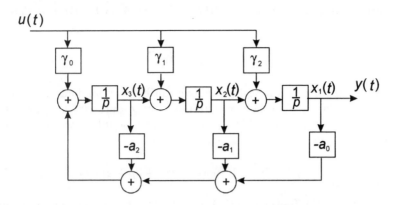

Figure 3.27: Observability canonical form

where the coefficients γ_1 and δ_1 are determined from the coefficients a_1, a_2, c_0, c_1, and c_2. It will be noted that the magnitude of the highest derivative on the right hand side of Eq. (3-46) has been reduced by a factor of one. Applying the above procedure once again to Eq. (3-46) and defining the state variable x_3 as

$$x_3(t) = px_2(t) - \gamma_1 u(t) \tag{3.47}$$

gives the following equation

$$px_3(t) = -a_0 x_1(t) - a_1 x_2(t) - a_2 x_3(t) + \gamma_0 u(t) \tag{3.48}$$

where γ_0 is obtained in a manner similar to the one used to obtain γ_1. The state equations Eqs. (3-45) to (3-48) can be written in the vector form given by Eq. (3-42) and (3-43), where the coefficient matrices are

$$\mathbf{x}(t) = \begin{bmatrix} x_1(t) \\ x_2(t) \\ x_3(t) \end{bmatrix}, \quad \mathbf{A} = \begin{bmatrix} 0 & 1 & 0 \\ 0 & 0 & 1 \\ -a_0 & -a_1 & -a_2 \end{bmatrix}, \quad \mathbf{b} = \begin{bmatrix} \gamma_2 \\ \gamma_1 \\ \gamma_0 \end{bmatrix}$$

$$\mathbf{c} = \begin{bmatrix} 1 & 0 & 0 \end{bmatrix}$$

The block diagram for the state equations given above is shown in Fig. 3-27. It will be noted that while the state variable signals are fed back and summed with the input $u(t)$ in a similar manner to the case shown in Fig. 3-26, the output time function $y(t)$ is no longer a linear combination of the state variables. Instead weighted values of the input time function $u(t)$ are summed with the state variables as shown in the figure.

Example 9 *Obtain the observability canonical form for the transfer function*

$$\frac{Y(s)}{U(s)} = \frac{(s+1)(s+2)}{s(s+3)(s+4)}$$

While the algorithm outlined above can be used for finding the coefficients γ_0, γ_1 and γ_2 we show an alternative procedure which simplifies the calculations considerably. Firstly, as shown in Example 8, we can write the transfer function in differential equation form as

$$(p^3 + 7p^2 + 12p)y(t) = (p^2 + 3p + 2)u(t) \qquad (3.49)$$

From Eqs. (3-45), (3-47), and (3-48) the following equation can be obtained

$$\{p\left[p\left(p+a_2\right)+a_1\right]+a_0\}\, y(t) = \{\gamma_2\left[p\left(p+a_2\right)+a_1\right]+\gamma_1\left(p+a_2\right)+\gamma_0\}\, u(t)$$

It will be noted that the terms contained in the braces on the left-hand and right-hand sides expand to

$$p^3 + a_2 p^2 + a_1 p + a_0$$

and

$$\gamma_2 p^2 + (a_2\gamma_2 + \gamma_1)p + (a_1\gamma_2 + a_2\gamma_1 + \gamma_0)$$

respectively. Comparing the terms in these two expansions with those shown in Eq. (3-49) shows that $a_0 = 0, a_1 = 12, a_2 = 7$, while γ_0, γ_1, and γ_2 are given by the solution of the equation

$$\begin{bmatrix} 1 & 0 & 0 \\ a_2 & 1 & 0 \\ a_1 & a_2 & 1 \end{bmatrix} \begin{bmatrix} \gamma_2 \\ \gamma_1 \\ \gamma_0 \end{bmatrix} = \begin{bmatrix} 1 \\ 3 \\ 2 \end{bmatrix}$$

Hence $\gamma_0 = 18, \gamma_1 = -4$, and $\gamma_2 = 1$.

In the general case the matrix $\mathbf{A}$, in the state equation, is the companion matrix and the coefficients $\gamma_0, \cdots, \gamma_{n-1}$ are given by the solution of the equation

$$\begin{bmatrix} 1 & 0 & \cdots & 0 & 0 \\ a_{n-1} & 1 & \cdots & 0 & 0 \\ \vdots & \vdots & & \vdots & \vdots \\ a_1 & a_2 & \cdots & a_{n-1} & 1 \end{bmatrix} \begin{bmatrix} \gamma_{n-1} \\ \gamma_{n-2} \\ \vdots \\ \gamma_0 \end{bmatrix} = \begin{bmatrix} c_{n-1} \\ c_{n-2} \\ \vdots \\ c_0 \end{bmatrix}$$

Observer Canonical Form

Taking Eq. (3-44), which is the differential equation form of the transfer function Eq. (3-35) we have

$$p^3 x_1(t) + p^2(a_2 x_1(t) - c_2 u(t)) + p(a_1 x_1(t) - c_1 u(t)) + (a_0 x_1(t) - c_0 u(t)) = 0 \tag{3.50}$$

where the output $y(t) = x_1(t)$. Eq. (3-50) can be re-written in the following form

$$p^2(p x_1(t) + a_2 x_1(t) - c_2 u(t)) + p(a_1 x_1(t) - c_1 u(t)) + (a_0 x_1(t) - c_0 u(t)) = 0 \tag{3.51}$$

and we can write

$$p x_1(t) + a_2 x_1(t) - c_2 u(t) = x_2(t) \tag{3.52}$$

Substituting the above expression in Eq. (3-51) we obtain

$$p(p x_2(t) + a_1 x_1(t) - c_1 u(t)) + (a_0 x_1(t) - c_0 u(t)) = 0$$

where we again write

$$p x_2(t) + a_1 x_1(t) - c_1 u(t) = x_3(t)$$

Finally we have the state equations,

$$\dot{x}_1(t) = x_2(t) - a_2 x_1(t) + c_2 u(t)$$

$$\dot{x}_2(t) = x_3(t) - a_1 x_1(t) + c_1 u(t)$$

$$\dot{x}_3(t) = -a_0 x_1(t) + c_0 u(t)$$

If the state equations are written in the vector form given by Eqs. (3-42) and (3-43), the coefficient matrices are

$$\mathbf{x}(t) = \begin{bmatrix} x_1(t) \\ x_2(t) \\ x_3(t) \end{bmatrix}, \quad \mathbf{A} = \begin{bmatrix} -a_2 & 1 & 0 \\ -a_1 & 0 & 1 \\ -a_0 & 0 & 0 \end{bmatrix}, \quad \mathbf{b} = \begin{bmatrix} c_2 \\ c_1 \\ c_0 \end{bmatrix}$$

$$\mathbf{c} = \begin{bmatrix} 1 & 0 & 0 \end{bmatrix}$$

where $\mathbf{A}$ is an alternative form of the **companion matrix**. The block diagram for the state equations given above is shown in Fig. 3-28.

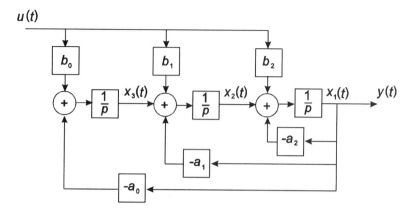

Figure 3.28: Observer canonical form

Controllability Canonical Form

For the last canonical form to be examined we re-write Eqs. (3-38) and (3-39) as

$$y(t) = c_2 p^2 x_1(t) + c_1 p x_1(t) + c_0 x_1(t) \qquad (3.53)$$

$$p^3 x_1(t) + a_2 p^2 x_1(t) + a_1 p x_1(t) + a_0 x_1(t) = u(t) \qquad (3.54)$$

Following the same arguments used in transforming Eq. (3-50) into a set of state equations, we find

$$\dot{x}_1(t) \;=\; x_2(t) - a_2 x_1(t)$$

$$\dot{x}_2(t) \;=\; x_3(t) - a_1 x_1(t)$$

$$\dot{x}_3(t) \;=\; -a_0 x_1(t) + u(t)$$

From Eq. (3-53) and the state equation for $\dot{x}_1(t)$ we obtain

$$y(t) \;=\; c_2(p x_2(t) - a_2 p x_1(t)) + c_1 p x_1(t) + c_0 x_1(t)$$

$$\;=\; c_2(x_3(t) - a_1 x_1(t) - a_2 p x_1(t)) + c_1 p x_1(t) + c_0 x_1(t)$$

Combining like terms in the above equation and taking $\beta_2 = c_2$ gives

$$y(t) = \beta_2 x_3(t) + \alpha_2 p x_1(t) + \delta_2 x_1(t)$$

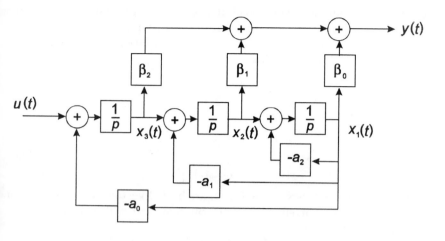

Figure 3.29: Controllability canonical form

Repeating the above procedure we find

$$y(t) = \beta_2 x_3(t) + \beta_1 x_2(t) + \beta_0 x_1(t)$$

If the state equations are written in vector form as given in Eqs. (3-42) and (3-43), then the coefficient matrices are

$$\mathbf{x}(t) = \begin{bmatrix} x_1(t) \\ x_2(t) \\ x_3(t) \end{bmatrix}, \quad \mathbf{A} = \begin{bmatrix} -a_2 & 1 & 0 \\ -a_1 & 0 & 1 \\ -a_0 & 0 & 0 \end{bmatrix}, \quad \mathbf{b} = \begin{bmatrix} 0 \\ 0 \\ 1 \end{bmatrix}$$

$$\mathbf{c} = \begin{bmatrix} \beta_0 & \beta_1 & \beta_2 \end{bmatrix}$$

The block diagram for the state equations given above is shown in Fig. 3-29.

Similarity Transformations Between State Equation Realizations

We have seen that a single-input single-output system whose transfer function has the form shown in Eq. (3-35) has a number of state equation realizations which can be written using vector notation as shown in Eqs. (3-26) and (3-27). Since these realizations depend upon how the state variables are defined their number is clearly unlimited.

Provided the state equation realizations for the transfer function given by Eq. (3-35) have three state variables, which will be the case for the canonical realizations given above, it can be shown that the state vector time functions are related by **non-singular similarity transformations**. To illustrate this suppose two state equation realizations for Eq. (3-35) are

$$\dot{\mathbf{x}}_a(t) = \mathbf{A}_a \mathbf{x}_a(t) + \mathbf{b}_a u(t)$$
$$y(t) = \mathbf{c}_a \mathbf{x}_a(t)$$

(3.55)

and

$$\dot{\mathbf{x}}_b(t) = \mathbf{A}_b \mathbf{x}_b(t) + \mathbf{b}_b u(t)$$
$$y(t) = \mathbf{c}_b \mathbf{x}_b(t)$$

(3.56)

where the state vectors $\mathbf{x}_a(t)$ and $\mathbf{x}_b(t)$ are both 3-vectors. It can be shown that there exists a non-singular transformation matrix $\mathbf{T}$ such that

$$\mathbf{x}_a(t) = \mathbf{T}\mathbf{x}_b(t), \text{ and } \mathbf{x}_b(t) = \mathbf{T}^{-1}\mathbf{x}_a(t)$$

(3.57)

Using this transformation we show that the coefficient matrices $\mathbf{A}$, $\mathbf{b}$, and $\mathbf{c}$ for the two realizations are related. From Eqs. (3-55) and (3-57)

$$\dot{\mathbf{x}}_b(t) = \mathbf{T}^{-1}\dot{\mathbf{x}}_a(t) = (\mathbf{T}^{-1}\mathbf{A}_a\mathbf{T})\mathbf{x}_b(t) + (\mathbf{T}^{-1}\mathbf{b}_a)u(t)$$
$$y(t) = (\mathbf{c}_a\mathbf{T})\mathbf{x}_b(t)$$

(3.58)

Equating coefficients in Eqs. (3-58) and (3-56) shows that $\mathbf{A}_b = \mathbf{T}^{-1}\mathbf{A}_a\mathbf{T}$, $\mathbf{b}_b = \mathbf{T}^{-1}\mathbf{b}_a$, and $\mathbf{c}_b = \mathbf{c}_a\mathbf{T}$.

It can be shown that in the case above the transformation $\mathbf{T}$ is given by

$$\mathbf{T} = \mathcal{C}\,(\mathbf{A}_a, \mathbf{b}_a)\,\mathcal{C}^{-1}\,(\mathbf{A}_b, \mathbf{b}_b)$$

(3.59)

where we define the matrices $\mathcal{C}\,(\mathbf{A}_a, \mathbf{b}_a) = \begin{bmatrix} \mathbf{b}_a & \mathbf{A}_a\mathbf{b}_a & \mathbf{A}_a^2\mathbf{b}_a \end{bmatrix}$ and $\mathcal{C}\,(\mathbf{A}_b, \mathbf{b}_b) = \begin{bmatrix} \mathbf{b}_b & \mathbf{A}_b\mathbf{b}_b & \mathbf{A}_b^2\mathbf{b}_b \end{bmatrix}$. Both these matrices are non-singular so that $\mathbf{T}$ also has an inverse.

Example 10 *We now demonstrate that the transformation* $\mathbf{T}$ *defined in Eq. (3-59) can be applied to transform the controller to the observability canonical form for the transfer function given in Example 8.*

Designating the coefficient matrices in Example 8 by $\mathbf{A}_a, \mathbf{b}_a, \mathbf{c}_a$, and those in Example 9 by $\mathbf{A}_b, \mathbf{b}_b, \mathbf{c}_b$, we use them to evaluate $\mathcal{C}(\mathbf{A}_a, \mathbf{b}_a)$ and $\mathcal{C}(\mathbf{A}_b, \mathbf{b}_b)$ which are defined above. Thus

$$\mathcal{C}\,(\mathbf{A}_a, \mathbf{b}_a) = \begin{bmatrix} 0 & 0 & 1 \\ 0 & 1 & -7 \\ 1 & -7 & 37 \end{bmatrix}, \text{ and } \mathcal{C}\,(\mathbf{A}_b, \mathbf{b}_b) = \begin{bmatrix} 1 & -4 & 18 \\ -4 & 18 & -78 \\ 18 & -78 & 330 \end{bmatrix}$$

Now evaluating Eq. (3-59) gives

$$\mathbf{T} = \begin{bmatrix} 1/2 & -1/4 & -1/12 \\ 0 & 3/2 & 1/3 \\ 0 & -4 & 5/6 \end{bmatrix}, \text{ and } \mathbf{T}^{-1} = \begin{bmatrix} 2 & 3 & 1 \\ 0 & -10 & -4 \\ 0 & 48 & 18 \end{bmatrix}$$

Simple numeric computation shows that

$$\mathbf{T}^{-1}\mathbf{b}_a = \begin{bmatrix} 1 \\ -4 \\ 18 \end{bmatrix} = \mathbf{b}_b$$

$$\mathbf{T}^{-1}\mathbf{A}_a\mathbf{T} = \begin{bmatrix} 0 & 1 & 0 \\ 0 & 0 & 1 \\ 0 & -12 & -7 \end{bmatrix} = \mathbf{A}_b$$

and

$$\mathbf{c}_a\mathbf{T} = \begin{bmatrix} 1 & 0 & 0 \end{bmatrix} = \mathbf{c}_b$$

Thus we see that the equation

$$\mathbf{x}_a(t) = \mathbf{T}\mathbf{x}_b(t)$$

where $\mathbf{T}$ is given above, defines the relationship between the corresponding state variables of $\mathbf{x}_a(t)$ and $\mathbf{x}_b(t)$.

3.8   Small Disturbance of Nonlinear Systems

We have seen in Chapter 2 that the equations of motion for many physical systems are non-linear, and for state equations, in particular, their right-hand side is a non-linear function of the state variables. It is often difficult to design systems for controlling non-linear processes by analytical methods so that we are often forced to use numerical methods for simulating their operation.

In many cases when these systems are subjected to small disturbances about some specified operating conditions their operation can be approximated by a linear system of state equations. To see this for a general system suppose the state equations are in vector form

$$\dot{\mathbf{x}}(t) = f(\mathbf{x}(t), \mathbf{u}(t)) \tag{3.60}$$

where $\mathbf{x}(t)$ is an n-dimensional state vector and $\mathbf{u}(t)$ is an r-dimensional control vector. The function $f(\mathbf{x}, \mathbf{u})$ is assumed to be a non-linear function of $\mathbf{x}$ and $\mathbf{u}$, which can be approximated at any point by a Taylor series.

Let $\mathbf{x}_0(t)$ be the nominal operating trajectory for the input $\mathbf{u}_0(t)$ and initial state $\mathbf{x}_0(t_0) = \mathbf{x}_0$. We now expand the non-linear state equation Eq. (3-60) about the nominal trajectory $\mathbf{x}_0(t)$ using Taylor's theorem and assuming the terms $x_i(t) - x_{i0}(t)$ and $u_i(t) - u_{i0}(t)$ are small we can neglect the remainder terms. Thus

$$\dot{x}_i(t) = f_i(\mathbf{x}_0(t), \mathbf{u}_0(t)) + \sum_{j=1}^{n} \frac{\partial f_i(\mathbf{x}_0(t), \mathbf{u}_0(t))}{\partial x_j}(x_j(t) - x_{j0}(t))$$

$$+ \sum_{j=1}^{r} \frac{\partial f_i(\mathbf{x}_0(t), \mathbf{u}_0(t))}{\partial u_j}(u_j(t) - u_{j0}(t)) \tag{3.61}$$

for $i = 1, 2, \cdots, n$. Defining $\Delta x_i(t) = x_i(t) - x_{i0}(t)$ and $\Delta u_i(t) = u_i(t) - u_{i0}(t)$, and noting that

$$\dot{x}_{i0}(t) = f_i(\mathbf{x}_0(t), \mathbf{u}_0(t)) \tag{3.62}$$

we find Eq. (3-61) becomes

$$\Delta \dot{x}_i(t) = \sum_{j=1}^{n} \frac{\partial f_i(\mathbf{x}_0(t), \mathbf{u}_0(t))}{\partial x_j} \Delta x_j(t) + \sum_{j=1}^{r} \frac{\partial f_i(\mathbf{x}_0(t), \mathbf{u}_0(t))}{\partial u_j} \Delta u_j \tag{3.63}$$

which can be written in vector equation form,

$$\Delta \dot{\mathbf{x}}(t) = \mathbf{A}_0 \Delta \mathbf{x}(t) + \mathbf{B}_0 \Delta \mathbf{u}(t) \tag{3.64}$$

The matrices $\mathbf{A}_0$ and $\mathbf{B}_0$, which are evaluated at $\mathbf{x}_0(t)$ and $\mathbf{u}_0(t)$, are

$$\mathbf{A}_0 = \begin{bmatrix} \dfrac{\partial f_1}{\partial x_1} & \dfrac{\partial f_1}{\partial x_2} & \cdots & \dfrac{\partial f_1}{\partial x_n} \\ \dfrac{\partial f_2}{\partial x_1} & \dfrac{\partial f_2}{\partial x_2} & \cdots & \dfrac{\partial f_2}{\partial x_n} \\ \vdots & \vdots & & \vdots \\ \dfrac{\partial f_n}{\partial x_1} & \dfrac{\partial f_n}{\partial x_2} & \cdots & \dfrac{\partial f_n}{\partial x_n} \end{bmatrix}, \quad \mathbf{B}_0 = \begin{bmatrix} \dfrac{\partial f_1}{\partial u_1} & \dfrac{\partial f_1}{\partial u_2} & \cdots & \dfrac{\partial f_1}{\partial u_n} \\ \dfrac{\partial f_2}{\partial u_1} & \dfrac{\partial f_2}{\partial u_2} & \cdots & \dfrac{\partial f_2}{\partial u_n} \\ \vdots & \vdots & & \vdots \\ \dfrac{\partial f_n}{\partial u_1} & \dfrac{\partial f_n}{\partial u_2} & \cdots & \dfrac{\partial f_n}{\partial u_n} \end{bmatrix}$$

Example 11 *The simplified block diagram of a phase-lock-loop motor speed control system is shown in Fig. 3-30. Find the linearized equations of motion for this system.*

From the block diagram we find

$$\dot{\phi}(t) = x(t)$$

$$\dot{x}(t) = -\frac{1}{\tau}x(t) + \frac{AK}{\tau}\sin(\phi_d(t) - \phi(t))$$

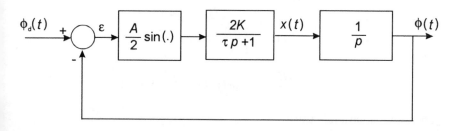

Figure 3.30: Phase lock loop motor speed control system

Let $x_0(t)$ and $\phi_0(t)$ describe the nominal operating trajectory for the input $\phi_{d0}(t)$. Expanding the right hand side of this equation in a Taylor series gives

$$\dot{\phi}(t) \;=\; x_0(t) + (x(t) - x_0(t))$$

$$\dot{x}(t) \;=\; -\frac{1}{\tau}x_0(t) + \frac{AK}{\tau}\sin(\phi_{d0}(t) - \phi_0(t)) + \Bigg\{ -\frac{1}{\tau}[x(t) - x_0(t)]$$

$$+ \left[-\frac{AK}{\tau}\cos(\phi_{d0}(t) - \phi_0(t)) \right](\phi(t) - \phi_0(t))$$

$$+ \left[\frac{AK}{\tau}\cos(\phi_{d0}(t) - \phi_0(t)) \right](\phi_d(t) - \phi_{d0}(t)) \Bigg\}$$

We note from above that

$$\dot{\phi}_0(t) \;=\; x_0(t)$$

$$\dot{x}_0(t) \;=\; -\frac{1}{\tau}x_0(t) + \frac{AK}{\tau}\sin(\phi_{d0}(t) - \phi_0(t))$$

Defining $\Delta\phi(t) = \phi(t) - \phi_0(t)$, $\Delta x(t) = x(t) - x_0(t)$, and $\Delta\phi_d(t) = \phi_d(t) -$

$\phi_{d0}(t)$ we obtain

$$\Delta\dot\phi(t) = \Delta x(t)$$

$$\Delta\dot x(t) = -\frac{1}{\tau}\Delta x(t) - \left[\frac{AK}{\tau}\cos(\phi_{d0}(t) - \phi_0(t))\right]\Delta\phi(t)$$

$$+ \left[\frac{AK}{\tau}\cos(\phi_{d0}(t) - \phi_0(t))\right]\Delta\phi_d(t)$$

The state equation is thus

$$\dot{\mathbf x}(t) = \mathbf A\mathbf x(t) + \mathbf b u(t),$$

where

$$\mathbf x(t) = \left[\begin{array}{c} \Delta\phi(t) \\ \Delta x(t) \end{array}\right], \quad \mathbf A = \left[\begin{array}{cc} 0 & 1 \\ -\dfrac{AK}{\tau}\cos(\phi_{d0} - \phi_0) & -\dfrac{1}{\tau} \end{array}\right]$$

$$\mathbf b = \left[\begin{array}{c} 0 \\ \dfrac{AK}{\tau}\cos(\phi_{d0} - \phi_0) \end{array}\right]$$

and $u(t) = \Delta\phi_d(t)$.

Example 12 *The angular attitude motions θ_x, θ_y, and θ_z of an artificial satellite are given by the following three equations,*

$$p^2\theta_x(t) + \delta_1 p\theta_y(t)p\theta_z(t) = f_x(t)$$

$$p^2\theta_y(t) + \delta_2 p\theta_x(t)p\theta_z(t) + \delta_4 p\theta_z(t) = f_y(t)$$

$$p^2\theta_z(t) + \delta_3 p\theta_x(t)p\theta_y(t) - \delta_5 p\theta_y(t) = f_z(t)$$

where it is assumed that $\dot\theta_y, \dot\theta_y$ are small and $\dot\theta_x = \omega + \dot\phi_x$ where ω is constant but $\dot\phi_x$ is small, and $\delta_1, \delta_2, \delta_3, \delta_4$, and δ_5 are all constants. Obtain the state representation for these equations involving $\dot\phi_x, \dot\theta_y$, and $\dot\theta_z$.

We first linearize the equations. Since $\dot\theta_y, \dot\theta_z, \dot\phi_x$ are small their products can be neglected, so that

$$p^2\phi_x(t) = f_x(t)$$

$$p^2\theta_y(t) + \alpha p\theta_z(t) = f_y(t)$$

$$p^2\theta_z(t) - \beta p\theta_y(t) = f_z(t)$$

where $\alpha = \delta_4 + \delta_2\omega$, and $\beta = \delta_5 - \delta_3\omega$. The state equation becomes

$$\dot{\mathbf{x}}(t) = \mathbf{A}\mathbf{x}(t) + \mathbf{b}u(t)$$

where

$$\mathbf{x(t)} = \begin{bmatrix} \dot{\phi}_x(t) \\ \dot{\theta}_y(t) \\ \dot{\theta}_z(t) \end{bmatrix}, \quad \mathbf{A} = \begin{bmatrix} 0 & 0 & 0 \\ 0 & 0 & -\alpha \\ 0 & \beta & 0 \end{bmatrix}, \quad \mathbf{b}u(t) = \begin{bmatrix} f_x(t) \\ f_y(t) \\ f_z(t) \end{bmatrix}$$

Note this yields three first-order differential equations since we are solving for the rates only, and also we note there are three inputs f_x, f_y, f_z.

3.9 Summary

In this chapter we have been concerned with various ways of representing control systems. Each block diagram was characterized by its own transfer function, i.e. the ratio of the output over the input signal when the signals are represented in the Laplace domain. This transfer function contains all the information we require about the behavior of the physical system represented by the block diagram. By a systematic handling of the transfer functions we were able to reduce complicated systems to a single overall transfer function.

The state representation involved the characterization of the entire system in matrix notation. It was seen that although the number of state variables is unique, any particular set is not. We described how the differential equation, transfer function, and the state equations of a system are related.

3.10 References

1. N. Balabanian, T. A. Bickart, *Linear Network Theory*, Matrix Publishers, Inc., Beaverton, Oregon, 1981.

2. S. Seshu, N. Balabanian, *Linear Network Analysis*, John Wiley and Sons, Inc., New York, 1959.

3. T. Kailath, *Linear Systems*, Prentice-Hall, Inc., Englewood Cliffs, N. J., 1980.

4. R. J. Mayhan, *Discrete-Time and Continuous-Time Linear Systems*, Addison Wesley Publ. Co., Reading, Mass., 1983.

5. R. W. Brockett, *Finite Dimensional Linear Systems*, John Wiley and Sons, Inc., New York, 1970.

6. N. Minorsky, *Non Linear Oscillations*, Robert E. Krieger Publ. Co., Inc., Malabar, Florida, 1987.

7. J. J. D'Azzo, C. H. Houpis, *Linear Control System Analysis and Design — Conventional and Modern*, Mc Graw Hill Book Co., New York, 1988.

8. J. Meisel, *Principles of Electromechanical Energy Conversion*, Mc Graw Hill Book Co., New York, 1966.

9. N. H. Mc Clamroch, *State Models of Dynamic Systems — A Case Study Approach*, Springer-Verlag, New York, 1980.

10. A. E. Bryson, Y. C. Ho, *Applied Optimal Control — Optimization, Estimation, and Control*, Ginn and Co., Waltham, Mass, 1969.

3.11 Problems

3-1 Evaluate the convolution integrals

(a) $H(t) * e^{-4t}H(t)$
(b) $(1 - e^{-10t})H(t) * e^{-3t}H(t)$
(c) $H(t)\cos t * [H(t) - H(t-3)]$

3-2 Evaluate the Du Hamel integral for the systems having step responses and inputs given below

(a) $g_s(t) = (1 - e^{-4t})H(t), \quad u(t) = H(t)$
(b) $g_s(t) = (1 - e^{-3t})H(t), \quad u(t) = (1 - e^{-10t})H(t)$
(c) $g_s(t) = H(t)\sin t, \quad u(t) = H(t) - H(t-3)$

Compare these results with those obtained in Problem 3-1.

3-3 Using graphical methods find the convolutions for the functions shown in Fig. P3-3.

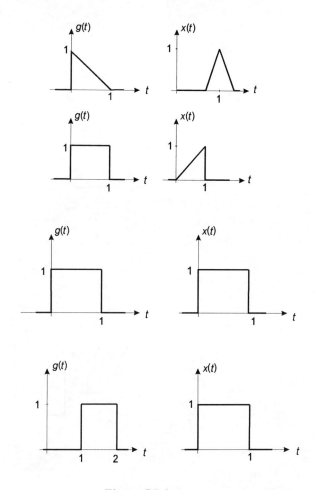

Figure P3-3

3-4 The impulse and step responses for the systems shown in Figs. P3-4 (a), (b), and (c) are

(a) $g(t) = e^{-10t/3}H(t)/3,\ g_s(t) = \left(1 - e^{-10t/3}\right)H(t)/10$
(b) $g(t) = 5(3e^{-3t} - 2e^{-2t})H(t),\ g_s(t) = 5(e^{-2t} - e^{-3t})H(t)$
(c) $g(t) = (2e^{-2t} - 3e^{-3t})H(t),\ g_s(t) = \left(1 - 3e^{-2t} + 2e^{-3t}\right)H(t)/6$

Find the response of each system to the input signals shown in Figs. P3-4 (d) and (e) using both the convolution and Du Hamel integrals.

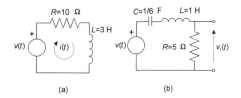

(a)   (b)

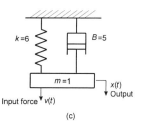

(c)

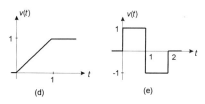

(d)   (e)

Figure P3-4

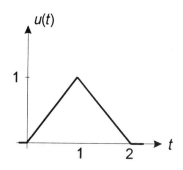

Figure P3-5

3-5 Suppose we apply the triangular pulse given in Fig. P3-5 to the system shown in Fig. 3-3. Using the convolution integral find its time response for all time.

3-6 Draw a functional block diagram of the working of the national economy. Assume that the input is the money spent by the government and the output is the national income. Comment on the validity of your assumptions.

3-7 Draw a functional block diagram of the behavior of your car. Consider the input to be the displacement of your accelerator and the output the velocity of your car.

3-8 Obtain $C(s)/R(s)$ and $E(s)/R(s)$ for the following control systems shown in Fig. P3-8.

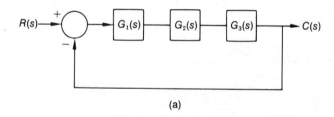

(a)

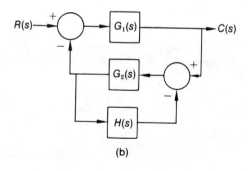

(b)

Figure P3-8

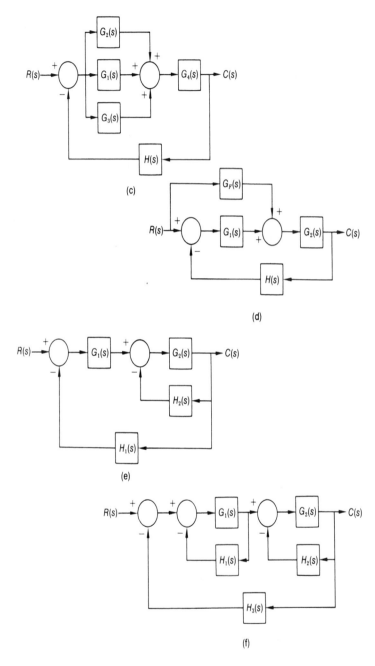

Figure P3-8 (contd.)

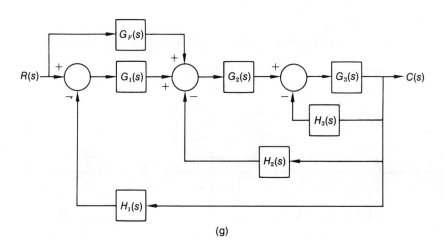

(g)

Figure P3-8 (contd.)

3-9 Obtain the output $C(s)$ and error $E(s)$ for the control systems shown in Fig. P3-9.

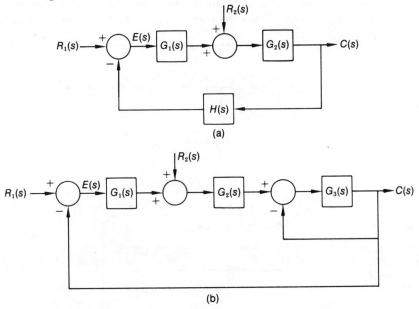

(a)

(b)

Figure P3-9

3-10 Using block diagram analysis techniques show that

$$\frac{C(s)}{R(s)} = \frac{G_1 G_2 G_3}{1 + (G_1 H_1 + G_2 H_2 + G_3 H_3) + G_1 G_3 H_1 H_3}$$

for the control system of Fig. P3-10.

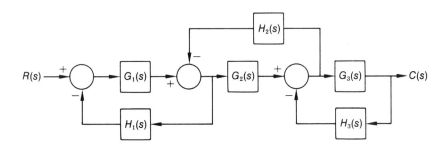

Figure P3-10

3-11 Consider the systems shown in Fig. P3-11. For each system use the state variables shown and find the resultant state and output equations. From these obtain the corresponding **A**, **b**, **c**, and **d** matrices.

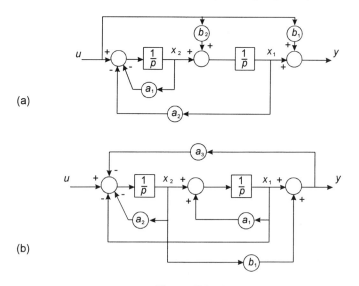

Figure P3-11

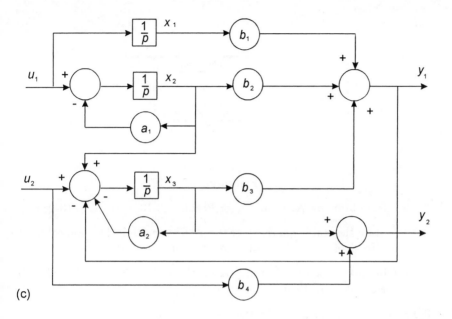

Figure P3-11 (contd.)

3-12 The differential equation of a linear system is given by

$$p^3 x(t) + 6p^2 x(t) + 11px(t) + 6x(t) = f(t)$$

Using each of the methods described in Sec. 3-7, transform the equation into state variable form.

3-13 For a closed loop transfer function

$$\frac{C(s)}{R(s)} = \frac{s+1}{s^3 + 3s^2 + 5s + 3}$$

(a) Determine the dynamical equations using three methods.

(b) Discuss the relative advantages of each.

3-14 A feedback control system is shown in Fig. P3-14. Assuming $u(t)$ and $y(t)$ to be the input and output variables, obtain the state equations.

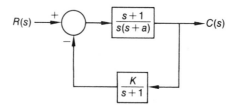

Figure P3-14

3-15 After obtaining **A**, **b**, and **c** for Problem 3-13, verify Eq. (3-28).

3-16 Obtain the state equations for the electromechanical servomechanism of Fig. 3-14. Having obtained **A**, **b**, and **c** verify Eq. (3-28).

3-17 Obtain the coefficient matrix **A** for the stable platform described by the equation in Example 5.

3-18 A system is described by the state equations

$$\left[\begin{array}{c} \dot{x}_1(t) \\ \dot{x}_2(t) \end{array} \right] = \left[\begin{array}{cc} -1 & 0 \\ 2 & 3 \end{array} \right] \left[\begin{array}{c} x_1(t) \\ x_2(t) \end{array} \right] + \left[\begin{array}{c} 1 \\ 0 \end{array} \right] u(t)$$

$$y(t) = \left[\begin{array}{cc} 3 & 1 \end{array} \right] \mathbf{x}(t)$$

Determine the transfer function for this system.

3-19 Find the Jordan, controller, observer, controllability, and observability canonical realizations for the transfer functions

$$G_1(s) = \frac{s^2 + 16s + 2}{(s + 4)(s + 10)(s + 1)}$$

$$G_2(s) = \frac{3s^3 + 5s^2 + 2s + 1}{(s + 2)(s + 3)(s + 5)}$$

For each realization construct a block diagram. **Hint**: In transforming $G_2(s)$ to state equation form, initially divide the numerator by the denominator so as to reduce it to a polynomial of second order. The state equations will thus be seen to be slightly modified from those discussed in Sec. 3-6, and will have the form

$$\dot{\mathbf{x}}(t) \quad = \quad \mathbf{A}\mathbf{x}(t) + \mathbf{b}u(t)$$

$$y(t) \quad = \quad \mathbf{c}\mathbf{x}(t) + \mathbf{d}u(t)$$

3-20 Consider the interconnected systems shown in Fig. P3-20, where the systems $\mathcal{S}_i$, $i = 1, 2$ have the state equation matrices $\mathbf{A}_i, \mathbf{b}_i, \mathbf{c}_i, \mathbf{d}_i$. Find the state equations for each of these systems in terms of the state equations of each sub-system. For each case find the resultant transfer function and deduce its relationship to the transfer functions of the respective sub-systems.

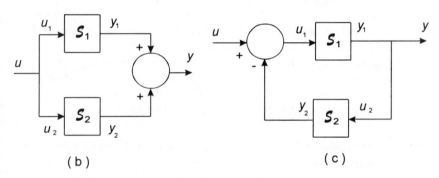

Figure P3-20

3-21 For each of the canonical realizations discussed in Sec. 3-7 show that the transfer function

$$c[s\mathbf{I} - \mathbf{A}]^{-1}\mathbf{b} = (c_2s^2 + c_1s + c_0)/(s^3 + a_2s^2 + a_1s + a_0).$$

Hint: You will need to use the matrix relationship $\mathbf{A}^{-1} = \operatorname{adj}\mathbf{A}/\det\mathbf{A}$, and for the companion matrix $\mathbf{A}_c$, $\det(s\mathbf{I} - \mathbf{A}_c) = s^3 + a_2s^2 + a_1s + a_0$.

3-22 Consider two state equation realizations for the systems $\mathcal{S}_a$ and $\mathcal{S}_b$ having the system matrices $\mathbf{A}_a$, $\mathbf{b}_a$, $\mathbf{c}_a$, and $\mathbf{A}_b$, $\mathbf{b}_b$, $\mathbf{c}_b$, which are linked by the similarity transformation $\mathbf{T}$, so that $\mathbf{A}_b = \mathbf{T}^{-1}\mathbf{A}_a\mathbf{T}$, etc. Show that both systems have the same transfer function.

3-23 Consider the state equation system $\dot{\mathbf{x}}_a(t) = \mathbf{A}_a\mathbf{x}_a(t) + \mathbf{b}_a u(t)$, $y(t) = \mathbf{c}_a\mathbf{x}_a(t)$, where

$$\mathbf{A}_a = \begin{bmatrix} -1 & 0 & 0 \\ 1 & -2 & 0 \\ -1 & -2 & -4 \end{bmatrix}, \quad \mathbf{b}_a = \begin{bmatrix} 1 \\ 0 \\ 1 \end{bmatrix}, \quad \mathbf{c}_a = \begin{bmatrix} 1 & 1 & 1 \end{bmatrix}$$

Find the transformation matrix $\mathbf{T}$ so that the transformation $\mathbf{x}_a = \mathbf{T}\mathbf{x}_b$ gives (a) the controller canonical form, and (b) the Jordan canonical form.

3-24 (Meisel) The equations of motion for an electromechanical system are

$$mp^2x(t) + Dpx(t) + \frac{x(t) - b}{K} + \frac{ai(t)^2}{(d + x(t))^2} = mg$$

$$\frac{2a}{d + x(t)}pi(t) + Ri(t) + \frac{2ai(t)}{(d + x(t))^2}px(t) = v(t)$$

where $x(t)$ is the armature displacement, $i(t)$ the coil current, and $v(t)$ is the input voltage. The remaining parameters $m, D, b, K, a, d, g,$ and R are system constants. Take the system state variables as $x_1(t) = x(t)$, $x_2(t) = px(t)$, $x_3(t) = i(t)$, obtain a set of nonlinear state equations, and linearize them about the operating conditions x_{10}, x_{20}, and x_{30} to obtain the small disturbance state equations.

3-25 (Mc Clamroch) The nonlinear state equations for three continuously fed stirred tank chemical reactors are

$$\dot{x}_1(t) = -\left(\frac{Q}{V}\right)x_1(t) + \left(\frac{Q}{V}\right)u_1(t)$$

$$\dot{x}_2(t) = -\left(\frac{Q}{V}\right)x_2(t) - kx_1(t)x_2(t) + \left(\frac{Q}{V}\right)u_2(t)$$

$$\dot{x}_3(t) = -\left(\frac{Q}{V}\right)x_3(t) + kx_1(t)x_2(t)$$

where $x_1(t)$, $x_2(t)$, $x_3(t)$ are the chemical concentrations in the three tanks, and $u_1(t)$, $u_2(t)$ are the inflow concentrations which are controllable. Supposing the system is in equilibrium for these variables having the values x_{10}, x_{20}, x_{30}, u_{10}, and u_{20}, find a set of linearized state equations for small disturbances about these equilibrium conditions.

3-26 (Bryson and Ho) The state equations for a constant thrust rocket operating in a fixed plane and undergoing an orbital transfer, are given by

$$\dot{r}(t) = u(t)$$

$$\dot{u}(t) = \frac{v(t)^2}{r(t)} - \frac{u(t)}{r(t)^2} + \frac{T \sin \phi(t)}{m}$$

$$\dot{v}(t) = -\frac{u(t)v(t)}{r(t)} + \frac{T \cos \phi(t)}{m}$$

where $\phi(t)$ is the controllable rocket thrust angle. Supposing the nominal rocket motion time history is given by the time functions $r_0(t)$, $u_0(t)$, $v_0(t)$, and $\phi_0(t)$, linearize the state equations about the trajectories and find the resultant small disturbance state equations.

Chapter 4

Time Response - Classical Method

4.1 Introduction

Having represented control systems using block diagrams as well as state variables, we turn our attention to determining the system response i.e. how does a system respond as a function of time when subjected to various types of stimuli? Here we are interested in the system output without regard to the behavior of variables inside the control systems. When this is the case, we can work with the system transfer function. If we desire $C(s)$ we can work with $C(s)/R(s)$ and specify $R(s)$. On the other hand, if we need $E(s)$ we would work with $E(s)/R(s)$ and specify $R(s)$.

In any event it is important to recognize that when the response to a single input is required without regard to the behavior of variables inside a control system, we speak of applying the classical approach. This can be most readily achieved by employing operational techniques. This technique involves representing the output (or desired variable) as the ratio of two polynomials and then expanding the expression into partial fractions. The constants of the partial fractions are calculated by the residue theorem. The output in the time domain is then obtained by taking the inverse Laplace transform. A detailed discussion of the Laplace transform is given in Appendix A and should be reviewed by those who do not have a good working knowledge of its use.

In general, the input excitation to a control system is not known ahead of time. However, for purposes of analysis it is necessary that we assume some simple types of excitation and obtain the system response to at least

133

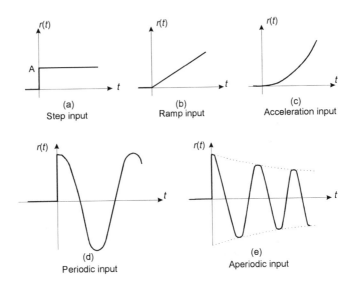

Figure 4.1: Test signals commonly used for linear feedback systems

these types of signals. In general, there are five types of excitations used in obtaining the response of linear feedback control systems. They are the step input, ramp input, parabolic input and the periodic and aperiodic inputs, which can be expressed in the phasor forms $e^{j\omega t}$ and e^{at} respectively. These are typical test or reference inputs. In practice, the input is never exactly specifiable.

Step Input

A step input consists of a sudden change of the reference input at $t = 0$. Mathematically it is

$$r(t) = \begin{cases} A & t \geq 0 \\ 0 & t < 0 \end{cases}$$

The function is shown in Fig. 4-1(a) and has a Laplace transform of A/s.

Ramp Input

A ramp input has a constant velocity and is represented as

$$r(t) = \begin{cases} At & t \geq 0 \\ 0 & t < 0 \end{cases}$$

The function is shown in Fig. 4-1(b) and has a Laplace transform of A/s^2.

Parabolic Input (Step Acceleration)

In this case the input has a constant acceleration and is represented as

$$r(t) = \begin{cases} At^2 & t \geq 0 \\ 0 & t < 0 \end{cases}$$

The function is shown in Fig. 4-1(c) and has a Laplace transform of $2A/s^3$.

Periodic Input

A periodic signal input is represented as

$$r(t) = \begin{cases} A \cos \omega t & t \geq 0 \\ 0 & t < 0 \end{cases}$$

This function can be represented in phasor notation by the function

$$r(t) = \begin{cases} \text{Re}[Ae^{j\omega t}] & t \geq 0 \\ 0 & t < 0 \end{cases}$$

In either case the function, which is shown in Fig. 4-1(d) has a Laplace transform of $As/(s^2 + \omega^2)$.

Aperiodic Input

A general sinusoid aperiodic input is

$$r(t) = \begin{cases} \text{Re}[Ae^{\alpha t}] & t \geq 0 \\ 0 & t < 0 \end{cases}$$

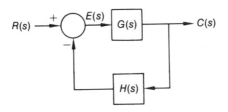

Figure 4.2: A closed loop system

where α is complex and can be written as $\alpha = a + jb$. The function is shown in Fig. 4-1(e) and has the Laplace transform $A(s - a)/[(s - a)^2 + b^2]$.

In studying the system response of a feedback control system there are three things we wish to know, viz. the **transient response** the **steady state** or **forced response**, and the **stability** of the system. The transient solution yields information on how much the system deviates from the input and the time necessary for the system response to settle to within certain limits. The steady state or forced response gives an indication of the accuracy of the system. Whenever the steady state output does not agree with the input, the system is said to have a **steady state error**. By **stability** we mean that the output does not become uncontrollably large.

4.2   Transient Response

Consider a closed loop system as shown in Fig. 4-2. The output and the error transfer functions are

$$\frac{C(s)}{R(s)} = \frac{G(s)}{1 + G(s)H(s)}$$

$$\frac{E(s)}{R(s)} = \frac{1}{1 + G(s)H(s)}$$

$$(4.1)$$

The transient response of the system, be it the error $E(s)$ or the output $C(s)$, depends upon the roots (also called zeros) of the **characteristic equation**

$$1 + G(s)H(s) = 0 \qquad (4.2)$$

The zeros of the characteristic equation are also the **poles** of the transfer functions given by Eqs. (4-1). These poles are known as the **closed loop poles**. It is interesting to note that the transient response does not depend upon the kind of input but depends only on the zeros of the characteristic equation. Now if the forward and feedback transfer functions are defined as

$$G(s) = \frac{P(s)}{Q(s)}, \quad H(s) = \frac{A(s)}{B(s)}$$

where $P(s), Q(s), A(s)$, and $B(s)$ are polynomials in s, then the characteristic equation becomes

$$\frac{Q(s)B(s) + P(s)A(s)}{Q(s)B(s)} = 0 \tag{4.3}$$

and the zeros which are obtained from

$$Q(s)B(s) + P(s)A(s) = 0 \tag{4.4}$$

corresponds to the poles of

$$\frac{C(s)}{R(s)} = \frac{P(s)B(s)}{Q(s)B(s) + P(s)A(s)} \tag{4.5}$$

The zeros of $P(s)$ and $A(s)$ are usually referred to as the **open loop zeros** for the feedback system while the zeros of $Q(s)$ and $B(s)$ are called the **open loop poles**. It will be noted from Eq. (4-5) that the zeros of the **closed loop transfer function** are simply given by the zeros of the transfer function $G(s)$ in the system forward path and the poles of the transfer function $H(s)$ in the feedback path. This should be contrasted with the complex relationship given in Eq. (4-4) from which the **closed loop poles** of Eq. (4-5) are determined. It is this complexity which generates many of the difficulties that occur in feedback control system design.

The right hand side of Eq. (4-5) can be written as a ratio of two polynomials where the degree of the denominator is equal to or higher than the degree of the numerator. Let us assume that the degrees of the denominator and numerator are n and v respectively, then if we factorize the first of the Eqs. (4-1) it can be written as

$$\frac{C(s)}{R(s)} = \frac{K \prod_{i=1}^{v}(s + z_i)}{\prod_{i=1}^{n}(s + p_i)} \tag{4.6}$$

Note that we began with the system transfer function to obtain this expression. It is perhaps interesting to observe that had we begun with a state equation representation, then

$$\frac{C(s)}{R(s)} = \mathbf{c}\,[s\mathbf{I} - \mathbf{A}]^{-1}\,\mathbf{b}$$

which would have been factored to obtain the right-hand side of Eq. (4-6). When we are interested only in the system transients we need not be concerned with the form of the input since the transient responses are *only* a function of the characteristic roots. It is therefore convenient to set $R(s) = 1$. (Since $R(s) = 1$ when $r(t)$ is an impulse, it follows that the transient response may be obtained by applying an impulse to the input of a system.) When the input is included, the transients will not only include the response due to the characteristic roots, but there will be terms in the output corresponding to the input and its derivatives. Only these terms survive as $t \to \infty$ and yield the steady state response. For obtaining the total transient response we will include the input term.

Returning to Eq.(4-6) we now assume that the input $r(t)$ is a unit step, then $R(s) = 1/s$ and the output becomes

$$C(s) = \frac{K \prod_{i=1}^{v}(s + z_i)}{s \prod_{i=1}^{n}(s + p_i)} \tag{4.7}$$

Let us now assume that of the n distinct poles, $2k$ poles are complex[1] and the remaining poles are real. If we denote the conjugates of s_m and K_m by $\bar{s}_m$ and $\bar{K}_m$, then Eq. (4-4) may be expanded in partial fractions and written as

$$C(s) = \frac{K_0}{s} + \sum_{m}^{k} \frac{K_m}{(s + s_m)} + \sum_{m}^{k} \frac{\bar{K}_m}{(s + \bar{s}_m)} + \sum_{i}^{n} \frac{K_i}{(s + s_i)} \tag{4.8}$$

where

$$K_0 = \lim_{s \to 0} sC(s)$$

$$K_m = \lim_{s \to -s_m} (s + s_m)C(s)$$

$$K_i = \lim_{s \to -s_i} (s + s_i)C(s)$$

If we denote $s_m = \sigma_m + j\omega_m$, then the output in the time domain is obtained by taking the inverse Laplace transform of Eq. (4-8),

$$
\begin{aligned}
c(t) &= K_0 + \sum_{m} K_m e^{-(\sigma_m + j\omega_m)t} + \sum_{m} \bar{K}_m e^{-(\sigma_m - j\omega_m)t} + \sum_{i} K_i e^{-s_i t} \\
&= K_0 + \sum_{m} |K_m| e^{-\sigma_m t} \cos(\omega_m t + \phi_m) + \sum_{i} K_i e^{-s_i t}
\end{aligned}
\tag{4.9}
$$

---

[1] Complex poles appear as complex conjugates.

where ϕ_m is the phase contribution of the constant K_m. Notice the second term of Eq. (4-9) is obtained by combining two terms.

If $C(s)$ has m poles that are equal (i.e. repeated), then

$$C(s) = \frac{K_0}{s} + \sum_{i=1}^{m} \frac{K_i}{(s + s_a)^i} \qquad (4.10)$$

where

$$K_i = \lim_{s \to -s_a} \frac{1}{(m - i)!} \frac{d^{(m-i)}}{ds^{(m-i)}} [(s + s_a)^m C(s)]$$

and

$$c(t) = K_0 + \sum_i \frac{K_i t^{i-1} e^{-s_a t}}{(i - 1)!} \qquad (4.11)$$

In the above i goes from 1 to m. In general, the response of a system contains terms of the type given in Eq. (4-9) as well as Eq. (4-11). The important fact here is that the form of the transient response is a function of the location of the closed loop poles, which are identical to the zeros of the characteristic equation, on the s-plane.

For a real, simple pole, the time response is simply an exponential which decays if the pole is in the left half s-plane, and increases with time if the pole is in the right half s-plane. The rate of this decay or growth is dependent upon the magnitude of the pole. Poles closer to the imaginary axis, but in the left half of the s-plane, are referred to as **dominant poles** since the decay due to them takes longer.

For complex poles the response is oscillatory with the peak magnitude varying exponentially with time. Again, if the real part is in the left half s-plane, the magnitude decreases with time. If the real part is positive, then the magnitude increases exponentially with time.

Finally, if the poles are real and of multiplicity m, then the time response is of the form $t^m e^{-s_a t}$. We have not shown the response if the poles are multiple and complex. It is left for the reader to show that for complex multiple poles the response is of the form

$$\left(\frac{t^{m-1}}{(m - 1)!} \right) e^{-\sigma_m t} |K_m| \cos(\omega_m t + \phi_m)$$

Our ideas of this section are consolidated and shown graphically in Fig. 4-3. We note that for well-behaved systems, i.e. systems exhibiting a stable response, it is reasonable to require that the closed loop poles of the control systems be located in the left half s-plane. If the poles exist on the imaginary axis they must be simple. Otherwise, the control system responds in such a way that the magnitude of the output becomes uncontrollably large for a step transient.

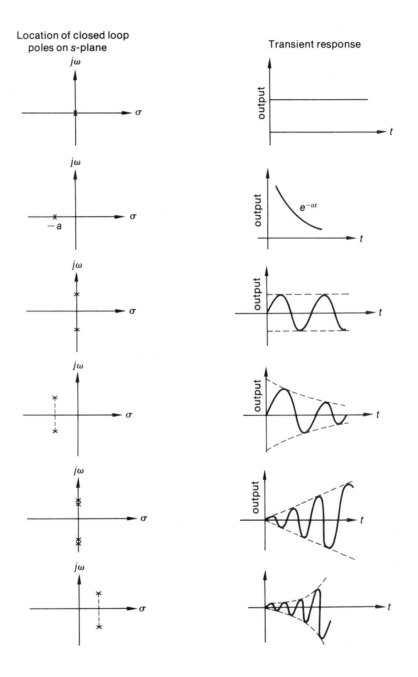

Figure 4.3: Transient response as a function of the closed loop poles on the s-plane

Example 1 *The forward loop transfer function of a unity feedback (i.e.*
$H(s) = 1$) *control system is given by* $G(s) = K/[s(s^2 + 19s + 118)]$. *Obtain*
$e(t)$ *if* $K = 240$ *and the input is* $r(t) = t$ *for* $t > 0$.

$$\frac{E(s)}{R(s)} = \frac{1}{1 + G(s)} = \frac{s(s^2 + 19s + 118)}{s(s^2 + 19s + 118) + K}$$

For $K = 240$ and $r(t) = tH(t)$

$$E(s) = \frac{(s^2 + 19s + 118)}{s(s + 5)(s + 6)(s + 8)}$$

Expanding this in partial fractions we obtain

$$E(s) = \frac{K_1}{s + s_1} + \frac{K_2}{s + s_2} + \frac{K_3}{s + s_3} + \frac{K_4}{s + s_4}$$

where $s_1 = 0, s_2 = 5, s_3 = 6$ and $s_4 = 8$. Now we evaluate the constants,

$$K_1 = \lim_{s \to 0}[sE(s)] = \left[\frac{(s^2 + 19s + 118)}{(s + 5)(s + 6)(s + 8)}\right]_{s=0} = 0.492$$

$$K_2 = \lim_{s \to -5}[(s + 5)E(s)] = -3.2$$

$$K_3 = \lim_{s \to -6}[(s + 6)E(s)] = 3.33$$

$$K_4 = \lim_{s \to -8}[(s + 8)E(s)] = -0.625$$

The inverse Laplace transform yields

$$e(t) = 0.492 - 3.2e^{-5t} + 3.33e^{-6t} - 0.625e^{-8t}$$

Example 2 *The forward loop of a unity feedback control system is given*
by $G(s) = K/s(s + 6)$. *It is desired to vary* K *from 8 to 13 for the case of*
a unit step input. Obtain the output for $K = 8$ *and 13 and determine the*
value of K *above which the system exhibits oscillatory behavior.*

The overall transfer function is given by

$$\frac{C(s)}{R(s)} = \frac{G(s)}{1 + G(s)}$$

Substituting for $G(s)$ and simplifying yields

$$C(s) = \frac{KR(s)}{s^2 + 6s + K}$$

The roots of the denominator are s_1 and s_2 where

$$s_1 = -3 + \sqrt{9 - K}, \quad s_2 = -3 + \sqrt{9 - K}$$

For $K \leq 9$ the roots are real and for $K > 9$ the roots are complex. The system therefore will exhibit oscillations for $K > 9$. (See next example). When $K = 8$ the roots become

$$s_1 = -2, \quad s_2 = -4$$

and for a step input

$$C(s) = \frac{8}{s(s+2)(s+4)} = \frac{K_1}{s} + \frac{K_2}{s+2} + \frac{K_3}{s+4}$$

The constants are

$$K_1 = \lim_{s \to 0}[sC(s)] = 1$$

$$K_2 = \lim_{s \to -2}[(s+2)C(s)] = -2$$

$$K_3 = \lim_{s \to -4}[(s+4)C(s)] = 1$$

The output for $K = 8$ becomes

$$c(t) = 1 - 2e^{-2t} + e^{-4t}$$

When $K = 13$ the roots become

$$s_1 = -3 + j2, \quad s_2 = -3 - j2$$

and for a step input

$$C(s) = \frac{13}{s(s+3-j2)(s+3+j2)} = \frac{K_1}{s} + \frac{K_2}{s+3-j2} + \frac{K_3}{s+3+j2}$$

The constants become

$$K_1 = \lim_{s \to 0}[sC(s)] = 1$$

$$K_2 = \lim_{s \to -3+j2}[(s+3-j2)C(s)] = -0.5 + j0.75 = 0.9e^{j\theta}$$

$$K_3 = \lim_{s \to -3-j2}[(s+3+j2)C(s)] = -0.5 - j0.75 = 0.9e^{-j\theta}$$

where $\theta = 123.7°$. We note that K_2 is the complex conjugate of K_3. Whenever the roots are complex conjugates we will find that the constants are also complex conjugates. The output for $K = 13$ becomes

$$c(t) = 1 + 0.9e^{j\theta}e^{(-3+j2)t} + 0.9e^{-j\theta}e^{(-3-j2)t}$$

$$= 1 + 0.9e^{-3t}[e^{j(2t+\theta)} + e^{-j(2t+\theta)}]$$

Noting that $2\cos(2t + \theta) = e^{j(2t+\theta)} + e^{-j(2t+\theta)}$

$$c(t) = 1 + 1.8e^{-3t}\cos(2t + \theta)$$

The output is seen to have an exponentially damped oscillatory term superimposed on a constant term.

Example 3 *Obtain the output for $K = 9$ for the system described in Example 2. Assume the input is a unit step.*

For $K = 9$, the overall transfer function becomes

$$C(s) = \frac{9}{s(s^2 + 6s + 9)} = \frac{9}{s(s + 3)^2}$$

Since we have repeated roots the partial fraction expansion becomes

$$C(s) = \frac{K_1}{s} + \frac{K_2}{s + 3} + \frac{K_3}{(s + 3)^2}$$

The constants are

$$K_1 = \lim_{s \to 0}[sC(s)] = 1$$

$$K_2 = \lim_{s \to -3}\left[\frac{d}{ds}(s + 3)^2 C(s)\right] = \left[\frac{-9}{s^2}\right]_{s=-3} = -1$$

$$K_3 = \lim_{s \to -3}[(s + 3)^2 C(s)] = -3$$

Substituting yields

$$C(s) = \frac{1}{s} - \frac{1}{s + 3} - \frac{3}{(s + 3)^2}$$

Taking the inverse Laplace transform, we obtain

$$c(t) = 1 - (1 + 3t)e^{-3t}$$

We note that as $t \to \infty$ the last term goes to zero.

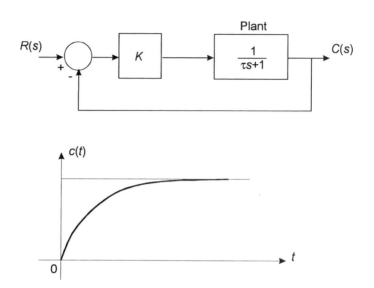

Figure 4.4: (a) A first-order unity feedback system (b) System time response for $K = 100, \tau = 1\,\mathrm{sec}$

First Order Systems

The plants of many control systems in process industries are dominated by a first order time constant and can be well modeled as first order systems even though there may, in addition, be other fast dynamic effects present. To illustrate this case let us examine the system shown in Fig. 4-4(a). From this figure and Eq. (4-1) we find

$$\frac{C(s)}{R(s)} = \frac{K}{\tau s + 1 + K} = \frac{K/(K+1)}{\left(\dfrac{\tau}{K+1}\right)s + 1} \tag{4.12}$$

For a unit step input $R(s) = 1/s$, and assuming the system is initially at rest, it can be shown by using the partial fraction expansion method that

$$c(t) = \begin{cases} \dfrac{K}{K+1}\left[1 - e^{-t/\tau_1}\right] & t \geq 0 \\[2mm] 0 & t < 0 \end{cases} \tag{4.13}$$

where the **time constant** $\tau_1 = \tau/(K+1)$. The time response of this system for $K = 100$ and $\tau = 1$ sec. is given in Fig. 4-4(b).

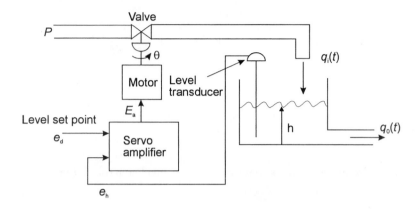

Figure 4.5: Fluid level control system

Two interesting observations can be made about the effect of feedback upon the operation of first order plants. First it will be noted that as K increases and becomes much larger than unity then the output will track the input quite accurately providing the input is changing slowly. Thus the transfer characteristic of the closed loop control system becomes practically independent of the controller gain K provided it is much larger than unity. Secondly it will be noted that feedback also speeds up the response time of the control system to transient inputs. Thus the feedback has the effect of improving the transient performance of the system, which can be of great benefit in practical applications.

Example 4 *Level control systems of the type shown in Fig. 4-5 are commonly used in many process and chemical industries. In this system the level h is measured by the level transducer and its output is used to control a servo motor controlling the valve position which in turn adjusts the inlet flow rate. Derive the equations of motion for the system and determine its dynamic behavior.*

From continuity conditions the differential equation for the tank fluid level is given by Eqs. (2-43) and (2-44)

$$Aph(t) = q_i(t) - q_o(t) = q_i(t) - \frac{\rho g}{R}h(t)$$

where ρ is the fluid density, and R is the outlet flow pipe resistance. Assuming zero initial conditions and taking the Laplace transform gives the

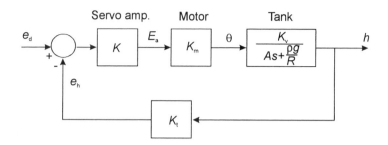

Figure 4.6: Block diagram for fluid level control system

transfer function

$$\frac{H(s)}{Q_i(s)} = \frac{1}{As + \rho g/R}$$

Analyzing the block diagram for the system shown in Fig. 4-6 shows the closed loop transfer function is

$$\frac{H(s)}{E_d(s)} = \frac{K K_m K_v}{K K_m K_v K_t + \rho g/R} \cdot \frac{1}{\left(\dfrac{A}{K K_m K_v K_t + \rho g/R}\right) s + 1}$$

The system is recognized as being the same as the first order system illustrated in Fig. 4-4. It will be noted that the tank time constant is proportional to the servo amplifier gain K and that it can be made short by making K sufficiently large. In the analysis the motor dynamic behavior has been ignored. This is usually acceptable because the response time of the motor and valve is generally much shorter than the tank time constant. Inclusion of this time constant in the analysis would result in a second order characteristic polynomial for the system.

Second-Order Systems

While many systems can be modeled with first order plants, to take into account their dynamic operation some are better modeled by plants having second order characteristics. These types of systems are most abundant and consequently play a very fundamental role in the physical world. We shall therefore dwell on a second-order system and analyze it a bit more thoroughly. Consider the system of Fig. 4-7. The output becomes

$$C(s) = \frac{K R(s)}{As^2 + Bs + K}$$

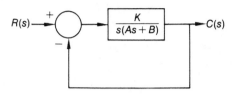

Figure 4.7: A second-order unity feedback system

and, if we define

$$\omega_n^2 = \frac{K}{A}, \quad \zeta = \frac{B}{2\sqrt{KA}}$$

then

$$\frac{C(s)}{R(s)} = \frac{\omega_n{}^2}{s^2 + 2\zeta\omega_n s + \omega_n{}^2} \tag{4.14}$$

Here ω_n is the **natural frequency** and ζ is the **damping ratio** of the control system. The transient response is dependent upon the roots of the **characteristic equation**

$$s^2 + 2\zeta\omega_n s + \omega_n{}^2 = 0 \tag{4.15}$$

The roots, for non-negative ζ, are

$$
\begin{array}{llll}
(\zeta < 1) & s_1, s_2 & = & -\zeta\omega_n \pm j\omega_n\sqrt{1 - \zeta^2} \quad \text{(Under damping)} \\
(\zeta = 1) & s_1, s_2 & = & -\omega_n \quad \text{(Critical damping)} \\
(\zeta > 1) & s_1, s_2 & = & -\zeta\omega_n \pm \omega_n\sqrt{\zeta^2 - 1} \quad \text{(Over damping)} \\
(\zeta = 0) & s_1, s_2 & = & \pm j\omega_n \quad \text{(No damping)}
\end{array}
$$

The term $\omega_n\sqrt{1 - \zeta^2}$ is often referred to as the **damped natural frequency** and denoted by ω_d. If we assume that ζ is positive, then the system has positive damping and the roots of the characteristic equation exist in the left half of the s-plane. This means that this system shall always exhibit a stable time response. The migration of the closed loop poles as a function of the damping ratio for this system is shown in Fig. 4-8 for constant natural frequency. In addition the coordinates of the characteristic roots on the s-plane for the case where $\zeta < 1$, together with the relationship of the angle ψ to the damping ratio ζ, are also shown.

Let us assume the input to be a unit step so that $R(s) = 1/s$. The time response of the output may be obtained by the techniques already discussed

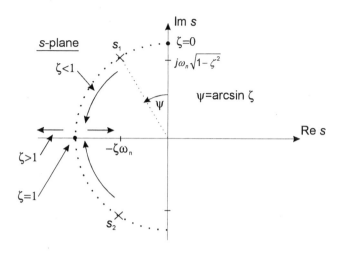

Figure 4.8: Closed loop pole migration as a function of damping ratio ζ for a second-order system

or read directly from tables. For the under damped case the time response to a unit step is

$$c(t) = 1 - \frac{e^{-\zeta \omega_n t}}{\sqrt{1 - \zeta^2}} \sin(\omega_d t + \phi) \tag{4.16}$$

where

$$\phi = \arctan \frac{\sqrt{1 - \zeta^2}}{\zeta}, \text{ and } \omega_d = \omega_n \sqrt{1 - \zeta^2}$$

Here, as mentioned above, ω_d is referred to as the **damped natural frequency**. A plot of $c(t)$ is shown in Fig. 4-9(a). We note that an under damped system exhibits an **overshoot**. The magnitude of this overshoot may be obtained by solving $dc(t)/dt = 0$. The first maximum value occurs at

$$t_{\max} = \frac{\pi}{\omega_n \sqrt{1 - \zeta^2}} \tag{4.17}$$

Then the magnitude of the overshoot is obtained by substituting $t_{\max}$ in Eq. (4-17) which yields

$$c(t)_{\max} = 1 + e^{-\pi \zeta / \sqrt{1 - \zeta^2}} \tag{4.18}$$

and is seen to be a function of the damping ratio only. This is shown in Fig. 4-9(b). The time t_s required for the oscillatory response to settle to

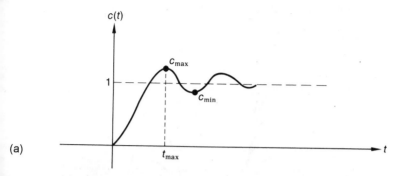

(a)

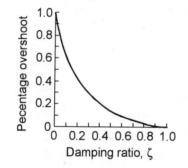

(b)

Figure 4.9: (a) Underdamped response of a second-order system (b) Overshoot as a function of damping ratio for a step input

within 2 percent of its steady state value is

$$t_s \approx \frac{4}{\zeta \omega_n}$$

When the damping ratio is decreased to zero the output becomes

$$c(t) = 1 - \cos \omega_n t$$

and we obtain pure oscillations. On the other hand if we increase the damping ratio to $\zeta = 1$, we obtain double roots from the characteristic equation so that

$$C(s) = \frac{\omega_n^2}{s(s + \omega_n)^2}$$

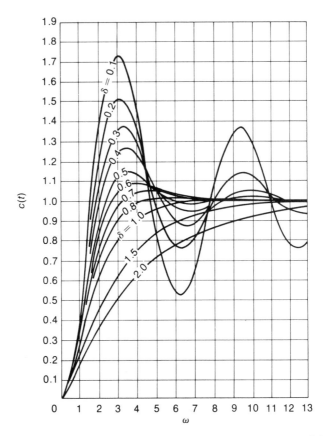

Figure 4.10: Transient response of a second-order system subjected to a unit step input

The inverse Laplace transform of this expression gives the output

$$c(t) = 1 - (1 + \omega_n t)e^{-\omega_n t}$$

which is seen to be non-oscillatory. Note the appearance of the term $\omega_n t e^{-\omega_n t}$ due to the multiplicity of the pole at $s = -\omega_n$. This system dampens the input without any overshoot.

The last case, over damping, has two unequal real poles. Again we expect to see non-oscillatory behavior. The response for different values of ζ is shown in Fig. 4-10. We observe that as ζ becomes small the system has larger overshoots and is faster acting.

As ζ is increased the system overshoot decreases and then vanishes as

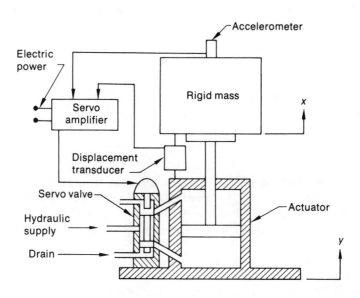

Figure 4.11: An electrohydraulic isolation system

we approach critical damping and $\zeta > 1$. In this case the system is sluggish. It is a characteristic of second-order physical systems that sluggish systems are over damped, whereas fast acting systems are under damped.

Example 5 *Consider the electrohydraulic vibration isolation system shown in Fig. 4-11. The system essentially consists of a rigid mass m vibrating as a result of the excitation input y(t). The mass m is connected to a hydraulic actuator rod having a negligible cross sectional area. By considering the acceleration, velocity, and relative displacement of the rigid body as feedback signals, we operate a servovalve to control the flow of a relatively incompressible fluid to and from the hydraulic actuator. Derive the equations of this systems and discuss the behavior of the system.*

Neglecting leakage, the flow $q(t)$ through the servovalve is that due to the motion across the actuator

$$q(t) = Ap(x(t) - y(t))$$

By processing the acceleration, relative displacement, and relative velocity signals, the flow through the valve can be made proportional to $p^2 x(t)$, $p(x(t) - y(t))$, and $(x(t) - y(t))$, namely

$$q(t) = -[C_a p^2 x(t) + C_v p(x(t) - y(t)) + C_d(x(t) - y(t))]$$

where C_a, C_v and C_d are proportionality constants. Equating the two equations for the flow and taking the Laplace transform, we obtain

$$\frac{X(s)}{Y(s)} = \frac{C_d + (C_v + A)s}{C_a s^2 + (C_v + A)s + C_d}$$

where the initial conditions are assumed to be zero. If we define the natural frequency $\omega_n = \sqrt{C_d/C_a}$ and the damping ratio $\zeta = (C_v + A)/\sqrt{4C_a C_d}$, then

$$\frac{X(s)}{Y(s)} = \frac{\omega_n{}^2 + 2\zeta\omega_n s}{s^2 + 2\zeta\omega_n s + \omega_n{}^2}$$

The characteristic equation is recognized as being identical to the second-order equation already discussed. Since the values of ω_n and ζ are dependent upon the feedback signals C_a, C_v and C_d, we have the ability of obtaining vibration isolation over a wide range of frequencies. In conventional passive isolation this is not possible since the properties of a passive isolator for any given system are fixed. Note also that since the mass m does not appear in the equations of motion, the isolation system performance characteristics are independent of the weight of the isolated body. Examination of the natural frequency and damping ratio definitions indicate that the relative velocity feedback affects only the damping ratio. This provides us with the facility of varying the damping ratio but with fixed natural frequency. In this problem if we employ only acceleration feedback gain, then the resulting characteristic equation becomes a polynomial of order one, i.e. the transient response exhibits an exponential decay. Another special case is obtained if in addition to acceleration, velocity, and displacement gains we have integral feedback of the displacement. In this case the system characteristic equation becomes of third-order.

Example 6 *Consider the third-order servomechanism shown in Fig. 4-12(a). Such a closed loop system can be used for positioning a large antenna or a mass having a large moment of inertia. The output potentiometer measures the output shaft position and converts this to voltage $e(t) = K_0 c(t)$ where K_0 has units of volts per radian and is the transfer function of the potentiometer. A particular position servo with known speed torque characteristics of the servomotor is shown in Fig. 4-12(b) and it is assumed that the gear ratio is unity. Derive the overall transfer function and obtain $c(t)$ for (a) $A = 2.625 \times 10^5$ and (b) $A = 9375$. Assume a unit step input.*

The output transfer function becomes

$$\frac{C(s)}{R(s)} = \frac{250A}{s(s + 200)(s + 62.5) + 250A}$$

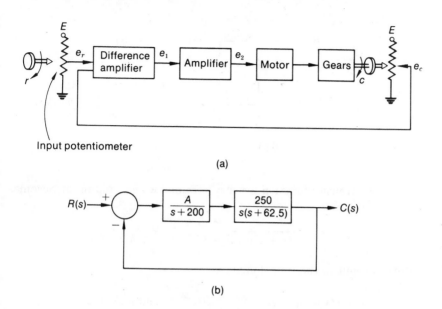

Figure 4.12: A position servomechanism

where the characteristic equation is a third-order polynomial.
   (a) The transfer function for an amplifier gain $A = 2.625 \times 10^5$ is

$$\frac{C(s)}{R(s)} = \frac{6.56 \times 10^7}{s^3 + (2.625 \times 10^2)s^2 + (1.25 \times 10^4)s + 6.56 \times 10^7}$$

Factoring the characteristic equation and substituting for a unit step $R(s) = 1/s$,

$$C(s) = \frac{6.56 \times 10^7}{s(s + 500)(s^2 - 237.5s + 1.312 \times 10^5)}$$

The output in the time domain becomes

$$c(t) = A_0 - A_1 e^{-500t} - A_2 e^{118.7t} \sin(1140t + \phi)$$

Examination of this equation indicates that the first two terms of $c(t)$ are
well behaved but the last terms becomes large, i.e. the magnitude of the
oscillations increase exponentially with time. It is clear that this is an undesirable characteristic and must be avoided. Actually when this happens
we speak of the system as being **unstable**.

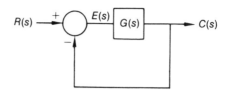

Figure 4.13: Unity feedback control system

(b) The transfer function for $A = 9375$ and a unit step input becomes

$$C(s) = \frac{2.35 \times 10^6}{s(s + 250)(s^2 + 12.5s + (9.375 \times 10^3))}$$

and the output in the time domain is

$$c(t) = B_0 - B_1 e^{-250t} - B_2 e^{-6.25t} \sin(96.6t + \phi)$$

The last term now is a positively damped sinusoid. The output of the position servomechanism now exhibits exponentially decaying oscillations, about a constant value B_0 which corresponds to some constant input. We speak of such a system as being **stable**.

4.3   Steady State Response

Unity Feedback Control System

In addition to the system transient response, we are also interested in its steady state response, i.e. the character of the system after a very long time has elapsed. The steady state value of the error, denoted by $e(\infty)$, can be obtained by using the final value theorem

$$e(\infty) = e(t)|_{t \to \infty} = \lim_{s \to 0} sE(s) \qquad (4.19)$$

provided a final value exists. Note that if a system is purely oscillatory, this theorem cannot be used. The steady state error, shown in Eq. (4-19), is important since it tells us the accuracy of the idealized system. The steady state performance of a system is best defined in terms of **error constants** when the inputs are simple aperiodic signals. These error constants for a unity feedback system are defined as

$$K_0 = \lim_{s \to 0} G(s)$$

$$K_1 = \lim_{s \to 0} sG(s) \tag{4.20}$$

$$K_2 = \lim_{s \to 0} s^2 G(s)$$

where K_0 is referred to as the **position error constant**, K_1 as the **velocity error constant**, and K_2 as the **acceleration error constant**. The relationship between the steady state error and the error constants may be readily shown.

The error for the unity feedback control system shown in Fig. 4-13 is

$$E(s) = \frac{R(s)}{1 + G(s)}$$

If the input is a unit step, then the steady state error becomes

$$e(\infty) = \lim_{s \to 0} sE(s) = \lim_{s \to 0} s \left[\frac{1}{s} \frac{1}{1 + G(s)} \right] = \frac{1}{1 + K_0}$$

If the input is a ramp, then $R(s) = 1/s^2$ and

$$e(\infty) = \lim_{s \to 0} s \left[\frac{1}{s^2} \frac{1}{1 + G(s)} \right] = \frac{1}{K_1}$$

For a parabolic input $R(s) = 1/s^3$ and

$$e(\infty) = \lim_{s \to 0} s \left[\frac{1}{s^3} \frac{1}{1 + G(s)} \right] = \frac{1}{K_2}$$

We conclude then that K_0, K_1 and K_2 are measures of the system error in following a step, ramp, or parabolic input.

Example 7 *The open loop transfer function of the unity feedback control system shown in Fig. 4-13 is given by*

$$G(s) = \frac{K}{s(As + B)}$$

Derive the steady state error and error coefficients for a step, ramp and parabolic input.

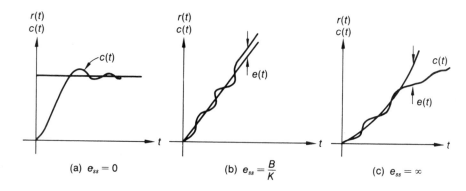

Figure 4.14: Steady state error for a step, ramp, and parabolic input, $G(s) = \dfrac{K}{s(As + B)}$, $H(s) = 1$

For a step input,
$$K_0 = \lim_{s \to 0} G(s) = \infty$$
and
$$e(\infty) = \frac{1}{1 + K_0} = 0$$

which suggests that such a system can follow a step input without any steady state error. If the input is a ramp, $R(s) = 1/s^2$, then

$$K_1 = \lim_{s \to 0} sG(s) = \lim_{s \to 0} \frac{sK}{s(As + B)} = \frac{K}{B}$$

and the steady state error becomes

$$e(\infty) = \frac{B}{K}$$

Since K cannot be increased without bound, there is a physical limitation as to how small the steady state error may become.

Finally, if the input to this control system is parabolic, $R(s) = 1/s^3$, then $K_2 = 0$ and
$$e(\infty) = \infty$$

i.e. this control system cannot follow a parabolic input as time becomes large. The physical interpretation of the steady state error for different inputs is illustrated in Fig. 4-14.

It is interesting to see how we can modify the control system of Example 7 so that it can follow even a parabolic input with only a finite error. A quick look at $G(s)$ indicates that if we introduce another pole at $s = 0$, then

$$G(s) = \frac{K}{s^2(As + B)}$$

and $K_2 = K/B$. This achieves two things. First, the error due to a ramp input goes to zero. Second, the error due to a parabolic input becomes finite. We may conclude, therefore, that the value of a specific error constant and the steady state error are dependent upon the number of poles at the origin of the open loop transfer function. The rest of the polynomial contributes nothing to the form of the error excepting of course for the constant term.

To clarify the effect of poles at $s = 0$ upon the steady state error for step, ramp, parabolic, etc., inputs we introduce the concept of **System Type**. In general $G(s)$ may be written as

$$G(s) = \frac{K(1 + T_1 s)(1 + T_2 s) \cdots (1 + T_m s)}{s^\ell (1 + T_a s)(1 + T_b s) \cdots (1 + T_n s)} \tag{4.21}$$

where K and all the coefficients T_j are constants. The **System Type** of a unity feedback control system is said to be the order of the pole of $G(s)$ at $s = 0$. Thus the system where $G(s)$ is defined by Eq. (4-21) is of **Type** ℓ, where $\ell = 0, 1, 2, \ldots$ The transfer function $G(s)$ given in Eq.(4-21) is said to be written in **gain form**. When $G(s)$ is written in this form only the system **Type** ℓ and the **gain constant** K affect the steady state error. For example suppose the system is of Type 1, then $\ell = 1$ and it can be seen from Eq. (4-20) that $K_1 = K$. Thus in evaluating the error constants it suffices to write the transfer function $G(s)$ in gain form, as the error constant can then be easily determined by inspection. The steady state error for a unit step, ramp, or parabola as a function of the **System Type** is given in Table 4-1.

Non-Unity Feedback Control System

The system error signal $e_s(t)$ for a non-unity feedback control system is defined in Fig. 4-15, where $G_m(s)$ is the idealized system model. As an example suppose the control system output is to track the input, but there is a constant non-unity feedback element $H(s) = S_m$. In this case we would

Table 4-1 Steady state error for unity feedback system

No of Open Loop Poles at $s = 0$	System Type	Steady State Error $e(\infty)$		
		Position	Ramp	Parabola
0	0	$1/(1+K_0)$	∞	∞
1	1	0	$1/K_1$	∞
2	2	0	0	$1/K_2$
3	3	0	0	0

take $G_m(s) = 1/S_m$, so that for a unit step input

$$
\begin{aligned}
e_s(\infty) &= \lim_{t \to \infty} e_s(t) = \lim_{s \to 0} s \left[\frac{1}{S_m} - \frac{G(s)}{1 + G(s)S_m} \right] \frac{1}{s} \\
&= \frac{1}{S_m} \cdot \frac{1}{1 + K_0 S_m}
\end{aligned}
\tag{4.22}
$$

Similarly for ramp and parabolic inputs the steady state system errors become

$$
e_s(\infty) = \frac{1}{S_m} \cdot \frac{1}{K_1 S_m}
$$

and

$$
e_s(\infty) = \frac{1}{S_m} \cdot \frac{1}{K_2 S_m}
$$

respectively.

Example 8 *A velocity servo has the plant and feedback transfer functions*

$$
G(s) = \frac{K(s + C_i)}{s^2(Ts + 1)}, \quad \text{and } H(s) = K_t s
$$

For this system the objective is to control the output velocity so the idealized model will be taken as $G_m(s) = 1/K_t s$, so that a unit step will yield a ramp of slope $1/K_t$ at its output. Find the steady state system error for step and ramp inputs.

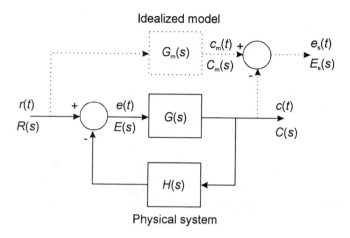

Figure 4.15: Feedback control system with idealized system model

For a unit step input,

$$e_s(\infty) = \lim_{s \to 0} s \left[\frac{1}{K_t s} - \frac{K(s + C_i)}{s(s(Ts + 1) + KK_t(s + C_1))} \right] \frac{1}{s}$$

$$= \frac{1}{K_t} \cdot \frac{1}{K_t K C_i} = \frac{1}{K_t} \cdot \frac{1}{K_t K_2}$$

Thus we see that under steady state conditions there is a constant error between the output of the idealized model and the system output $c(t)$.

For a ramp input we find

$$e_s(\infty) = \lim_{s \to 0} s \left[\frac{1}{K_t s} - \frac{K(s + C_i)}{s(s(Ts + 1) + KK_t(s + C_1))} \right] \frac{1}{s^2} = \infty$$

This result shows that the velocity servo is incapable of following a ramp speed command without an ever increasing system error.

It will be noted for the case of the tracking servo that the system error $e_s(t)$ can be related to the error signal $e'(t)$ for an equivalent unity feedback system. Using block diagram reduction methods upon the system shown in Fig. 4-15 we obtain the system shown in Fig. 4-16. As can be seen the error signal $E'(s)$ for the equivalent unity feedback system equals the system error signal $E_s(s)$, whereas the signal $E(s)$ in Fig. 4-15 differs from these variables by the scale factor $1/S_m$. It will be left to the reader to determine the corresponding equivalent feedback system for a velocity

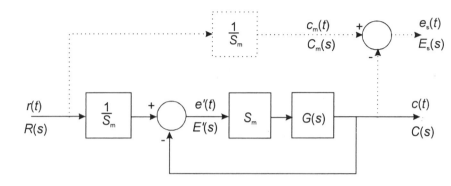

Figure 4.16: Equivalent unity feedback system for a tracking servo

servo, and the relationship between its error signal and the system error signal $E_s(s)$.

Error Series

When the inputs are not simple aperiodic signals, the steady state error can be expressed as a function of more general error coefficients. These error coefficients are meaningful for unity feedback systems or more general systems that can be represented as equivalent unity feedback systems.

Consider a unity feedback control system where the error in the Laplace domain is

$$E(s) = W_e(s)R(s) \qquad (4.23)$$

with the error transfer function $W_e(s)$ defined as

$$W_e(s) = \frac{1}{1 + G(s)} \qquad (4.24)$$

Since $W_e(s)$ is a quotient of polynomials it can be expanded in a Taylor series in a small neighborhood about $s = 0$,

$$W_e(0) = W_e(0) + \frac{dW_e}{ds}(0)s + \frac{1}{2!}\frac{d^2W_e}{ds^2}(0)s^2 + \frac{1}{3!}\frac{d^3W_e}{ds^3}(0)s^3 + \cdots \qquad (4.25)$$

Provided the infinite sum in Eq. (4-25) converges, we can write $E(s)$ as

$$\begin{aligned} E(s) &= W_e(0)R(s) + \frac{dW_e}{ds}(0)[sR(s)] + \frac{d^2W_e}{ds^2}(0)[s^2R(s)/2!] \\ &\quad + \frac{d^3W_e}{ds^3}(0)[s^3R(s)/3!] + \cdots \end{aligned} \qquad (4.26)$$

Taking the inverse Laplace transform of Eq. (4-26) gives

$$e(t) = W_e(0)r(t) + \frac{dW_e}{ds}(0)[pr(t)] + \frac{1}{2!}\frac{d^2W_e}{ds^2}(0)[p^2r(t)] + \cdots \quad (4.27)$$

Thus the error signal $e(t)$ can be expressed as a weighted sum of the input signal $r(t)$ and its derivatives provided this series converges and $t \gg 0$.
    Defining the **generalized error coefficients** as

$$C_0 = \lim_{s \to 0} W_e(s)$$

$$C_1 = \lim_{s \to 0} \frac{dW_e}{ds}(s) \quad (4.28)$$

$$\vdots$$

$$C_n = \lim_{s \to 0} \frac{d^n W_e}{ds^n}(s)$$

the **error series** given in Eq. (4-27) becomes

$$e(t) = C_0 r(t) + C_1[pr(t)] + \frac{C_2}{2!}[p^2r(t)] + \cdots \quad (4.29)$$

and the coefficients $C_0, C_1, C_2, \ldots$ are the **generalized error coefficients**.
    Denoting the inverse Laplace transform of $W_e(s)$ by $w_e(t)$, let us examine the definitions of the generalized error coefficients a little further. From the definition of the Laplace transform,

$$W_e(s) = \int_0^\infty w_e(t)e^{-st}dt \quad (4.30)$$

and taking the limit of Eq. (4-30) as $s \to 0$ we have

$$C_0 = \lim_{s \to 0} W_e(s) = \int_0^\infty w_e(t)dt$$

From the complex differentiation property of the Laplace transform we also have

$$C_1 = \lim_{s \to 0} \frac{dW_e}{ds}(s) = -\int_0^\infty t w_e(t)dt$$

$$\vdots \quad (4.31)$$

$$C_n = \lim_{s \to 0} \frac{d^n W_e}{ds^n}(s) = (-1)^n \int_0^\infty t^n w_e(t)dt$$

Thus we see that the generalized error coefficient C_n is the nth moment of the impulse response $w_e(t)$.

From the above analysis we note that given any input, the steady state system error for a unity feedback system may be obtained without solving for the complete transient response of the system. For example, if the input is a step of magnitude A, then all derivatives of $r(t)$ go to zero and the error series becomes

$$e(t) = C_0 A, \text{ for } t \gg 0$$

If the input $r(t) = At$, a ramp, then $pr(t) = A$, and

$$e(t) = C_0 At + C_1 A, \text{ for } t \gg 0$$

We note that for this control system the steady state error becomes infinite as $t \to \infty$ unless $C_0 = 0$. Since C_0 is determined by the time function $w_e(t)$, its value is a characteristic of the control system.

Example 9 *Obtain the error series for the second-order control system with the following transfer function*

$$\frac{E(s)}{R(s)} = W_e(s) = \frac{s^2}{s^2 + 2\zeta\omega_n s + \omega_n^2}$$

From the definition of the error coefficients, we have

$$C_0 = \lim_{s \to 0} W_e(s) = 0$$

$$C_1 = \lim_{s \to 0} \frac{dW_e(s)}{ds} = 0$$

$$C_2 = \lim_{s \to 0} \frac{d^2 W_e(s)}{ds^2} = \frac{2}{\omega_n^2}$$

$$C_3 = \lim_{s \to 0} \frac{d^3 W_e(s)}{ds^3} = -\frac{12\zeta}{\omega_n^3}$$

Substituting the error coefficient in Eq. (4-29), the steady state error becomes

$$e(t) = \left(\frac{1}{\omega_n^2}\right) p^2 r(t) - \left(\frac{2\zeta}{\omega_n^3}\right) p^3 r(t) + \cdots$$

Since the steady state error is a function of the second and higher derivatives of $r(t)$, the only way the steady state error can be zero or constant is if the input is at *most* parabolic.

The error series is seen to be a fairly general expression capable of yielding the steady state error for *any* class of inputs.

Since the error constants are a subset of the more general error coefficient, it is only natural that we see how they are related. For a Type 0 system and unit step input, the error series yields

$$e(t) = C_0$$

and since we established that the steady state error for $t \to \infty$ is related to the position error constant by

$$e(t) = \frac{1}{1 + K_0}$$

we have

$$C_0 = \frac{1}{1 + K_0}$$

Similarly, for a Type 1 system and a ramp input

$$e(t) = C_1$$

and again since

$$e(t) = \frac{1}{K_1}$$

we have

$$C_1 = \frac{1}{K_1}$$

Following this procedure we may relate as many of the error coefficients to the error constants as we desire. Substituting for the error coefficients in Eq. (4-29)

$$e(t) = \frac{1}{1 + K_0} r(t) + \frac{1}{K_1} pr(t) + \frac{1}{K_2} p^2 r(t) + \cdots \text{ for } t \gg 0$$

Although this series looks general enough, we must caution ourselves with the understanding that its utility depends upon its convergence properties.

Example 10 *Consider a unity feedback control system having a Type 0 plant whose transfer function is*

$$G(s) = \frac{K}{s + 1}$$

Find the error for unit step and ramp inputs.

From Eq. (4-24)

$$W_e(s) = \frac{s+1}{s+1+K}$$

and

$$C_0 = \frac{1}{1+K}$$

$$C_1 = \frac{K}{(1+K)^2}$$

Thus for $t \gg 0$, Eq. (4-29) becomes

$$e(t) = \frac{1}{1+K}r(t) + \frac{K}{(1+K)^2}[pr(t)] + \cdots$$

For a step input, $r(t) = 1$ and $pr(t) = 0$ for $t \gg 0$, so that

$$e(t) = \frac{1}{1+K}$$

for $t \gg 0$. This agrees with the result obtained using the steady state error constant method. For a ramp input $r(t) = t$ and $pr(t) = 1$, so the error

$$e(t) = \frac{1}{1+K}t + \frac{K}{(1+K)^2}$$

for $t \gg 0$. The error series thus gives the time dependence of the error signal $e(t)$, and shows that it becomes infinite as $t \to \infty$, which again agrees with the results of the error constant method.

Dynamic Error for Periodic Inputs

The response of control systems to periodic, and in particular, sinusoidal input signals is of considerable practical and theoretical importance. As unity feedback systems are not able to faithfully follow sinewave inputs of all frequencies significant **dynamic errors** will exist under some conditions. The error series method discussed above may be used for estimating the magnitude and phase of these dynamic errors.

Example 11 *A position servo, having unity feedback, is used in a radar tracking system and has the open loop transfer function*

$$G(s) = \frac{100}{s(0.1s+1)}$$

Find the dynamic error for the system when the servo is required to track an input signal $r(t) = 0.1\sin \pi t$ rad.

Using Eq. (4-24)

$$W_e(s) = \frac{s^2 + 10s}{s^2 + 10s + 1000}$$

and

$$C_0 = \lim_{s \to 0} W_e(s) = 0$$

$$C_1 = \lim_{s \to 0} \frac{dW_e(s)}{ds} = 0.01$$

$$C_2 = \lim_{s \to 0} \frac{d^2 W_e(s)}{ds^2} = 0.0018$$

For the system input $r(t)$ given above

$$pr(t) = 0.1\pi \cos \pi t$$

$$p^2 r(t) = -0.1\pi^2 \sin \pi t$$

The error series is thus

$$e(t) = (0)(0.1 \sin \pi t) + (0.01)(0.1\pi \cos \pi t) - \left(\frac{0.0018}{2}\right)(0.1\pi^2 \sin \pi t) + \cdots$$

$$\approx 0.003142 \cos \pi t - 0.000888 \sin \pi t$$

$$\approx 0.003265 \sin(\pi t - 1.846) \text{ for } t \gg 0$$

4.4 Response to Periodic Inputs

Periodic and aperiodic signals can be decomposed into the sum of a large number of sinusoidal functions or components by using Fourier series or transform methods. The key information for each component is its frequency and its amplitude, and the collection of all such frequencies and amplitudes of the components is termed the signal **frequency spectrum**. As the control systems under consideration in this text are assumed to be linear, superposition allows us to find the control system output signal and its frequency spectrum by computing the system response to each sinusoidal input component independently of the others. This approach leads us to introduce the concept of the **frequency response function** which is analogous to the transfer function introduced in Chapter 3.

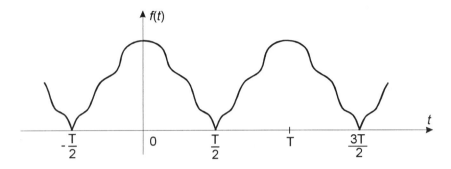

Figure 4.17: Periodic wave signal

Fourier Series Expansion of Periodic Signals

A signal $f(t)$ is **periodic** if there exists a constant T so that $f(t) = f(t+T)$ for all t, and the smallest such constant is called the **period**. The periodic signal $f(t)$ of period T shown in Fig. 4-17 can under certain conditions be represented by the complex **Fourier series**

$$f(t) = \frac{1}{T} \sum_{n=-\infty}^{\infty} F(jn\Delta\omega)e^{jn\Delta\omega t} \tag{4.32}$$

where $\Delta\omega = 2\pi/T$ is the **fundamental frequency** of the periodic signal, and $\{F(jn\Delta\omega)\}_{n=-\infty}^{\infty}$ are the Fourier coefficients. While it is common to write these coefficients as F_n, we have used the more explicit notation $F(jn\Delta\omega)$ to simplify our subsequent analysis. It can be easily shown that the Fourier coefficients are given by

$$F(jn\Delta\omega) = \int_{-T/2}^{T/2} f(t)e^{-jn\Delta\omega t}dt \tag{4.33}$$

for $n = 0, \pm1, \pm2, \cdots$.

Example 12 *Consider the rectangular pulse train signal given in Fig. 4-18, where the period is T, the pulse duration is τ, and its amplitude is $1/\tau$. Find the Fourier coefficients for this waveform.*

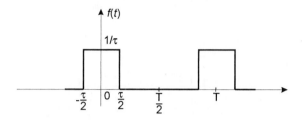

Figure 4.18: Rectangular pulse train signal

From Eq. (4-33) we obtain

$$
\begin{aligned}
F(jn\Delta\omega) &= \frac{1}{\tau}\int_{-\tau/2}^{\tau/2} e^{-jn\Delta\omega t}dt \\[2mm]
&= -\frac{1}{jn\Delta\omega\tau}\left[e^{-jn\Delta\omega t}\right]_{-\tau/2}^{\tau/2} \\[2mm]
&= \frac{\sin(n\Delta\omega\tau/2)}{n\Delta\omega\tau/2}
\end{aligned}
$$

for $n = 0, \pm1, \pm2, \cdots$.

The Frequency Spectrum

It will be noted from Eqs. (4-32) and (4-33) that the Fourier coefficients $F(jn\Delta\omega)$ give the magnitude of the respective sinusoidal components in the Fourier series expansion of $f(t)$. The sinusoidal term in the Fourier series for $f(t)$ corresponding to $n = \pm1$ is termed the **fundamental component**, while those corresponding to $n = \pm2, \pm3, \cdots$ are called the **harmonic components**. Thus a periodic signal is constituted of its discrete fundamental and harmonic components which may be represented graphically by the coefficients $F(jn\Delta\omega)$ as a **frequency spectrum**. This is shown in Fig. 4-19 for the pulse train of Example 12. In this figure each spectral line represents a harmonic frequency component of the periodic signal and its length is proportional to the corresponding Fourier coefficient $F(jn\Delta\omega)$. The distance between spectral lines is $\Delta\omega = 2\pi/T$ rad/sec. For the signal analyzed in Example 12 the spectral lines are enveloped by the function

$$
F(\omega) = \frac{\sin \omega\tau/2}{\omega\tau/2}
$$

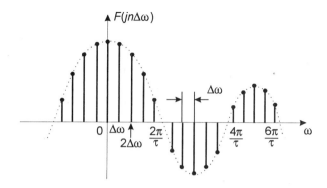

Figure 4.19: Frequency spectrum of pulse train

The frequency spectrum $\{F(jn\Delta\omega)\}$ for $-\infty < n < \infty$ of a signal is often referred to as the **frequency domain representation** of the signal while the graph of $f(t)$, for $-\infty < t < \infty$ is referred to as its **time domain representation**. Because the frequency spectrum uniquely represents a signal it can be considered to be a dual representation of it so that any changes in the spectrum will be reflected by changes in its time domain behavior and vice-versa.

A difficulty arises when we try to find the Fourier series representation for an aperiodic signal as it is not immediately clear whether it should be able to be decomposed into constituent sinusoidal components. However by using limiting arguments we show that the physical concept of a frequency spectrum for these signals can be retained. However the spectrum will now become continuous rather than having discrete spectral lines as was the case for periodic signals.

The Fourier Transform

To develop these new concepts consider the periodic signal $f(t)$ shown in Fig. 4-17 but let us assume that the period $T \to \infty$. In this case $\Delta\omega \to 0$, so that the spectral line separation approaches zero. This means that the frequency spectrum appears to become a continuous curve as $T \to \infty$. Let us now examine what happens to the Fourier series expansion of $f(t)$ as $T \to \infty$. Eq. (4-32) can be re-written as

$$f(t) = \frac{1}{2\pi} \sum_{n=-\infty}^{\infty} F(jn\Delta\omega)e^{jn\Delta\omega t}\Delta\omega \qquad (4.34)$$

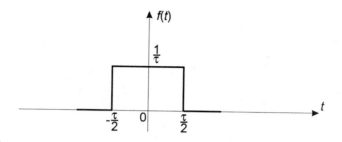

Figure 4.20: Aperiodic rectangular pulse signal

The summation on the right side of Eq. (4-34) can be viewed as a Riemann sum, so that as $T \to \infty$ and $\Delta\omega \to 0$, the sum becomes an integral giving

$$f(t) = \frac{1}{2\pi} \int_{-\infty}^{\infty} F(j\omega)e^{j\omega t}\,d\omega \qquad (4.35)$$

Similarly Eq. (4-33) becomes

$$F(j\omega) = \int_{-\infty}^{\infty} f(t)e^{-j\omega t}\,dt \qquad (4.36)$$

where $F(j\omega)$ is called the **Fourier transform** of the signal $f(t)$ and is often denoted by $F(j\omega) = \mathcal{F}\{f(t)\}$. Similarly we can write $f(t) = \mathcal{F}^{-1}\{F(j\omega)\}$.

It can be noted that as $T \to \infty$ the discretely varying variable $n\Delta\omega$ will transform to the smoothly varying variable ω so that when we plot $F(j\omega)$ as a function of ω we obtain a continuous function which is the **frequency spectrum** of $f(t)$. Now, however, the spectrum of an aperiodic signal contains energy at all frequencies, not just at the harmonic frequencies as is the case for a periodic signal.

Example 13 *Consider the rectangular pulse signal given in Fig. 4-20. Find the Fourier transform for this signal and plot its frequency spectrum.*

From Eq. (4-36)

$$
\begin{aligned}
F(j\omega) &= \frac{1}{\tau} \int_{-\tau/2}^{\tau/2} e^{-j\omega t}\,dt \\[2mm]
&= \frac{\sin \omega\tau/2}{\omega\tau/2} = \operatorname{sinc}\left(\frac{\omega\tau}{2}\right)
\end{aligned}
$$

The frequency spectrum for the signal $f(t)$ is plotted in Fig. 4-21.

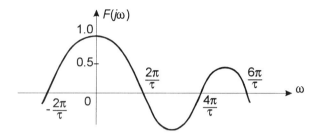

Figure 4.21: Frequency spectrum for rectangular pulse signal

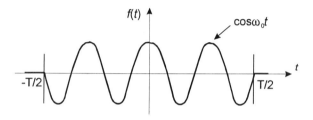

Figure 4.22: Pulse burst cosine wave of frequency ω_o and duration T

Example 14 *Consider the rectangular burst cosine wave shown in Fig. 4-22. Find the Fourier transform for this signal and plot its frequency spectrum. Supposing the period T of the burst becomes of infinite duration find the spectrum for a cosine wave of infinite duration.*

From Eq. (4-36)

$$
\begin{aligned}
F(j\omega) &= \frac{1}{2} \int_{-T/2}^{T/2} (e^{j\omega_0 t} + e^{-j\omega_0 t}) e^{-j\omega t} dt \\
&= \frac{T}{2} \frac{\sin(\omega + \omega_0)T/2}{(\omega + \omega_0)T/2} + \frac{T}{2} \frac{\sin(\omega - \omega_0)T/2}{(\omega - \omega_0)T/2}
\end{aligned}
\tag{4.37}
$$

The frequency spectrum for the signal $f(t)$ given in Fig. 4-22 is plotted in Fig. 4-23(a). We find that the spectrum consists of two components centered about the frequencies at $\pm\omega_0$ rad/sec, and having peak amplitudes of $T/2$.

Now as $T \to \infty$, the cosine wave will become of infinite duration. Each of the terms in Eq. (4-37) will converge to a Dirac δ-function of weight π.

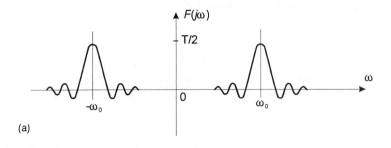

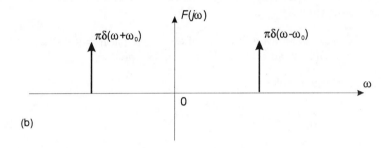

Figure 4.23: (a) Frequency spectrum for pulse burst cosine wave (b) Frequency spectrum for a cosine wave

Thus in the limit as $T \to \infty$ we find

$$F(j\omega) = \pi\delta(\omega + \omega_0) + \pi\delta(\omega - \omega_0)$$

The spectrum of a cosine wave of infinite duration is plotted in Fig. 4-23(b).

Frequency Response Function

The spectrum of the output signal of a system $\mathcal{S}$ can be determined from the system input spectrum by the use of the **frequency response function**. From Eq. (3-5) the output signal $y(t)$ is given by the convolution of the input signal $u(t)$ and the impulse $g(t)$, thus

$$y(t) = \int_{-\infty}^{t} g(t - x)u(x)dx \qquad (4.38)$$

if we assume the input signal exists for all $t > -\infty$. Taking the Fourier transform of the output signal and utilizing the convolution property we find that output spectrum is given by

$$Y(j\omega) = G(j\omega)U(j\omega) \tag{4.39}$$

where $G(j\omega)$ is the Fourier transform of the impulse response function $g(t)$, and is termed the **frequency response function** for system $\mathcal{S}$. The frequency response function $G(j\omega)$ is a complex function of ω and can be written as

$$G(j\omega) = |G(j\omega)|\, e^{j\,\phi(\omega)} \tag{4.40}$$

where $|G(j\omega)|$ and $\phi(\omega)$ will be referred to as the **magnitude** and **phase responses** of $G(j\omega)$ respectively. Because $g(t)$ is a real function of time we can show that,

$|G(j\omega)|$ is an even function of ω, i.e. $|G(j\omega)| = |G(-j\omega)|$, and $\phi(\omega)$ is an odd function of ω, i.e. $\phi(\omega) = -\phi(-\omega)$.

To show this we observe

$$
\begin{aligned}
G^*(j\omega) &= \int_{-\infty}^{\infty} [g(t)e^{-j\omega t}]^* dt \\[2mm]
&= \int_{-\infty}^{\infty} g(t)e^{j\omega t} dt = G(-j\omega)
\end{aligned}
$$

From Eq. (4-40) we obtain

$$|G(j\omega)|\, e^{-j\,\phi(\omega)} = |G(-j\omega)|\, e^{j\,\phi(-\omega)}$$

so that the result follows immediately.

To illustrate the meaning of the frequency response function $G(j\omega)$ suppose the system is excited by a unit cosine signal of frequency ω_0. In this case the frequency spectrum of the input is

$$U(j\omega) = \pi[\delta(\omega + \omega_0) + \delta(\omega - \omega_0)] \tag{4.41}$$

From Eqs. (4-39) and (4-41) we find the spectrum of the output signal to be

$$
\begin{aligned}
Y(j\omega) &= \pi[G(-j\omega_0)\delta(\omega + \omega_0) + G(j\omega_0)\delta(\omega - \omega_0)] \\[2mm]
&= \pi|G(j\omega_0)|[e^{-\phi(\omega_0)}\delta(\omega + \omega_0) + e^{\phi(\omega_0)}\delta(\omega - \omega_0)]
\end{aligned}
\tag{4.42}
$$

Taking the inverse Fourier transform of Eq. (4-42) gives

$$y(t) = |G(j\omega_0)| \cos(\omega_0 t + \phi(\omega_0))$$

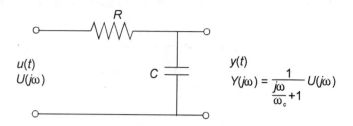

Figure 4.24: RC network

We see that at all frequencies the magnitude of the output cosine wave is given by $|G(j\omega)|$, while its phase is given by $\phi(\omega)$.

We have shown two useful interpretations of the frequency response function of a system $\mathcal{S}$. In the first case we take the system viewpoint where the system input signal spectrum $U(j\omega)$ is treated as an entity, so that the frequency response function can be viewed as modifying its spectrum so as to give the output signal spectrum. This viewpoint is summarized by Eq. (4-39). From the second viewpoint we assume the system is excited by a cosine wave of frequency ω. In this case the magnitude and phase of the frequency response function at the excitation frequency ω determines the magnitude $|G(j\omega)|$ and phase $\phi(\omega)$ of the output cosine wave.

Bode Response Plot

From the discussion above it can be seen that a method of graphically presenting the frequency response function information is to plot the magnitude and phase responses as a function of frequency. The amplitudes and frequencies in these plots often vary over a wide range of values, so it is convenient in this case to use logarithmic scales for amplitude and frequency while using a linear scale for phase. These plots of the frequency response function are referred to as the **Bode amplitude** and **phase plots**. In Chapter 7 we will examine the construction of Bode plots from transfer functions using asymptotic methods.

Example 15 *The simple RC network shown in Fig. 4-24 has the frequency response function*

$$G(j\omega) = \frac{1}{j\omega/\omega_c + 1}$$

where $\omega_c = RC$. Find the Bode amplitude and phase response plots for this frequency response function.

Resolving $G(j\omega)$ into the magnitude and phases responses as given in Eq. (4-40) we find

$$|G(j\omega)| = \frac{1}{\sqrt{\left(\dfrac{\omega}{\omega_c}\right)^2 + 1}}$$

and

$$\phi(\omega) = -\arctan\frac{\omega}{\omega_c}$$

As suggested above we plot $\log |G(j\omega)|$ and $\phi(\omega)$ as functions of $\log\ \omega$. Thus

$$\log |G(j\omega)| = -\frac{1}{2}\log\left[\left(\frac{\omega}{\omega_c}\right)^2 + 1\right]$$

and

$$\phi(\omega) = \arctan 10^{[\log\ \omega/\omega_c]}$$

which are plotted in Fig. 4-25.

The magnitude $|G(j\omega)|$ can be expressed in decibels, and is then denoted as $|G(j\omega)|_{dB}$, where we define

$$|G(j\omega)|_{dB} = 20\log |G(j\omega)|$$

It has become conventional to plot the magnitude $|G(j\omega)|$ in decibels rather than plotting $\log |G(j\omega)|$ as a function of $\log\ \omega$. The scale of $|G(j\omega)|_{dB}$ is also plotted on Fig. 4-25 for comparison.

Comparison Between Frequency Response Function and Transfer Function

From above the frequency response function $G(j\omega)$ is the Fourier transform of the system impulse response $g(t)$, while from Eq. (3-11) we observe that the transfer function $G(s)$ is the Laplace transform of $g(t)$. Providing the number of poles of the transfer function $G(s)$ exceeds the number of its zeros, comparison of the definitions of the Fourier and Laplace transforms of $g(t)$ shows that

$$G(j\omega) = G(s)|_{s=j\omega} \tag{4.43}$$

As a consequence of this equivalence many of the block diagram properties of transfer functions carry over to frequency response functions, so that block diagram algebra may also be applied in the frequency domain.

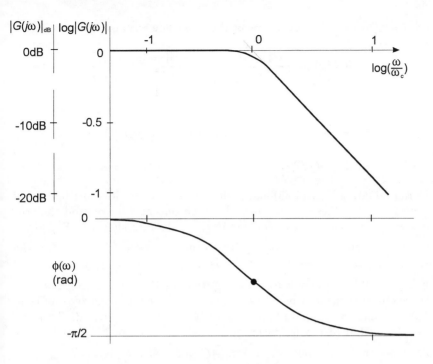

Figure 4.25: Bode amplitude and phase diagrams for $G(s) = \dfrac{1}{j\omega/\omega_c + 1}$

Frequency Response Function and the Pole-Zero Plot

While, as shown above, the graphical presentation of the frequency response function as Bode plots may be obtained from the system transfer function $G(s)$ by analytical means, a graphical interpretation of the pole-zero diagram and its relationship to the Bode diagrams gives valuable additional insight.

Suppose a system has the transfer function $G(s)$ as given in Eq. 4-21. We can easily re-write this transfer function as

$$G(s) = \frac{K'(s + s_1)(s + s_2)\cdots(s + s_m)}{s^\ell (s + s_a)(s + s_b)\cdots(s + s_n)} \qquad (4.44)$$

where $s_1 = 1/T_1, s_2 = 1/T_2, \cdots, s_m = 1/T_m,\ s_a = 1/T_a, \cdots, s_n = 1/T_n$, and $K' = KT_1T_2\cdots T_m/T_aT_b\cdots T_n$. When a transfer function is written as given in Eq. (4-44) we will say it is in **loop sensitivity form**. From

Eq. (4-43) we see the magnitude of the frequency response function is

$$|G(j\omega)| = \frac{|K'|\,|j\omega + s_1|\,|j\omega + s_2|\cdots|j\omega + s_m|}{|j\omega|^{\ell}\,|j\omega + s_a|\,|j\omega + s_b|\cdots|j\omega + s_n|} \tag{4.45}$$

and its phase is

$$\begin{aligned}
\phi(\omega) \;=\; & \arg K' + \arg(j\omega + s_1) + \cdots + \arg(j\omega + s_m) \\
& -\ell\pi/2 - \arg(j\omega + s_a) - \cdots - \arg(j\omega + s_n)
\end{aligned} \tag{4.46}$$

In evaluating Eqs. (4-45) and (4-46) let us examine the contributions of typical terms in these equations such as the zero at $s = -s_1$ and the pole at $s = -s_a$ which might be located in the s-plane as shown in Fig. 4-26. The vectors $\overrightarrow{AC}$ and $\overrightarrow{BC}$ correspond to the zero term $(j\omega + s_1)$ and the pole term $(j\omega + s_a)$ respectively. Thus the terms $|j\omega + s_1|$ and $|j\omega + s_a|$ in Eq. (4-45) correspond to the magnitude of the vectors $\overrightarrow{AC}$ and $\overrightarrow{BC}$. These observations may be repeated for each of the poles and zeros of $G(s)$ thus enabling the magnitude $|G(j\omega)|$ of $G(j\omega)$ to be computed. Similarly the phase $\phi(\omega)$ of the frequency response function can be computed by observing that the angles of vectors $\overrightarrow{AC}$ and $\overrightarrow{BC}$ are given by ϕ_{AC} and ϕ_{BC}. Repeated application of these observations for each pole and zero, where special note is made that the angles due to zeros must be added while those due to poles are subtracted, will yield the phase $\phi(\omega)$ of the frequency response function.

Example 16 *A system has the transfer function*

$$G(s) = \frac{10(s + 10)}{(s + 20 + j15)(s + 20 - j15)}$$

Find the magnitude and phase responses for the frequency response function $G(j\omega)$ using the pole-zero diagram.

It should be noted that the transfer function $G(s)$ is in loop sensitivity form. (It is always necessary to transform the transfer function to this form if the pole-zero plot is to be used for evaluating the frequency response function.) From the pole-zero diagram shown in Fig. 4-27 it can be easily seen that the magnitude response is given by

$$|G(j\omega)| = \frac{10\sqrt{\omega^2 + 10^2}}{\sqrt{(\omega + 15)^2 + 20^2}\sqrt{(\omega - 15)^2 + 20^2}}$$

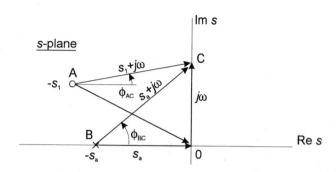

Figure 4.26: Pole-zero diagram showing typical pole and zero locations of $G(s)$ and the construction for computing the frequency response function $G(j\omega)$

Similarly the phase response can be easily deduced from Fig. 4-27 to be

$$\phi(\omega) = \arctan\left(\frac{\omega}{10}\right) - \arctan\left(\frac{\omega + 15}{20}\right) - \arctan\left(\frac{\omega - 15}{20}\right)$$

The Bode magnitude and phase responses are plotted in Fig. 4-28.

Example 17 *The transfer function for a second order system is*

$$G(s) = \frac{\omega_n^2}{s^2 + 2\zeta\omega_n s + \omega_n^2}$$

Find the Bode magnitude and phase response for this system, together with the maximum gain M_m and the frequency ω_m at which it occurs.

From the pole-zero diagram shown in Fig. 4-29 we find that the magnitude response is

$$|G(j\omega)| = \frac{\omega_n^2}{\sqrt{\left(\omega - \omega_n\sqrt{1 - \zeta^2}\right)^2 + \zeta^2\omega_n^2}} \cdot \frac{1}{\sqrt{\left(\omega + \omega_n\sqrt{1 - \zeta^2}\right)^2 + \zeta^2\omega_n^2}}$$

Similarly the phase response is

$$\phi(\omega) = -\arctan\left(\frac{\omega - \omega_n\sqrt{1 - \zeta^2}}{\zeta\omega_n}\right) - \arctan\left(\frac{\omega + \omega_n\sqrt{1 - \zeta^2}}{\zeta\omega_n}\right)$$

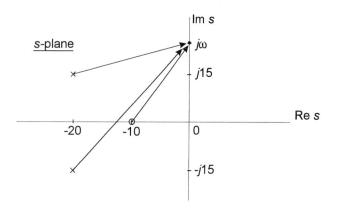

Figure 4.27: Pole zero diagram for the transfer function $G(s) = \dfrac{10(s+10)}{(s+20+j15)(s+20-j15)}$

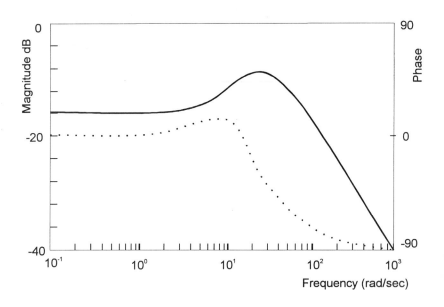

Figure 4.28: Bode magnitude and phase response for the frequency response function $G(j\omega) = \dfrac{10(j\omega + 10)}{(j\omega + 20 + j15)(j\omega + 20 - j15)}$

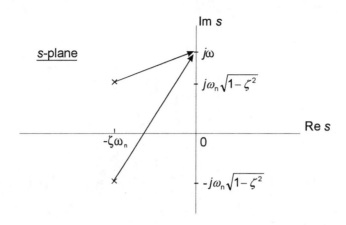

Figure 4.29: Pole-zero diagram for the transfer function $G(s) = \dfrac{\omega_n^2}{s^2 + 2\zeta\omega_n s + \omega_n^2}$

By simple algebraic manipulation

$$|G(j\omega)| = \frac{\omega_n^2}{\left\{[\omega^2 - \omega_n^2(1 - \zeta^2)]^2 + \zeta^4\omega_n^4 + 2\zeta^2\omega_n^2\,[\omega^2 + \omega_n^2(1 - \zeta^2)]\right\}^{1/2}}$$

(4.47)

To find the stationary point of $|G(j\omega)|$ the equation $d\,|G(j\omega)|\,/d\omega = 0$ is solved. The frequency at which the stationary point occurs is

$$\omega_m = \omega_n \sqrt{1 - 2\zeta^2}$$

(4.48)

Substituting ω_m into Eq. (4-47) gives the peak value of $G(j\omega)$ which we denote by M_m,

$$M_m = \frac{1}{2\zeta\sqrt{1 - \zeta^2}}$$

(4.49)

The Bode amplitude and phase responses for a range of damping factors ζ are shown in Fig. 4-30.

4.5 Approximate Transient Response

We have seen in Section 4-2 that the transient response can be characterized by a single time constant for first order systems, and by two parameters,

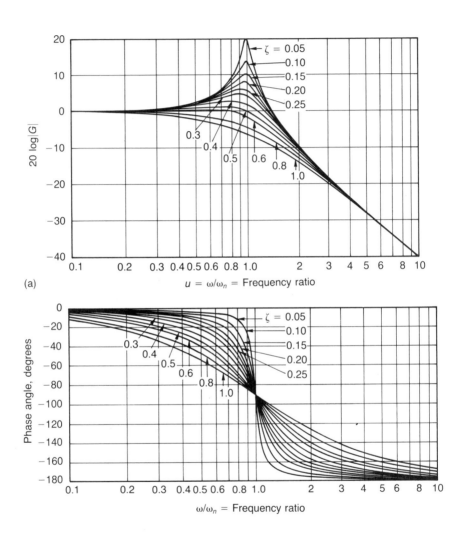

Figure 4.30:   Bode amplitude and phase response for $G(s)$ = $\dfrac{\omega_n^2}{s^2 + 2\zeta\omega_n s + \omega_n^2}$

the undamped natural frequency ω_n and the damping factor ζ, for second order systems. For high order systems the number of parameters required to characterize the transient response will multiply rapidly, and in general the transient response will become a complex function of time. Fortunately for many practical systems considerable simplification of the transient response of complex systems can be achieved, and they can be approximated by the responses of equivalent first or second order systems transfer functions.

Consider a system whose transfer function $G(s)$ is given by

$$G(s) = \frac{K(s^m + b_{m-1}s^{m-1} + \cdots + b_0)}{s^n + a_{n-1}s^{n-1} + \cdots + a_0} \tag{4.50}$$

We can factorize the denominator and expand $G(s)$ into partial fractions giving

$$G(s) = \frac{A_1}{s + s_1} + \frac{A_2}{s + s_2} + \cdots + \frac{A_3 s + A_4}{s^2 + 2\zeta_3\omega_{n3}s + \omega_{n3}^2} + \frac{A_5 s + A_6}{s^2 + 2\zeta_5\omega_{n5}s + \omega_{n5}^2} + \cdots \tag{4.51}$$

For a unit step input $U(s) = 1/s$, and we find

$$y(t) = A_1 \left(1 - e^{-s_1 t}\right) + A_2 \left(1 - e^{-s_2 t}\right) + \cdots$$

$$+ A_3' \left[1 - \frac{1}{\sqrt{1 - \zeta_3^2}} e^{-\zeta_3\omega_{n3}t} \sin\left(\omega_{d3}t + \phi_3\right) \right] + \cdots \tag{4.52}$$

The time response for this system is in general a very complex function of time, and it is quite difficult to easily characterize it in terms of a small number of parameters.

In many practical cases we can make considerable simplifications to the transfer function $G(s)$ by observing that at least one of the following situations exists:

1. A single real pole or a complex conjugate pair of poles is considerably closer to the imaginary axis of the s-plane than all the remaining poles of $G(s)$.

2. Some poles, either real or complex, close to the imaginary axis of the s-plane are also near corresponding zeros of $G(s)$, and condition (1) above exists amongst the remaining poles and zeros.

When either of these conditions exists we say the single real pole or complex conjugate poles are **dominant** because they largely determine the transient response of the system so that it may be well characterized by

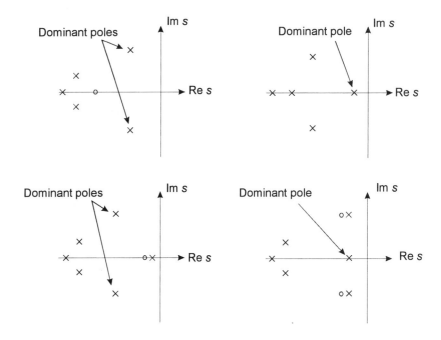

Figure 4.31: Pole-zero distribution and dominant poles

either one or possibly two parameters. To illustrate this situation a number of examples of various pole-zero distributions are shown, and the dominant pole or poles in each case is indicated in Fig. 4-31. The explanation for their dominance of the transient response can be seen by referring to the partial fraction expansion shown in Eq. (4-51). We will find that the magnitude of the partial fraction coefficients for all terms other than the dominant poles are small compared with those for the dominant poles. As a consequence the transient responses of the dominant pole terms will largely determine the system time response in each case.

Example 18 *Suppose a system has the transfer function*

$$G(s) = \frac{25a}{(s^2 + 6s + 25)(s + a)}$$

Find the transient response to a unit step for this system with $a = 0.3, 3$ *and 30, and determine the approximate transfer function in each case.*

Plots of the response to a unit step for $a = 0.3, 3$ and 30 are shown in Fig. 4-32. As can be seen when $a = 0.3$ the transient response is dominated

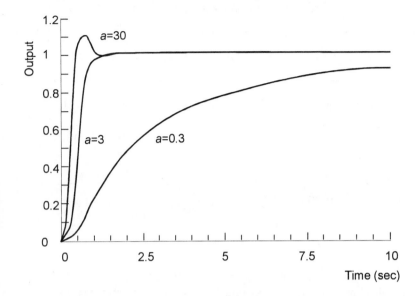

Figure 4.32: Transient response of $G(s)$ for $a = 0.3, 3$ and 30

by the time response of the term due to the pole at $s = -0.3$, while for the case where $a = 30$ the transient response is dominated by the term due to the poles at $s = -3 \pm j4$. To see that these results are as to be expected, let use expand $G(s)$ into its partial fractions,

$$G(s) = \frac{A_1 s + A_2}{\dfrac{1}{25}s^2 + \dfrac{6}{25}s + 1} + \frac{A_3}{\dfrac{1}{a}s + 1}$$

Evaluating the coefficients A_1, A_2 and A_3 for each case we obtain the results shown in Table 4-2. It can be seen that when $a = 0.3$ the coefficient $A_3 = 1.073$ which is much larger than the remaining two terms so that the pole at $s = -0.33$ is dominant, as observed from the time response plots. When $a = 30$ examination of the table shows that coefficient $A_2 = 0.966$ is much larger than either A_1 or A_3 so that the complex poles at $s = -3 \pm j4$ are dominant, which is confirmed by an examination of the time response plots shown in Fig. 4-32. In each of these cases the small partial fraction coefficients can, to a first approximation, be neglected so that the approximate transfer functions shown in the table are obtained. For the intermediate case where $a = 3$ all partial fraction coefficients are of comparable magnitude so that no simplified transfer function appears

Table 4-2 Partial fraction coefficients

a	A_1	A_2	A_3	Approx $G(s)$	Remarks
0.3	-0.013	-0.073	**1.073**	$\frac{1.073}{3.33s+1}$	Dominant real pole
3	-0.188	-0.562	1.563		Transfer function cannot be simplified
30	-0.040	**0.966**	0.034	$\frac{0.966}{s^2/25+6s/25+1}$	Dominant complex poles

possible. This conclusion is confirmed by the plot in Fig. 4-32 for $a = 3$ which lies between the two extreme cases for $a = 0.3$ and $a = 30$.

We see from this example that the real pole or complex poles closest to the imaginary axis in the s-plane essentially determine the transient response of the system because the partial fraction coefficients of the other terms in the expansion given in Eq. (4-51) are small and can be neglected.

Example 19 *Suppose a system has the transfer function*

$$G_1(s) = \frac{24.19(s + 0.31)}{(s^2 + 6s + 25)(s + 0.3)} \tag{4.53}$$

Find the transient response of this system for a unit step input, and show that its transfer function can be approximated by

$$G_2(s) = \frac{1}{\frac{1}{25}s^2 + \frac{6}{25}s + 1}$$

The computed transient response for $G_1(s)$ is shown in Fig. 4-33 together with the response of $G_2(s)$ for comparison. It will be noted from this figure that the time response for $G_1(s)$ conforms very closely to the one for $G_2(s)$. Expanding Eq. (4-53) into its partial fractions we find

$$G_1(s) = \frac{A_1 s + A_2}{\frac{1}{25}s^2 + \frac{6}{25}s + 1} + \frac{A_3}{\frac{1}{0.3}s + 1}$$

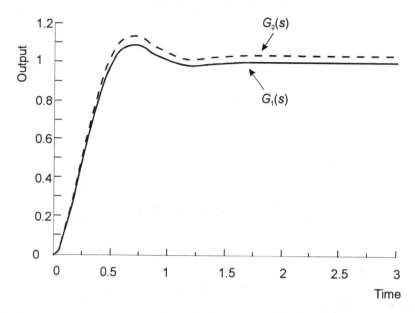

Figure 4.33: Transient response of system transfer function $G_1(s)$ and approximate transfer function $G_2(s)$

with $A_1 = -0.000415$, $A_2 = 0.965$, and $A_3 = 0.0346$, where it will be noted that $|A_1|, |A_3| \ll |A_2|$. Because of this observation we conclude that $G_1(s) \approx G_2(s)$.

We see from this example that even though a system pole may be close to the imaginary axis in the s-plane the presence of a zero close-by may cancel its effect upon the system transient response to a step input. This result is characterized by the fact that the partial fraction coefficient corresponding to this pole in the partial fraction expansion is small compared to other coefficients.

Estimating Transient and Frequency Response Performance

From the above discussions when a systems has either a dominant real pole or a complex conjugate pair of dominant poles then the transient response to a step input and the frequency response of the system is largely determined by the responses of the respective approximate system transfer functions. Thus for the system transfer function discussed in Example 18, when $a = 30$ the approximate transient response is given in Fig. 4-10 while the frequency response is given in Fig. 4-30. On the other hand when

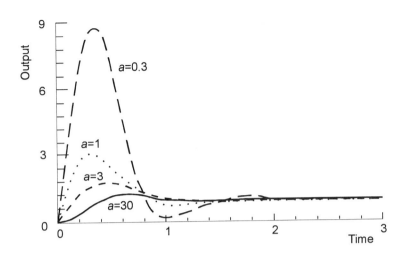

Figure 4.34:   Transient response for a unit step input of $G(s)$ $=$ $\dfrac{25(s/a+1)}{s^2+6s+25}$, where $a = 0.3$, 1, 3, and 30

$a = 0.3$ the time response will have the exponential form shown in Fig. 4-4(b), while the frequency response will have the form shown in Fig. 4-25.

Effect of System Zeros on Transient Overshoot

The presence of an uncanceled zero in a system transfer function, especially when it is close to the imaginary axis of the s-plane, can have a profound effect on the system transient response. We find in this case that there is often a large increase in the magnitude of the overshoot, and at the same time a significant decrease in the transient response time.

Example 20 *Suppose a system has the transfer function*

$$G(s) = \frac{25(s/a+1)}{s^2+6s+25}$$

Find the transient response of this system for a unit step input, when $a = 0.3, 1, 3,$ *and* 30, *and compare the resultant transient overshoot, rise-time and settling time for the four cases.*

The computed transient responses are shown in Fig. 4-34. It can be seen from this figure and from Table 4-3 that as the zero at $s = -a$ approaches

Table 4-3 Response characteristics of $G(s) = 25(s/a + 1)/(s^2 + 6s + 25)$

a	t_s(sec)	c_{max}
0.3	2.17	8.73
1.0	1.53	2.95
3.0	1.45	1.41
30.0	1.15	1.10

the imaginary axis of the s-plane the overshoot increases extremely rapidly, while at the same time there is a moderate increase in the settling time. However the presence of the zero can be advantageous because it will be noted that it leads to a significant improvement in the system response time.

4.6   Summary

This chapter has dwelt upon the analysis of control systems in the time domain via the classical method.

The transient response was seen to depend upon the roots of the characteristic equation, or the closed loop poles, and not upon the type of input. The form of response was directly related to the location of these poles on the complex s-plane. It was seen that the response became uncontrollably large if the poles were in the right half of the s-plane or repeated on the imaginary axis.

The steady state response of control systems was studied and the concepts of error coefficients and error constants were introduced. Following this the concept of dynamic errors for periodically excited systems was studied, and methods of approximately calculating their value using error coefficients was examined.

Fourier series expansion of periodic signals and their extension to the Fourier transform was introduced. The important concepts of frequency spectrum, and frequency response function were presented. It was shown that the frequency response function has many properties similar to the transfer function introduced in Chapter 3. It was also shown that the Bode response plot is an extremely useful way of representing the frequency response function graphically.

The concept of dominant poles for a transfer function was introduced and it was shown to be useful in approximating the transient and periodic responses of systems.

4.7   References

1. J. J. D'Azzo, C. H. Houpis, *Linear Control System Analysis and Design — Conventional and Modern*, Mc Graw Hill Book Co., New York, 1988.

2. J. J. Di Stefano, A. R. Stubberud, I. J. Williams, *Feedback and Control Systems (Schaum's Outline Series)*, Mc Graw Hill Book Co., New York, 1967.

3. M. F. Gardner, J. L. Barnes, *Transients in Linear Systems*, John Wiley and Sons, Inc., New York, 1942.

4. E. Kreysig, *Advanced Engineering Mathematics*, John Wiley and Sons, Inc., New York, 1988.

5. T. J. Viersma, *Analysis, Synthesis, and Design of Hydraulic Servosystems and Pipelines*, Elsevier Scientific Publ. Co., Amsterdam, 1980.

6. C. J. Savant, *Control System Design*, Mc Graw Hill Book Co., New York, 1964.

7. H. Chestnut, R. W. Mayer, *Servomechanism and Regulating System Design*, Mc Graw Hill Book Co., 1959.

8. H. M. James, N. B. Nichols, R. S. Philips, *Theory of Servomechanisms*, Mc Graw Hill Book Co., New York, 1947.

9. B. C. Kuo, *Automatic Control Systems*, Prentice-Hall, Inc., Englewood Cliffs New Jersey, 1987.

10. M. Schwartz, *Information Transmission, Modulation, and Noise: A Unified Approach to Communications*, Mc Graw Hill Book Co., New York, 1981.

11. R. J. Mayhan, *Discrete-Time and Continuous-Time Linear Systems*, Addison Wesley Publ. Co., Reading, Mass., 1983.

12. H. W. Bode, *Network Analysis and Feedback Amplifier Design*, Van Nostrand, Princeton N. J., 1945.

13. R. C. Dorf, *Modern Control Systems*, Addison Wesley Publ. Co., Reading, Mass., 1989.

14. N. H. McClamroch, *State Models of Dynamic Systems — A Case Study Approach*, Springer-Verlag, New York, 1980.

15. J. Meisel, *Principles of Electromechanical Energy Conversion*, Mc Graw Hill Book Co., New York, 1966.

16. A. Papoulis, *The Fourier Integral and Its Applications*, Mc Graw Hill Book Co., New York, 1962.

17. R. B. Bracewell, *The Fourier Integral and Its Applications*, Mc Graw Hill Book Co., New York, 1986.

4.8 Problems

4-1 The block diagram of a simplified servo is shown in Fig. P4-1. Obtain $c(t)$ if $r(t) = H(t)$ and $K = 2$. At what time does $c(t)$ reach its first peak? What is the percent overshoot of the response?

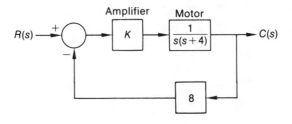

Figure P4-1

4-2 For what values of amplifier gain does the system shown in Fig. 4-12 becomes unstable?

4-3 A d.c. position servomechanism is shown in Fig. P4-3. Derive the overall transfer function if $T_1 \approx 0$, $K = 1.0$, $K_1 = 0.5$, and $T_2 = 0.04$. Obtain $\theta(t)$ if $\theta_1(t)$ is a unit step input. Obtain the peak overshoot.

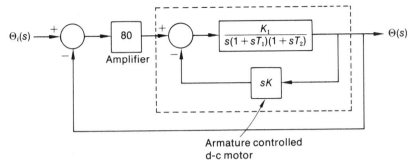

Armature controlled
d-c motor

Figure P4-3

4-4 Find the partial fraction expansion for:

(a) $F(s) = \dfrac{1}{(s+1)(s+2)(s+3)}$

(b) $F(s) = \dfrac{1}{s(s+1)^2}$

(c) $F(s) = \dfrac{10(s+2)(s+10)^2}{(s+4)^3(s^2+6s+25)}$

(d) $F(s) = \dfrac{2(s+3)}{s^2(s+5)(s+10)}$

4-5 Obtain $f(t)$ for the following:

(a) $F(s) = \dfrac{1}{(s^2+s+3)^2}$

(b) $F(s) = \dfrac{6s^2+1}{s^2(s+3)(s^2+2s+3)}$

(c) $F(s) = \dfrac{s^2+s+1}{s^2+3s+8}$

(d) $F(s) = \dfrac{20}{s(s^2+2s+5)(s+6)}$

(e) $F(s) = \dfrac{10}{(s+3)^2(s^2+2s+2)(s^2+6s+25)}$

4-6 The block diagram of the autopilot for the DC3 aircraft is shown in Fig. P4-6. Obtain the transient response of the aircraft pitch angle $\theta(t)$ for a unit step change in the input $e_{\theta_i}(t)$ when $S_\theta S_{\delta_e} = 2$.

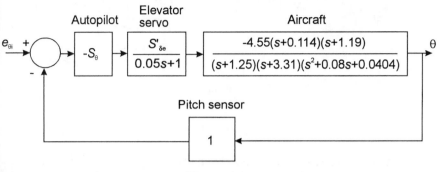

Figure P4-6

4-7 The electrohydraulic vibration isolation system discussed in Example 5 has been shown to have the transfer function

$$\frac{X(s)}{Y(s)} = \frac{\omega_n(2\zeta s + \omega_n)}{s^2 + 2\zeta\omega_n s + \omega_n^2}$$

Supposing $\omega_n = 10$ and $\zeta = 0.5$, find the response of the system to the step input $x(t) = H(t)$.

4-8 (McClamroch) Consider two continuously stirred mixing tanks, Fig. P4-8, with reflux flow, of a type used frequently in process industries. In the system the tanks are held filled at constant volumes V_1 and V_2. The constant through- and reflux-flow rates are q_1 and q_2 respectively, and the output salt concentration $c_o(t)$ is controlled by varying the input concentration $c_i(t)$.

(a) Show the transfer function of this system is

$$\frac{C_o(s)}{C_i(s)} = \frac{q_1(q_1 - q_2)}{V_1 V_2 s^2 + (q_1 V_2 + q_2 V_1)s + q_1(q_1 - q_2)}$$

(b) Suppose the initial concentration in each tank is zero, and there is a step change in the input concentration from zero to $c_{i\ max}$. Find the transient response of the output concentration as a function of time.

(c) If the input concentration varies sinusoidally

$$c_i(t) = c_{i\ max} \sin \omega t$$

find the steady state variation of the output concentration $c_o(t)$. From this result draw a conclusion about the effect of the tanks upon the output concentration variation for different frequencies of the input concentration $c_i(t)$.

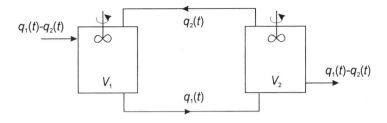

Figure P4-8

4-9 A model of the suspension system for a car wheel is shown in Fig. P4-9, where m_1 is the equivalent car body mass referred to the wheel, and m_2 is the unsprung mass of the wheel. Coupling the masses m_1 and m_2 is the suspension spring of stiffness k_1, and the shock absorber having damping coefficient b. The compliance of the tire is represented by the spring with stiffness k_2. The displacements of the car body and the wheel relative to their equilibrium points when $x_3(t) = 0$ are denoted by $x_1(t)$ and $x_2(t)$ respectively.

(a) Find the transfer function $X_1(s)/X_3(s)$.

(b) Calculate the transient response of the car body if there is a step change in the roadway position given by $x_3(t) = x_{30}H(t)$.

(c) Find the steady state response of the car body to a sinusoidally varying road surface having a frequency ω rad/sec.

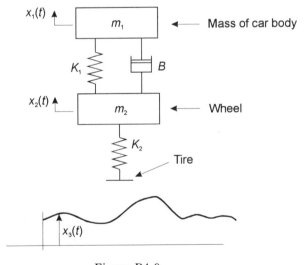

Figure P4-9

4-10 (Meisel) A simplified diagram of a force balance accelerometer is shown in Fig P4-10(a). Here mass M is free to move along the sense axis, and the position is detected by a sensitive inductive displacement transducer. The mass is forced to return to its null displacement position by a current driven actuator incorporated into a feedback loop as shown in Fig. 4-10(b). The transfer function for this system is

$$\frac{I(s)}{\ddot{X}(s)} = \frac{K S_d (1 + C_d s)}{s^2 + s \left(\dfrac{C_d S_d S_f K}{M} \right) + \dfrac{S_d S_f K}{M}}$$

Find the transient response of the accelerometer to a unit step change of the acceleration $\ddot{x}(t)$.

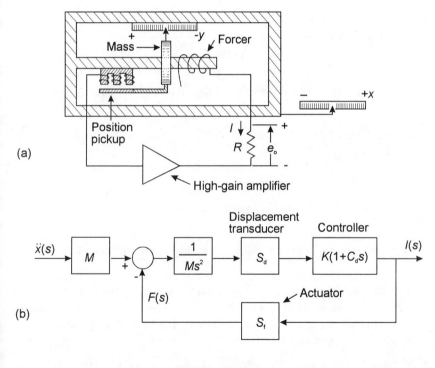

(a)

(b)

Figure P4-10

4-11 Obtain the steady state output and error for a step input to the control system shown in Fig. P4-11.

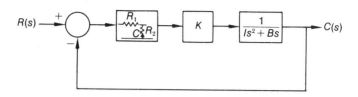

Figure P4-11

4-12 The overall transfer function for a control system is given by

$$\frac{C(s)}{R(s)} = \frac{s^2 + 2}{(s+1)^2(s^2 + 2s + 4)}$$

Obtain the steady state and transient response $c(t)$ if $r(t) = 2H(t)$.

4-13 A closed loop unity feedback systems has the following forward loop transfer function

$$G(s) = \frac{K}{s^2(s + A)}$$

What is the steady state value of the error and the output if the input is:

(a) Unit ramp

(b) Unit step

(c) $\sin \omega t$

If you cannot obtain the answer to any of the above three questions, explain why.

4-14 What is the steady state error and output for Problem 4-3?

4-15 The attitude stabilization control system for a vertical takeoff aircraft (VTOL) is shown in Fig. P4-15. Obtain the transient response to a unit step input. What is the steady state error to a constant disturbance if $a = 0.5$, $b = 10$, $c = 3$, $K_A = 1$, $K_B = 17$, $K_C = 1$, $u(t) = 0.1$, and $r(t) = H(t)$.

4-16 How do changes in K_C affect the output of the VTOL described in Problem 4-15?

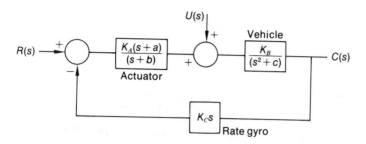

Figure P4-15

4-17 A simplified control system for the speed setting of a gasoline engine is shown in Fig. P4-17. The lags τ_1 and τ_2 occur at the carburetor and the engine itself. The lag τ_3 is the time constant associated with the speed measurement. If a steady state error of 10% of the reference setting is permissible, what must the gain be? Suggest some typical values of τ_1, τ_2, and τ_3.

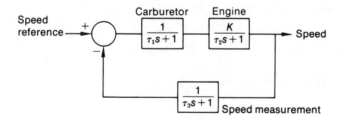

Figure P4-17

4-18 A control system for controlling the temperature of a liquid container is shown in Fig. P4-18. The controller operates the solenoid valve which allows refrigerant to flow into the evaporator coil. Assume that the valve takes τ seconds to open fully and that the percentage opening varies linearly with time. Obtain the overall transfer function and obtain $T(t)$ if the desired temperature is suddenly *decreased* to T_L. State your assumptions. Does the response agree with the answer you arrive at intuitively?

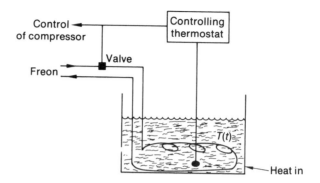

Figure P4-18

4-19 For the aircraft described in Problem 4-6 find the generalized error coefficients C_0 and C_1, and determine the pitch angle error for $t \gg 0$ if $e_{\theta_i}(t) = 0.1H(t)$.

4-20 For the force balance accelerometer discussed in Problem 4-10 find the generalized error coefficients C_0, C_1, and C_2. Using these coefficients determine the dynamic error when $t \gg 0$, for ramp, parabolic, and sinusoidal variations of the acceleration $\ddot{x}(t)$.

4-21 For a unity feedback system whose open loop transfer function is

$$G(s) = \frac{100}{s(0.05s + 1)}$$

find the dynamic error of the system for $t \gg 0$ when $r(t) = 1 + 2t + 0.3 \sin t$.

4-22 Consider the non-unity feedback position control system, where the transfer functions in the forward and feedback paths are

$$G(s) = \frac{75}{s(0.1s + 1)}, \quad \text{and } H(s) = 2 + \frac{0.08s^2}{0.2s + 1}$$

Supposing the idealized model for this control system is $G_m(s) = 1/2$, find the steady-state error for a unit ramp input.

4-23 Find the Fourier series for the periodic waveforms $f(t)$ shown in Fig. P4-23. Plot and compare the amplitude spectra for these two waveforms. In addition examine and compare the phase components of the spectra and give a physical interpretation of their differences.

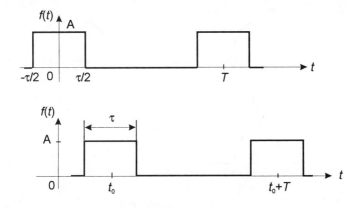

Figure P4-23

4-24 Find the Fourier series for the periodic waveforms shown in Fig. P4-24.

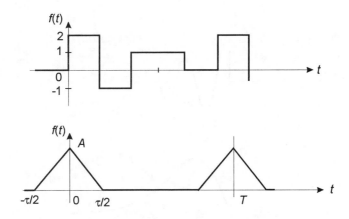

Figure P4-24

4-25 Find the Fourier spectrum of the periodic waveform consisting of the half-cosine pulses shown in Fig. P4-25(a), and show that as the period $T \to \infty$ the amplitude spectra for the periodic signal approaches the Fourier transform of the half-cosine pulse shown in Fig. P4-25(b).

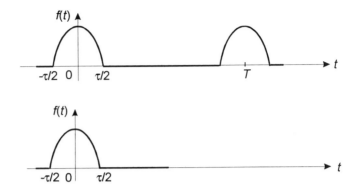

Figure P4-25

4-26 Find the Fourier transform of the time functions shown in Fig. P4-26.

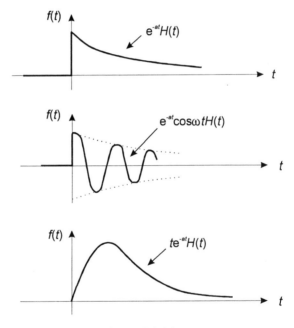

Figure P4-26

(a) Show the Fourier transform of the Dirac delta function $\delta(t)$ is given by

$$\mathcal{F}[\delta(t)] = 1$$

so that by the uniqueness property of the Fourier transform we can conclude $\mathcal{F}^{-1}[1] = \delta(t)$.

(b) Similarly show that if the signum function sgn t is as defined in
Fig. P4-27 then

$$\mathcal{F}^{-1}[2/j\omega] = \text{sgn } t$$

so that by the uniqueness property $\mathcal{F}[\text{sgn } t] = 2/j\omega$.

(c) Find the Fourier transform of the unit step $H(t)$ by observing
that

$$H(t) = \frac{1}{2} + \frac{1}{2}\text{sgn } t$$

and utilizing the results of (a) and (b) above show

$$\mathcal{F}[H(t)] = \pi\delta(t) + \frac{1}{j\omega}$$

Plot the magnitude of this spectrum.

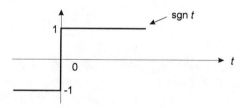

Figure P4-27

4-28 If $F(j\omega) = \mathcal{F}[f(t)]$ then using the definition of the Fourier transform
show that

$$\mathcal{F}[f(t)\sin \omega_0 t] = [F(j\omega - j\omega_0) - F(j\omega + j\omega_0)]/2j.$$

(a) Supposing $F_1(j\omega) = \mathcal{F}[f_1(t)]$ and $F_2(j\omega) = \mathcal{F}[f_2(t)]$ show using
Eqs. 4-38, 4-39, and 4-36, that

$$\int_{-\infty}^{\infty} f_1(\tau)f_2(t-\tau)d\tau = \frac{1}{2\pi}\int_{-\infty}^{\infty} F_1(j\omega)F_2(j\omega)e^{-j\omega t}\,d\omega$$

and so if $t = 0$

$$\int_{-\infty}^{\infty} f_1(\tau)f_2(-\tau)d\tau = \frac{1}{2\pi}\int_{-\infty}^{\infty} F_1(j\omega)F_2(j\omega)d\omega$$

(b) From the definition of the Fourier transform show that if $\mathcal{F}[f(t)] = F(j\omega)$ then $\mathcal{F}[f^*(t)] = F^*(-j\omega)$, and $\mathcal{F}[f(-t)] = F(-\omega)$ where
$f^*(t)$ and $F^*(j\omega)$ are complex conjugates of $f(t)$ and $F(j\omega)$ respectively.

(c) Taking $g^*(t) = f_2(-t)$ show that

$$\int_{-\infty}^{\infty} f(t)g^*(t)dt = \frac{1}{2\pi} \int_{-\infty}^{\infty} F(j\omega)G^*(j\omega)d\omega$$

and so deduce Parseval's theorem

$$\int_{-\infty}^{\infty} |f(t)|^2 \, dt = \frac{1}{2\pi} \int_{-\infty}^{\infty} |F(j\omega)|^2 \, d\omega$$

[If for example the signal $f(t)$ is the potential across a one ohm resistor then it can be seen that

$$E = \int_{-\infty}^{\infty} |f(t)|^2 \, dt$$

is the total energy dissipated in the resistor, and the Parseval identity shows this is given by the area under the $|F(j\omega)|^2/2\pi$ curve. Since $F(j\omega)$ is the spectrum of $f(t)$ the quantity $|F(j\omega)|^2$ is often referred to as the **energy spectral density function** of $f(t)$.]

4-30 For the transfer functions shown in Problem 4-4, plot the respective amplitude and phase responses using the pole-zero plotting method discussed in Section 4-4.

4-31 For the transfer functions given below determine whether they have dominant poles, and if so determine their approximate transfer functions. Verify your conclusions by expanding these transfer functions into partial fractions and identifying which of their coefficients can be neglected.

(a) $\quad G(s) \quad = \quad \dfrac{492000\,(s+50)}{[(s+50)^2 + 40^2]\,(s+30)\,(s+4)}$

(b) $\quad G(s) \quad = \quad \dfrac{1.372 \times 10^7}{\left[(s+60)^2 + 50^2\right]\left[(s+3)^2 + 4^2\right](s+90)}$

(c) $G(s) \;\; = \;\; \dfrac{8.984 \times 10^7 \, (s + 0.011)}{\left[(s + 60)^2 + 50^2 \right] \left[(s + 6)^2 + 12^2 \right] (s + 90)(s + 0.01)}$

(d) $G(s) \;\; = \;\; \dfrac{6.069 \times 10^5 \left[(s + 5.95)^2 + 4.99^2 \right]}{\left[(s + 80)^2 + 60^2 \right] \left[(s + 6)^2 + 5^2 \right] (s + 120)(s + 3)}$

(e) $G(s) \;\; = \;\; \dfrac{224 \, (s + 40)}{\left[(s + 80)^2 + 10^2 \right] \left[(s + 4)^2 + 4^2 \right] (s + 3)}$

4-32 Consider the position servomechanism used in a numerically controlled laser cutting machine shown in Fig. P4-32. Find the closed loop transfer function for this system and show that it can be approximated by the transfer function

$$\frac{C(s)}{R(s)} = \frac{10.7}{(s + 10.7)}$$

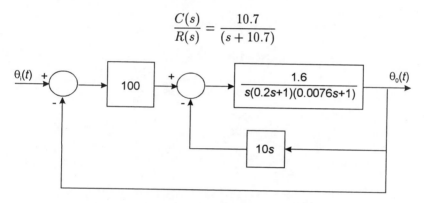

Figure P4-32

Chapter 5

Time Response - State Equation Method

5.1 Introduction

We have dealt with the analysis of systems that were represented by higher-order linear differential equations with constant coefficients. The response of the system was obtained via classical techniques that rely on operational mathematics. Such an approach is the direct outgrowth of techniques developed in mechanical vibration theory and electrical network theory. This method is very attractive when we study a single output of a system subjected to a single input. However, modern systems tend to get quite complex and it becomes not only necessary to study several inputs and outputs simultaneously but also the behavior of variables inside the control system. This can be most effectively achieved by the use of state space techniques where the system is represented by first-order differential equations and the analysis exploits matrix theory which is briefly reviewed in Appendix C. Since the characterization of systems using matrix methods requires much computational time, the use of a digital computer becomes imperative. With the advent of faster computers the state space technique has enabled us to tackle some very complex but interesting systems.

The vector-matrix differential equation representation was introduced in Chapter 3 and it was also shown how it is related to the system transfer function. Here we shall be concerned with obtaining the time response of the system represented using state space notation. We will also establish the stability of the system by investigating the nature of the eigenvalues of its coefficient matrix. Finally we shall examine the concepts of controllability and observability and the important role they play in relating the transfer function and the state equations of a system.

5.2   Solution of the State Equation

We consider the state equations for a system $\mathcal{S}$, which was discussed in Chapter 3

$$\dot{\mathbf{x}}(t) = \mathbf{A}\mathbf{x}(t) + \mathbf{b}u(t) \qquad (5.1)$$

$$y(t) = \mathbf{c}\mathbf{x}(t) \qquad (5.2)$$

Its general solution contains two parts, these being the **zero state** (or forced) **response** and the **zero input** (or transient) **response**. Since the total system response is dependent upon a knowledge of the zero input response we first treat this case and then show how it can be used to find the system response when the input is non-zero. There are several methods of finding the solution of these equations but we shall content ourselves with examining two. The first is based upon using the **Laplace transform**, often together with the Fadeeva algorithm, to evaluate the relevant coefficients, while the second is based upon the use of the **Cayley-Hamilton theorem**, so we will refer to it by this appellation.

The Laplace Transform Method

Taking the Laplace transform of Eq. (5-1) and assuming $\mathbf{x}(0)$ is the initial state vector (i.e. the initial condition) we obtain

$$s\mathbf{X}(s) - \mathbf{x}(0) = \mathbf{A}\mathbf{X}(s) + \mathbf{b}U(s) \qquad (5.3)$$

where

$$\mathbf{x}(0) = \begin{bmatrix} x_1(0) \\ \vdots \\ x_n(0) \end{bmatrix} \text{ and } \mathbf{X}(s) = \begin{bmatrix} X_1(s) \\ \vdots \\ X_n(s) \end{bmatrix}$$

Rewriting this

$$[s\mathbf{I} - \mathbf{A}]\mathbf{X}(s) = \mathbf{x}(0) + \mathbf{b}U(s)$$

where $\mathbf{I}$ is the identity matrix. Solving the matrix equation

$$\mathbf{X}(s) = [s\mathbf{I} - \mathbf{A}]^{-1}[\mathbf{x}(0) + \mathbf{b}U(s)] \qquad (5.4)$$

The inverse Laplace transform of Eq. (5-4) yields

$$\mathbf{x}(t) = [\mathcal{L}^{-1}(s\mathbf{I} - \mathbf{A})^{-1}]\mathbf{x}(0) + \mathcal{L}^{-1}[(s\mathbf{I} - \mathbf{A})^{-1}\mathbf{b}U(s)] \qquad (5.5)$$

Taking the Laplace transform of the output given by Eq. (5-2)

$$Y(s) = \mathbf{c}\mathbf{X}(s)$$

Substituting Eq. (5-4)

$$Y(s) = \mathbf{c}[s\mathbf{I} - \mathbf{A}]^{-1}\mathbf{x}(0) + \mathbf{c}[s\mathbf{I} - \mathbf{A}]^{-1}\mathbf{b}U(s) \qquad (5.6)$$

where the first term represents the zero input response, and the second term represents the zero state response. Here we shall concern ourselves with the first term which can be rewritten if we observe that

$$[s\mathbf{I} - \mathbf{A}]^{-1} = \frac{\text{adj}[s\mathbf{I} - \mathbf{A}]}{\det[s\mathbf{I} - \mathbf{A}]} \qquad (5.7)$$

where **adj** is the adjoint and **det** is the determinant. The zero input response becomes

$$\mathbf{x}(t) = \mathcal{L}^{-1}\left[\frac{\text{adj}[s\mathbf{I} - \mathbf{A}]}{\det[s\mathbf{I} - \mathbf{A}]}\right]\mathbf{x}(0) \qquad (5.8)$$

$$y(t) = \mathbf{c}\mathcal{L}^{-1}\left[\frac{\text{adj}[s\mathbf{I} - \mathbf{A}]}{\det[s\mathbf{I} - \mathbf{A}]}\right]\mathbf{x}(0) \qquad (5.9)$$

Here the equation

$$\det[s\mathbf{I} - \mathbf{A}] = 0 \qquad (5.10)$$

is the **characteristic equation** of the system.

Example 1 *A system is characterized by*

$$\dot{x}_1(t) \;=\; 0$$

$$\dot{x}_2(t) \;=\; ax_1(t) - ax_2(t)$$

and the output vector is $\mathbf{c} = \begin{bmatrix} 1 & 2 \end{bmatrix}$. *If* $x_1(0) = 1$ *and* $x_2(0) = 2$, *determine the output time function.*

The coefficient matrix becomes

$$\mathbf{A} = \begin{bmatrix} 0 & 0 \\ a & -a \end{bmatrix}$$

and

$$[s\mathbf{I} - \mathbf{A}]^{-1} = \frac{1}{s(s+a)}\begin{bmatrix} s+a & 0 \\ a & s \end{bmatrix}$$

The characteristic equation is $s(s + a) = 0$. Therefore by using the partial fraction expansion the above equation may be written as

$$[s\mathbf{I} - \mathbf{A}]^{-1} = \frac{1}{as} \begin{bmatrix} a & 0 \\ a & 0 \end{bmatrix} - \frac{1}{a(s + a)} \begin{bmatrix} 0 & 0 \\ a & -a \end{bmatrix}$$

so that

$$\mathbf{x}(t) = \mathcal{L}^{-1}[s\mathbf{I} - \mathbf{A}]^{-1}\mathbf{x}(0) = \begin{bmatrix} 1 & 0 \\ 1 - e^{-at} & e^{-at} \end{bmatrix} \begin{bmatrix} x_1(0) \\ x_2(0) \end{bmatrix}$$

Carrying out the matrix multiplication and letting $x_1(0) = 1, x_2(0) = 2$

$$\begin{bmatrix} x_1(t) \\ x_2(t) \end{bmatrix} = \begin{bmatrix} 1 \\ 1 + e^{-at} \end{bmatrix}$$

The output is

$$\begin{aligned} y(t) &= \mathbf{cx}(t) = x_1(t) + 2x_2(t) \\ &= 3 + 2e^{-at} \end{aligned}$$

Example 2 *Obtain the characteristic equation for the system described by the following set of differential equations*

$$\begin{aligned} \dot{x}_1(t) &= x_2(t) \\ \dot{x}_2(t) &= x_3(t) \\ \dot{x}_3(t) &= -6x_1(t) - 11x_2(t) - 6x_3(t) \end{aligned}$$

The coefficient matrix becomes

$$\mathbf{A} = \begin{bmatrix} 0 & 1 & 0 \\ 0 & 0 & 1 \\ -6 & -11 & -6 \end{bmatrix}$$

Forming $[s\mathbf{I} - \mathbf{A}]^{-1}$ we find

$$[s\mathbf{I} - \mathbf{A}]^{-1} = \frac{1}{P(s)} \begin{bmatrix} s^2 + 6s + 11 & s + 6 & 1 \\ -6 & s^2 + 6s & s \\ -6s & -11s - 6 & s^2 \end{bmatrix}$$

where the characteristic equation $P(s)$ is

$$P(s) = s^3 + 6s^2 + 11s + 6 = 0$$

and it can be shown it has zeros equal to $-1, -2$, and -3.

The procedure of taking the inverse Laplace transform may be formalized as we did for polynomials in Chapter 4. Let us define the **resolvent matrix** $\Phi(s)$ as

$$\Phi(s) = \frac{\text{adj}[s\mathbf{I} - \mathbf{A}]}{\det[s\mathbf{I} - \mathbf{A}]} = [s\mathbf{I} - \mathbf{A}]^{-1} \tag{5.11}$$

The characteristic equation here is $\det[s\mathbf{I} - \mathbf{A}] = 0$ and its roots are $s_1, s_2, s_3, \ldots, s_n$. Therefore the $\Phi(s)$ matrix may be expressed as

$$\Phi(s) = \frac{[\mathbf{K}]_1}{s - s_1} + \frac{[\mathbf{K}]_2}{s - s_2} + \cdots + \frac{[\mathbf{K}]_n}{s - s_n} \tag{5.12}$$

where $[\mathbf{K}]_n$ is an undetermined coefficient matrix. If the roots of the characteristic equation $\det[s\mathbf{I} - \mathbf{A}] = 0$ are distinct, then the undetermined coefficient matrix may be determined by

$$[\mathbf{K}]_i = \lim_{s \to s_i} \left[[s\mathbf{I} - \mathbf{A}]^{-1} (s - s_i) \right] \tag{5.13}$$

If however there is a multiple root, then the matrix becomes

$$\Phi(s) = \frac{[\mathbf{K}]_1}{s - s_1} + \frac{[\mathbf{K}]_2}{s - s_2} + \frac{[\mathbf{K}]_3}{(s - s_2)^2} + \cdots + \frac{[\mathbf{K}]_n}{s - s_{n-1}} \tag{5.14}$$

In this case all the matrices, except $[\mathbf{K}]_2$ and $[\mathbf{K}]_3$, may be evaluated using the previous method. $[\mathbf{K}]_2$ and $[\mathbf{K}]_3$ are evaluated from

$$[\mathbf{K}]_2 = \lim_{s \to s_2} \frac{d}{ds} \left[[s\mathbf{I} - \mathbf{A}]^{-1} (s - s_2)^2 \right]$$

$$[\mathbf{K}]_3 = \lim_{s \to s_2} \left[[s\mathbf{I} - \mathbf{A}]^{-1} (s - s_2)^2 \right]$$

This method is referred to as a generalization of **Heaviside's partial fraction expansion**.

Once the constants in Eq. (5-12) or Eq. (5-14) are evaluated the inverse Laplace transform yields

$$\phi(t) = [\mathbf{K}]_1 e^{s_1 t} + \cdots + [\mathbf{K}]_n e^{s_n t} \tag{5.15}$$

or for multiple roots it yields

$$\phi(t) = [\mathbf{K}]_1 e^{s_1 t} + [\mathbf{K}]_2 e^{s_2 t} + [\mathbf{K}]_3 t e^{s_2 t} + \cdots \tag{5.16}$$

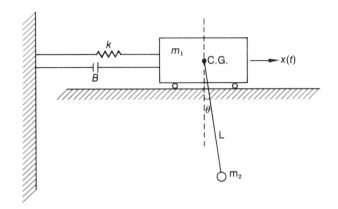

Figure 5.1: A mechanical system

Example 3 *Consider the mechanical system shown in Fig. 5-1. The governing differential equations are*

$$\ddot{x}(t) + \frac{B}{m_1}\dot{x}(t) + \frac{k}{m_1}x(t) - \frac{m_2 g}{m_1}\theta(t) = 0$$

$$\ddot{\theta}(t) - \frac{B}{m_1 L}\dot{x}(t) - \frac{k}{m_1 L}x(t) + \frac{m_1 + m_2}{m_1 L}g\theta(t) = 0$$

We wish to obtain the zero input response $\mathbf{x}(t)$.

The state equation can be obtained by first defining the state variables $x_1(t) = x(t)$, $x_2(t) = \dot{x}(t)$, $x_3(t) = \theta(t)$, $x_4(t) = \dot{\theta}(t)$ and then substituting, we obtain

$$\dot{x}_1(t) = x_2(t)$$

$$\dot{x}_2(t) = -\frac{k}{m_1}x_1(t) - \frac{B}{m_1}x_2(t) + \frac{m_2 g}{m_1}x_3(t)$$

$$\dot{x}_3(t) = x_4(t)$$

$$\dot{x}_4(t) = \frac{k}{m_1 L}x_1(t) + \frac{B}{m_1 L}x_2(t) - \frac{m_1 + m_2}{m_1 L}g x_3(t)$$

which can be written as

$$\dot{\mathbf{x}}(t) = \mathbf{A}\mathbf{x}(t)$$

where

$$\mathbf{A} = \begin{bmatrix} 0 & 1 & 0 & 0 \\ \dfrac{-k}{m_1} & \dfrac{-B}{m_1} & \dfrac{m_2 g}{m_1} & 0 \\ 0 & 0 & 0 & 1 \\ \dfrac{k}{m_1 L} & \dfrac{B}{m_1 L} & \dfrac{-(m_1 + m_2)g}{m_1 L} & 0 \end{bmatrix}$$

The characteristic equation is

$$\det[s\mathbf{I} - \mathbf{A}] = 0$$

Substituting and carrying out the algebra we obtain

$$s^4 + \beta_3 s^3 + \beta_2 s^2 + \beta_1 s + \beta_0 = 0$$

where

$$\beta_0 = \frac{kg}{m_1 L} \qquad\qquad \beta_1 = \frac{Bg}{m_1 L}$$

$$\beta_2 = \frac{(m_1 + m_2)g + kL}{m_1 L} \qquad \beta_3 = \frac{B}{m_1}$$

The solution of the state equation depends upon the constants β_0, β_1, β_2, β_3. Let us assume that the physical system is such that the characteristic polynomial becomes

$$P(s) = s^4 + 5s^3 + 10s^2 + 10s + 4$$

The roots of this equation are $s = -1, -2, -1 - j, -1 + j$ and the solution becomes

$$\mathbf{x}(t) = \left[[\mathbf{K}]_1 \, e^{-t} + [\mathbf{K}]_2 \, e^{-2t} + [\mathbf{K}]_3 \, e^{-t} \sin(t + \alpha) \right] \mathbf{x}(0)$$

where the constant matrices can be easily evaluated using the method shown previously.

Example 4 *A control system is characterized by the following coefficient matrix*

$$\mathbf{A} = \begin{bmatrix} 0 & 1 \\ 0 & -2 \end{bmatrix}$$

where the time response is $\mathbf{x}(t) = \phi(t)\mathbf{x}(0)$. *Evaluate* $\phi(t)$.

The roots of the characteristic equation $s(s+2) = 0$ are $s = 0, -2$. The $\mathbf{\Phi}(s)$ matrix becomes

$$\mathbf{\Phi}(s) = \frac{[\mathbf{K}]_1}{s} + \frac{[\mathbf{K}]_2}{s + 2}$$

and the undetermined coefficients may be directly evaluated.

$$[\mathbf{K}]_1 = \lim_{s \to 0} [s\mathbf{I} - \mathbf{A}]^{-1} s = \frac{1}{2} \begin{bmatrix} 2 & 1 \\ 0 & 0 \end{bmatrix}$$

$$[\mathbf{K}]_2 = \lim_{s \to -2} [s\mathbf{I} - \mathbf{A}]^{-1} (s + 2) = \frac{1}{2} \begin{bmatrix} 0 & 1 \\ 0 & -2 \end{bmatrix}$$

Substituting in the expression for $\mathbf{\Phi}(s)$

$$\mathbf{\Phi}(s) = \frac{1}{2s} \begin{bmatrix} 2 & 1 \\ 0 & 0 \end{bmatrix} - \frac{1}{2(s+2)} \begin{bmatrix} 0 & 1 \\ 0 & -2 \end{bmatrix}$$

Taking the inverse Laplace transform and adding we obtain $\phi(t)$

$$\phi(t) = \begin{bmatrix} 1 & \frac{1}{2}(1 - e^{-2t}) \\ 0 & e^{-2t} \end{bmatrix}$$

Example 5 *A control system is characterized by* $\dot{\mathbf{x}}(t) = \mathbf{A}\mathbf{x}(t)$ *where*

$$\mathbf{A} = \begin{bmatrix} -1 & K \\ -1 & -3 \end{bmatrix} \quad and \quad \mathbf{x}(0) = \begin{bmatrix} 1 \\ 0 \end{bmatrix}$$

Obtain $\mathbf{x}(t)$ *for* $K = 2$.

For $K = 2$,

$$[s\mathbf{I} - \mathbf{A}]^{-1} = \frac{1}{s^2 + 4s + 5} \begin{bmatrix} s+3 & 2 \\ -1 & s+1 \end{bmatrix}$$

The roots of the characteristic equation are $s_1 = -2 + j, s_2 = -2 - j$. The $\mathbf{\Phi}(s)$ matrix becomes

$$\mathbf{\Phi}(s) = \frac{[\mathbf{K}]_1}{s + 2 - j} + \frac{[\mathbf{K}]_2}{s + 2 + j}$$

where

$$[\mathbf{K}]_1 = \lim_{s \to -2+j} [s\mathbf{I} - \mathbf{A}]^{-1}(s + 2 - j)$$

$$= \frac{1}{2} \begin{bmatrix} \sqrt{2}e^{-j\pi/4} & 2e^{-j\pi/2} \\ e^{j\pi/2} & \sqrt{2}e^{j\pi/4} \end{bmatrix}$$

$$[\mathbf{K}]_2 = \lim_{s \to -2-j} [s\mathbf{I} - \mathbf{A}]^{-1}(s + 2 + j)$$

$$= \frac{1}{2} \begin{bmatrix} \sqrt{2}e^{j\pi/4} & 2e^{j\pi/2} \\ e^{-j\pi/2} & \sqrt{2}e^{-j\pi/4} \end{bmatrix}$$

We note, as we did in the classical method, that $[\mathbf{K}]_1$ and $[\mathbf{K}]_2$ are complex conjugates. Substituting these constants and taking the inverse Laplace transform and simplifying yields,

$$\phi(t) = \left[\begin{array}{cc} \sqrt{2}\cos\left(t - \dfrac{\pi}{4}\right) & 2\cos\left(t - \dfrac{\pi}{2}\right) \\[2mm] \cos\left(t + \dfrac{\pi}{2}\right) & \sqrt{2}\cos\left(t + \dfrac{\pi}{4}\right) \end{array} \right] e^{-2t}$$

Using the initial condition gives the result

$$\mathbf{x}(t) = \left[\begin{array}{c} \sqrt{2}e^{-2t}\cos\left(t - \dfrac{\pi}{4}\right) \\[2mm] e^{-2t}\cos\left(t + \dfrac{\pi}{2}\right) \end{array} \right]$$

The Transition Matrix

As we have seen above in Eq.(5-5) the complete solution of the state equations involves the sum of the zero input and the zero state responses. In the following discussion we will initially restrict the discussion to the zero input response so that Eq. (5-1) simplifies to the **homogeneous equation**

$$\dot{\mathbf{x}}(t) = \mathbf{A}\mathbf{x}(t) \tag{5.17}$$

where the initial condition is assumed to be $\mathbf{x}(0)$. In seeking a time domain solution for (5-17) we introduce the exponential matrix $\exp[\mathbf{A}t]$, which is defined by

$$\exp[\mathbf{A}t] = \mathbf{I} + t\mathbf{A} + \frac{t^2}{2!}\mathbf{A}^2 + \frac{t^3}{3!}\mathbf{A}^3 + \cdots \tag{5.18}$$

and which is frequently denoted by $e^{\mathbf{A}t}$. Writing

$$\mathbf{x}(t) = e^{\mathbf{A}t}\mathbf{x}(0) \tag{5.19}$$

and noting that

$$\frac{d}{dt}(e^{\mathbf{A}t}) = \mathbf{A}e^{\mathbf{A}t} \tag{5.20}$$

it can be easily shown by direct substitution in Eq. (5-17) that Eq. (5-19) is a solution of the homogeneous equation.

By direct term-by-term multiplication we can show

$$e^{\mathbf{A}(t+\tau)} = e^{\mathbf{A}t}e^{\mathbf{A}\tau} \tag{5.21}$$

If the state at time $t_0 > 0$ is $\mathbf{x}(t_0)$ then

$$\begin{aligned} \mathbf{x}(t) &= e^{\mathbf{A}(t-t_0)}e^{\mathbf{A}t_0}\mathbf{x}(0) \\ &= e^{\mathbf{A}(t-t_0)}\mathbf{x}(t_0) \end{aligned}$$

The matrix exponential in the above equation is referred to as the **transition matrix** or the **fundamental matrix** $\phi(t - t_0)$, where we take

$$\phi(t - t_0) = e^{\mathbf{A}(t-t_0)} \qquad (5.22)$$

The state vector and the output in terms of the transition matrix become

$$\mathbf{x}(t) = \phi(t - t_0)\mathbf{x}(t_0) \qquad (5.23)$$

$$y(t) = \mathbf{c}\phi(t - t_0)\mathbf{x}(t_0) \qquad (5.24)$$

When the input is non-zero we can use the state transition matrix to find the total system response. The resulting equation is called the **variation of parameters formula**. To see this result we note that

$$\frac{d}{dt}[e^{-\mathbf{A}t}\mathbf{x}(t)] = e^{-\mathbf{A}t}[\dot{\mathbf{x}}(t) - \mathbf{A}\mathbf{x}(t)] \qquad (5.25)$$

Substituting Eq. (5-1) into the right hand side of Eq. (5-25) gives

$$\frac{d}{dt}[e^{-\mathbf{A}t}\mathbf{x}(t)] = e^{-\mathbf{A}t}\mathbf{b}u(t)$$

Integrating both sides between the initial time t_0 and time t gives

$$\int_{t_0}^{t} d(e^{-\mathbf{A}t}\mathbf{x}(t)) = \int_{t_0}^{t} e^{-\mathbf{A}\tau}\mathbf{b}u(\tau)d\tau$$

which after some manipulation becomes

$$\mathbf{x}(t) = e^{\mathbf{A}(t-t_0)}\mathbf{x}(t_0) + \int_{t_0}^{t} e^{-\mathbf{A}(t-\tau)}\mathbf{b}u(\tau)d\tau$$

Writing this in terms of the transition matrix $\phi(t)$ gives the state **variation of parameters formula**

$$\mathbf{x}(t) = \phi(t - t_0)\mathbf{x}(t_0) + \int_{t_0}^{t} \phi(t - \tau)\mathbf{b}u(\tau)d\tau \qquad (5.26)$$

The system output $y(t)$ is

$$y(t) = \mathbf{c}\phi(t - t_0)\mathbf{x}(t_0) + \int_{t_0}^{t} \mathbf{c}\phi(t - \tau)\mathbf{b}u(\tau)d\tau \qquad (5.27)$$

An examination of Eq. (5-22) indicates that the transition matrix transfers the state of the system from t_i to t_{i+1}. As a matter of fact this matrix contains all the information necessary to define the state of a system at any time. Indeed, if the transition matrix defined by Eq. (5-22) is obtained, then the system may be analyzed directly in the time domain. This is by far the most important reason why the transition matrix is particularly useful. Some of the more useful properties of the transition matrix are:

1. The transition matrix is non-singular and its inverse exists for all t.

2. It is the solution of the matrix differential equation

$$\frac{d}{dt}[\phi(t - t_0)] = \mathbf{A}\phi(t - t_0)$$

3. It has the property that $\phi(t_0 - t_0) = \mathbf{I}$.

4. It has a simple inverse, i.e. $\phi^{-1}(t) = \phi(-t)$. (Such matrices are called **symplectic**.)

5. The transition matrix has a sequential property, i.e.

$$\phi(t_{i+2} - t_i) = \phi(t_{i+2} - t_{i+1})\phi(t_{i+1} - t_i)$$

The sequence of finite transitions will help when we make manipulations with $\phi(t - t_0)$.

6. It is related to the resolvent matrix, i.e. $\phi(t) = \mathcal{L}^{-1}[(s\mathbf{I} - \mathbf{A})^{-1}]$.

Having obtained the solution of the matrix differential equation in terms of the transition matrix, let us see how we may evaluate this transition matrix. The state response in terms of the transition matrix and the initial condition is

$$\mathbf{x}(t) = \phi(t)\mathbf{x}(0)$$

This equation may be expanded and written as

$$
\begin{aligned}
x_1(t) &= \phi_{11}x_1(0) + \cdots + \phi_{1n}x_n(0)\\[4pt]
x_2(t) &= \phi_{21}x_1(0) + \cdots + \phi_{2n}x_n(0)\\
&\;\;\vdots\\
x_n(t) &= \phi_{n1}x_1(0) + \cdots + \phi_{nn}x_n(0)
\end{aligned}
$$

where ϕ_{ij} are, in general, functions of time and the terms $x_j(0)$ represent initial conditions. If $x_j(0)$ is unity, then ϕ_{ij} may be interpreted as the response of the ith state variable to a unit initial condition of the jth state variable.

Example 6 *Obtain the transition matrix for the system discussed in Example 1.*

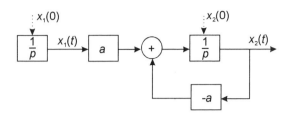

Figure 5.2: Block diagram representing $\dot{x}_1(t) = 0$, $\dot{x}_2(t) = a[x_1(t) - x_2(t)]$

The solution to the state equation can be expanded and written as

$$\left[\begin{array}{c} x_1(t) \\ x_2(t) \end{array} \right] = \left[\begin{array}{cc} \phi_{11}(t) & \phi_{12}(t) \\ \phi_{21}(t) & \phi_{22}(t) \end{array} \right] \left[\begin{array}{c} x_1(0) \\ x_2(0) \end{array} \right]$$

where each element $\phi_{ij}(t)$ of the transition matrix must be evaluated.

Re-calling that $\phi_{ij}(t)$ is the response of the ith state variable to a unit initial condition of the jth state variable, we see from Fig. 5-2 that

ϕ_{11} = 1 (response measured at x_1 for unit initial condition at integrator x_1)

ϕ_{12} = 0 (response measured at x_1 for unit initial condition at integrator x_2)

ϕ_{21} = $\mathcal{L}^{-1}[a/s(a+s)] = 1 - e^{-at}$ (response measured at x_2 for unit initial condition at integrator x_1)

ϕ_{22} = $\mathcal{L}^{-1}[1/(s+a)] = e^{-at}$ (response measured at x_2 for unit initial condition at integrator x_2)

In evaluating each $\phi_{ij}(t)$ we are simply evaluating the gain between the ith state variable and the initial condition of the jth state variable and then taking the inverse Laplace transform. Substituting the above results we obtain the transition matrix

$$\phi(t) = \left[\begin{array}{cc} 1 & 0 \\ 1 - e^{-at} & e^{-at} \end{array} \right]$$

which is identical to the result obtained in Example 5-1.

Example 7 *A control system has the following transfer function*

$$\frac{Y(s)}{U(s)} = \frac{1}{s^2(s+1)}$$

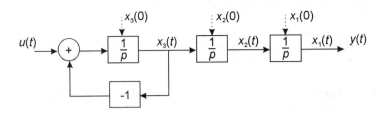

Figure 5.3: Block diagram representing the state equation for $G(s) =$
$\dfrac{1}{s^2(s+1)}$

Obtain the transition matrix $\phi(t)$, assuming that the state equations are in observability canonical form as given in Chapter 3.

The state vector becomes

$$\mathbf{x}(t) = \begin{bmatrix} x_1(t) \\ x_2(t) \\ x_3(t) \end{bmatrix}$$

and the block diagram representation of the system is shown in Fig. 5-3. The elements of the transition matrix may be evaluated as in the previous problem

$$\phi_{11} = 1$$

$$\phi_{12} = t$$

$$\phi_{13} = \mathcal{L}^{-1}\left[\frac{1}{s^2(s+1)}\right] = t - 1 - e^{-t}$$

$$\phi_{31} = \phi_{21} = \phi_{32} = 0$$

$$\phi_{22} = 1$$

$$\phi_{23} = \mathcal{L}^{-1}\left[\frac{1}{s(s+1)}\right] = 1 - e^{-t}$$

$$\phi_{33} = \mathcal{L}^{-1}\left[\frac{1}{s+1}\right] = e^{-t}$$

The entire matrix becomes

$$\phi(t) = \begin{bmatrix} 1 & t & t-1-e^{-t} \\ 0 & 1 & 1-e^{-t} \\ 0 & 0 & e^{-t} \end{bmatrix}$$

As an exercise (in convincing yourself) you should form $[s\mathbf{I} - \mathbf{A}]^{-1}$ and take its inverse Laplace transform and see that it is equivalent to the transition matrix shown in the previous example.

Before continuing further, several observations are in order. First, the ijth element of the transition matrix is zero if there is no signal flow from the jth state variable to the ith state variable. Second, the term $\phi_{ij}(t)$ is the inverse Laplace transform of the overall gain from the jth state variable to the ith state variable. Although the above method appears difficult initially, its clarity and overall simplicity cannot be overstated. This method is particularly suited for hand computation.

It is not necessary to obtain $\phi(t)$ by the above method. There are a number of additional techniques for obtaining the transition matrix, one of which we shall consider here. This method deals directly with the matrix $\mathbf{A}$ and operates on the eigenvalues of $\mathbf{A}$.

The transition matrix in the s-domain expressed in terms of an adjoint matrix determinant is

$$\phi(s) = \frac{\text{adj}[s\mathbf{I} - \mathbf{A}]}{\det[s\mathbf{I} - \mathbf{A}]} \tag{5.28}$$

However this is identical to the resolvent matrix $\mathbf{\Phi}(s)$ previously defined in Eq. (5-11). Therefore the Heaviside expansion may be used to evaluate the transition matrix. In using this method you will recall we had to invert a matrix, which is not always desirable when the matrix $\mathbf{A}$ is large. The inverse Laplace transform then needs to be taken. An alternative approach using the Cayley-Hamilton theorem (see Appendix C) gives the matrix function $\exp[\mathbf{A}t]$ directly without having to compute the inverse matrix function and then find its inverse Laplace transform. To develop this method we need to recall that the matrix $\mathbf{A}$ has the **characteristic equation** defined as

$$\Delta(\lambda) = \det(\lambda\mathbf{I} - \mathbf{A}) = 0 \tag{5.29}$$

and the values of λ for which this equation is satisfied are termed the **eigenvalues** of $\mathbf{A}$. If $\mathbf{A}$ is an $(n \times n)$ matrix there are in general n eigenvalues. Writing $\Delta(\lambda)$ in polynomial form as

$$\Delta(\lambda) = \lambda^n + a_{n-1}\lambda^{n-1} + \cdots + a_1\lambda + a_0 = 0 \tag{5.30}$$

the Cayley-Hamilton theorem shows that every $(n \times n)$ matrix $\mathbf{A}$ satisfies its own characteristic equation so that

$$\Delta(\mathbf{A}) = \mathbf{A}^n + a_{n-1}\mathbf{A}^{n-1} + \cdots + a_1\mathbf{A} + a_0\mathbf{I} = 0 \qquad (5.31)$$

Finding $e^{\mathbf{A}t}$ Using the Cayley-Hamilton Theorem

In the following discussion we will show how the **exponential matrix** can be obtained from the fundamental definition given in Eq. (5-18). To do this we first examine how the Cayley-Hamilton theorem can be used for evaluating powers of the matrix $\mathbf{A}$ of the form $\mathbf{A}^k$, where $k \geq 0$, and will then show the form that Eq. (5-18) takes when this result is taken into account.

To find $\mathbf{A}^k$ we need to consider two cases. When $k < n$ the matrix $\mathbf{A}^k$ can be evaluated by iteration in a straightforward manner. For $k \geq n$, we can show by repeated use of the Cayley-Hamilton theorem that

$$
\begin{aligned}
\mathbf{A}^k &= \alpha_0(k)\mathbf{I} + \cdots + \alpha_{n-1}(k)\mathbf{A}^{n-1} \\
&= \sum_{i=0}^{n-1} \alpha_i(k)\mathbf{A}^i
\end{aligned}
\qquad (5.32)
$$

where the functional dependence of the coefficients α_i, $0 \leq i \leq n-1$ on k is shown explicitly. Since the eigenvalues λ_i of $\mathbf{A}$ also satisfy the characteristic equation given by Eq. (5-30) we can show by the repeated use of Eq. (5-30) that λ_i^k also satisfies

$$
\begin{aligned}
\lambda_i^k &= \alpha_0(k) + \alpha_1(k)\lambda_i + \cdots + \alpha_{n-1}(k)\lambda_i^{n-1} \\
&= \sum_{i=0}^{n-1} \alpha_i(k)\lambda^i
\end{aligned}
\qquad (5.33)
$$

for $k \geq n$ where the coefficients are the same as those shown in Eq. (5-32). When there are n distinct eigenvalues this equation suggests a simple method of finding the n coefficients $\alpha_0(k), \ldots, \alpha_{n-1}(k)$ for each k, by substituting the n distinct eigenvalues into Eq. (5-33) to give n equations with n unknowns, as below

$$
\begin{aligned}
\lambda_1^k &= \alpha_0(k) + \cdots + \alpha_{n-1}(k)\lambda_1^{n-1} \\
&\ \ \vdots \\
\lambda_n^k &= \alpha_0(k) + \cdots + \alpha_{n-1}(k)\lambda_n^{n-1}
\end{aligned}
\qquad (5.34)
$$

Example 8 *Consider the matrix* **A** *defined by*

$$\mathbf{A} = \begin{bmatrix} 1 & -1 & 0 \\ 0 & 2 & 0 \\ 0 & 1 & 3 \end{bmatrix}$$

Find a general relationship for the matrix $\mathbf{A}^k, k > 0$.

The matrix **A** has the eigenvalues $\lambda_1 = 1$, $\lambda_2 = 2$, $\lambda_3 = 3$. From Eq. (5-32) and (5-33) we have

$$\mathbf{A}^k = \alpha_0(k)\mathbf{I} + \alpha_1(k)\mathbf{A} + \alpha_2(k)\mathbf{A}^2 \tag{5.35}$$

for $k \geq 3$, and

$$\lambda^k = \alpha_0(k) + \alpha_1(k)\lambda + \alpha_2(k)\lambda^2 \tag{5.36}$$

for $\lambda = \lambda_1, \lambda_2, \lambda_3$, and $k \geq 3$. Substituting λ_1, λ_2 and λ_3 into Eq. (5-36) gives

$$1^k = \alpha_0 + \alpha_1 + \alpha_2$$

$$2^k = \alpha_0 + 2\alpha_1 + 4\alpha_2$$

$$3^k = \alpha_0 + 3\alpha_1 + 9\alpha_2$$

Solving for α_1, α_2 and α_3 gives

$$\alpha_0 = 1 + (3^k - 1) - 3(2^k - 1)$$

$$\alpha_1 = -\frac{3}{2}(3^k - 1) + 4(2^k - 1)$$

$$\alpha_2 = \frac{1}{2}(3^k - 1) - (2^k - 1)$$

Substituting for α_0, α_1 and α_2 into Eq. (5-35) gives

$$\mathbf{A}^k = \begin{bmatrix} 1 & 1 - 2^k & 0 \\ 0 & 2^k & 0 \\ 0 & 3^k - 2^k & 3^k \end{bmatrix}$$

for $k \geq 3$. For $k = 0, 1, 2$, $\mathbf{A}^k$ is obtained by direct observation or multiplication.

To find $\exp[\mathbf{A}t]$ we refer to the definition given by Eq. (5-18), where

$$e^{\mathbf{A}t} = \sum_{k=0}^{\infty} \frac{t^k}{k!} \mathbf{A}^k \tag{5.37}$$

We also note that the scalar exponential function is given by

$$e^{\lambda t} = \sum_{k=0}^{\infty} \frac{t^k}{k!} \lambda^k \tag{5.38}$$

From the analysis above we have seen that $\mathbf{A}^k$ and λ_i^k can be expressed by the finite sums shown in Eq. (5-32) and (5-33) for $i = 1,\ldots,n$ and $k \geq n$, where the λ_i are the distinct eigenvalues of $\mathbf{A}$. These results also apply trivially for $0 \leq k < n$ so that we may conclude that they apply for all $k \geq 0$. We now substitute these equations in Eq. (5-37) and (5-38) respectively to obtain

$$
\begin{aligned}
e^{\mathbf{A}t} &= \sum_{k=0}^{\infty} \frac{t^k}{k!} \sum_{j=0}^{n-1} \alpha_j(k) \mathbf{A}^j \\
&= \sum_{j=0}^{n-1} \left(\sum_{k=0}^{\infty} \frac{t^k}{k!} \alpha_j(k) \right) \mathbf{A}^j
\end{aligned}
$$

and

$$
\begin{aligned}
e^{\lambda t} &= \sum_{k=0}^{\infty} \frac{t^k}{k!} \sum_{j=0}^{n-1} \alpha_j(k) \lambda^j \\
&= \sum_{j=0}^{n-1} \left(\sum_{k=0}^{\infty} \frac{t^k}{k!} \alpha_j(k) \right) \lambda^j
\end{aligned}
$$

for $\lambda = \lambda_1, \lambda_2, \ldots, \lambda_n$. Defining

$$\beta_j(t) = \sum_{k=0}^{\infty} \frac{t^k}{k!} \alpha_j(k)$$

we obtain

$$e^{\mathbf{A}t} = \sum_{j=0}^{n-1} \beta_j(t) \mathbf{A}^j \tag{5.39}$$

and

$$e^{\lambda t} = \sum_{j=0}^{n-1} \beta_j(t) \lambda^j \tag{5.40}$$

for $\lambda = \lambda_1, \lambda_2, \ldots, \lambda_n$.

To find $\exp[\mathbf{A}t]$ we need to determine the functions $\beta_j(t)$, $j = 0, 1, \ldots,$ $n - 1$. Fortunately, we have n equations with the n unknowns $\beta_0(t), \ldots,$

$\beta_{n-1}(t)$, given by

$$
\begin{aligned}
e^{\lambda_1 t} &= \beta_0(t) + \beta_1(t)\lambda_1 + \cdots + \beta_{n-1}(t)\lambda_1^{n-1} \\
&\vdots \\
e^{\lambda_n t} &= \beta_0(t) + \beta_1(t)\lambda_n + \cdots + \beta_{n-1}(t)\lambda_n^{n-1}
\end{aligned}
\tag{5.41}
$$

so that provided the eigenvalues $\lambda_i, i = 1, \ldots, n$ are all distinct the coefficients $\beta_j(t)$, $j = 0, \ldots, n-1$ can be obtained.

Example 9 *Suppose the matrix* **A** *is given by*

$$
\mathbf{A} = \left[\begin{array}{cc} 0 & 6 \\ -1 & -5 \end{array} \right]
$$

Find the exponential matrix $e^{\mathbf{A}t}$.

The characteristic equation for **A** is

$$
\Delta(\lambda) = \det(\lambda I - \mathbf{A}) = \lambda^2 + 5\lambda + 6 = 0
$$

so the eigenvalues are $\lambda_1 = -2$ and $\lambda_2 = -3$. From Eqs. (5-39) and (5-40) we have

$$
e^{\mathbf{A}t} = \beta_0(t)\mathbf{I} + \beta_1(t)\mathbf{A}
\tag{5.42}
$$

and

$$
e^{\lambda t} = \beta_0(t) + \beta_1(t)\lambda \text{ for } \lambda = -2, -3
$$

As shown in Eq. (5-41)

$$
\begin{aligned}
\beta_0 - 2\beta_1 &= e^{-2t} \\
\beta_0 - 3\beta_1 &= e^{-3t}
\end{aligned}
$$

Solving these equations gives

$$
\begin{aligned}
\beta_0(t) &= 3e^{-2t} - 2e^{-3t} \\
\beta_1(t) &= e^{-2t} - e^{-3t}
\end{aligned}
$$

Substituting $\beta_0(t)$ and $\beta_1(t)$ into Eq. (5-42) results in the exponential matrix

$$
e^{\mathbf{A}t} = \left[\begin{array}{cc} 3e^{-2t} - 2e^{-3t} & 6e^{-2t} - 6e^{-3t} \\ -e^{-2t} + e^{-3t} & -2e^{-2t} + 3e^{-3t} \end{array} \right]
$$

When the matrix **A** has repeated eigenvalues the procedure given above must be modified as Eq. (5-40) fails to yield sufficient independent equations to allow $\beta_0(t), \ldots, \beta_n(t)$ to be uniquely determined. The approach we take is to differentiate Eq. (5-40) with respect to λ as many times as is necessary to obtain n independent equations.

Example 10 *Let the matrix* **A** *be*

$$\mathbf{A} = \left[\begin{array}{cc} -2 & 1 \\ 0 & -2 \end{array} \right]$$

Find the exponential matrix $e^{\mathbf{A}t}$.

The characteristic equation is

$$\Delta(\lambda) = (\lambda + 2)^2 = 0$$

so the eigenvalues are $\lambda_1 = \lambda_2 = -2$. Thus

$$e^{\mathbf{A}t} = \beta_0 \mathbf{I} + \beta_1 \mathbf{A}$$

and

$$e^{\lambda t} = \beta_0 + \beta_1 \lambda \text{ for } \lambda = -2 \tag{5.43}$$

Since **A** has repeated eigenvalues Eq. (5-43) does not uniquely determine $\beta_0(t)$ and $\beta_1(t)$. Differentiating Eq. (5-43) with respect to λ gives

$$te^{\lambda t} = \beta_1 \text{ for } \lambda = -2.$$

Solving these equations yields

$$\beta_0 = (2t + 1)e^{-2t}$$

$$\beta_1 = te^{-2t}$$

and

$$e^{\mathbf{A}t} = \left[\begin{array}{cc} e^{-2t} & te^{-2t} \\ 0 & e^{-2t} \end{array} \right]$$

5.3 Eigenvalues of Matrix A and Stability

In Chapter 4 Example 6 where we studied a third-order servomechanism it was observed that the magnitude of one term of the output response increased exponentially with time. In this case we say the system response is **unstable**.

Similarly in the state space method it can be seen that the system response, which is given by the variation of parameters formula Eq. (5-27), is a function of the exponential matrix $\exp[\mathbf{A}t]$. The form of the matrix is dependent upon the roots of the characteristic equation

$$\Delta(\lambda) = \det[\lambda \mathbf{I} - \mathbf{A}] = 0$$

where $\mathbf{A}$ is the coefficient matrix of the system. *For the system to be stable we require that the roots of the above characteristic equation (eigenvalues of $\mathbf{A}$ matrix) either have no positive real parts, or, if they are imaginary they cannot be repeated on the imaginary axis.*

From the above discussion it can be seen that, to determine stability we need to find the eigenvalues of the system $\mathbf{A}$ matrix. When these eigenvalues are distinct and the state equations are in Jordan canonical form, as given in Section 3-7, the $\mathbf{A}$ matrix will be diagonal and the eigenvalues can be found from this matrix by inspection. However, when the matrix $\mathbf{A}$ has repeated eigenvalues it cannot be written in diagonal form, so that their determination is no longer straight forward.

Fadeeva's Method for Finding the Characteristic Polynomial of Matrix A

We now present Fadeeva's method of obtaining the characteristic equation for the $\mathbf{A}$ matrix of a general state equation representation. The characteristic equation for matrix $\mathbf{A}$ is given by Eq. (5-30), and we define the trace of matrix $\mathbf{A}$ as

$$\mathrm{tr}(\mathbf{A}) = \sum_{i=1}^{n} a_{ii}$$

where $\{a_{ii}\}$ are the diagonal elements of the $(n \times n)$ matrix $\mathbf{A}$. For this matrix the coefficients of its characteristic equation are determined iteratively by the algorithm

$$
\begin{aligned}
\mathbf{B}_{n-1} &= \mathbf{I} & a_{n-1} &= -\mathrm{tr}(\mathbf{B}_{n-1}\mathbf{A}) \\
\mathbf{B}_{n-2} &= \mathbf{B}_{n-1}\mathbf{A} + a_{n-1}\mathbf{I} & a_{n-2} &= -\tfrac{1}{2}\mathrm{tr}(\mathbf{B}_{n-2}\mathbf{A}) \\
\mathbf{B}_{n-3} &= \mathbf{B}_{n-2}\mathbf{A} + a_{n-2}\mathbf{I} & a_{n-3} &= -\tfrac{1}{3}\mathrm{tr}(\mathbf{B}_{n-3}\mathbf{A}) \\
&\ \ \vdots & &\ \ \vdots
\end{aligned}
\tag{5.44}
$$

$$\mathbf{B}_{n-k} = \mathbf{B}_{n-k+1}\mathbf{A} + a_{n-k+1}\mathbf{I} \quad a_{n-k} = -\frac{1}{k}\mathrm{tr}(\mathbf{B}_{n-k}\mathbf{A})$$

$$\vdots \qquad\qquad\qquad \vdots$$

$$\mathbf{B}_1 = \mathbf{B}_2\mathbf{A} + a_2\mathbf{I} \qquad\qquad a_1 = -\frac{1}{n-1}\mathrm{tr}(\mathbf{B}_1\mathbf{A})$$

$$\mathbf{B}_0 = \mathbf{B}_1\mathbf{A} + a_1\mathbf{I} \qquad\qquad a_0 = -\frac{1}{n}\mathrm{tr}(\mathbf{B}_0\mathbf{A})$$

$$0 = \mathbf{B}_0\mathbf{A} + a_0\mathbf{I}$$

Once the coefficients of the characteristic equation are determined numerical techniques can be used to find its roots. These correspond to the eigenvalues of **A**.

While Fadeeva's method is unaffected when **A** has repeated eigenvalues it is, however, numerically ill-conditioned when the order of **A** becomes large, because **A** is raised to the $(n-1)$th power so that small errors in elements of **A** may cause large changes in the coefficients $a_0, a_1, \cdots, a_{n-1}$. Much work has been done on the numerical calculation of eigenvalues and there are many powerful computer library routines such as the QR algorithm for their computation. The reader should consult some of the standard references and control system computational packages for appropriate algorithms.

Fadeeva's method can be used not only for recursively computing the coefficients of the characteristic equation, but it can also be shown that the matrix $\mathrm{adj}(\lambda\mathbf{I} - \mathbf{A})$ is given by

$$\mathrm{adj}(\lambda\mathbf{I} - \mathbf{A}) = \lambda^{n-1}\mathbf{B}_{n-1} + \lambda^{n-2}\mathbf{B}_{n-2} + \cdots + \lambda\mathbf{B}_1 + \mathbf{B}_0 \qquad (5.45)$$

where the matrix coefficients $\mathbf{B}_0, \mathbf{B}_1, \cdots, \mathbf{B}_{n-1}$ are derived in Eq. (5-44). Consequently it can be seen that Fadeeva's method gives a recursive method of determining the resolvent matrix $\phi(s)$ since Eq. (5-11) shows

$$\phi(s) = [s\mathbf{I} - \mathbf{A}]^{-1} = \left.\frac{\mathrm{adj}(\lambda\mathbf{I} - \mathbf{A})}{\Delta(\lambda)}\right|_{\lambda=s}$$

This approach gives an alternative method to the direct evaluation of the inverse matrix in finding the resolvent matrix $\phi(s)$, and is therefore useful in the Laplace transform method of finding the exponential matrix $e^{\mathbf{A}t}$, as discussed in Section 5-2.

Example 11 *Consider a system characterized by the following* **A** *matrix*

$$\mathbf{A} = \begin{bmatrix} 0 & 1 & 0 \\ 0 & -2 & 1 \\ -2 & -3 & -1 \end{bmatrix}$$

Find the characteristic equation and hence the eigenvalues.

Because of the simplicity of the matrix $\mathbf{A}$ we can evaluate $\Delta(\lambda)|_{\lambda=s}$ directly, thus

$$\Delta(s) = \det(s\mathbf{I} - \mathbf{A}) = \begin{bmatrix} s & -1 & 0 \\ 0 & s+2 & -1 \\ 2 & 3 & s+1 \end{bmatrix}$$

Consequently

$$\Delta(s) = s\lambda^3 + 3s^2 + 5s + 2 = 0$$

Solving the characteristic equation gives the eigenvalues: $-0.0547, -1.227 \pm j1.467$. When a system is represented in state space form, the stability is studied either graphically or using more advanced analytical techniques. We shall consider some of these ideas in more detail in a later chapter.

5.4 Two Examples

The governing equations of satellite attitude motion possess properties that make them attractive candidates for examples. This is true because the equations are simply derived using familiar concepts and can be linearized in some applications to constant coefficient linear differential equations. Also, the dynamics of the motion is such that the three rotational modes are uncoupled from the three translational motions. Finally, the three rotational motions generally appear in such a way that at least one degree is uncoupled from the other two. Such a fortunate circumstance is rare!

For our first example we will solve for the motion of two satellites connected by a tether. The arrangement consists of two satellites connected by a massless and inelastic tether so that the center of mass of the system moves on a given circular orbit. Furthermore, the motion of the satellites is planar, i.e. restricted to the plane of the orbit.

Treating the two satellites as identical point masses the equations of motion, of one of the masses, are derived via Newtonian mechanics and are

$$\frac{d^2z(t)}{dt^2} = 2\omega \frac{dy(t)}{dt} + \omega^2 f_z(t)$$

$$\frac{d^2y(t)}{dt^2} = -2\omega \frac{dz(t)}{dt} + 3\omega^2(t) + \omega^2 f_y(t)$$

(5.46)

where ω is the mean motion of the orbit, and $\omega^2 f_z(t)$ and $\omega^2 f_y(t)$ are control forces. The motion of the other mass is obtained by symmetry. If we introduce the independent variable $\tau = \omega t$, then the equations become

$$\ddot{z}(\tau) = 2\dot{y}(\tau) + f_z(\tau)$$

$$\ddot{y}(\tau) = -2\dot{z}(\tau) + 3y(\tau) + f_y(\tau)$$

where $(\ddot{z})$ denotes the second derivative with respect of ωt.

We now define the state vector

$$\mathbf{x}(\tau) = \begin{bmatrix} x_1(\tau) \\ x_2(\tau) \\ x_3(\tau) \\ x_4(\tau) \end{bmatrix} = \begin{bmatrix} z(\tau) \\ \dot{z}(\tau) \\ y(\tau) \\ \dot{y}(\tau) \end{bmatrix}$$

so that the state equation becomes

$$\dot{\mathbf{x}}(\tau) = \mathbf{A}\mathbf{x}(\tau) + \mathbf{b}u(\tau)$$

where

$$\mathbf{A} = \begin{bmatrix} 0 & 1 & 0 & 0 \\ 0 & 0 & 0 & 2 \\ 0 & 0 & 0 & 1 \\ 0 & -2 & 3 & 0 \end{bmatrix}, \quad \mathbf{b}u(\tau) = \begin{bmatrix} 0 \\ f_z(\tau) \\ 0 \\ f_y(\tau) \end{bmatrix}$$

The characteristic equation yields the following roots, $\lambda_1 = 0, \lambda_2 = 0$, $\lambda_3 = j, \lambda_4 = -j$. Using the Cayley-Hamilton method we form the following equation

$$e^{\mathbf{A}\tau} = \beta_0 \mathbf{I} + \beta_1 \mathbf{A} + \beta_2 \mathbf{A}^2 + \beta_3 \mathbf{A}^3$$

Since there are four eigenvalues, this leads to the following four equations

$$\left[\frac{d}{d\lambda} e^{\lambda \tau} \right]_{\lambda=0} = \beta_1$$

$$\left[e^{\lambda \tau} \right]_{\lambda=0} = \beta_0$$

$$\left[e^{\lambda \tau} \right]_{\lambda=j} = \beta_0 + j\beta_1 + j^2\beta_2 + j^3\beta_3$$

$$\left[e^{\lambda \tau} \right]_{\lambda=-j} = \beta_0 - j\beta_1 + j^2\beta_2 - j^3\beta_3$$

Solving these equations for $\beta_0, \beta_1, \beta_2, \beta_3$ and substituting we obtain

$$e^{\mathbf{A}\tau} = \phi(\tau) = \begin{bmatrix} 1 & 4\sin\tau - 3\tau & 6(\tau - \sin\tau) & 2(1 - \cos\tau) \\ 0 & 4\cos\tau - 3 & 6(1 - \cos\tau) & 2\sin\tau \\ 0 & -2(1 - \cos\tau) & 4 - 3\cos\tau & \sin\tau \\ 0 & -2\sin\tau & 3\sin\tau & \cos\tau \end{bmatrix} \quad (5.47)$$

where $\tau = \omega t$. We can determine system stability by observing the eigenvalues of $\mathbf{A}$. Since there are two eigenvalues at the origin, the system is unstable. This is also obvious from the solution given by Eq. (5-47). This means that the satellites will continue to move away from each other as long as the tether is slack. In practice, the control forces $f_z(\tau)$ and $f_y(\tau)$ are used to control the motion of the tethered satellites.

Our next example involves a spacecraft whose attitude motion is to be controlled by jets mounted orthogonally on the vehicle. We would like to obtain the response and stability of this satellite. Although this problem can get quite complex, we shall make several assumptions in order to simplify it without detracting from the salient points. If we define ϕ, θ, and ψ as the pitch, roll and yaw motion, the linearized dynamical equations can be derived from Newtonian mechanics. We simply write down the equations as

$$I_y \ddot{\phi}(t) + 3\omega^2(I_x - I_z)\phi(t) = M_y(t)$$

$$I_x \ddot{\theta}(t) + 4\omega^2(I_y - I_z)\theta(t) - \omega(I_y - I_x - I_z)\dot{\psi}(t) = M_x(t) \qquad (5.48)$$

$$I_z \ddot{\psi}(t) + \omega^2(I_y - I_x)\psi(t) + \omega(I_y - I_x - I_z)\dot{\theta}(t) = M_z(t)$$

where I_x, I_y, I_z are the principle moments of inertia, ω is the orbital velocity, and $M_x(t), M_y(t), M_z(t)$ are applied torques. We note that the first equation is a second-order undamped differential equation and is uncoupled from the following two equations. Since we have analyzed such equations previously, we shall investigate only the coupled equations. We would like to investigate the transient behavior of this satellite.

Let us first establish the physical characteristics of the satellite as follows:

$$\begin{aligned} I_x &= 100 \text{ kg m}^2 \\ I_y &= 200 \text{ kg m}^2 \\ I_z &= 50 \text{ kg m}^2 \end{aligned}$$

Also, if we set $\tau = \omega t$ and then substitute for the above parameters

$$\ddot{\theta}(\tau) + 6\theta(\tau) - 0.5\dot{\psi}(\tau) = 0$$

$$\ddot{\psi}(\tau) + 2\psi(\tau) + \dot{\theta}(\tau) = (M_z(\tau)/I_z)$$

where we have assumed only one input. Now if we define the state variables as $x_1(\tau) = \theta(t)$, $x_2(\tau) = \dot{\theta}(\tau)$, $x_3(\tau) = \psi(\tau)$, $x_4(\tau) = \dot{\psi}(\tau)$ then we obtain

$$\dot{\mathbf{x}}(\tau) = \mathbf{A}\mathbf{x}(\tau) + \mathbf{b}u(\tau)$$

where

$$\mathbf{A} = \begin{bmatrix} 0 & 1 & 0 & 0 \\ -6 & 0 & 0 & 0.5 \\ 0 & 0 & 0 & 1 \\ 0 & -1 & -2 & 0 \end{bmatrix}, \quad \mathbf{b} = \begin{bmatrix} 0 \\ 0 \\ 0 \\ 1 \end{bmatrix}, \quad u(\tau) = (M_z(\tau)/I_z)$$

The output is given by

$$\theta(\tau) = \mathbf{c}_1 \mathbf{x}(\tau)$$

$$\psi(\tau) = \mathbf{c}_2 \mathbf{x}(\tau)$$

where

$$\mathbf{c}_1 = \begin{bmatrix} 1 & 0 & 0 & 0 \end{bmatrix}$$

$$\mathbf{c}_2 = \begin{bmatrix} 0 & 0 & 1 & 0 \end{bmatrix}$$

We shall investigate the time response of $\theta(\tau)$ only, since the results for $\psi(\tau)$ are similar. The characteristic equation becomes

$$s^4 + 8.5s^2 + 12 = 0$$

having the following roots: $s = j2.59, -j2.59, j1.33, -j1.33$. Since all the roots are imaginary and simple, the system is stable. The response of the homogeneous equation is

$$\mathbf{x}(\tau) = \mathcal{L}^{-1}\left[\frac{\text{adj}(s\mathbf{I} - \mathbf{A})}{\det(s\mathbf{I} - \mathbf{A})} \right] \mathbf{x}(0)$$

where

$$\text{adj}[s\mathbf{I} - \mathbf{A}] =$$

$$\begin{bmatrix} s\left(s^2 + 2.5\right) + 0.5s & s^2 + 2 & -1 & 0.5s \\ -6\left(s^2 + 2\right) & s\left(s^2 + 2\right) & -s & 0.5s^2 \\ 6 & -s & s(s^2 + 6.5) + 6s & s^2 + 6 \\ 6s & -s^2 & -2(s^2 + 6) & s(s^2 + 6) \end{bmatrix}$$

The time response $\theta(\tau)$ is

$$\theta(\tau) = x_1(\tau)$$

$$= \mathcal{L}^{-1}\left(\frac{[s(s^2 + 2) + 0.5s]x_1(0) + (s^2 + 2)\,x_2(0) - x_3(0) + 0.5sx_4(0)}{(s - j2.59)(s + j2.59)(s - j1.33)(s + j1.33)} \right)$$

$$(5.49)$$

Let us assume that the only non-zero initial condition is $x_4(0)$, then

$$\theta(\tau) = 0.5x_4(0)[K_1 \cos 2.59\tau + K_2 \sin 1.33\tau]$$

where K_1 and K_2 are constants.

5.5   Controllability and Observability

Two concepts which have proven to have far-reaching significance in the modern theory of control systems are **controllability** and, its complementary concept, **observability**. These ideas infringe on many aspects of system operation including the relationship between the transfer function and state space realizations of a system, and the behavior of state space trajectories when a system is excited by a defined class of inputs. In this section we shall introduce these concepts from a geometric view point and then develop algebraic tests for determining when a system is controllable or observable respectively. Their role in linking the transfer function and state variable realization of a system will also be discussed.

Controllability

While controllability can be introduced from an algebraic or system structure direction, we will take a geometric approach because it demonstrates its close links with control systems and because of its concrete nature when considered from this standpoint. When examined from this viewpoint it is natural to consider what conditions a system needs to satisfy so that its state trajectory will be reached or can be steered to any given final state from an initial state. Examples that can be considered are: An aircraft which must be flown from London to New York, or a rocket being used to launch a satellite into earth orbit, or lastly, but not exhaustively, a turbo-generator starting from rest and being controlled so as to reach a given set of operating conditions which would include its terminal voltages, its frequency and its power output. In each of these examples the system S starts from known, but often arbitrary, initial conditions, which define an initial state at time t_0 in the state space, and it is required to manipulate the control variable(s) so that at the final time $t_f > t_0$ the system state has assumed a desired final value.

This concept can be interpreted geometrically by examining Fig. 3-21 which illustrates the trajectory for a system having a 3-dimensional state space. For the system S described by Eq. (5-1) and (5-2) we have shown that the state space trajectory $\mathbf{x}(t)$ is given by the variation of parameters formula. Examining Eq. (5-26), which is repeated here for convenience with $\phi(t)$ replaced by the exponential matrix

$$\mathbf{x}(t) = e^{A(t-t_0)}\mathbf{x}(t_0) + \int_{t_0}^{t} e^{A(t-\tau)}\mathbf{b}u(\tau)d\tau \qquad (5.50)$$

it can be seen that the state space trajectory is the sum of the zero-input response and the zero-state response as shown in Fig. 5-4. Once the initial

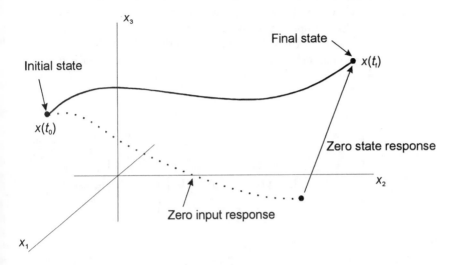

Figure 5.4: State space trajectory from initial state $\mathbf{x}(t_0)$ to final state $\mathbf{x}(t_f)$ of system $\mathcal{S}$ given by Eq. 5-1

and final states are determined the control variable $u(t)$ must be manipulated so as to steer the system from its initial to final state.

We can formalize these ideas by the following definition. A system $\mathcal{S}$ is said to be **completely state controllable** if, for any time t_0, each initial state $\mathbf{x}(t_0)$ can be steered to any final state $\mathbf{x}(t_f)$ by a permissible input $u(t)$, $t_0 \leq t \leq t_f$.

Many extensions of this definition of controllability can be stated, such as output controllability and function space controllability, but here we will restrict our discussions to the definition given above. The reader is referred to references at the end of the chapter for a discussion of these alternative concepts.

We now derive an algebraic test of controllability for the system whose response is described by Eq. (5-50). Transposing the zero input response term to the left hand side and multiplying both sides by $\exp[-\mathbf{A}t_f]$ gives

$$e^{-\mathbf{A}t_f}\mathbf{x}(t_f) - e^{-\mathbf{A}t_0}\mathbf{x}(t_0) = \int_{t_0}^{t} e^{-\mathbf{A}\tau}\mathbf{b}u(\tau)d\tau \qquad (5.51)$$

Now suppose the system is completely state controllable then $\mathbf{x}(t_f)$ and $\mathbf{x}(t_0)$ can be any points in the n-dimensional state space. Since $\exp[\mathbf{A}t]$ is non-singular the term $(\exp[-\mathbf{A}t_f]\mathbf{x}(t_f) - \exp[-\mathbf{A}t_0]\mathbf{x}(t_0))$ on the left hand side of Eq. (5-51) will correspond to any point $\mathbf{x}$ in this n-dimensional state space, which in turn implies the integral term on the right hand side must range over all points in the same space. We known from Eq. (5-39) that $\exp[\mathbf{A}t]$ can be expanded into a finite sum, and when it is substituted into Eq. (5-51) we obtain

$$
\begin{aligned}
\mathbf{x} &= \sum_{j=0}^{n-1} \mathbf{A}^j \mathbf{b} \int_{t_0}^{t_f} \beta_j(\tau)u(\tau)d\tau \\
\\
&= \begin{bmatrix} \mathbf{b} & \mathbf{Ab} & \cdots & \mathbf{A}^{n-1}\mathbf{b} \end{bmatrix}
\begin{bmatrix} \alpha_0 \\ \alpha_1 \\ \vdots \\ \alpha_{n-1} \end{bmatrix}
\end{aligned}
\tag{5.52}
$$

where we define

$$
\alpha_j = \int_{t_0}^{t_f} \beta_j(\tau)u(\tau)d\tau
$$

Since $\mathbf{x}$ can be any point in the n-dimensional state space we see that $\mathcal{C}(\mathbf{A},\mathbf{b}) = \begin{bmatrix} \mathbf{b} & \mathbf{Ab} & \cdots & \mathbf{A}^{n-1}\mathbf{b} \end{bmatrix}$ must have rank n. As Eq. (5-1) describes a single input system this statement is equivalent to $\mathcal{C}(\mathbf{A},\mathbf{b})$ being non-singular.

Let us now suppose the system is not controllable. Then it follows that the integral on the right hand side of Eq. (5-51) cannot span the n-dimensional state space, so there exists a non-zero vector η, such that

$$
\int_{t_0}^{t_f} \eta^T e^{-\mathbf{A}\tau} \mathbf{b} u(\tau)d\tau = 0
$$

for all inputs $u(t)$, $t_0 \leq t \leq t_f$. Taking

$$
u(t) = [\eta^T e^{-\mathbf{A}t} \mathbf{b}]^T
\tag{5.53}
$$

we find

$$
\int_{t_0}^{t_f} (\eta^T e^{-\mathbf{A}\tau} \mathbf{b}\mathbf{b}^T e^{\mathbf{A}^T \tau} \eta)d\tau = 0
$$

so that

$$
\eta^T e^{-\mathbf{A}t} \mathbf{b} = 0
\tag{5.54}
$$

for $t_0 \leq t \leq t_f$ Differentiating Eq. (5-54) $(n-1)$ times and setting $t = 0$, gives

$$
\eta^T \mathbf{b} = \eta^T \mathbf{Ab} = \cdots = \eta^T \mathbf{A}^{n-1}\mathbf{b} = 0
$$

Thus it follows that the rank of matrix $C(\mathbf{A}, \mathbf{b})$ is less than n, or in other words $C(\mathbf{A}, \mathbf{b})$ is singular.

Consequently we have shown that the system $\mathcal{S}$ described by Eq. (5-1) is completely state controllable if and only if the matrix $C(\mathbf{A}, \mathbf{b})$ has rank n. In the case where $C(\mathbf{A}, \mathbf{b})$ has rank n we say that the matrix $\mathbf{A}$ and vector $\mathbf{b}$ are a **controllable pair**.

It will be noted we have not said anything about the output of the system. This is so because state controllability is independent of the output of a system.

Example 12 *Consider the dynamical system characterized by*

$$\dot{x}_1 = x_2$$

$$\dot{x}_2 = x_3$$

$$\dot{x}_3 = -12x_2 - 7x_3 + u(t)$$

Then

$$\mathbf{A} = \begin{bmatrix} 0 & 1 & 0 \\ 0 & 0 & 1 \\ 0 & -12 & -7 \end{bmatrix}, \quad \mathbf{b} = \begin{bmatrix} 0 \\ 0 \\ 1 \end{bmatrix}$$

Determine whether the system is controllable.

We establish system controllability by requiring that $\det C(\mathbf{A}, \mathbf{b}) \neq 0$. First we form the vectors,

$$\mathbf{Ab} = \begin{bmatrix} 0 & 1 & 0 \\ 0 & 0 & 1 \\ 0 & -12 & -7 \end{bmatrix} \begin{bmatrix} 0 \\ 0 \\ 1 \end{bmatrix} = \begin{bmatrix} 0 \\ 1 \\ -7 \end{bmatrix}$$

and similarly

$$\mathbf{A}^2\mathbf{b} = \begin{bmatrix} 0 \\ -7 \\ 37 \end{bmatrix}$$

The matrix $C(\mathbf{A}, \mathbf{b})$ becomes

$$C(\mathbf{A}, \mathbf{b}) = \begin{bmatrix} \mathbf{b} & \mathbf{Ab} & \mathbf{A}^2\mathbf{b} \end{bmatrix} = \begin{bmatrix} 0 & 0 & 1 \\ 0 & 1 & -7 \\ 1 & -7 & 37 \end{bmatrix}$$

and

$$\det C(\mathbf{A}, \mathbf{b}) = -1$$

Since $\det C(\mathbf{A}, \mathbf{b})$ non zero the system is controllable.

Observability

Just as we approached the definition of controllability from a geometric viewpoint we do the same for observability. For the system $\mathcal{S}$, described by Eq. (5-1) and (5-2) with output $y(t)$ it is natural to consider under what conditions all initial states can be deduced by observing the output $y(t)$ over a finite period. When all initial states can be deduced the system is said to be **observable**. In simple terms a system will be observable if all the state variables have an influence on the output $y(t)$, as if one state variable had no influence then it would not be possible to determine its value at time t_0.

We formalize these ideas by the following definition. A system $\mathcal{S}$ is said to be **completely state observable** if every initial state $\mathbf{x}(t_0)$ can be determined from the observation of the output $y(t)$ over the finite interval $t_0 \leq t \leq t_f$, assuming the input variable $u(t)$ is known over this interval.

In a manner similar to our discussion of controllability we develop an algebraic test for observability of the system $\mathcal{S}$ whose output $y(t)$ is given by Eq. (5-51). In this analysis it is possible to assume the input $u(t) \equiv 0$ since the component of the output which depends upon the initial conditions is only the zero input response. Consequently Eq. (5-51) becomes

$$y(t) = \mathbf{c}e^{\mathbf{A}(t-t_0)}\mathbf{x}(t_0), \quad t_0 \leq t \leq t \tag{5.55}$$

Differentiating this equation $(n-1)$ times and taking $t = t_0$ we find

$$\begin{bmatrix} y(t_0) \\ y^{(1)}(t_0) \\ \vdots \\ y^{(n-1)}(t_0) \end{bmatrix} = \begin{bmatrix} \mathbf{c} \\ \mathbf{cA} \\ \vdots \\ \mathbf{cA}^{n-1} \end{bmatrix} \mathbf{x}(t_0) \tag{5.56}$$

We thus see that if $\mathcal{O}(\mathbf{c}, \mathbf{A}) = \begin{bmatrix} \mathbf{c}^T & \mathbf{A}^T\mathbf{c}^T & \cdots & \left(\mathbf{A}^T\right)^{n-1}\mathbf{c}^T \end{bmatrix}^T$ has rank n then the initial condition $\mathbf{x}(t_0)$ can be deduced from the output signal and its $(n-1)$ derivatives evaluated at t_0. For a single output system $\mathcal{O}(\mathbf{c}, \mathbf{A})$ has rank n if and only if it is non-singular. Alternatively if the rank of $\mathcal{O}(\mathbf{c}, \mathbf{A})$ is less than n then there exists a non-zero vector η for which $\mathcal{O}(\mathbf{c}, \mathbf{A})\eta = \mathbf{0}$. Thus

$$\mathbf{c}\eta = \mathbf{cA}\eta = \cdots = \mathbf{cA}^{n-1}\eta = 0$$

From Eqs. (5-39) and (5-55)

$$y(t) = \left(\sum_{j=0}^{n-1} \mathbf{cA}^j \beta_j(t) \right) \mathbf{x}(t_0)$$

If we take $\mathbf{x}(t_0) = \eta \neq \mathbf{0}$ then $y(t) \equiv 0$. That is there exists a non-zero initial condition which cannot be detected by observing the output $y(t)$, so that the system is not observable.

From the above discussion it can be seen that the systems $\mathcal{S}$ described by Eqs. (5-1) and (5-2) is completely state observable if and only if the matrix $\mathcal{O}(\mathbf{c}, \mathbf{A})$ has rank n.

The system give in Eqs. (5-1) and (5-2) can be characterized by the notation $\mathcal{S}(\mathbf{A}, \mathbf{b}, \mathbf{c})$. Using this notation there is a **duality** between controllability and observability for it is easy see that $\mathcal{S}(\mathbf{A}, \mathbf{b}, \mathbf{c})$ is completely controllable (observable) if and only if the system $\mathcal{S}(\mathbf{A}^T, \mathbf{c}^T, \mathbf{b}^T)$ is completely observable (controllable). In addition it is easy to see for a system $\mathcal{S}$ whose $\mathbf{A}$ matrix is in diagonal form that it is controllable or observable if and only if $\mathbf{b}$ or $\mathbf{c}$ respectively have no zero entries.

Example 13 *Consider a system characterized by*

$$\mathbf{A} = \begin{bmatrix} 0 & 1 & 0 \\ 0 & 0 & 1 \\ 0 & -12 & -7 \end{bmatrix}, \quad \mathbf{c}^T = \begin{bmatrix} 2 \\ 3 \\ 1 \end{bmatrix}$$

Determine whether the system is observable.

The elements of the $\mathcal{O}(\mathbf{c}, \mathbf{A})$ matrix are formed

$$\mathbf{cA} = \begin{bmatrix} 2 & 3 & 1 \end{bmatrix} \begin{bmatrix} 0 & 1 & 0 \\ 0 & 0 & 1 \\ 0 & -12 & -7 \end{bmatrix} = \begin{bmatrix} 0 & -10 & -4 \end{bmatrix}$$

Similarly

$$\mathbf{cA}^2 = \begin{bmatrix} 0 & 48 & 18 \end{bmatrix}$$

The $\mathcal{O}(\mathbf{c}, \mathbf{A})$ matrix becomes

$$\mathcal{O}(\mathbf{c}, \mathbf{A}) = \begin{bmatrix} 2 & 3 & 1 \\ 0 & -10 & -4 \\ 0 & 48 & 18 \end{bmatrix}$$

and

$$\det \mathcal{O}(\mathbf{c}, \mathbf{A}) = 24$$

Since $\det \mathcal{O}(\mathbf{c}, \mathbf{A}) \neq 0$ the system is observable.

We note that a system can be controllable and stable but not observable. In general the requirements of stability, controllability and observability need to be independently determined. An important point to remember is that when there is a loss of information, possibly due to pole and zero cancellations inside the system, then the system will not be observable as shown in the next example.

Example 14 *The closed loop transfer function of a control system is given*

$$\frac{C(s)}{R(s)} = \frac{s+2}{s^3 + 5s^2 + 9s + 6}$$

Determine if the system is stable, controllable and observable.

The state equations for the system are

$$\dot{x}_1 = x_2$$

$$\dot{x}_2 = x_3$$

$$\dot{x}_3 = -6x_1 - 9x_2 - 5x_3 + u(t)$$

The coefficient matrix is

$$\mathbf{A} = \begin{bmatrix} 0 & 1 & 0 \\ 0 & 0 & 1 \\ -6 & -9 & -5 \end{bmatrix}$$

and the input and output vectors are

$$\mathbf{b} = \begin{bmatrix} 0 \\ 0 \\ 1 \end{bmatrix}, \quad \mathbf{c}^T = \begin{bmatrix} 2 \\ 1 \\ 0 \end{bmatrix}$$

For establishing system observability we form the elements of the matrix $\mathcal{O}(\mathbf{c}, \mathbf{A})$ which are

$$\mathcal{O}(\mathbf{c}, \mathbf{A}) = \begin{bmatrix} 2 & 1 & 0 \\ 0 & 2 & 1 \\ -6 & -9 & -3 \end{bmatrix}$$

and

$$\det \mathcal{O}(\mathbf{c}, \mathbf{A}) = 0$$

The system is therefore not observable. Examination of the transfer function indicates that the denominator has a zero at $s = -2$ which cancels the one in the numerator thereby reducing the transfer function to

$$\frac{C(s)}{R(s)} = \frac{1}{s^2 + 3s + 3}$$

Since we are only looking at the output we have no idea what happens inside the system owing to a pole and zero cancellation. This lack of a unique transfer function renders the system unobservable. Let us see, however, if the system is stable and controllable. Since after canceling $(s + 2)$ it is a second-order system, and all coefficients of s are positive the system is stable. For controllability we form the $\mathcal{C}(\mathbf{A}, \mathbf{b})$ matrix,

$$\mathcal{C}(\mathbf{A}, \mathbf{b}) = \begin{bmatrix} 0 & 0 & 1 \\ 0 & 1 & -5 \\ 1 & -5 & 16 \end{bmatrix}$$

From this matrix we find $\det \mathcal{C}(\mathbf{A}, \mathbf{b}) = 1$, so that the system is controllable. This means that if we specify an arbitrary final state the system can always be driven to this point by an appropriate input. The system under consideration is therefore controllable and stable but not observable.

Example 15 *It is desired to control, by means of on-board jets, the attitude motion of the tethered satellite discussed in Section 5-4. Obtain the controllability and observability of the roll $\theta(t)$ and yaw motions $\psi(t)$ of the satellite if the governing equations are*

$$\ddot{\theta} + 6\theta - 0.5\dot{\psi} = 0$$

$$\ddot{\psi} + 2\psi + \dot{\theta} = T(t)$$

where $T(t)$ is a torque input. It is assumed that the pitch motion is completely uncoupled from the other two motions and is stable, controllable and observable.

Let the state variables be $x_1 = \theta$, $x_2 = \dot{\theta}$, $x_3 = \psi$, $x_4 = \dot{\psi}$. Then, as before, the state equation is

$$\dot{\mathbf{x}}(t) = \mathbf{A}\mathbf{x}(t) + \mathbf{b}u(t)$$

where

$$\mathbf{A} = \begin{bmatrix} 0 & 1 & 0 & 0 \\ -6 & 0 & 0 & 0.5 \\ 0 & 0 & 0 & 1 \\ 0 & -1 & -2 & 0 \end{bmatrix} \qquad \mathbf{b} = \begin{bmatrix} 0 \\ 0 \\ 0 \\ 1 \end{bmatrix} \qquad u(t) = T(t)$$

The output is given by

$$\theta(t) = \mathbf{c}_1 \mathbf{x}$$

$$\psi(t) = \mathbf{c}_2 \mathbf{x}$$

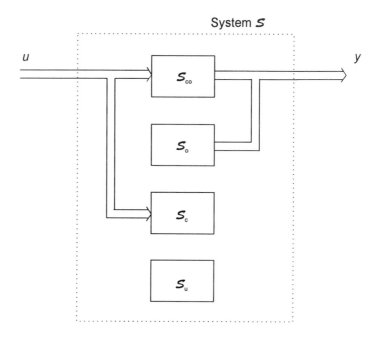

Figure 5.5: Canonical decomposition of a system $\mathcal{S}$ into four subsystems

where

$$\mathbf{c}_1^T = \begin{bmatrix} 1 \\ 0 \\ 0 \\ 0 \end{bmatrix}, \quad \mathbf{c}_2^T = \begin{bmatrix} 0 \\ 0 \\ 1 \\ 0 \end{bmatrix}$$

The question of controllability can be answered by forming the matrix $\mathcal{C}(\mathbf{A}, \mathbf{b})$

$$\mathcal{C}(\mathbf{A}, \mathbf{b}) = \begin{bmatrix} \mathbf{b} & \mathbf{Ab} & \mathbf{A}^2\mathbf{b} & \mathbf{A}^3\mathbf{b} \end{bmatrix} = \begin{bmatrix} 0 & 0 & 0.5 & 0 \\ 0 & 0.5 & 0 & -4.25 \\ 0 & 1 & 0 & -2.5 \\ 1 & 0 & -2.5 & 0 \end{bmatrix}$$

Since by calculation $\det \mathcal{C}(\mathbf{A}, \mathbf{b}) = 1.5$, the system is controllable. This means that the state variables may be controlled by controlling the input torque.

The system observability is determined from the $\mathcal{O}(\mathbf{A}, \mathbf{c})$ matrix. For the output vector c_1, the $\mathcal{O}(\mathbf{A}, c_1)$ becomes

$$\mathcal{O}(\mathbf{A}, c_1) = \begin{bmatrix} 1 & 0 & 0 & 0 \\ 0 & 1 & 0 & 0 \\ -6 & 0 & 0 & 0.5 \\ 0 & -6.5 & -1 & 0 \end{bmatrix}$$

The determinant of $\mathcal{O}(\mathbf{A}, c_1)$ is found to be 0.5. Since it is non-zero the system is observable. Thus the system under consideration, in addition to being stable, is also controllable and observable.

Canonical Decomposition

We saw in Section 3-7 that similarity transformations exist between various state realizations of a systems $\mathcal{S}$. Gilbert and Kalman have shown that a similarity transformation can always be found which decomposes the system $\mathcal{S}$ into four subsystem realizations which we will denote by $\mathcal{S}_{co}$, $\mathcal{S}_o$, $\mathcal{S}_c$, $\mathcal{S}_u$. This **canonical decomposition**, while not unique, leads to the block diagram shown in Fig. 5-5. Here $\mathcal{S}_{co}$ is a completely controllable and observable subsystem, $\mathcal{S}_o$ is only a completely observable subsystem, $\mathcal{S}_c$ is only a completely observable subsystem and $\mathcal{S}_u$ is a totally unobservable and uncontrollable subsystem.

The transfer function of a system $\mathcal{S}$ is defined by its terminal behavior, as discussed in Chapter 3. From Fig. 5-5 it can be seen that the subsystem $\mathcal{S}_{co}$ is the only one relating the input to the output of system $\mathcal{S}$ so that the transfer function is identified with this subsystem. As a consequence the state space description of a system $\mathcal{S}$ is more general than its transfer function description as it describes the internal structure of a system whereas the transfer function only relates to the terminal behavior. When a state realization of a system only includes the subsystem $\mathcal{S}_{co}$, with the others being absent, it is referred to as a **minimal state realization** of the transfer function.

It can be shown that the minimal realization for a system transfer function has a state equation of the lowest possible order. By this we mean there is no other system representation of the transfer function whose $\mathbf{A}$ matrix is of a lower dimension than the minimal realization. Many further interesting and useful results relating state equation realizations and transfer functions have been obtained and the interested reader should refer to Kailath for an extensive discussion.

5.6   Summary

A control system is represented in state space form by the vector-matrix differential equations

$$\dot{\mathbf{x}}(t) = \mathbf{A}\mathbf{x}(t) + \mathbf{b}u(t)$$

$$y(t) = \mathbf{c}\mathbf{x}(t)$$

The solution of these equations given in terms of the transition matrix $\phi(t)$ is

$$\mathbf{x}(t) = \phi(t)\mathbf{x}(0) + \int_0^t \phi(t-\tau)\mathbf{b}u(t)d\tau$$

$$y(t) = \mathbf{c}\phi(t)\mathbf{x}(0) + \int_0^t \mathbf{c}\phi(t-\tau)\mathbf{b}u(t)d\tau$$

where the first term is the **zero input response** and the second is the **zero state response**.

We investigated several methods for obtaining the transition matrix. Although all the methods gave similar results, their relative attraction lays with the computational aids available. While the Laplace transform method is well suited for hand computation, the Cayley-Hamilton method is better suited for machine computation. The Fadeeva method can also be used for machine computation for low order systems.

The stability of the system is dependent upon the eigenvalues of the **A** matrix obtained by solving the equation $\det[s\mathbf{I} - \mathbf{A}] = 0$.

As you progress further, it is important that you realize that the techniques developed in the present chapter are not to be considered as just another way of obtaining a solution to problems outlined in earlier chapters. Instead, these techniques are to be interpreted as new tools useful for not only observing variables inside a control system, but handling a very large number of input-output combinations. As systems get more complex (and interesting), we shall be more dependent upon the computer as a computational aid and consequently rely more heavily on state space techniques.

Finally the concepts of controllability and observability are introduced, and simple algebraic tests for determining whether systems are completely controllable or observable are developed. In addition some simple properties of such systems are presented. Following this it is shown that systems may be canonically decomposed into four types of sub-systems.

5.7   References

1. V. N. Fadeeva, *Computational Methods in Linear Algebra,* Dover Publications, Inc., New York 1959.

2. L. A. Zadeh, C. A. Desoer, *Linear System Theory,* Mc Graw Hill Book Co. New York, 1963.

3. R. J. Mayhan, *Discrete-Time and Continuous-Time Linear Systems,* Addison Wesley Publ. Co., Reading, Mass., 1984.

4. F. R. Gantmacher, *The Theory of Matrices,* Chelsea Publ. Co., New York, 1960.

5. J. M. Wilkinson, *The Algebraic Eigenvalue Problem,* Oxford University Press, 1965.

6. C. J. Swet, J. M. Whisnant, "Deployment of a Tethered Orbiting Interferometer", *The J. Astronautical Sciences,* Vol. 17, No. 1 (1969) pp. 44-59.

7. R. W. Brockett, *Finite Dimensional Linear Systems,* John Wiley and Sons, Inc., New York, 1970.

8. R. E. Kalman, "Mathematical Description of Linear Dynamical Systems", *SIAM J. Control,* Vol. 1, No. 2 (1963) pp. 152-192.

9. E. G. Gilbert, "Controllability and Observability in Multivariable Control Systems", *SIAM J. Control,* Vol. 1, No. 2 (1963) pp. 128-151.

10. T. Kailath, *Linear Systems,* Prentice-Hall, Inc., New Jersey, 1980.

5.8   Problems

5-1 Beginning with the state representation of Problem 3-13, obtain $c(t)$.

5-2 The coefficient matrix of a physical systems is given by

$$\mathbf{A} = \begin{bmatrix} -1 & 3 & -2 \\ 0 & -1 & -1 \\ 0 & -2 & -4 \end{bmatrix}$$

Derive the transition matrix $\phi(t)$ using the Laplace transform technique.

5-3 Solve Problem 5-2 using the Cayley-Hamilton method and compare $\phi(t)$ at $t = 0, 1, 5$, and 10 seconds.

5-4 Obtain the solution to $\dot{\mathbf{x}}(t) = \mathbf{A}\mathbf{x}(t) + \mathbf{b}u(t)$ using the state variable diagrams discussed in Section 3-6 for evaluating the transition matrix

$$\mathbf{A} = \begin{bmatrix} -1 & 3 & -2 \\ 0 & -1 & 0 \\ 0 & 2 & -3 \end{bmatrix}, \quad \mathbf{b} = \begin{bmatrix} 1 \\ 0 \\ 0 \end{bmatrix}, \quad u(t) = H(t)$$

Assume the initial state $\mathbf{x}(0) = \mathbf{0}$.

5-5 Obtain $\mathbf{x}(t)$ for Example 4 by solving for $\phi(t)$ using the Cayley-Hamilton and Laplace transform methods.

5-6 A system has a coefficient matrix given by

$$\mathbf{A} = \begin{bmatrix} -1 & -1 & 0 \\ 2 & 1 & -1 \\ 3 & 0 & 1 \end{bmatrix}$$

Is it stable?

5-7 For what values of α, β, γ is the system with the following coefficient matrix $\mathbf{A}$ stable?
$$\mathbf{A} = \begin{bmatrix} 0 & -1 & 0 \\ \alpha & 0 & 1 \\ 0 & -\beta & \gamma \end{bmatrix}$$

5-8 Express the attitude stabilization control system in Problem 4-15 in state form and obtain the characteristic equation. Do these agree with the results in Problem 4-15?

5-9 Solve for β_0, β_1, β_2, β_3 and verify $\theta(\tau)$ shown in Eq. (5-47).

5-10 Obtain the solution in Example 12, Chapter 3, for the case where $f_x = f_y = f_z = 0$ (i.e. torque free environment), $\alpha = -\beta = 1$ and initial conditions

$$\mathbf{x}(0) = \begin{bmatrix} 1 \\ 2 \\ 0 \end{bmatrix}$$

5-11 It is desired that mass m_1 shown in Fig. P5-11 be stabilized by moving the base. If we assume that $\theta(t)$ and $x(t)$ are small and also that $m_2 \ll m_1$, the resulting equations are seen to be linear but the

system itself is unstable. The system can be stabilized if we apply a
force $f(t)$ on the cart such that

$$f(t) = k_1 \left(\theta - \frac{x}{L}\right) + k_2 \left(\dot{\theta} - \frac{\dot{x}}{L}\right)$$

Write the system equations in state space form and find the values of
k_1 and k_2 such that the system is stable.

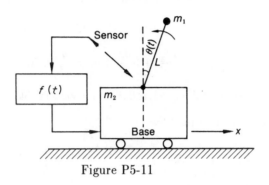

Figure P5-11

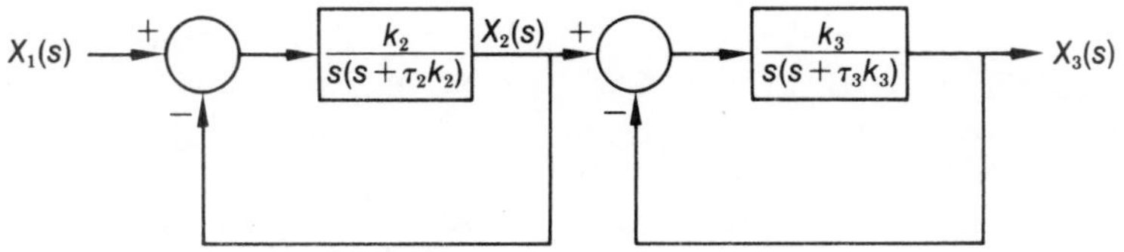

Figure P5-12

5-12 In studying the flow of automobile traffic it was noticed that the
acceleration of a car was determined by the distance between two cars,
the velocity of the lead car, and natural constants that described the
car and driver. If we consider the first (lead) car's position as $x_1(t)$,
that of the second car as $x_2(t)$, and so on, the control system that
models the automobile traffic is shown in Fig. P5-12. The constant
k_i is the natural frequency of the car and driver and τ_i is a combined
time constant. We wish to use this model to study the start up and
stopping of traffic involving three cars. We assume that the length of
the cars is neglected. Obtain an expression for the position of each
car.

5-13 Under what condition is the system defined in Problem 5-7 observable and controllable? Let

$$\mathbf{b} = \begin{bmatrix} 1 \\ 0 \\ 0 \end{bmatrix}, \quad \mathbf{x}(0) = \begin{bmatrix} 0 \\ 1 \\ 0 \end{bmatrix}, \quad \text{and } \mathbf{c} = \begin{bmatrix} 1 & 0 & 0 \end{bmatrix}$$

5-14 Show that ψ is observable and controllable in Example 15.

5-15 A system is described by the following equation,

$$\frac{d^2x}{dt^2} + 5\frac{dx}{dt} + 6x = \frac{dy}{dt} + 3y$$

Is this system stable, controllable, and observable?

5-16 A system is described by

$$\mathbf{A} = \begin{bmatrix} \alpha & -1 \\ 2 & 1 \end{bmatrix}, \quad \mathbf{b} = \begin{bmatrix} 1 \\ 0 \end{bmatrix}$$

For what values of α is this system controllable and stable?

Chapter 6

Performance Criteria

6.1 Introduction

When a specific system is proposed for a given application, it must satisfy certain requirements. This may involve the optimization of the system or the system response in a specified way. The requirements that a control system must meet are generally called performance specifications. The analysis of the system, by methods presented in the previous chapters, yields results that must be compared to the desired performance specifications. If they do not compare favorably, the response can be altered by either introducing a controller into the control system or by introducing other compensating elements.

The first part of this chapter is concerned with control system performance specifications in the time domain. We will then introduce the concept of the performance index. Also, the effects of parameter variation on system error will be investigated. This is important as the ability of a control system to suppress the system error below a certain level is sometimes vital. The chapter also includes consideration of various types of controllers suitable for altering system performance.

6.2 Control System Specification

The performance of a control system can be considered in three parts. The first part pertains to the specifications as they directly relate to the system response. The second has to do with a performance index that is a function of the error or output. The last part is concerned with system error caused by parameter variations.

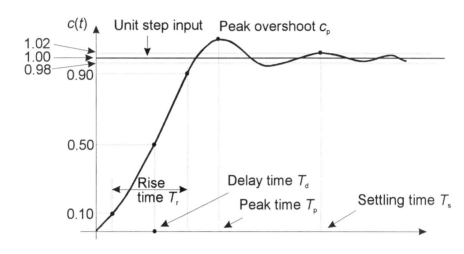

Figure 6.1: Time domain specification of control system performance

Control system specifications can be directly related to system response as shown in Fig. 6-1. This type of information is germane to second-order systems or higher-order systems which have a pair of characteristic roots that are complex and dominate the transient behavior. For example, a system with characteristic roots at $-5, -10 \pm j2$ and $-0.5 \pm j2$ is a fifth-order system but the dominant roots are $-0.5 \pm j2$. The commonly used terms to describe system specifications are peak overshoot, peak time, rise time, delay time, settling time, bandwidth, damping ratio and undamped natural frequency. These are briefly discussed below.

Peak Overshoot (c_p)

This is measured when the response has a maximum value at the peak time, T_p. It is an indication of the largest error between input and output during the transient state. For the systems considered in Chapter 4 we observed that the peak overshoot increased as the damping ratio decreased. The concept of peak overshoot is not limited to only second-order systems. It is often used for higher-order systems that have a dominant pair of complex poles. These poles are those located nearest the imaginary axis. It is often convenient in practice to express the peak overshoot c_p as a fraction M_p

of the output steady state value, and in most well-designed systems, peak overshoots are lower than 30 percent.

Rise Time (T_r)

The rise time is a measure of the speed of response. It is defined as the time necessary for the response to rise from 10% to 90% of its final steady state value. Sometimes an equivalent measure is to represent the rise time as the reciprocal of the slope of the response at the instant the response is 50% of its final steady state value. For second-order under-damped systems, the time to reach the peak overshoot is also a good measure of the speed of response.

Delay Time (T_d)

The time necessary for the response to reach some value (usually 50 percent) of its steady state value is called delay time.

Settling Time (T_s)

The settling time is defined as the time necessary for the response to decrease to and stay within a specified range of its final value. Two or five percent is often stated as the tolerable range. The number of oscillations necessary to reach this condition is also a useful index.

Bandwidth (ω_b)

The bandwidth is defined as the frequency at which the output magnitude is 0.707 as compared to the output magnitude at low (or zero) frequency when the system is subjected to sinusoidal inputs. We shall consider this in more detail in the next chapter.

Damping Ratio (ζ)

This is a ratio of the system damping to the critical damping for a second-order system. It measures the damping of a complex pole pair. Higher-order systems may have more than one damping ratio although the damping measured by the most dominant complex pole pair is of greatest importance. The damping ratio is an important parameter in determining the transient performance and stability of a system.

Undamped Natural Frequency (ω_n)

This is directly related to the "springiness" of a system. Like the damping ratio it can be applied to second-order systems or higher-order systems possessing dominant poles.

For second-order systems, simple formulas can be derived relating the parameters T_r, T_d, T_s, and T_p, to the system damping ratio ζ and undamped natural frequency ω_n. These are summarized in Table 6-1. For higher-order systems having a dominant pair of complex conjugate poles these formulas will give approximate values for the parameters mentioned above. As a consequence these parameters are often used for specifying the time domain performance of such higher-order systems.

System specifications are also given in terms of the **error constants** (or coefficients) as well as the **System Type**. The error constants are used to relate the system gain and time constants to the system errors of a unity feedback system. They measure directly the minimum ideal steady state error of a system for a step, ramp, and parabolic input. The error coefficients are identical, in concept, to the error constants although generalized for any type of input. In addition to this, system performance is also often given in the frequency domain. This is considered in Chapter 7.

Example 1 *Consider a control systems whose output is given by*

$$C(s) = \frac{K\omega_n^2}{s(s^2 + 2\zeta\omega_n s + \omega_n^2)}$$

where $\omega_n = 8$ *rad/sec and* $\zeta = 0.5$. *Determine* ω_d, $c(t)$, M_p, T_p, T_s *and the number of oscillations to reach* T_s.

The damped frequency becomes

$$\omega_d = \omega_n \sqrt{1 - \zeta^2} = 6.93 \text{ rad/sec}$$

The response is

$$c(t) = K[1 - 1.155e^{-4t}\sin(6.928t + \phi)]$$

where $\cos \phi = 0.5$. The overshoot M_p is given by

$$M_p = \exp\left(-\frac{\zeta\pi}{\sqrt{1 - \zeta^2}}\right) = 0.163$$

Table 6-1 Formulas for computing time domain parameters of second order systems assuming $\zeta \leq 1$.

1.	Peak Time	$T_p = \dfrac{\pi}{\omega_n \sqrt{1 - \zeta^2}}$	
2.	Rise Time	$T_r \approx \dfrac{1}{\omega_n}[1 + 1.4\zeta]$	$0 \leq \zeta \leq 0.7$
3.	Delay Time	$T_d \approx \dfrac{1}{\omega_n}[1.1 + 1.2\zeta]$	
4.	Settling Time	$T_s \approx \dfrac{4}{\zeta\omega_n}$	(within 2% range)
		$T_s \approx \dfrac{3}{\zeta\omega_n}$	(within 5% range)
5.	Number of Oscillations	$N \approx \dfrac{4\sqrt{1 - \zeta^2}}{2\pi\zeta}$	(within 2% range)
		$N \approx \dfrac{3\sqrt{1 - \zeta^2}}{2\pi\zeta}$	(within 5% range)
6.	Peak Overshoot	$c_p = 1 + \exp\left(-\dfrac{\pi\zeta}{\sqrt{1 - \zeta^2}}\right)$	
		$M_p = \exp\left(-\dfrac{\pi\zeta}{\sqrt{1 - \zeta^2}}\right)$	

This 16 percent overshoot occurs at

$$T_p = \frac{\pi}{\omega_n \sqrt{1 - \zeta^2}} = 0.453 \text{ sec}$$

The 5 percent settling time T_s (as given in Table 6-1) is approximately three time constants of the envelope of the damped sinusoidal oscillation

$$T_s = \frac{3}{\zeta \omega_n} = 0.75 \text{ sec}$$

The number of oscillations necessary to reach this time is similarly obtained from

$$N = \frac{3\sqrt{1 - \zeta^2}}{2\pi\zeta} = 0.827 \text{ oscillations.}$$

Although the above specifications are quite popular, an increasing amount of stress is being laid on the mathematical representation of the performance of a control system. This is given in terms of the single performance index which is to be considered next.

6.3   Dynamic Performance Indices

A performance index is a quantitative measure of system performance. We must of course be clear as to what we mean by system performance. For example, do we mean that the error must be a minimum or constant? Or would we prefer to measure system performance by the square of the error? Clearly, the performance index will depend upon the specific criterion that we wish to invoke. It is therefore important that the design engineer must know *a priori* what he would like to optimize.

In general the **performance index** J is represented as

$$J = \int_0^T f(e(t)) \, dt \tag{6.1}$$

where $e(t)$ may be the system error, output, input or some combination of these quantities. For many control systems, the performance index based upon the system error is quite useful. Four different indices that are in common usage are

$$J_1 = \int_0^T e^2(t) dt \tag{6.2}$$

$$J_2 = \int_0^T |e(t)| \, dt \tag{6.3}$$

Table 6-2 Coefficients of characteristic equation based on optimizing ITAE, for a step input

$$s + \omega_n$$

$$s^2 + 1.4\omega_n s + \omega_n^2$$

$$s^3 + 1.75\omega_n s^2 + 2.15\omega^2 s + \omega_n^3$$

$$s^4 + 2.1\omega_n s^3 + 3.4\omega_n^2 s^2 + 2.7\omega_n^3 s + \omega_n^4$$

$$J_3 = \int_0^T |te(t)|\, dt \qquad (6.4)$$

$$J_4 = \int_0^T te^2(t)\, dt \qquad (6.5)$$

The upper time limit is arbitrary. Quite often the settling time is used. Other times it is replaced by infinity. J_1 is the performance index based on the **integral of the square error** and designated by ISE. The integral of the absolute error, IAE, is given by J_2. The integral of time t multiplied by the absolute error $|e(t)|$ is J_3 and designated by ITAE. Finally, J_4 is designated by ITSE. If it is desired to have a minimum performance index, the indices most useful are ITAE and ITSE, in that order.

The ITAE index has been developed into a standard form that provides optimum coefficients for various systems. This is shown in Table 6-2. Consider, for example, the forward loop transfer function of a control system, where it is desired to optimize ITAE.

$$G(s) = \frac{K}{s^2 + bs + a}$$

The closed loop transfer function becomes

$$\frac{C(s)}{R(s)} = \frac{K}{s^2 + bs + (a + K)}$$

From Table 6-2 we note that for optimizing ITAE we must have $(a + K) = \omega_n^2$, and $b = 1.4\omega_n$, where ω_n can be selected from other additional requirements. If the above system had an integrator K_1/s in the forward loop, the closed loop transfer function would then become

$$\frac{C(s)}{R(s)} = \frac{KK_1}{s^3 + bs^2 + (a + K)s + KK_1}$$

Now from Table 6-2, the constants for optimum ITAE are

$$b = 1.75\omega_n, \quad (a + K) = 2.15\omega_n^2, \quad KK_1 = \omega_n^3$$

Again we can select ω_n from other requirements.

Example 2 *The output of a second-order control system subjected to a unit input is given by*

$$c(t) = 1 - \frac{e^{-\zeta\omega_n t}}{\sqrt{1 - \zeta^2}} \sin(\omega_d t + \phi)$$

where

$$\omega_d = \omega_n\sqrt{1 - \zeta^2}, \quad \phi = \arctan\frac{\sqrt{1 - \zeta^2}}{\zeta}$$

Obtain a value or values of ζ that minimizes J_1 for a fixed value of ω_n.

The departure from unity may be considered as the error,

$$e(t) = \frac{e^{-\zeta\omega_n t}}{\sqrt{1 - \zeta^2}} \sin(\omega_d t + \phi)$$

from which we compute ISE,

$$J_1 = \int_0^\infty \frac{e^{-2\zeta\omega_n t}}{1 - \zeta^2} \sin^2(\omega_d t + \phi) dt$$

where the upper limit has been selected as infinity. Carrying out the integration,

$$J_1 = \frac{1 + 4\zeta^2}{4\omega_n\zeta}$$

For minimizing J_1 we set $dJ_1/d\zeta = 0$ and obtain $\zeta = 0.5$.

In general, it is necessary to plot J_1 versus ζ to obtain the optimum value. Various performance indices are available in the form of tables for aiding the design of control systems.

When the system is represented in state form, the performance index is written as

$$J = \int_0^T f(\mathbf{x}) dt \tag{6.6}$$

where $\mathbf{x}$ is the state vector and a function of time. In general, $f(\mathbf{x})$ will be a combination of several state variables. A performance index based on the sum of the square of the state variables is obtained if

$$f(\mathbf{x}) = \mathbf{x}^T \mathbf{x}$$

$$= \begin{bmatrix} x_1 & x_2 & \cdots & x_n \end{bmatrix} \begin{bmatrix} x_1 \\ x_2 \\ \vdots \\ x_n \end{bmatrix}$$

$$= x_1^2 + x_2^2 + \cdots + x_n^2$$

so that J becomes

$$J = \int_0^T (\mathbf{x}^T \mathbf{x}) dt \tag{6.7}$$

All of the indices given by Eq. (6-2) through Eq. (6-5) may be written in a more general form involving state variables. A more general form of Eq. (6-7) is considered in Chapter 9.

Example 3 *The state vector of a control system is given by*

$$\mathbf{x}(t) = \begin{bmatrix} x_1(t) \\ x_2(t) \end{bmatrix} = \begin{bmatrix} \sigma e^{-\sigma t} \\ 2e^{-\sigma t} \end{bmatrix}$$

Determine the optimum value σ for minimizing J based on Eq. (6-7).

The performance index is

$$J = \int_0^\infty (\mathbf{x}^T \mathbf{x}) dt$$

$$= \int_0^\infty (\sigma^2 e^{-2\sigma t} + 4e^{-2\sigma t}) dt$$

$$= \frac{\sigma^2 + 4}{2\sigma}$$

For minimizing J we set $dJ/d\sigma = 0$, which yields $\sigma = 2.0$

Example 4 *The state vector of a control system is given by*

$$\mathbf{x}(t) = \begin{bmatrix} x_1(t) \\ x_2(t) \end{bmatrix} = \begin{bmatrix} \dfrac{\zeta}{\sqrt{1-\zeta^2}} e^{-\zeta t} \sin\left(\sqrt{1-\zeta^2}\,t\right) \\ 0.5 e^{-\zeta t} \end{bmatrix}$$

Determine the optimum value of ζ for minimizing J based on Eq. (6-7).

The performance index is

$$J = \int_0^\infty \left(\mathbf{x}^T \mathbf{x}\right) dt$$

$$J = \int_0^\infty \left(\frac{\zeta^2 e^{-2\zeta t}}{(1-\zeta^2)} \sin^2\left(\sqrt{1-\zeta^2}t\right) + 0.25 e^{-2\zeta t}\right) dt$$

$$J = \frac{0.5 + \zeta^2}{4\zeta}$$

For minimizing J we set $dJ/d\zeta = 0$, which yields $\zeta = 0.707$.

6.4   Steady State Performance

In addition to the transient performance criteria for a control system discussed in Section 6-2 it is often necessary to specify its steady state performance. We saw in Chapter 4 that the steady state error performance of a unity feedback control system is related to the magnitude of the gain constant K and the number of open loop poles ℓ at the origin of the s-plane, where the terms K and ℓ are defined in Eq. (4-21). The number of poles at the origin is usually referred to as the **System Type**. As an example it may be specified that a unity feedback control system must have a steady state error less than some specified value for a unit ramp input. From Table 4-1 we see that the transfer function $G(s)$ must be Type-1 if the error is to be finite for a ramp input. It can also be seen that the error magnitude is inversely related to the velocity error constant K_1. In addition, it will be noted that for this case the steady state error is zero for a step input and infinite for a parabolic input, irrespective of the value for K_1.

When the operation of a control system fails to satisfy the specified transient or steady state performance requirements we may modify either, or both, of these characteristics by introducing a **controller** $G_c(s)$ into the control system as shown in Fig. 6-2. We shall show that an appropriate controller does indeed alter the response of the system although it also brings about additional changes that may be undesirable. It is not always possible to optimize everything and the particular price that is to be paid must be weighed against the advantages to be gained.

The output signal $M(s)$ of the controller, shown in Fig. 6-2, is the actuating signal that is employed for making necessary corrections so that the output $C(s)$ better corresponds to the input in some manner. The relationship of the actuating signal $M(s)$ to the error signal $E(s)$ is directly

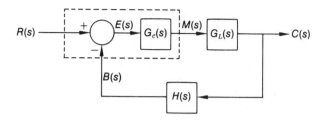

Figure 6.2: General feedback control system

dependent upon $G_c(s)$,

$$\frac{M(s)}{E(s)} = G_c(s)$$

$$\frac{C(s)}{E(s)} = G_c(s)G_p(s)$$

$$\frac{C(s)}{R(s)} = \frac{G_c(s)G_p(s)}{1 + G_c(s)G_p(s)H(s)}$$

We note that the response $C(s)$ will depend upon the nature of $G_c(s)$. Since we can select $G_c(s)$ we have some control upon $C(s)$ for a given input. There are, in general, three basic types of controls that we may select,

1. Proportional control

2. Derivative control

3. Integral control.

We may of course, select any combination of the three basic types. In the remaining section the effect of different controllers on system response is investigated.

Proportional Control

With this control the actuating signal is proportional to the error signal,

$$M(s) = K_p[R(s) - B(s)]$$

$$= K_p E(s) \tag{6.8}$$

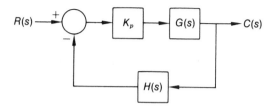

Figure 6.3: Proportional control

A feedback control system with proportional control is shown in Fig. 6-3. In such a system a compromise is often necessary in selecting a proper gain so that the steady state error and maximum overshoot are within acceptable limits. Practically, however, a compromise cannot always be reached since an optimum value of K may satisfy the steady state error but may cause excessive overshoot or even instability. This problem can be overcome if we employ proportional control in conjunction with some other type of control.

Derivative Control

When it is necessary to have an actuating signal which is proportional to the time derivative of the error signal, then we employ derivative control,

$$M(s) = K_d s E(s) \qquad (6.9)$$

where K_d is the gain of the controller. Although this form of control is very useful when it is necessary to increase system damping, it cannot be used alone since it does not respond if the error is constant. Therefore, it is used in combination with other controls. Let us consider here derivative and proportional control of a second-order system shown in Fig. 6-4. The output becomes

$$\frac{C(s)}{R(s)} = \frac{K_d s + 1}{As^2 + (B + K_d)s + 1} \qquad (6.10)$$

We immediately note that the effective damping of the system has increased. Since the overshoot is directly dependent upon the damping ratio, derivative control enables us to control system overshoot. The use of derivative control for improving system damping may be extended to the output signal as well. The inclusion of derivative control in the feedback[1] path of

---

[1] This often referred to as Rate-Feedback or Tachometric Control since a tachometer generator in the feedback yields the necessary derivative action.

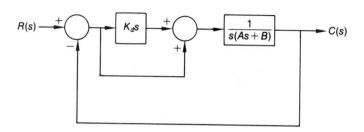

Figure 6.4: Control system with derivative and proportional control

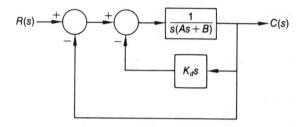

Figure 6.5: Derivative control in feedback loop

a second-order control system is shown in Fig. 6-5. The output becomes

$$\frac{C(s)}{R(s)} = \frac{1}{As^2 + (B + K_d)s + 1} \tag{6.11}$$

and the characteristic equation is identical to that of Eq. (6-10) and therefore we obtain the same control of the output overshoot. What then is the
difference between these two types of derivative control? The difference is
in the rise time of the response. Since Eq. (6-14) has an open loop zero,
whereas Eq. (6-15) does not, we see from Section 4-5 that the output rise-
time of the system shown in Fig. 6-4 is faster than for the one shown in
Fig. 6-5, when the input $r(t)$ is a step signal.

Integral Control

Sometimes it is desirable to completely eliminate positional error as in a
servomotor. In such cases a signal proportional to the integral of the error
is used. Consider the control system employing integral control in Fig. 6-6.

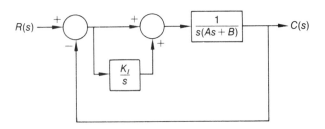

Figure 6.6: Feedback control system with integral control

The output becomes

$$\frac{C(s)}{R(s)} = \frac{s + K_1}{As^3 + Bs^2 + s + K_1} \qquad (6.12)$$

and the error becomes

$$E(s) = \frac{s^2(As + B)}{As^3 + Bs^2 + s + K_1}R(s)$$

We notice immediately that in applying integral control to a second-order system we have converted it to a third-order system. Whenever this is done, the system is no longer stable over all ranges of gain, i.e. if the gain is made sufficiently large, the system will become unstable. Although this is undesirable it is the price we pay for eliminating the steady state error, under load, which is

$$e_{ss}(t) = \lim_{s \to 0} sE(s)$$

and is zero for step and ramp inputs. For a unit parabolic input the steady state error is B/K_1. Without integral control the steady state error is a non-zero constant for a ramp input but infinite for a parabolic input.

Proportional, Derivative, and Integral Control

Knowing the advantages of individual types of control action, let us consider an example employing all three control actions. The second-order system shown in Fig. 6-7 has proportional, derivative, and integral control (PID).

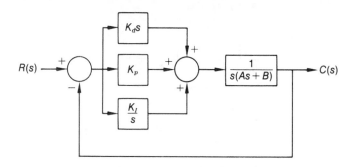

Figure 6.7: Control system with proportional, derivative, and integral control

The output and error become

$$C(s) = \frac{K_d s^2 + K_p s + K_i}{A s^3 + s^2(B + K_d) + K_p s + K_i} R(s)$$

$$E(s) = \frac{s^2(As + B)}{A s^3 + s^2(B + K_d) + K_p s + K_i} R(s)$$

(6.13)

With a proper choice of K_d and K_i we can get an oscillatory system with acceptable damping ratio. The error is zero for a step or ramp input. When selecting the parameters K_d and K_i we need to be careful about system stability; a topic which will be discussed in great detail in Chapter 7. Suffice for now to note that the Routh-Hurwitz criterion which is discussed in this chapter shows that the system is stable when the gain constants K_p, K_d and K_i satisfy the inequalities

$$K_d \geq 0, \quad K_i \geq 0, \quad K_d > \frac{A K_i}{K} - B$$

(6.14)

Examining the last inequality we can see that integral control has a destabilizing effect upon the system because we need to increase the damping coefficient K_d as K_i increases. So, in conclusion we observe derivative control enhances system damping, while integral control eliminates steady state errors but decreases the margin of stability.

Example 5 *A proportional plus derivative controller of the form shown in Fig. 6-8 is used in the forward loop of a unity feedback control system having*

$$G(s) = \frac{1}{s^2 + bs + a}$$

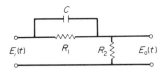

Figure 6.8: Proportional plus derivative controller

Obtain expressions for damping ratio and natural frequency after the controller is introduced.

The transfer function of the controller can be shown to be

$$G_c(s) = K_1 \left[\frac{1 + sT}{1 + sTK_1} \right]$$

where $T = R_1 C$ and $K_1 = R_2/(R_1 + R_2)$. If K_1 is small (say less than 0.1) and T is small (say about 0.05 sec.) then the controller transfer function can be approximated by

$$G_c = K_1(1 + Ts)$$

The new closed loop transfer function becomes

$$\frac{C(s)}{R(s)} = \frac{(1 + sT)K_1}{s^2 + (b + K_1 T)s + (a + K_1)}$$

The damping ratio and natural frequency become

$$\zeta = \frac{b + K_1 T}{2\sqrt{a + K_1}}, \quad \omega_n = \sqrt{a + K_1}$$

Example 6 *The control systems shown in Figs. 6-4 and 6-5 have the coefficients $A = 0.1$ and $B = 1$. Determine and compare the step responses of these systems for $K_d = 10$.*

The closed loop transfer functions for the systems having cascade and feedback compensation as shown in Figs. 6-4 and 6-5 are

$$\frac{C(s)}{R(s)} = \frac{10s + 1}{0.1s^2 + 11s + 1}$$

and

$$\frac{C(s)}{R(s)} = \frac{1}{0.1s^2 + 11s + 1}$$

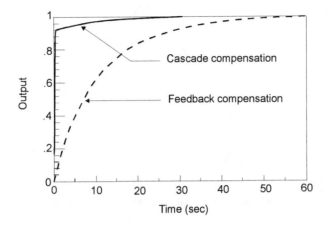

Figure 6.9: Step response of feedback systems having cascade and feedback compensation

respectively. The step responses for these systems are plotted in Fig. 6-9. It will be observed that the effect of the zero at $s = -0.1$ for the system with cascade compensation, as shown in Fig. 6-4, is to produce a significantly faster response time than the comparable system using feedback compensation as shown in Fig. 6-5. This result is consistent with the observations made, about the effects of zeros on a system transfer function, in Section 4-5. The effect on the closed loop transfer function of a control system by placing a compensator, in either the forward or feedback paths, can be exploited advantageously at times and will be discussed further in Chapter 9.

Example 7 *The liquid-level in the tank shown in Fig. 6-10 is controlled by an integral controller which consists of a hydraulic valve operated by a signal representing the height of the liquid. Obtain the overall transfer function* $H(s)/Q_i(s)$ *of this control system.*

For an incompressible fluid the mass balance yields

$$q_i + q - q_o = A\frac{dh}{dt}$$

where h is a deviation about a steady state position. We also have

$$q_o = \frac{\rho g h}{R}$$

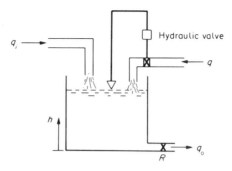

Figure 6.10: Fluid storage tank

and for the hydraulic valve the output signal x is given by

$$x = -\int K_v h\, dt$$

as discussed in Chapter 2. The negative sign accounts for the fact that h is measured positive in the opposite direction. The flow through the control valve is assumed to be a linear function of x, so that

$$q = -K\int K_v h\, dt$$

Substituting and rearranging we obtain

$$q_i = A\frac{dh}{dt} + \frac{\rho g h}{R} + K K_v \int h\, dt$$

where the integral controller term appears due to the hydraulic valve. The overall transfer function becomes

$$\frac{H(s)}{Q_i(s)} = \frac{Rs}{RAs^2 + \rho g s + K K_v R}$$

A system such as this is capable of maintaining a given level and eliminating steady state error. Note that if $K_v = 0$, i.e. no hydraulic controller, the transfer function becomes

$$\frac{H(s)}{Q_i(s)} = \frac{R}{RAs + \rho g}$$

and the order of the system drops to a first-order.

Figure 6.11: Sensitivity of open loop system output to changes in $G(s)$

6.5 Sensitivity Functions and Robustness

The behavior of a control system changes as its system parameters change, as the reference input varies, or due to loading effects. These parameters change due to environmental conditions, manufacturing defects, aging, etc. Since these changes generally go uncorrected, they affect the response of a system. If, however, the system has feedback, the effect of changes in system parameters which cause output changes will be reduced by feedback action. Since the quality of a control system depends upon its error reducing capability, we shall investigate its sensitivity to these disturbance functions. In addition to the effects of parameter variations, disturbance and measurement error signal inputs occur at various locations within a control system and these can also affect its operation. As will be seen it is often necessary, for satisfactory system operation, to seek to maximize the system response to reference inputs, while at the same time to minimize the sensitivity to parameter and disturbance inputs. One of the most challenging aspects of control system design is the need to seek acceptable compromises between these conflicting requirements.

Sensitivity Function

If a change in parameter K causes a change in M, then the fractional change in M, due to K, divided by the fractional change in K is defined as the **sensitivity function**

$$S_K^M = \frac{dM/M}{dK/K} = \frac{K}{M}\frac{dM}{dK} \tag{6.15}$$

If M and K are defined for the **open loop** system shown in Fig. 6-11, by

$$M = \frac{C(s)}{R(s)} = G(s), \quad K = G(s)$$

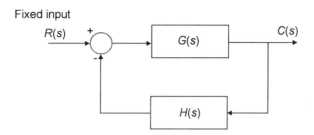

Figure 6.12: Sensitivity of feedback system output to changes in $G(s)$ and $H(s)$

then

$$S_G^M = \frac{G}{M}\frac{dM}{dG} = 1 \tag{6.16}$$

This relationship states that a change in $G(s)$ and a change in the open loop system output have a one-to-one ratio provided the input is fixed. It follows that there is also a one-to-one ratio in output changes caused by input changes.

The sensitivity with respect to the transfer function $G(s)$ for the **closed loop** system shown in Fig. 6-12 is obtained by defining

$$M = \frac{G(s)}{1 + G(s)H(s)}, \quad K = G(s)$$

Now since

$$\frac{dM}{dG} = \frac{1}{[1 + G(s)H(s)]^2}$$

the sensitivity function becomes

$$
\begin{aligned}
S_G^M &= \frac{G}{M}\cdot\frac{dM}{dG} = G\frac{(1+GH)}{G}\cdot\frac{1}{(1+GH)^2} \\
&= \frac{1}{1 + G(s)H(s)}
\end{aligned} \tag{6.17}
$$

Thus we see that the output of the system is less affected by changes in $G(s)$ than for the **open loop** case by the factor $(1 + G(s)H(s))$, provided its magnitude is much greater than unity. The effect on the closed loop system output due to parameter changes in the feedback element can be obtained by defining the sensitivity function

$$S_H^M = \frac{H}{M}\frac{dM}{dH}$$

where

$$M = \frac{G}{1 + GH}$$

Taking the derivative and substituting yields

$$S_H^M = \frac{-G(s)H(s)}{1 + G(s)H(s)} \tag{6.18}$$

Provided the magnitude of the factor $(1 + G(s)H(s))$ is much larger than unity we note that

$$S_H^M \approx -1$$

This result implies that there is a one-to-one ratio between changes in the output and changes in $H(s)$, indicating that the system is sensitive to changes in $H(s)$. This should be contrasted with the effect of changes in $G(s)$ which were shown to be small, so long as the magnitude of $(1 + G(s)H(s))$ is large. Because the term $(1 + G(s)H(s))$ arises frequently we define the **return difference**

$$F(s) = 1 + G(s)H(s) \tag{6.19}$$

so

$$S_G^M = 1/F(s), \quad S_H^M = -G(s)H(s)/F(s)$$

Disturbance and Measurement Noise Sensitivity

Not only does feedback decrease the effects of parameter changes on the system output but it also affects the influence of disturbances and measurement noise. To see this we examine the block diagram of the control system shown in Fig. 6-13 which shows the disturbance signal $D(s)$ and measurement noise $N(s)$, in addition to the reference input $R(s)$. From Fig. 6-13 we see that

$$C(s) = P(s)\frac{G(s)}{1 + G(s)H(s)}R(s) + \frac{1}{1 + G(s)H(s)}D(s) - \frac{G(s)H(s)}{1 + G(s)H(s)}N(s) \tag{6.20}$$

Denoting $S_G^M(s)$ and $-S_H^M(s)$, given in Eqs. (6-17) and (6-18), by $S(s)$ and $T(s)$ respectively, we will refer to $S(s)$ as the **system sensitivity**

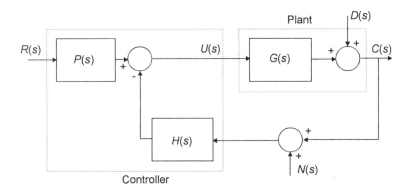

Figure 6.13: Feedback control system showing disturbance signal $D(s)$ and measurement noise $N(s)$

and to $T(s)$ as the **complementary sensitivity**. Taking the control ratio $C(s)/R(s)$ as $G_c(s)$, Eq. (6-20) becomes

$$C(s) = G_c(s)P(s)R(s) + S(s)D(s) - T(s)N(s) \qquad (6.21)$$

It is useful to note that if the control system is to be insensitive to disturbance inputs and measurement error signals then the system and complementary sensitivities must both be small. However examination of Eqs. (6-17) and (6-18) show that

$$S(s) + T(s) = 1 \qquad (6.22)$$

so if $S(s)$ is to be nearly zero then $T(s)$ must be approximately unity. This important relationship shows that a trade-off between the system and complementary sensitivities must be achieved for any feedback control system design. Since the disturbance $D(s)$ and input reference signal $R(s)$ have their frequency spectra concentrated at low frequencies while the measurement noise $N(s)$ is wideband, the conflicting requirements mentioned above can be resolved by taking

$$S(j\omega) \approx \begin{cases} 0 & \text{for} \quad 0 < \omega < \omega_o \\ \\ 1 & \text{for} \quad \omega_c < \omega \end{cases} \qquad (6.23)$$

and $T(j\omega)$ to be small at high frequencies, and approximately unity at low frequencies. This means the system best rejects the effects of the disturbance $D(s)$ and minimizes the effect of the measurement noise $N(s)$. A

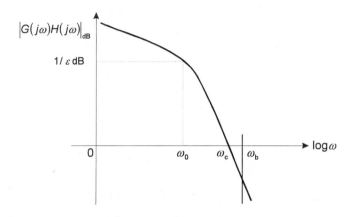

Figure 6.14: Open loop frequency response for control

typical open loop frequency response for the open loop system $G(j\omega)H(j\omega)$ which achieves these objectives is shown in Fig. 6-14. From Eq. (6-17) we see $|S(j\omega)| < \varepsilon$ for $0 < \omega < \omega_o$, and for $\omega > \omega_c$ we see $|S(j\omega)| \approx 1$. Plots of the corresponding system sensitivity $S(j\omega)$ and complementary sensitivity $T(j\omega)$ are shown in Fig. 6-15.

Now if $P(s) = H(s)$, which is the case for cascade compensation, the bandwidth ω_b of the complementary sensitivity also determines the bandwidth of the control ratio $C(s)/R(s)$. Thus since the system must also respond sufficiently fast to reference inputs $R(s)$ we see that we do not have complete freedom over the choice of the bandwidth for the complementary sensitivity. The interplay of these various limiting requirements determines in each case the final design outcome.

Example 8 *The output of a control system is given by*

$$C(s) = \frac{K}{s\left[(A+s)(B+s) + K\right]} = G$$

where it is known that K varies slowly. Obtain the effect of this change on system output.

The change in output is

$$\Delta C(s) = G(s)\left(1 - sG(s)\right)\frac{\Delta K}{K}$$

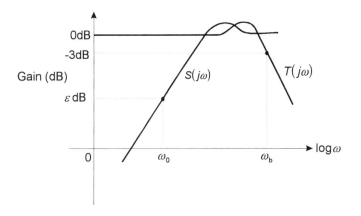

Figure 6.15: Gain characteristic for system sensitivity $S(j\omega)$ and complementary sensitivity $T(j\omega)$

The steady state value can be obtained from the final value theorem,

$$\Delta c(\infty) = \lim_{s \to 0} sG(s)(1 - sG(s))\frac{\Delta K}{K} = \frac{AB}{(AB + K)^2}\Delta K$$

Therefore if K changes by 1, the output varies by $AB/(AB + K)^2$ and this is the benefit of feedback. If we form the sensitivity function, we have

$$S_K^C = \frac{K}{C} \cdot \frac{\Delta C}{\Delta K} = 1 - sG(s)$$

For steady state values

$$\left[\frac{\Delta c}{\Delta K}\right](\infty) = \lim_{s \to 0} \frac{sC(s)}{K}[1 - sG(s)]$$

which yields the same result.

Example 9 *Obtain the sensitivity of the closed loop transfer function to changes in feedback gain K. Assume the forward loop transfer function to be G.*

The closed loop transfer function is

$$M = \frac{G}{1 + GK}$$

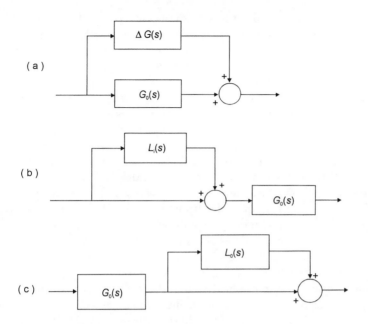

Figure 6.16: Methods of modelling systems having uncertainty in the plant tranfer function, (a) additive uncertainty, (b) input multiplicative uncertainty, (c) output multiplicative uncertainty

Then, the sensitivity becomes

$$S_K^M = \frac{dM/M}{dK/K} = \frac{K}{M}\frac{dM}{dK}$$

where

$$\frac{dM}{dK} = -\frac{G^2}{(1+KG)^2} = -M^2$$

so that

$$S_K^M = -MK$$

If KG is much larger than one, then the sensitivity function approaches -1.

System Uncertainty and Performance Robustness

As we have seen the uncertainty of a plant transfer function can lead to a change in the system output even when the reference input remains unchanged. In the analysis presented above we used differential methods to calculate these changes but these methods only give valid results when the plant uncertainties are very small. We now examine how finite plant uncertainties may be modeled for single-output systems and will use them to investigate the effect of plant uncertainties upon system performance.

The uncertainty of a plant may be represented by any of the models shown in Fig. 6-16, where in each case $G_o(s)$ is the nominal plant transfer function around which the actual plant transfer function is perturbed by $\Delta G(s)$, $L_i(s)$ or $L_o(s)$ respectively. The model shown in Fig. 6-16(a), where the variation in the plant is given by the additive term $\Delta G(s)$, is said to represent **additive uncertainty** in the plant, while the models shown in Figs. 6-16(b) and (c) represent **multiplicative uncertainties**. In these latter cases $L_i(s)$ and $L_o(s)$ represent multiplicative uncertainty relative to the plant input and output respectively. Let us now examine the **performance sensitivity** or **robustness** of a feedback design when uncertainties in the plant of the type shown in Fig. 6-16 are present. Knowledge of performance robustness gives insight into the accuracy of the plant model which is needed for a design to be successful.

The additive and multiplicative uncertainties of the plant also need to be characterized. We do this by defining an upper and lower bound on these uncertainties. Thus for the additive uncertainty $\Delta G(s)$ we suppose there is a real bounding function $v(s)$. The set of additive uncertainties denoted by $\mathcal{G}$ is defined as

$$\mathcal{G} = \{\Delta G : |\Delta G(s)| < v(s)\} \tag{6.24}$$

The set of multiplicative uncertainties can be characterized in a similar way.

In the following discussion we restrict our attention to additive plant uncertainties in the feedback system shown in Fig. 6-17. The analysis of systems with plant variations represented by multiplicative uncertainties follows along similar lines to the results given below.

The closed loop transfer function of the uncertain system is

$$\frac{C(j\omega)}{R(j\omega)} = G_c(j\omega) = \frac{K(j\omega)[G_o(j\omega) + \Delta G(j\omega)]}{1 + K(j\omega)[G_o(j\omega) + \Delta G(j\omega)]} \tag{6.25}$$

and we assume that $\Delta G(j\omega)$ belongs to the set $\mathcal{G}$. From Eq. (6-25) we find

$$|G_c(j\omega)| = \frac{|K(j\omega)||G_o(j\omega) + \Delta G(j\omega)|}{|1 + K(j\omega)[G_o(j\omega) + \Delta G(j\omega)]|} \tag{6.26}$$

Using the triangle inequality $|a + b| < |a| + |b|$, we find that the term $(G_o(j\omega) + \Delta G(j\omega))$ satisfies the inequalities

$$\max\{0, |G_o(j\omega)| - v(\omega)\} < |G_o(j\omega) + \Delta G(j\omega)| < |G_o(j\omega)| + v(\omega)$$

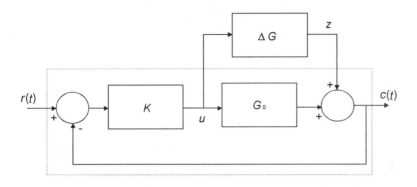

Figure 6.17: Feedback control system having additive plant uncertainty

and the denominator of Eq. (6-26) satisfies the inequalities

$$|1 + K(j\omega)[G_o(j\omega) + \Delta G(j\omega)]| < 1 + |K(j\omega)|[|G_o(j\omega)| + v(\omega)]$$

and

$$|1 + K(j\omega)[G_o(j\omega) + \Delta G(j\omega)]| >$$

$$\max\{0, 1 - |K(j\omega)|[|G_o(j\omega)| + v(\omega)], |K(j\omega)||[|G_o(j\omega)| - v(\omega)]| - 1\}$$

Substituting these results into Eq. (6-26) we find that the magnitude of the closed loop frequency response function $|G_c(j\omega)|$ satisfies the inequalities

$$|G_c(j\omega)| < G_{c,\max}(\omega) =$$

$$\frac{|K(j\omega)|[|G_o(j\omega)| + v(\omega)]}{\max\{0, 1 - |K(j\omega)|(|G_o(j\omega)| + v(\omega)), |K(j\omega)||[|G_o(j\omega)| - v(\omega)]| - 1\}} \tag{6.27}$$

and

$$|G_c(j\omega)| > G_{c,\min}(\omega) = \frac{|K(j\omega)|\max\{0, |G_o(j\omega)| - v(\omega)\}}{1 + |K(j\omega)|(|G_o(j\omega)| + v(\omega))} \tag{6.28}$$

These results may be easily used to graphically plot the bounds of the closed loop frequency response function $G_c(j\omega)$.

Example 10 *Consider the feedback control system shown in Fig. 6-17 where the nominal plant transfer function is*

$$G_o(s) = \frac{1}{5s + 1}$$

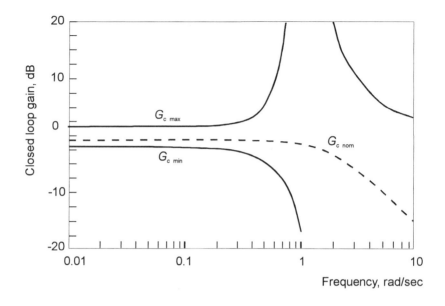

Figure 6.18: Performance robustness of position servo control system

Suppose it has a proportional controller with gain $K = 10$. Find the bounds of the closed loop frequency response function $G_c(j\omega)$ and graphically plot the system performance robustness, assuming the uncertainty in $G_o(j\omega)$ is defined by

$$v(\omega) = \frac{0.03\sqrt{(5\omega)^2 + 1}}{\sqrt{(0.5\omega)^2 + 1}}$$

From the inequalities (6-27) and (6-28) the upper and lower bounds for the closed frequency response are plotted in Fig. 6-18 for frequencies ω in the range 0.01 to 10 rad/s. Also plotted for reference is the frequency response function $|G_{c,\text{nom}}(j\omega)|$ for the nominal plant. From this figure it can be seen that for frequencies in the region $\omega = 1$ there is considerable uncertainty in the closed loop frequency response. However at low frequencies the closed loop system frequency response function band is quite narrow due to the fact that the uncertainty in the plant frequency response function is small at these frequencies. As a consequence it can be seen that the system has good performance robustness at low frequencies, but this robustness rapidly degrades at frequencies above $\omega = 0.3$ rad/s.

We have briefly examined the robustness of single-input single-output systems and their relationship to sensitivity functions here. More advanced discussions of these ideas are in the references.

6.6 Summary

In this chapter we defined some commonly used terms to describe control system specifications and directly related them to the system response in the time domain. The performance index was seen to be a quantitative measure of the error of a system. This index can be mathematically represented and is useful for optimizing system performance, which generally involves the minimization of some form of the error.

The effects of different types of controllers upon the transient and steady state responses of control systems have been examined. We noted how different types of controllers could be employed to enhance system damping and to eliminate steady state errors.

In the final section, the sensitivity to variations of inputs as well as system parameters was obtained. It was also shown how the sensitivity function can be related to the performance robustness of single-input single-output systems. Finally, it was observed that feedback tends to reduce the influence of parameter variations.

6.7 References

1. E. O. Doeblin, *Control System Principles and Design*, John Wiley and Sons, Inc., New York, 1985.

2. J. J. D'Azzo and C. H. Houpis, *Linear Control System Analysis and Design - Conventional and Modern*, Mc Graw-Hill Book Co., New York, 1988.

3. S. M. Shinners, *Modern Control System Theory and Application*, Addison Wesley Publ. Co., Reading, Mass., 1978.

4. R. E. Dorf, *Modern Control Systems*, Addison Wesley Publ. Co., Reading, Mass.,1989.

5. J. C. Doyle and G. Stein, "Multivariable Feedback Design; Concepts for Classical/Modern Synthesis," *IEEE Trans. Auto. Control*, Vol. AC-26, No. 1(1981), pp.4-16.

6. J. M. Maciejowski, *Multivariable Feedback Design*, Addison Wesley Publ. Co., Wokinghan, England, 1989.

7. W. H. Rosenbrock, *Computer-Aided Control System Design*, Academic Press, London, 1974.

6.8   Problems

6-1 An airplane is idealized as shown in Fig. P6-1. Derive the equation for the vertical motion of the plane as it lands, assuming that as it touches the ground it experiences a step force input. Obtain $x(t)$ and determine the natural and damped frequency, damping ratio, peak overshoot and the time at which it occurs, settling time and the number of oscillations to reach it. Neglect the weight of the wheels and assume

$$m = 10,000 \text{ kg}, \quad B = 10^5 \text{ N/m/sec}, \quad k = 10^6 \text{ N/m}$$

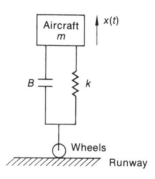

Figure P6-1

6-2 For the system shown in Fig. P4-1, calculate the maximum overshoot and rise time for a step input.

6-3 For the system investigated in Example 1, if the natural frequency is 20 rad/sec, then what damping ratio is necessary for the settling time to be 0.5 sec? How much overshoot results?

6-4 Compute J_1 for Example 1. Obtain the value of ζ that minimizes the performance index.

6-5 For the system shown in Example 1, plot J_4 as a function of ζ. Obtain the value of ζ that minimizes this performance index.

6-6 Show that for a second-order system, the minimization of J_3 yields $\zeta = 0.7$.

6-7 The error for a second-order system is given by

$$e(t) = [e^{-\zeta \omega_n t}\sin(\omega_d t + \phi)]/\sqrt{1 - \zeta^2}$$

where $\tan \phi = \sqrt{(1 - \zeta^2)}/\zeta$. Plot J_1 for this system if $\omega_n = 1$ for ζ between 0 and 1. What is the minimum value of J_1?

6-8 If B, k were to gradually change, due to aging of the shock absorbers and springs of the landing gear, what effect does this have on the steady state output of the airplane of Problem 6-1?

6-9 (a) What is the effect on the error and the output due to a change in $G_1(s)$ in Problems 3-8(a). (b) What is the effect on the output due to a change in $H_3(s)$ in Problems 3-8(f).

6-10 A control system employing proportional and derivative feedback control for stability is shown in Fig. P6-10. What must K be if the overshoot to a step input must not exceed the steady state value by more than 20%?

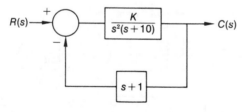

Figure P6-10

6-11 A second-order system is given by Fig. P6-11(a). For the same system, derivative plus proportional control is shown added in Fig. P6-11(b), integral plus proportional control is shown in Fig. P6-11(c), and all three in Fig. P6-11(d). For each case obtain $c(t)$ for a step input. Compare the results and discuss the merits of derivative and proportional control.

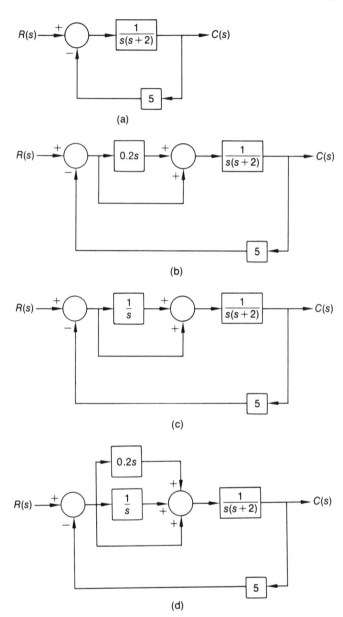

Figure P6-11

6-12 Derivative control is shown in the forward loop in Fig. P6-12(a) and in the feedback loop in Fig. P6-12(b). What is the essential difference? Obtain $c(t)$ for each and compare the results.

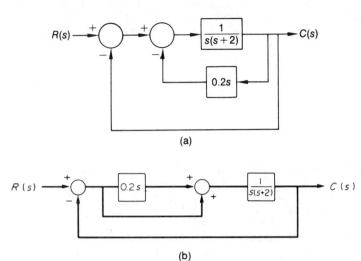

(a)

(b)

Figure P6-12

6-13 The plant of the system shown in Fig. P6-13, has the transfer function

$$G(s) = \frac{400}{s^2 + 20s + 400}$$

Find the steady state output $c(\infty)$ when $R(s) = 0$ and $D(s) = d/s$. When $R(s) = 1/s$, $D(s) = 0$, and $K_1 = 10$ find the range of permissible values of K_2 such that the change in the steady state output $c(\infty)$ is less than 5 percent.

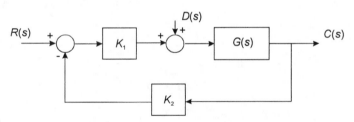

Figure P6-13

6-14 A feedback control system is shown in Fig. P6-14. Suppose the motor gain constant $K_m = 2 \pm 20\%$ and the tachometer sensitivity $K_t = 0.1 \pm 0.2\%$. Assume the command signal V_w is held constant. Determine the minimum value of the controller gain K which ensures that the changes in the output steady state speed $\omega(\infty)$ due to changes in K_m and K_t is less than 0.5%.

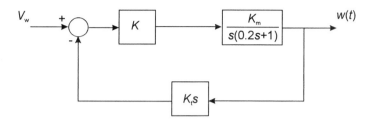

Figure P6-14

6-15 An operational amplifier circuit is shown in Fig. P6-15, where

$$A_o(s) = \frac{K}{\tau s + 1}$$

Because of manufacturing tolerances the amplifier gain K lies in the range $30,000$ to $60,000$ and the amplifier open loop bandwidth $\omega_b \ (= 1/\tau)$ lies in the range 1 to 5 rad/sec. Find bounds on the closed loop frequency response for the operational amplifier and graphically plot its performance robustness.

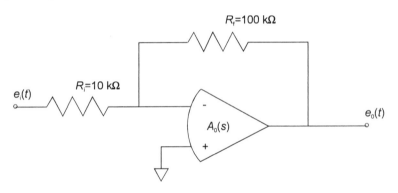

Figure P6-15

Chapter 7

Assessing Stability and Performance

7.1 Introduction

So far, the analysis of control systems has relied exclusively on several analytical techniques. System response and **stability** obtained via classical or state space methods, is dependent upon the closed loop poles. It is evident that even for relatively simple systems, the amount of algebra can be quite formidable. This is further aggravated when the characteristic polynomial is large. But that is not all. Assume that a control system has a characteristic polynomial of the form

$$s^3 + 4s^2 + (5 + K)s + 5K = 0$$

and we would like to obtain system performance for six values of K so that we can select a value that best suits our requirements. This requires that we obtain six sets of three closed loop poles! Additionally, we may not only want to know whether the system is absolutely stable but also about its **relative stability** as well as ways of improving system stability. Although the techniques developed so far can answer some of these questions, the amount of trial and error work as well as hand computation that must necessarily accompany these techniques becomes discouraging. Clearly, the motivation for looking for quicker ways becomes strong.

We can get around the difficulties posed in the above paragraph by either using a computer[1] for faster computation or by turning to analyt-

---

[1] A brief list of computer software packages for control system analysis which can operate on desk-top computers and engineering workstations is given in Appendix D.

ical and graphical methods for investigating control system stability and operation. Of the analytical techniques the Routh-Hurwitz criterion is particularly valuable as it can be used for deriving algebraic conditions for system stability in terms of the system coefficients and gains. Also, since the information obtained from graphical methods complements that obtained from time domain analysis, it is important that we understand the fundamentals of these methods. Generally, a complete analysis of a control system includes time domain analysis, as discussed above, in addition to the graphical approach to be discussed now. The power of the time domain as well as the graphical approach is greatly enhanced by relatively simple interactive computer programs mentioned above.

Although analysis based on a step input provides a uniform basis of comparison, the analysis of a system subjected to a sinusoidal input is quite common. While the system response will also be sinusoidal, its magnitude and its phase will differ from the input. The frequency response technique is a graphical method which considers the input to be a sinusoidally varying function. It is an easy, quick and powerful technique for obtaining the steady state response at specified frequencies. Furthermore, when the transfer function of a system is not known, it can be approximated using an experimentally determined frequency response of the system. This technique can also be extended to include a certain class of nonlinear problems as we shall show in a later chapter. We will also investigate the system stability in the real frequency domain, i.e. in terms of the frequency response of a system. We shall see that the frequency method of analysis enables us not only to study the relative stability of a system but is also a useful tool for modifying the system in order to vary this relative stability.

While the frequency response method dwells on the frequency response function of the system, another very useful graphical technique deals directly with the roots of the characteristic equation. This second method is called the root locus method. It shows graphically how the poles of the closed loop system migrate on the s-plane as the overall gain of the system is varied. Here it is possible to see directly what the dynamic performance of the system will be, and how it may be modified. Since we find the closed loop poles we can obtain the transient behavior of the system as well. Before we consider details of these graphical approaches it is perhaps useful to reiterate the following important points:

1. The response of a system is dependent upon the input, initial conditions, and the closed loop eigenvalues.

2. The absolute stability of a system is determined by the location of the closed loop eigenvalues.

3. The closed loop eigenvalues equal the closed loop poles, if there are no pole-zero cancellations.

4. The relative stability is very important and is dependent upon the closeness of the closed loop poles to the $j\omega$-axis.

5. Obtaining closed loop poles is generally difficult and often they depend upon design parameters.

6. The graphical methods, to be considered here, allow us to quickly determine the relative stability as well as response. Although a function of the hard to find closed loop poles, we shall determine response and relative stability from open loop poles which are known.

7.2 Stability via Routh-Hurwitz Criterion

We have seen how the response of a system is dependent upon the **characteristic equation zeros**. If the order of the characteristic polynomial is small, then it is generally not too difficult to obtain the zeros. However, if the order of the characteristic equation becomes large, then obtaining the zeros becomes nontrivial. The problem gets worse if some of the coefficients of the polynomial can vary and thereby cause some of the roots of the characteristic equation to move to the right half s-plane, as in the case of the positional servomechanism example considered in Section 4.2. It is therefore necessary that we have some means of knowing the admissible values of the coefficients of the characteristic equation that would insure system stability, i.e. to avoid roots that have real parts in the right half s-plane or multiple zeros on the imaginary axis. In this section we consider this problem and introduce a stability criterion that allows us to say whether the characteristic equation roots lie in the right half s-plane.

The **Routh-Hurwitz stability criterion** is a method for determining system stability from the characteristic equation.

Routhian Array

Consider the characteristic equation[2] of a control system

$$a_n s^n + a_{n-1} s^{n-1} + \cdots + a_0 = 0 \qquad (7.1)$$

---

[2]From the theory of equations we know that if any coefficient a_i is negative, then there must be roots with positive real parts. Also if any coefficient except a_0 is zero, then we either have roots with positive real parts or roots exist on the imaginary axis.

where a_n is positive and we form the **Routh array** as shown below

$$
\begin{array}{c|ccccccc}
s^n & a_n & a_{n-2} & \cdots & & & \\
s^{n-1} & a_{n-1} & a_{n-3} & \cdots & & \cdots & a_0 \\
\hline
s^{n-2} & b_1 & b_2 & \cdots & b_3 & \cdots & 0 \\
s^{n-3} & c_1 & c_2 & \cdots & 0 & & \\
s^{n-4} & d_1 & d_2 & \cdots & & & \\
\vdots & \vdots & & & & & \\
s^0 & a_0 & & & & &
\end{array}
$$

where

$$
b_1 = -\frac{\begin{vmatrix} a_n & a_{n-2} \\ a_{n-1} & a_{n-3} \end{vmatrix}}{a_{n-1}} \qquad b_2 = -\frac{\begin{vmatrix} a_n & a_{n-4} \\ a_{n-1} & a_{n-5} \end{vmatrix}}{a_{n-1}} \qquad b_3 = -\frac{\begin{vmatrix} a_n & a_{n-6} \\ a_{n-1} & a_{n-7} \end{vmatrix}}{a_{n-1}}
$$

$$
c_1 = -\frac{\begin{vmatrix} a_{n-1} & a_{n-3} \\ b_1 & b_2 \end{vmatrix}}{b_1} \qquad c_2 = -\frac{\begin{vmatrix} a_{n-1} & a_{n-5} \\ b_1 & b_3 \end{vmatrix}}{b_1} \qquad \cdots
$$

$$
d_1 = -\frac{\begin{vmatrix} b_1 & b_2 \\ c_1 & c_2 \end{vmatrix}}{c_1} \qquad d_2 = -\frac{\begin{vmatrix} b_1 & b_3 \\ c_1 & c_3 \end{vmatrix}}{c_1} \qquad \cdots
$$

$$(7.2)$$

The coefficients in the first two rows of the array are obtained from the characteristic polynomial, whereas the other coefficients are evaluated as indicated. In the course of evaluating a row, the terms of any row may be multiplied or divided by a positive quantity without altering the results. The coefficients in each row are evaluated until zero terms are obtained. Once the table is complete, the **Routh criterion** states that:

> **the number of zeros of the characteristic equation with positive real parts is equal to the number of sign changes in the first column of the coefficients in the Routh array.**

Consider the characteristic equation given by

$$F(s) = s^3 + 6s^2 + 12s + 8 = 0$$

The Routh array becomes

$$
\begin{array}{c|cc}
s^3 & 1 & 12 \\
s^2 & 6 & 8 \\
\hline
s^1 & 32/3 & 0 \\
3s^1 & 32 & 0 \\
s^0 & 8 &
\end{array}
$$

The coefficients in the first column are 1, 6, 32/3 , 8 and since there are no sign changes, then $F(s)$ has no zeros with positive real parts.

As another example consider the characteristic equation

$$s^3 + 3s^2 + 3s + 11 = 0$$

The Routh array becomes

s^3	1	3
s^2	3	11
s^1	$-2/3$	0
s^0	11	

There are two sign changes in the first column, therefore the characteristic equation has zeros that have positive real parts and the system will naturally be unstable. When there are sign changes, the stability criterion states that:

the number of zeros having positive real parts is equal to the number of sign changes in the first column.

Therefore in this example, the characteristic equation has two zeros which have positive real roots.

Since the coefficients in a characteristic equation are dependent upon control system parameters, the Routh test is very useful in ascertaining values of these parameters which insure stability. We see this in the next example.

Example 1 *For the servomechanism in Chapter 4, Example 6 we obtained the value of the amplifier gain A that ensures a stable response. The characteristic equation was*

$$F(s) = s^3 + 262.5s^2 + 12500s + 250A = 0$$

The Routh array becomes

s^3	1	12500	
s^2	262.5	$250A$	$b_1 = \dfrac{(12500)(262.5) - 250A}{262.5}$
s^1	b_1	b_2	$b_2 = 0$
s^0	c_1		$c_1 = 250A$

Since b_1 and c_1 must be positive for stability, we know that A must be positive and

$$[(262.5)(12500) - 250A] > 0 \text{ or } A < 13125$$

which is why in Example 6 of Chapter 4 the response was unstable, since we selected $A = 26.25 \times 10^4$. The application of the Routh-Hurwitz criterion therefore yields the range of system parameters that ensure stability.

Zero Coefficient in First Column

In constructing the Routh array, there is one problem that is often encountered which prevents us from completing the array. Referring back to the original array we observe that if any of the coefficients in the first column go to zero (except the last), then the next coefficient cannot be evaluated. Consider the following characteristic equation,

$$F(s) = a_6 s^6 + a_5 s^5 + a_4 s^4 + a_3 s^3 + a_2 s^2 + a_1 s + a_0 = 0$$

The Routh array is formed until we reach the fourth row. At this point the coefficient c_1 vanishes but $c_2 \neq 0$. We replace $c_1 = 0$ by $c_1 = \epsilon > 0$ in the array below and continue

s^6	a_6	a_4	a_2	a_0
s^5	a_5	a_3	a_1	
s^4	b_1	b_2	b_3	
s^3	$c_1 = 0$	c_2		
s^3	$c_1 \rightarrow \epsilon$	c_2		
s^2	d_1	d_2		
s^1	e_1	e_2		
s^0	f_1			

$$d_1 = \frac{\epsilon b_2 - b_1 c_2}{\epsilon}$$

$$d_2 = \frac{\epsilon b_3 - 0}{\epsilon} = b_3$$

As an example, consider the characteristic equation

$$F(s) = s^4 + 2s^3 + 4s^2 + 8s + 10 = 0$$

The array becomes

s^4	1	4	10
s^3	2	8	
s^2	0	10	
s^2	ϵ	10	
s^1	b_1	b_2	
s^0	c_1		

$$b_1 = \frac{8\epsilon - 20}{\epsilon}$$

$$b_2 = 0$$

$$c_1 = 10$$

Since ϵ is very small and positive, $(8\epsilon - 20)/\epsilon$ is negative. This means that there are two sign changes in the first column and therefore $F(s)$ has two roots with positive real parts.

All Coefficients of a Row are Zero

The previous method of replacing the coefficient cannot be used if *all* the coefficients of a row go to zero. Let us assume that the unfinished Routh array looks like

s^6	a_6	a_4	a_2	a_0
s^5	a_5	a_3	a_1	
s^4	b_1	b_2	b_3	
s^3	0	0	0	

The presence of such a row indicates the existence of two roots having the same magnitude but opposite sign. Any one of the following will cause a row to vanish: a pair of purely imaginary roots, four complex roots or a pair of real roots symmetrically located about the origin of the s-plane. We proceed with the construction of the Routh array by forming what is known as the **auxiliary equation**,

$$A_1(s) = b_1 s^4 + b_2 s^2 + b_3$$

Notice that the row containing the constants b_1, b_2 and b_3 appears above the row of zeros. The derivative of $A(s)$ is used to form a new s^3 row in the array

s^4	b_1	b_2	b_3
s^3	0	0	0
s^3	$4b_1$	$2b_2$	
s^2	c_1	$\cdots$	

$$\frac{dA_1(s)}{ds} = (4b_1) s^3 + (2b_2) s$$

$$c_1 = \frac{4b_1 b_2 - 2b_1 b_2}{4b_1} = \frac{b_2}{2}$$

and the procedure is continued as before. If no additional rows go to zero and the first column has no sign changes, the system is stable. However, if another row goes to zero as indicated in the following array

s^4	b_1	b_2	b_3
s^3	0	0	0
s^3	$(4b_1)$	$(2b_2)$	
s^2	c_1	c_2	0
s^1	0	0	

then we not only have another pair of roots that are the negative of each other but they are also equal to the previous roots. The auxiliary equation now becomes

$$A_2(s) = c_1 s^2 + c_2$$

The array is completed as follows

$$
\begin{array}{c|cc}
s^2 & c_1 & c_2 \\
\hline
s^1 & (2c_1) & 0 \\
s^0 & c_2
\end{array}
\qquad \frac{dA_2(s)}{ds} = (2c_1)\, s
$$

Although there are no sign changes we do have repeating roots. The roots of the characteristic equation may be obtained by solving the auxiliary equation,

$$A_2(s) = c_1 s^2 + c_2 = 0$$

$$s = \pm j \sqrt{\frac{c_2}{c_1}}$$

The two sets of roots, that are the negative of each other, are $\pm j\sqrt{c_2/c_1}$, and $\pm j\sqrt{c_2/c_1}$. When roots of this form appear, we expect to see a response in the time domain having the form $t\sin\left(\sqrt{c_2/c_1}\,t\right)$, i.e. oscillatory motion whose magnitude increases linearly. Clearly this is an unstable phenomena. We therefore modify our Routh criterion to state that a **stable system requires no sign changes in the first column and at most one vanishing row.**

Example 2 *The forward loop transfer function of a unity feedback control system is given by*

$$G(s) = \frac{K}{s(s+4)(s+10)}$$

Obtain the value of K for marginal stability and the frequency of oscillations at this value of K.

The characteristic equation is given by

$$1 + G(s) = 1 + \frac{K}{s(s+4)(s+10)} = 0$$

This yields

$$s^3 + 14s^2 + 40s + K = 0$$

The Routh array is formed as follows

$$
\begin{array}{c|cc}
s^3 & 1 & 40 \\
s^2 & 14 & K \\
\hline
s^1 & b_1 & b_2 \\
s^0 & c_1
\end{array}
\qquad
\begin{aligned}
b_1 &= \frac{560-K}{14} \\
b_2 &= 0 \\
c_1 &= K
\end{aligned}
$$

For the system to be marginally stable $b_1 = 0$ so that $K = 560$. The frequency of oscillations can be obtained from the auxiliary equation

$$A(s) = 14s^2 + K = 0$$

Substituting $K = 560$ we obtain $s = \pm j6.32$. The frequency of oscillation is 6.32 rad/s when the system is marginally stable.

Example 3 *A control system has the following characteristic equation*

$$F(s) = s^7 + 2s^6 + s^5 + 2s^4 - s^3 - 2s^2 - s - 2 = 0$$

Construct the Routh array and determine system stability.

The Routh array becomes

s^7	1	1	-1	-1
s^6	2	2	-2	-2
s^6	1	1	-1	-1
s^5	0	0	0	0

(Dividing by 2) appears beside the s^6 row.

The auxiliary equation and its derivative are

$$A_1(s) = s^6 + s^4 - s^2 - 1; \quad \frac{dA_1(s)}{ds} = 6s^5 + 4s^3 - 2s$$

The array is continued as

s^6	1	1	-1	-1
s^5	6	4	-2	
s^5	3	2	-1	
s^4	1/3	$-2/3$	-1	
s^4	1	-2	-3	
s^3	8	8		
s^2	-3	-3		
s^2	-1	-1		
s^1	0	0		

(New row) and (Divide by 2) appear beside the s^5 rows; (Multiply by 3) beside the s^4 row; (Divide by 3) beside the s^2 row.

We have another vanishing row indicating a set of equal and opposite roots. Again forming the auxiliary equation

s^2	-1	-1
s^1	-2	0
s^0	-1	

$$A_2(s) = -s^2 - 1$$
$$\frac{dA_2(s)}{ds} = -2s$$

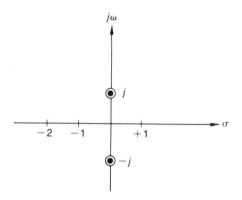

Figure 7.1: Roots of $s^7 + 2s^6 + s^5 + 2s^4 - s^3 - 2s^2 - s - 2$

The roots that appear as a set may be obtained by solving the auxiliary equation

$$A_2(s) = -s^2 - 1 = 0; \quad \text{so } s = \pm j$$

All the roots of the polynomial are shown in Fig. 7-1. Since there was one sign reversal in the first column one root exists in the right half s-plane. The system is unstable.

Stability of a State Space System

In the state space method discussed in Chapter 5 it was seen that the system response and its stability was a function of the **characteristic equation**,

$$\Delta(s) = \det[s\mathbf{I} - \mathbf{A}] = 0$$

As the above is a polynomial equation in s the **Routh-Hurwitz criterion** can be used for investigating whether the state space system is stable by ascertaining whether any of the eigenvalues of the $\mathbf{A}$ matrix have positive real parts or are repeated on the imaginary axis.

Example 4 *A state space system has the following coefficient matrix* $\mathbf{A}$

$$\mathbf{A} = \begin{bmatrix} 0 & 1 & 0 \\ 0 & -2 & 1 \\ -2 & -3 & -1 \end{bmatrix}$$

Determine whether this system is stable by using the Routh-Hurwitz criterion.

The characteristic equation becomes

$$\det[s\mathbf{I} - \mathbf{A}] = \begin{bmatrix} s & -1 & 0 \\ 0 & s+2 & -1 \\ 2 & 3 & s+1 \end{bmatrix} = 0$$

or

$$s^3 + 3s^2 + 5s + 2 = 0$$

Constructing the Routhian array

s^3	1	5
s^2	3	2
s^1	13/3	0
s^0	2	0

Since there are no sign reversals of the coefficients in the first column, by Routh's criterion the system is stable.

Stability Diagram

The Routh-Hurwitz criterion can be used to determine the influence of various system parameters on the stability of a feedback control system. It is quite useful for performing parametric studies during the system design stage when the interactions of the system parameters upon stability are being investigated. When the interaction of two parameters is being considered they may be taken as the coordinates of a **stability diagram** which is a plot of the regions of stability as the parameters vary.

Example 5 *The open loop transfer function for a disk drive memory head positioning control system is*

$$G(s)H(s) = \frac{K(s+a)}{s(s+130)(s^2 + 140s + 10^4)}$$

Find the conditions needing to be satisfied by the gain K and the compensation coefficient a for the system to be stable and plot the regions of stability on a stability diagram.

The characteristic equation of the system is

$$1 + G(s)H(s) = 1 + \frac{K(s+a)}{s(s+130)(s^2 + 140s + 10^4)} = 0$$

Simple algebraic manipulation shows that the zeros of the characteristic equation are given by the roots of the polynomial equation

$$s^4 + 270s^3 + 28200s^2 + (K + 1.3 \times 10^6)s + Ka = 0$$

Constructing the Routhian array

s^4	1	28200	Ka
s^3	270	$K + 1.3 \times 10^6$	0
$270s^2$	$6.314 \times 10^6 - K$	$270Ka$	
s^1	c_3	0	
s^0	$270Ka$		

Here

$$c_3 = \frac{(6.314 \times 10^6 - K)(K + 1.3 \times 10^6) - 270^2 Ka}{6.314 \times 10^6 - K}$$

For the control system to be stable there must be no sign reversals of the coefficients in the first column, thus we must have

$$K < 6.314 \times 10^6$$

$$(6.314 \times 10^6 - K)(K + 1.3 \times 10^6) > 72900Ka$$

and

$$Ka > 0$$

These inequalities are plotted on the stability diagram shown in Fig. 7-2, where the stable regions are indicated and the remaining regions are unstable. When the intersection of the ordinates a and K lie inside *stable* regions then the control system is stable. However it is not possible to draw any conclusions about the degree of stability of the system from this diagram.

7.3   Frequency Response Method

In Section 4-4 the frequency response of a system was introduced, and it was shown that the frequency response function $G(j\omega)$ can be derived from the system transfer function $G(s)$ by using Eq. (4-43). As shown in Eq. (4-40), apart from representing the frequency response function analytically it can be equally represented by its magnitude and phase responses $|G(j\omega)|$ and $\phi(\omega)$ respectively. Their graphical representation, either as **Bode response plots** which were discussed briefly in Chapter 4, or as a **Nyquist response plot**, have proven to be particularly useful in the analysis of control systems.

Bode Response Plot

We saw in Section 4-4 that the Bode response plot can be easily constructed using algebraic expressions derived from $G(s)$. It was also seen that the

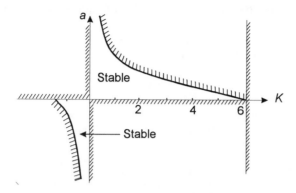

Figure 7.2: Stability diagram for a disk drive memory head positioning control system

same results can be obtained from the pole-zero plot on the s-plane. The Bode method will now be further discussed, and we will examine graphical methods for the construction of the Bode response plots.

The Bode response plot is also referred to as the **corner plot** or the **logarithmic plot**. This method of plotting frequency response information employs logarithms of functions so that multiplication and division are reduced to addition and subtraction and the work to obtain the response is largely graphical instead of analytical. This graphical approach makes the analysis of complex transfer functions with many poles and zeros straightforward. Varying the system gain on the Bode response plot simply involves the raising or lowering of the entire magnitude versus frequency plot, with respect to a reference, while the phase shift versus frequency remains unchanged.

The Bode response plot consists of two plots, viz. phase shift and the logarithm of the magnitude plotted as functions of frequency. The logarithm of the magnitude of a transfer function $G(j\omega)$ or for short the **log magnitude** of $G(j\omega)$, when expressed in decibels is defined in Section 4-4 and will be denoted by $|G(j\omega)|_{dB}$.

Since the log magnitude is also a function of frequency, a convenient way to express frequency bands is necessary. When the frequency varies[3] from ω_1 to ω_2 where $\omega_2 = 10\omega_1$, then the frequency band is referred to as a **decade**. The band from 1 Hz to 10 Hz or from 3.14 to 31.4 Hz is one

---

[3] The octave band is one where $\omega_2 = 2\omega_1$. This is not widely used.

decade. The number of decades in a given frequency band is given by

$$\frac{\log \omega_2/\omega_1}{\log 10} = \log \frac{\omega_2}{\omega_1} \text{ decades}$$

We observe that if $G(j\omega)$ increases by tenfold, or one decade, then the log magnitude increases by 20 dB.

In general, the transfer function of a system consists of four basic types of terms in the numerator and denominator. These are

$$K$$
$$(j\omega)^{\pm n}$$
$$(1 + j\omega T)^{\pm m}$$
$$(1 + j2\omega\zeta T + (j\omega T)^2)^{\pm p}$$

If we construct the curves of the log magnitude and angle versus frequency for each individual factor, then the Bode response plot of the entire transfer function can be obtained by adding the contribution of each term.

Constant or K Factor   The constant K is independent of frequency, and

$$K_{\text{dB}} = 20 \log K \text{ decibels} \tag{7.3}$$

appears as a horizontal line that raises or lowers the log magnitude curve of the complete transfer function by a fixed amount. There is no contribution on the phase shift since K is real and positive.

$(j\omega)^{\pm n}$ **Factor**   The log magnitude and phase shift of $(j\omega)^{\pm n}$ is

$$\begin{aligned}
\left| (j\omega)^{\pm n} \right|_{\text{dB}} &= \pm 20n \log \omega \\
\arg (j\omega)^{\pm n} &= \pm \frac{n\pi}{2}
\end{aligned} \tag{7.4}$$

The magnitude plot consists of a straight line whose slope $is \pm 20n$ dB/decade and passes through 0 dB at $\omega = 1$. The phase shift is a constant $\pm n\pi/2$.

$(1 + j\omega T)^{\pm m}$ **Factor**   As a first case let us assume that $m = 1$ and consider the negative exponent. Then

$$\begin{aligned}
\left| (1 + j\omega T)^{-1} \right|_{\text{dB}} &= -20 \log \sqrt{1 + \omega^2 T^2} \\
\arg (1 + j\omega T)^{-1} &= -\arctan \omega T
\end{aligned} \tag{7.5}$$

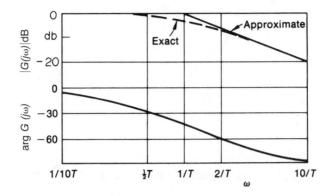

Figure 7.3: Bode plot for $(1 + j\omega T)^{-1}$

For very small values of ωT the log magnitude becomes

$$\left|(1 + j\omega T)^{-1}\right|_{\text{dB}} = -20 \log 1 = 0$$

i.e. for small frequencies the log magnitude is represented by the 0 dB line. When $\omega T \gg 1$,

$$\left|(1 + j\omega T)^{-1}\right|_{\text{dB}} \approx \left|(j\omega T)^{-1}\right|_{\text{dB}} \approx -20 \log \omega T$$

which is zero for $\omega = 1/T$ and a straight line with -20 dB/decade slope for $\omega T > 1$. The point defined by the intersection of the line obtained for $\omega T > 1$ and $\omega T < 1$ is $\omega = 1/T$. This point is called the **corner (break) frequency** and is shown in Fig. 7-3. The exact curve is shown by a dotted line. The error between the exact curve and the straight lines (or the **asymptotes**) is greatest at the corner frequency. In practice, the actual curve is drawn freehand with corrections inserted at $\omega = 1/T$, $\omega = 1/2T$, and $\omega = 2/T$. The log magnitude correction is shown in Fig. 7-4.

Before considering the next type of factor, we return to the case where $(1 + j\omega T)^{\pm m}$ occurs. The corner frequency is still at $\omega = 1/T$ except that the asymptote defined for $\omega T > 1$ has a slope of $\pm 20m$ dB/decade. This is the only difference. When two or more factors of this form appear, we simply add the contribution of each. For example, consider the transfer function

$$G(s) = (1 + sT_1)^{-1}(1 + sT_2)^{-1}$$

The magnitude and phase are

$$|G(j\omega)|_{\text{dB}} = -20 \, \log \sqrt{1 + \omega^2 T_1{}^2} - 20 \, \log \sqrt{1 + \omega^2 T_2{}^2}$$

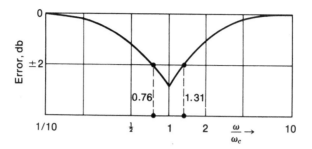

Figure 7.4: Log magnitude correction for $(1 + j\omega T)^{\pm 1}$

$$\arg G\left(j\omega\right) = -\arctan \omega T_1 - \arctan \omega T_2$$

If we assume that $T_2 < T_1$, then the contribution of both terms is 0 dB until $\omega = 1/T_1$ is reached. Now the first term contributes -20 dB/decade until $\omega = 1/T_2$ is reached. Up to this point the second factor has no contribution to the asymptotic behavior of the plot. At $\omega = 1/T_2$ (another corner frequency) the second term is approximated by $-20 \log |\omega T_2|$ and is represented by a straight line of -20 dB/decade slope. However, since this reinforces the first factor, the magnitude becomes a straight line of -40 dB/decade slope. Clearly, if the exponent of the first factor of $G(j\omega)$ had been positive, then the line would have been a horizontal line, or 0 dB/decade, after the second corner frequency. The log magnitude plot and phase shift plot for corner frequencies at $1/T_1$ and $1/T_2$ is shown in Fig. 7-5.

$\left(1 + j2\zeta\omega T + \left(j\omega T\right)^2\right)^{\pm p}$ **Factor**   As before, the case of $p = -1$ is considered without loss of generality. If the quadratic term can be factored yielding real roots, then we may use the technique developed in the previous section. However, if the roots are complex conjugates, then the entire factor should be plotted without factoring. In either case the gain and the phase shift are given by

$$
\begin{aligned}
|G(j\omega)|_{\mathrm{dB}} &= -20 \ \log[(1 - \omega^2 T^2)^2 + (2\zeta\omega T)^2]^{1/2} \\
\arg G\left(j\omega\right) &= -\arctan \frac{2\zeta\omega T}{1 - \omega^2 T^2}
\end{aligned}
\tag{7.6}
$$

Now if ω is very small, the log magnitude is zero. As ω increased the log magnitude asymptote has a slope of -40 dB/decade. The point of

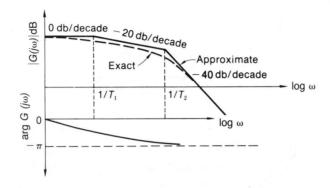

Figure 7.5: Bode plot for $G(j\omega) = \dfrac{1}{(j\omega T_1 + 1)(j\omega T_2 + 1)}$

intersection of the 0 dB line and the -40 dB/decade line occurs at $\omega = 1/T$ and this is the **corner frequency**. The phase shift is zero at $\omega = 0$, increases to $-\pi/2$ at the corner frequency, and finally goes to $-\pi$ as ω approaches infinity. This is shown in Fig. 7-6.

In the above case, when $\omega = \omega_n$, resonant conditions tend to be set up. This is why ω_n is often referred to as the **natural frequency**. Depending upon the value of the damping ratio ζ, considerable error may be introduced when comparing the actual Bode response plot to the straight line asymptotes. The exact plot with ζ as a parameter is shown in Figs. 4-30 and 7-7. The maximum value of the magnitude seen in Fig. 7-7 is often referred to as the **resonant peak** or **gain** M_m and has been shown in Eqs. (4-49) and (4-48) to be

$$M_m = \frac{1}{2\zeta\sqrt{1 - \zeta^2}}$$

and to occur at frequency ω_m, given by,

$$\omega_m = \omega_n\sqrt{1 - 2\zeta^2}$$

When the exponent for this type of factor becomes positive, the log magnitude and phase shift simply change sign.

Example 6 *Construct the Bode response plot for the following transfer function*

$$G(s) = \frac{0.25(1 + 0.5s)}{s(1 + 2s)(1 + 4s)}$$

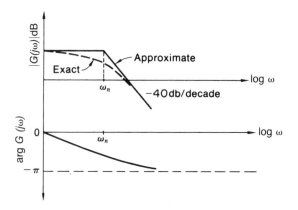

Figure 7.6: Bode plot for Eq. (7.6)

The log magnitude and phase shift are

$$|G(j\omega)|_{dB} = 20 \log 0.25 + 20 \log \sqrt{1 + 0.25\omega^2} - 20 \log \omega$$

$$-20 \log \sqrt{1 + 4\omega^2} - 20 \log \sqrt{1 + 16\omega^2}$$

$$\arg G(j\omega) = \arctan 0.5\omega - \frac{\pi}{2} - \arctan 2\omega - \arctan 4\omega$$

There are three corner frequencies at $\omega_1 = 1/4$, $\omega_2 = 1/2$ and $\omega_3 = 2$. The magnitude and phase plots are seen to vary as follows:

For $\quad 0 \leq \omega \leq \omega_1 \quad |G(j\omega)|_{dB} = -20 \log \omega + 20 \log 0.25 = \text{Lm}_1$

$$\arg G(j\omega) = -\pi/2 = \arg G_1$$

For $\quad \omega_1 \leq \omega \leq \omega_2 \quad |G(j\omega)|_{dB} = \text{Lm}_1 - 20 \log 4\omega = \text{Lm}_2$

$$\arg G(j\omega) = \arg G_1 - \arctan 4\omega = \arg G_2$$

For $\quad \omega_2 \leq \omega \leq \omega_3 \quad |G(j\omega)|_{dB} = \text{Lm}_2 - 20 \log 2\omega = \text{Lm}_3$

$$\arg G(j\omega) = \arg G_2 - \arctan 2\omega = \arg G_3$$

For $\quad \omega_3 \leq \omega \leq \infty \quad |G(j\omega)|_{dB} = \text{Lm}_3 + 20 \log 0.5\omega$

$$\arg G(j\omega) = \arg G_3 + \arctan 0.5\omega$$

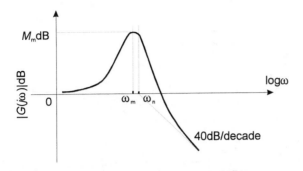

Figure 7.7: Bode amplitude response for an under-damped second-order system

The Bode plot is shown in Fig. 7-8.

System Type and Low Frequency Response

Assuming the transfer function of a system is in the form of Eq. (4-21), the **System Type** ℓ and gain constant K can be deduced from its Bode response plot, which can be of great value if this plot is measured experimentally. If the system is of Type ℓ then it can be easily seen that the Bode amplitude response will have a slope of -20ℓ dB/decade at low frequencies. To illustrate this consider the Type 0, 1 and 2 systems

$$G(j\omega) = \frac{K_0}{1 + j\omega T_a} \quad \text{(type 0)} \tag{7.7}$$

$$G(j\omega) = \frac{K_1}{(j\omega)(1 + j\omega T_a)} \quad \text{(type 1)} \tag{7.8}$$

$$G(j\omega) = \frac{K_2}{(j\omega)^2(1 + j\omega T_a)} \quad \text{(type 2)} \tag{7.9}$$

The asymptotic Bode response plots are shown in Fig. 7-9, and it will be noted that for frequencies much less than $\omega_a = 1/T_a$ the amplitude response has a slope of 0, -20, and -40 dB/decade respectively. It will also be noted that when $\omega << \omega_a$ the low frequency gain of the Type 0 system equals K_0, which can be easily read from an experimental plot. For the Type 1 and 2 systems the extensions of the low frequency asymptotes having slopes of -20 and -40 dB/decade intersect the 0 dB axis at frequencies ω_1 and ω_2

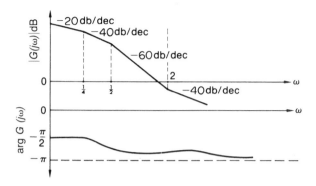

Figure 7.8: Bode plot for $G(s) = \dfrac{0.25(1 + 0.5s)}{s(1 + 2s)(1 + 4s)}$

respectively. From Eqs. (7-8) and (7-9) it can be seen that $K_1 = \omega_1$ and $K_2 = \omega_2^2$.

These observations concerning the initial frequency response slope and gain are not only useful when the response is obtained experimentally, but are also useful when we are constructing the Bode response plots from the system transfer function.

Relationship Between Amplitude and Phase Responses

The transfer function analyzed in Example 6 is said to be of minimum phase. In general the transfer function of any stable system which has no zeros or poles in the right half of the s-plane is said to be a **minimum phase transfer function**, while if it has one or more zeros in the right half then we say it is **non-minimum phase**. Bode studied the properties of stable systems having minimum phase transfer functions, and showed that the amplitude and phase responses of their frequency response functions are uniquely related. This is stated in the celebrated **phase-area theorem**.

Let the transfer function $G(s)$ of a system be of minimum-phase. The corresponding frequency response function $G(j\omega)$ can be written in polar form as

$$G(j\omega) = M(\omega)e^{j\,\phi(\omega)} \tag{7.10}$$

where $M(\omega) = |G(j\omega)|$ is said to be the gain or amplitude response of $G(j\omega)$. Taking the logarithm of $G(j\omega)$ we have

$$\ln G(j\omega) = \ln M(\omega) + j\phi(\omega) \tag{7.11}$$

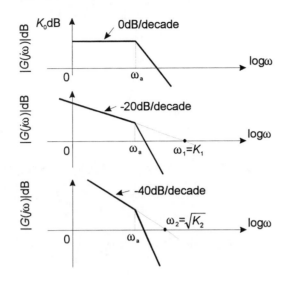

Figure 7.9: Bode response plots for Type 0, 1, and 2 systems

It can be shown that the phase response $\phi(\omega)$ is related to the amplitude response $M(\omega)$ by the relation

$$\phi(\omega) = \frac{1}{\pi} \int_{-\infty}^{\infty} \frac{d \ln M(y)}{du} \ln \coth \frac{|u|}{2} du \qquad (7.12)$$

where $u = \ln(y/\omega)$ or $y = \omega e^u$. Since the term $\ln M(y)$ is proportional to the log magnitude of $G(j\omega)$ Eq. (7-12) states that the phase response at any frequency depends upon the slope of the log-magnitude curve of $G(j\omega)$ plotted against $\log \omega$. The relative importance of the difference frequencies is determined by the weighting function $\ln \coth(|u|/2)$ plotted in Fig. 7-10.

Example 7 *Suppose the magnitude of a frequency response function $G(j\omega)$ varies as ω^n for all frequencies. Show that the phase response has the constant value $n\pi/2$ at all frequencies.*

We can write the gain $M(\omega)$ of $G(j\omega)$ as

$$M(\omega) = M(1)\omega^n$$

In this case $\ln M(\omega) = n \ln \omega + \ln M(1)$ and as a consequence $d \ln M(y)/du = n$. From Eq. (7-12)

$$\phi(\omega) = \frac{n}{\pi} \int_{-\infty}^{\infty} \ln \coth \frac{|u|}{2} du = n\frac{\pi}{2}$$

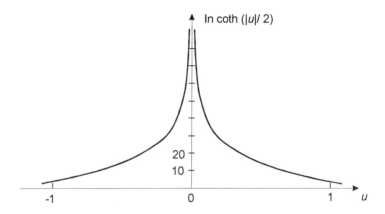

Figure 7.10: Plot of weighting factor $\ln \coth \dfrac{|u|}{2}$

If $M(\omega) = M(1)\omega^n$ for all frequencies, then from above the slope of $M(\omega)$ is $20n$ dB/decade if plotted in decibels along the ordinate and $\log_{10}\omega$ along the abscissa. As a consequence we see that if the amplitude response has a slope of $20n$ dB/decade then the phase response will be $n\pi/2$.

In general the Bode amplitude response for a frequency response function does not have a constant slope for all frequencies. Because however the weighting function in Eq. (7-12) is concentrated about $u = 0$ the phase response approaches $n\pi/2$ even if the amplitude response is approximately $20n$ dB/decade for only a limited range of frequencies.

Example 8 *The Bode amplitude response for a control system plant has been obtained experimentally and is shown in Fig. 7-11. Find the approximate phase response.*

We approximate the response curve by a series of straight lines having slopes of $20n$ dB/decade, where n is an integer. The break frequencies for the plant poles and zeros are at the intersections of the straight line segments. Taking the approximate phase response as $n\pi/2$ in the interval between each break point we obtain the discontinuous function shown by the broken lines. The actual phase response may be as shown by the superimposed smooth curve.

Example 9 *Obtain the transfer function for the experimentally derived frequency response shown in Fig. 7-12. Assume that the system is minimum phase.*

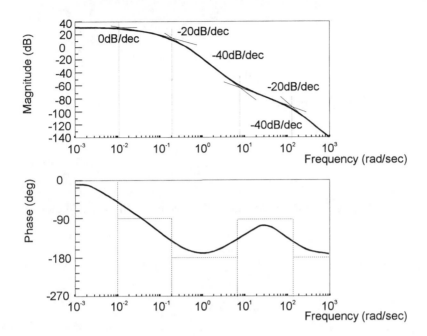

Figure 7.11: Approximate phase response for experimentally obtained minimum-phase amplitude response

We construct asymptotes as shown in the figure. The intersections of these asymptotes occur at the corner frequencies $\omega = 2, 4, 15, 20$, and 36. A positive slope indicates the existence of a zero and a negative slope indicates the existence of a pole. The factor that would yield -20 dB/decade between $\omega = 0$ and 2 is s^{-1}. Between $\omega = 2$ and 4, $(s + 2)$ would give a $+20$ dB/decade but this when added to the contribution of s^{-1} would yield a line having a slope of 0 dB/decade. The above argument can be carried through for the remaining corner frequencies. Finally we note the system is of Type 1, so comparing the low frequency response with the plot shown in Fig. 7-9, we see that the transfer function gain $K = 12.5$. Thus the overall transfer function becomes

$$G(s) = \frac{12.5(s/2 + 1)(s/4 + 1)^2}{s(s/15 + 1)(s/20 + 1)(s/36 + 1)}$$

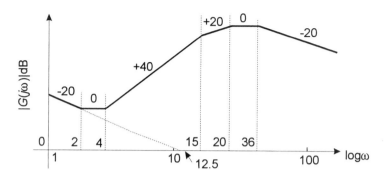

Figure 7.12: An experimentally derived log-magnitude plot

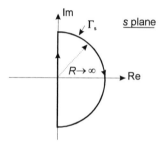

Figure 7.13: The Nyquist contour on the s-plane

Contour Mapping and the Nyquist Plot

The plot of $G(s)$ on the complex plane for values of s along the Nyquist contour Γ_s in a clockwise direction, as shown in Fig. 7-13, defines a contour plot of the function which will be referred to as a **polar** or **Nyquist plot**. If the number of poles $G(s)$ exceeds the number of zeros the points s on the semicircular path of infinite radius all map into the origin on the image plane. In this case the plots may be generated either by obtaining the magnitude and phase shift from the Bode response plot or by computing directly from the transfer function. If the Nyquist plot is obtained from the Bode response plot, care must be exercised in the conversion of decibels to the magnitude.

Plots of $G(j\omega)$ on the complex plane are referred to as direct plots, whereas those of $G(j\omega)^{-1}$ are referred to as inverse plots. The technique

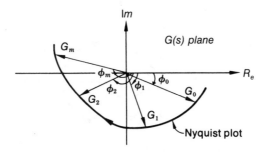

Figure 7.14: A Nyquist plot of $G\left(j\omega\right)$

of generating either the direct or the inverse plot is identical, therefore we shall only consider direct plots.

In general, the magnitude and phase shift of $G(j\omega)$ is

$$|G(j\omega)| \;=\; \frac{K \prod_z (1+\omega^2 T_z{}^2)^{1/2}}{\omega^m \prod_p (1+\omega^2 T_p{}^2)^{1/2}} \tag{7.13}$$

$$\arg G\left(j\omega\right) \;=\; \sum_z \arctan\omega T_z - \frac{m\pi}{2} - \sum_p \arctan\omega T_p$$

The magnitude and phase are computed for specific frequencies and plotted as shown in Fig. 7-14. A line joining all the permissible values of $|G(j\omega)|$ is the required plot. In general, $G(s) \to 0$ as $|s| \to \infty$, so we can obtain the Nyquist plot by varying the frequency from $-\infty$ to $+\infty$ as will be seen later. However, since the Nyquist plot is symmetrical about the real axis, the limit $0 \le \omega \le \infty$ suffices.

Example 10 *Construct the Nyquist plot for*

$$G(s) = \frac{K}{1+sT}$$

The magnitude and phase shift become

$$|G(j\omega)| = \frac{K}{\sqrt{1+\omega^2 T^2}}$$

$$\arg G\left(j\omega\right) = -\arctan\omega T$$

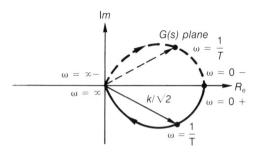

Figure 7.15: Nyquist plot for $G(s) = \dfrac{K}{1+sT}$

For specific values of ω the following table is constructed,

ω	$|G(j\omega)|$	$\arg G(j\omega)$
$-\infty$	0	$\pi/2$
$\vdots$	$\vdots$	$\vdots$
$-2/T$	$K/\sqrt{5}$	
$-1/T$	$K/\sqrt{2}$	$\pi/4$
0	K	0
$1/T$	$K/\sqrt{2}$	$-\pi/4$
$2/T$	$K/\sqrt{5}$	
$\vdots$	$\vdots$	$\vdots$
∞	0	$-\pi/2$

The Nyquist plot is shown in Fig. 7-15.

Consider next, when a transfer function has a pole at $s = 0$. In this case the Nyquist plot is constructed by taking the contour map of the Nyquist contour. However since there is a pole at $s = 0$ we need to deform the contour on the s-plane in the neighborhood of this pole.

Example 11 *Construct the Nyquist plot for*

$$G(s) = \frac{K}{s(1 + Ts)}$$

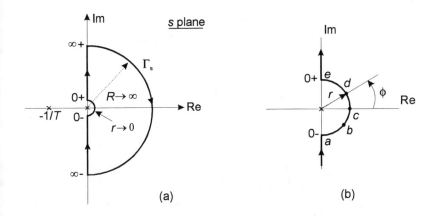

Figure 7.16: Nyquist contour with deformed semi-circular arc in neighborhood of origin

As mentioned above, to construct the Nyquist plot we need to deform the Nyquist contour in the neighborhood of $s = 0$ as shown in Fig. 7-16(a), where the radius r of the small semicircular path is assumed to approach zero.

For points along the imaginary axis where $0 < |\omega| < \infty$ the magnitude and phase shift of $G(j\omega)$ are given by

$$|G(j\omega)| = \frac{K}{\omega\sqrt{1 + \omega^2 T^2}}$$

$$\arg G(j\omega) = -\frac{\pi}{2} - \arctan \omega T$$

Consider now points along the small semi-circular arc in the neighborhood of the origin as shown in Fig. 7-16(a). An expanded view of the contour in this region is shown in Fig. 7-16(b) and points on this arc are given by $s = re^{j\phi}$. Substituting in the transfer function and assuming $r \to 0$ gives

$$G(s) \approx \frac{K}{r} e^{-j\phi}$$

for $-\pi/2 < \phi < \pi/2$. The points a, b, c, d, and e on this arc map into corresponding points at infinity as shown in Fig. 7-17.

Finally since the number of poles of $G(s)$ exceeds the number of zeros, all points along the semicircular path where $R \to \infty$ map into the origin in the $G(s)$ plane. The complete Nyquist plot for $G(s)$ is shown in Fig. 7-17.

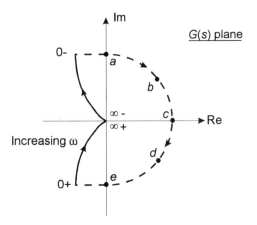

Figure 7.17: Nyquist plot for $G(s) = \dfrac{K}{s(1 + Ts)}$

We note in the above example that at $\omega = 0+$ the magnitude goes to infinity and the phase shift is $-\pi/2$. If we have n poles at the origin, then the initial phase shift will be $-n\pi/2$ and the final phase shift will be $-(n + 1)\pi/2$ for this type of transfer function. In each of the previous examples we saw that the gain K had no effect on the overall shape of the polar plot since it is a common factor to all the points. As K is varied the plot simply contracts (decreasing K) or dilates (increasing K) as shown in Fig. 7-18. Examination of the plots obtained so far show that they often cross the real and/or the imaginary axis. The point where the crossover occurs is a function of the frequency. Consider the open loop transfer function

$$G(s) = \frac{K}{s(1 + sT_1)(1 + sT_2)(1 + sT_3)}$$

Substitution of $s = j\omega$ and writing $G(j\omega)$ as a complex number,

$$
\begin{aligned}
G(j\omega) &= \frac{K}{j\omega(1 + j\omega T_1)(1 + j\omega T_2)(1 + j\omega T_3)} \\[2mm]
&= \frac{1}{P(\omega)}[R(\omega) + jI(\omega)]
\end{aligned}
$$

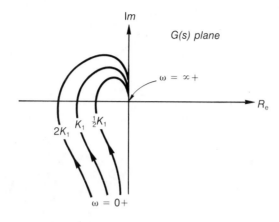

Figure 7.18: Effect of varying system gain

where

$$R(\omega) \;=\; \omega^2 K[T_1 T_2 T_3 \omega^2 - (T_1 + T_2 + T_3)]$$

$$I(\omega) \;=\; \omega K[\omega^2 (T_1 T_2 + T_1 T_3 + T_2 T_3) - 1]$$

The solution of the equations $R(\omega) = 0$ and $I(\omega) = 0$ will yield the frequencies at which the imaginary and real axes are crossed. In this problem the imaginary axis intercept is

$$R(\omega) = 0; \quad \omega = \omega_R = \pm \left[\frac{T_1 + T_2 + T_3}{T_1 T_2 T_3} \right]^{\frac{1}{2}}$$

and the real intercept occurs at

$$I(\omega) = 0; \quad \omega = \omega_I = \pm \left[\frac{1}{T_1 T_2 + T_1 T_3 + T_2 T_3} \right]^{\frac{1}{2}}$$

Since the limit under investigation is $0 \leq \omega \leq \infty$ we select the positive roots. Note that the points of real and imaginary axis intercepts are independent of K but are functions of the system time constants as shown in Fig. 7-19. In all the examples so far there have been no zeros in $G(s)$. An interesting Nyquist plot occurs for a third-order system having $G(s)$ with a zero at the origin

$$G(s) = \frac{Ks}{(1 + sT_1)(1 + sT_2)(1 + sT_3)}$$

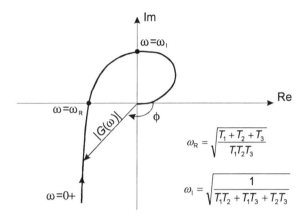

Figure 7.19: Plot of $G(s) = \dfrac{K}{s(1+sT_1)(1+sT_2)(1+sT_3)}$

The magnitude of $G(j\omega)$ is zero at $\omega = 0$ and $\omega = \infty$. The phase shift is $\pi/2$ at $\omega = 0$ and $-\pi$ at $\omega = \infty$. If the zero is at $s = -A$, then the magnitude is KA at $\omega = 0$ and zero at $\omega = \infty$. The phase in this case varies from zero to $-\pi$. This is shown in Fig. 7-20. The real and imaginary axis intercepts may be computed by setting the imaginary and real part of $G(s)$ equal to zero as before.

The closed loop response can be obtained from the Nyquist plot. This is discussed in a later section when we speak of relative stability and performance in the frequency domain. But first we will consider the system stability via a criterion applied on the Nyquist plot.

The Nyquist Stability Criterion

A stability criterion for application in the frequency domain was developed by H. Nyquist and is based upon a theorem in the theory of complex variables. This theorem is Cauchy's theorem and it is concerned with the mapping of contours in the s-plane. We shall present the criterion but without any formal proof.

The stability of a system may be studied by investigating the **characteristic equation**

$$F(s) = 1 + P(s) = 0 \qquad (7.14)$$

where $P(s)$ is a rational function of s. For a system to be stable no zeros of $F(s)$ can be in the right hand half of the s-plane. This may be ascertained

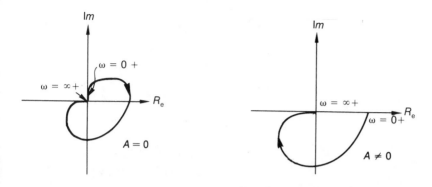

Figure 7.20: Nyquist plot for $G(s) = \dfrac{Ks(s+A)}{(1+sT_1)(1+sT_2)(1+sT_3)}$

if the right half s-plane is mapped on to the $F(s)$-plane and the Nyquist criterion is invoked.

The Nyquist criterion can be derived using the Cauchy theorem which is commonly called the **principle of the argument** in complex variable theory. Fortunately, the consequences of the theorem may be understood without reviewing the formal proofs.

In the discussion below when a contour Γ is traversed in the **clockwise** direction the points **inside** the contour are taken as those on the **right-hand side** of Γ. (The convention described here is the one commonly used in control system theory, although it is the opposite of the one normally adopted in complex variable theory. Either convention is valid so long as it is consistently applied). We say a point P is **encircled** by a contour Γ if in tracing out its complete path, the net number of rotations made by a vector from P to Γ is non-zero. Otherwise it is **not encircled**. Examples are shown in Fig. 7-21.

Let us now assume $F(s)$ is a single valued rational function. It must also be analytic in the s-plane except at some finite number of points. You will recall that these points are the poles of $F(s)$. Now suppose that we select a closed path Γ_s on the s-plane such that $F(s)$ is **analytic everywhere on this enclosed path.** We can now generate a corresponding locus on the $F(s)$-plane, i.e. for every value s_i on the closed path Γ_s we compute and plot a corresponding point $F(s_i)$ on the $F(s)$-plane. The resulting contour on the $F(s)$-plane is also a closed path and is labeled Γ_F as shown in Fig. 7-22. The locus Γ_F is called the contour map of Γ_s.

Now we use the **principle of the argument**, which states that:

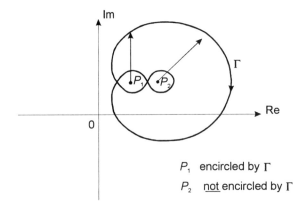

Figure 7.21: Encirclement of a point P by a contour Γ

If the contour Γ_s encircles Z zeros and P poles of $F\left(s\right)$ on the s-plane in a clockwise direction then the $F\left(s\right)$ locus Γ_F encircles the origin on the $F\left(s\right)$-plane $N = Z - P$ times in the clockwise direction if N is positive. If N is negative the encirclement in the $F\left(s\right)$-plane is counter-clockwise.

Depending upon Z and P, N may be zero, negative, or positive. If the contour on the s-plane encircles more zeros than poles, then $N > 0$. If the number of poles is equal to the number of zeros, then $N = 0$. Finally if the number of poles is greater than the number of zeros then $N < 0$. When N is negative, the $F(s)$ contour encircles the origin N times in a counterclockwise direction.

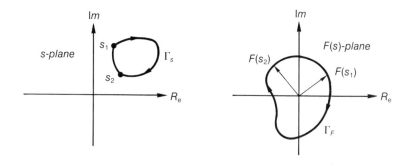

Figure 7.22: Conformal mapping

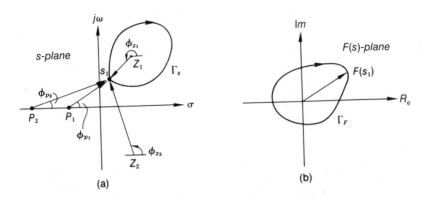

Figure 7.23: $F(s)$ encloses the origin $Z - P$ times

Consider the function $F(s)$ given by

$$F(s) = \frac{(s + z_1)(s + z_2)}{(s + p_1)(s + p_2)}$$

Writing this as magnitude and phase,

$$|F(s)| = \frac{|s + z_1| \cdot |s + z_2|}{|s + p_1| \cdot |s + p_2|}$$

$$\arg F(s) = \phi_{z1} + \phi_{z2} - \phi_{p1} - \phi_{p2}$$

where ϕ_{z1} is the angle due to the zero at z_1 and so on. Now let us select the path shown on the s-plane in Fig. 7-23(a). For $s = s_1$, the vector $F(s_1)$ is shown in Fig. 7-23(b). As we move around the contour on the s-plane along the indicated path, beginning from s_1 and returning to s_1, the net angle change is zero. However, as Γ_s traverses 2π on the s-plane, ϕ_{z1} also traverses 2π in the same direction. If the path is enlarged to include Z_2, then ϕ_{z2} also traverses 2π radians. The two zeros together traverse $2(2\pi)$ radians. In general if all the zeros are enclosed, then for Z zeros we have $Z(2\pi)$ traversals. Following the same argument, if the poles are enclosed we may say that the poles contribute $P(2\pi)$ traversals but in the opposite direction. The net traversals that the $F(s)$ contour experiences becomes

$$N(2\pi) = Z(2\pi) - P(2\pi)$$

or

$$N = Z - P \qquad\qquad (7.15)$$

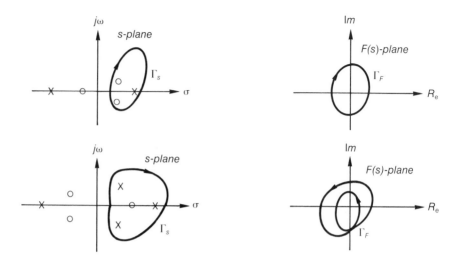

Figure 7.24: $F(s)$ enclosing the origin for different pole-zero configurations

i.e. the number of enclosures is equal to the difference of the number of zeros and poles in the right half s-plane.

As an example consider a function $F(s)$ having the pole zero configuration shown in Fig. 7-24(a). Since there is one pole and two zeros inside Γ_s, the origin on the $F(s)$-plane is encircled once by Γ_F. The pole-zero configuration of $F(s)$ in Fig. 7-24(b) indicates one zero and three poles inside Γ_s, therefore, Γ_F encircles the origin twice but in the opposite direction.

We investigate stability of a control system by considering the characteristic equation

$$F(s) = 1 + G(s)H(s) = 0 \qquad (7.16)$$

and requiring that the roots of this characteristic equation may not exist in the right half s-plane, i.e. to the right of the imaginary axis. Therefore, if we select Γ_s to include the entire right half s-plane, then we may employ the principle of argument to determine whether any poles or zeros of $F(s)$ exist in this region. Let us define Γ_s which we referred to above as the **Nyquist path**, to consist of four paths shown in Fig. 7-25. These paths are

1. From $-j\infty$ to $-j0$ along $j\omega$ axis

2. From $-j0$ to $+j0$ along the semicircle with infinitesimal radius

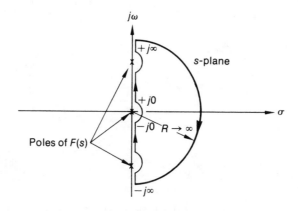

Figure 7.25: The Nyquist path

3. From $+j0$ to $+j\infty$ along $j\omega$ axis

4. From $+j\infty$ to $-j\infty$ along the semicircle with infinite radius.

We note that since the Nyquist path may not pass through any poles of $F(s)$ we must avoid them on the path by deforming it as shown in the figure.

The Nyquist method is concerned with the mapping of $F(s)$

$$F(s) = 1 + G(s)H(s) = 1 + \frac{K \prod_z (s + z_z)}{\prod_p (s + p_p)} \qquad (7.17)$$

and the consequent encirclement of the origin on the $F(s)$-plane. It is rather bothersome to plot the function $F(s) = 1 + G(s)H(s)$, whereas the **open loop transfer function** $G(s)H(s)$ is generally readily available either as a rational function in s, or as Bode amplitude and phase plots which may have been derived from the open loop transfer function or obtained experimentally. We define the conformal transformation $f(s)$ as

$$f(s) = F(s) - 1 = G(s)H(s)$$

and map the contour Γ_F onto the $f(s)$-plane, as the Γ_f contour. It is clear from Fig. 7-26 that the contour Γ_F encircles the origin if and only if Γ_f encircles the point $(-1, 0)$. With these observations in mind, we state the Nyquist stability criterion:

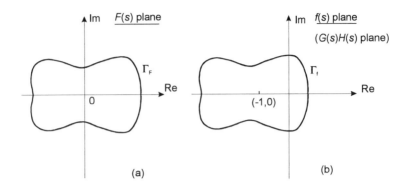

Figure 7.26: Mapping from the $F(s)$ to the $f(s)$ planes using the mapping function $f(s) = F(s) - 1$

> **A control system is stable if the number of counter-clockwise encirclements of the $(-1,0)$ point by the $GH(j\omega)$ plot is equal to the number of poles of $GH(s)$ with positive real parts.**

If the system is to be stable and the number of poles of GH in the right half s-plane is zero, then $N = 0$, i.e. the $GH(j\omega)$ plot may not encircle the $(-1,0)$ point. If the number of poles with positive real parts is 1, then

$$N = Z - 1$$

and for the stability we require that $Z = 0$, therefore

$$N = -1$$

For a stable system, the Nyquist criterion requires that $N = -P$ where N is the number of encirclement of the $(-1,0)$ point by the GH plot. Because of the central role the point $(-1,0)$, on the $GH(j\omega)$-plane, plays in the statement of the Nyquist criterion, it is usually referred to as the **critical point**.

Application of the Nyquist Criterion

The application of the Nyquist criterion to the stability analysis of linear control systems is best understood by considering several examples.

As shown earlier the Nyquist plots are constructed for all points along the Nyquist contour shown in Fig. 7-13, which is equivalent to constructing

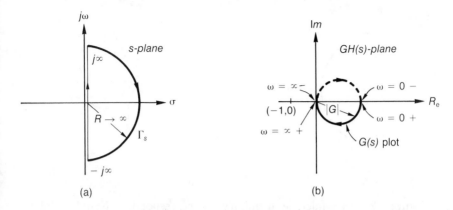

Figure 7.27: Nyquist plot of $G(s) = K/(s+a)$

it for both negative and positive frequencies. This is quite straightforward, especially if we note that the plot for negative frequencies will be a mirror image about the real axis of the positive frequency plot.

Example 12 *Determine the stability of the unity feedback control system with the forward loop transfer function given by*

$$G(s) = \frac{K}{s+a}$$

The magnitude and phase are given as

s	$|G(s)|$	$\arg G(s)$
0	K/a	0
js_1	$K/\sqrt{s_1^2 + a^2}$	$-\arctan\dfrac{s_1}{a}$
$j\infty$	0	$-\dfrac{\pi}{2}$

The Nyquist plot is shown in Fig. 7-27. Note that the plot for negative frequencies is simply the image about the real axis. Since $G(s)$ does not have poles with positive real parts, the $G(s)$ plot should not enclose the point $(-1,0)$ for stability. This in fact is the case and the system is stable for all gain.

Next we consider the transfer function with a pole at $s = 0$

$$G(s) = \frac{K}{s(s+a)}$$

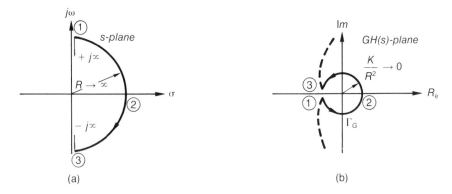

Figure 7.28: Γ_s semicircle at infinity and corresponding Nyquist plot Γ_G

The origin on the s-plane must now be avoided by the Γ_s contour as shown in Example 11. The magnitude and phase are computed,

s	$|G(s)|$	$\arg G(s)$
0	∞	$-\dfrac{\pi}{2}$
js_1	$k/s_1\sqrt{s_1^2 + a^2}$	$-\dfrac{\pi}{2} - \arctan\dfrac{s_1}{a}$
$j\infty$	-0	$-\pi$

We may use this information to construct the Nyquist plot as s varies from $-j\infty$ to -0 and $+0$ to $+j\infty$. In order to complete the Nyquist plot we need two additional paths, viz. one as s varies from $-j0$ to $+j0$ and the second as s varies from $+j\infty$ to $-j\infty$.

As shown in Example 11 we construct a semi-circular deformation of the Nyquist contour about the origin as shown in Fig. 7-16, where the radius $r \to 0$, so that points along this contour map into points at infinity as shown in Fig. 7-17.

We now consider the semicircle having infinite radius on the s-plane. Representing this as a vector

$$s = Re^{+j\phi}$$

Substituting this in $G(s)$

$$G(s) = \frac{K}{Re^{+j\phi}(Re^{+j\phi} + a)}$$

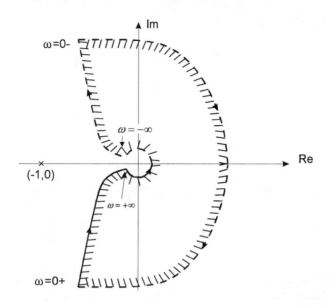

Figure 7.29: Nyquist plot for $G(s) = \dfrac{K}{s(s+a)}$

$$G(s) \approx \frac{K}{R^2 e^{+j2\phi}} = \frac{K}{R^2} e^{-j2\phi}$$

for $-\frac{\pi}{2} < \phi < \frac{\pi}{2}$. The last approximation is valid since $R >> a$. Now as we follow Γ_s along the semicircle with infinite radius, ϕ varies from $+\pi/2$ to $-\pi/2$ through zero. Meanwhile $G(s)$ varies from $-\pi$ to $+\pi$ through zero while having zero magnitude. Notice that the angle variation is twice that of s on the s-plane. This is shown in Fig. 7-28. Combining the two paths just considered we obtain the complete Nyquist plot shown in Fig. 7-29. We notice that the $(-1,0)$ point is not encircled. Since $P = 0$ and $N = 0$, the system is stable. Although it is interesting to observe the behavior of the $GH(j\omega)$ plot around the origin, we can generally ascertain system stability without completing this part of the plot.

The Nyquist plots of the above examples do not cross the negative real axis as the phase shift always exceeds $-\pi$ except at $\omega = \infty$. For this reason the point $(-1,0)$ is not encircled.

Example 13 *Determine the stability of a control system with the following* *open loop transfer function*

$$G(s)H(s) = \frac{K}{s(sT_1 + 1)(sT_2 + 1)}$$

The magnitude and phase are given by

s	$|G(s)H(s)|$	$\arg(G(s)H(s))$
0	∞	$-\dfrac{\pi}{2}$
js_1	$K/s_1\sqrt{(s_1^2T_1^2 + 1)(s_1^2T_2^2 + 1)}$	$-\dfrac{\pi}{2} - \arctan s_1 T_1 - \arctan s_1 T_2$
$j\infty$	0	$-\dfrac{3\pi}{2}$

and the Nyquist plot appears in Fig. 7-30. Note that as Γ_s goes around the origin in the s-plane, $GH(j\omega)$ goes from $+\pi/2$ to $-\pi/2$ with infinite radius. Since $P = 0$, the system is stable if $(-1, 0)$ is not enclosed as shown in Fig. 7-30(a). However, when K is sufficiently increased the point $(-1, 0)$ is enclosed and the systems becomes unstable as seen in Fig. 7-30(c). When $K = K_c$ the system is marginally stable and oscillates.

The effect of K on the system stability may be investigated using the Routh-Hurwitz criterion. The characteristic equation is

$$1 + \frac{K}{s(sT_1 + 1)(sT_2 + 1)} = 0$$

or

$$\frac{[s(sT_1 + 1)(sT_2 + 1) + K]}{(sT_1 + 1)(sT_2 + 1)} = 0$$

The zeros of the characteristic equation are determined by

$$s(sT_1 + 1)(sT_2 + 1) + K = 0$$

or

$$s^3 T_1 T_2 + s^2(T_1 + T_2) + s + K = 0$$

Forming the Routh array

s^3	$T_1 T_2$	1	
s^2	$(T_1 + T_2)$	K	$b_1 = \dfrac{T_1 + T_2 - KT_1 T_2}{T_1 + T_2}$
s^1	b_1	b_2	$b_2 = 0$
s^0	c_1		$c_1 = K$

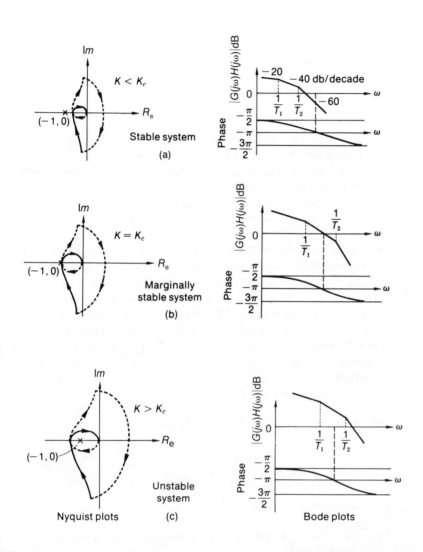

Figure 7.30: Stability of minimum phase function $G(s)H(s) = \dfrac{K}{s(sT_1 + 1)(sT_2 + 1)}$ on Nyquist plot and Bode plot

For system stability $b_1 \geq 0$, therefore

$$K \leq K_c; \quad K_c = (T_1 + T_2)/T_1 T_2$$

If $b_1 = 0$, we form the auxiliary equation to investigate the purely oscillatory behavior of the system,

$$s^2(T_1 + T_2) + K_c = 0$$

$$s = j\omega_\phi = \pm j\sqrt{\frac{1}{T_1 T_2}}$$

We conclude then that when $K < K_c$ the system is stable, when $K = K_c$ the system exhibits pure oscillations with a frequency ω_ϕ, and finally when $K > K_c$ the system becomes unstable. We note that all the plots have the same shape and when the plot goes through $(-1, 0)$ the system exhibits pure oscillations. When this happens, the system is called a marginally stable system.

Since the Nyquist plots for different values of K have the same shape it is a considerable nuisance to reconstruct them for each K when investigating system stability. It is more convenient to write the open loop transfer function as an explicit function of K and to restate the Nyquist criterion in terms of a moving critical point. We observe that the open loop transfer function can be written as $G(s)H(s) = K[G(s)H(s)]'$ where the term in brackets is independent of K. Referring to Eq. 7-17 we could write $F(s) = 1/K + [G(s)H(s)]'$, in which case it is easy to see the Nyquist criterion can be stated in terms of encirclements by the contour $[G(s)H(s)]'$ of the point $(-1/K, 0)$ rather than the point $(-1, 0)$.

Example 14 *Determine the stability of the control system with the open loop transfer function*

$$G(s)H(s) = \frac{K}{s(sT_1 + 1)(sT_2 + 1)}$$

using the modified Nyquist plot.

We can write the **modified open loop transfer function** as

$$[G(s)H(s)]' = \frac{1}{s(sT_1 + 1)(sT_2 + 1)}$$

and then plot its Nyquist contour as shown in Example 13. This is simply equivalent to plotting $G(s)H(s)$ with $K = 1$. From our discussion above the critical point is at $(-1/K, 0)$. A typical Nyquist plot for the above

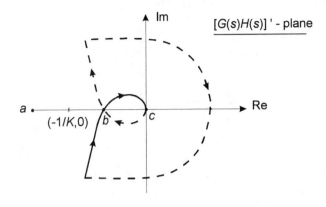

Figure 7.31: Modified Nyquist plot with critical point at $(-1/K, 0)$

system is shown in Fig. 7-31. If the critical point lies in the region ab then the system is stable because the number of clockwise encirclements of $(-1/K, 0)$ is zero. On the other hand if the critical point lies in the region bc, there are two clockwise encirclements of the critical point, so that the control system is unstable with two closed loop poles in the right half of the s-plane.

Let us for a moment review the results obtained so far. We began with the forward loop transfer function, $G(s) = K/s(s+a)$, and saw that the system was stable for all K. Then we added a pole to the transfer function and the system was no longer stable for all K. If the gain is increased without limit, the system becomes unstable. We may conclude therefore that the addition of poles to the open loop transfer function tends to have a destabilizing influence on the system response.

When the number of poles of a control system becomes large the system exhibits many stable and unstable responses depending upon system gain. Consider the transfer function of such a control system,

$$G(s)H(s) = \frac{N_1(s)}{N_2(s)} = \frac{K(s\tau_5 + 1)(s\tau_6 + 1)}{s(s\tau_1 + 1)(s\tau_2 + 1)(s\tau_3 + 1)(s\tau_4 + 1)}$$

The Nyquist plot is shown in Fig. 7-32. Since $P = 0$, we want no enclosures of $(-1/K, 0)$ to assure system stability. As the gain K varies the critical point $(-1/K, 0)$ will move along the real axis, and consequently there are four regions where this point might lie which need to be considered. For low gains the system is stable. As the gain is raised it becomes unstable. If the gain is raised some more it again becomes stable and finally gets unstable

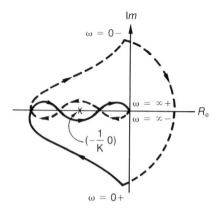

Figure 7.32: Nyquist plot of the open loop transfer function $G(s)H(s) = \dfrac{K(s\tau_5 + 1)(s\tau_6 + 1)}{s(s\tau_1 + 1)(s\tau_2 + 1)(s\tau_3 + 1)(s\tau_4 + 1)}$, with $K = 1$

for very high gain.

We have been concerned with Nyquist plots of systems having rational polynomials up to this point. However, many control systems have a time delay in the loop and this affects the system stability. Examples of systems having time delay are motors, valve delay, transmission and transportation delays and so on. In the next example we consider a control system with delay and show how the Nyquist criterion may be used to investigate its stability.

Consider the control system shown in Fig. 7-33 where $e^{-2\pi\tau s}$ charac-

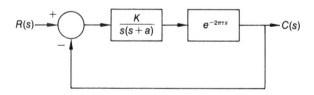

Figure 7.33: System with time delay

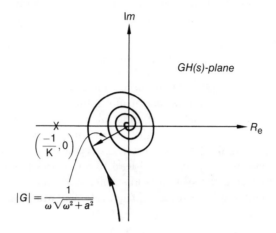

Figure 7.34: Nyquist plot of system with time delay, and $K = 1$

terizes the time delay. The open loop transfer function becomes

$$G(s) = \frac{K}{s(s+a)}e^{-2\pi\tau s} \tag{7.18}$$

This is rewritten as

$$G(j\omega) = M(\omega)e^{-j\phi}$$

where

$$M(\omega) = \frac{K}{\omega\sqrt{\omega^2 + a^2}}; \quad \phi = -2\pi\tau\omega - \arctan\frac{\omega}{a} - \frac{\pi}{2}$$

and note that the delay contributes only to the phase shift as it has no effect upon the magnitude. As ω varies from 0 to $1/\tau$, the time delay contributes 2π to the phase. As ω varies further from $1/\tau$ to $2/\tau$, the time delay contributes another 2π to the phase as shown in Fig. 7-34. If the point $(-1/K, 0)$ is not enclosed the first time the negative real axis is crossed, it is highly unlikely to be enclosed at any later time, because as ω increases the magnitude generally decreases. The acceptable value of τ is one which prevents $(-1/K, 0)$ from being enclosed the first time as seen in Fig. 7-34. We note that since $G(s) = K/s(s + a)$ was stable for all gain, the addition of time delay has had a destabilizing influence on system responses.

We conclude this section by observing that for a stable system the number of times the Nyquist plot of $G(s)H(s)$ must encircle $(-1, 0)$ must be equal to the number of poles of $G(s)(Hs)$ in the right half s-plane. The direction of this encirclement must be negative. Several typical Nyquist plots for various transfer functions are shown in Table 7-1.

System Stability from Bode Response Plot

The stability of a system can be ascertained via the conventional Bode response plot if the system is one of minimum phase as we have seen in Example 13. If we construct the Nyquist plot for a system and we find the Nyquist contour just passes through the critical point we say the system is **marginally (neutrally) stable**. From the Nyquist criterion it can be seen that any increase in the gain K will lead to an encirclement of the critical point thus indicating the system is unstable. A decrease in K will lead to the system being stable. Supposing the system is **marginally stable** we will observe that

$$G(j\omega_1)H(j\omega_1) = -1$$

for some frequency ω_1. We can write this as

$$|G(j\omega_1)H(j\omega_1)|_{dB} = 0 \quad \text{and} \quad \arg G(j\omega_1)H(j\omega_1) = -\pi$$

Thus if a system is marginally stable the phase shift of the open loop frequency response function will be $-\pi$ radians at the 0 dB cross-over frequency. If the phase shift at this frequency is less than $-\pi$ radian the system will be unstable, while if it is greater than $-\pi$ radian it will be stable.

Consider the third-order system with the following transfer function,

$$G(s)H(s) = \frac{K}{s(1 + T_1 s)(1 + T_2 s)}$$

The three possible Bode response plots are shown in Fig. 7-30. These plots indicate that the system is stable for $K < K_c$, neutrally stable (i.e. oscillatory) for $K = K_c$, and unstable for $K > K_c$. Notice that when $K > K_c$, the magnitude is larger than 0 dB when the phase is $-\pi$. When this happens the zeros of the characteristic equation have assumed positive real parts.

Although the Bode response plot refers to the open loop transfer function we can also obtain the closed loop response. For a specific frequency ω_c, we can relate the closed loop frequency response to $G(j\omega_c)$ as follows,

$$F(j\omega_c) = \frac{C(j\omega_c)}{R(j\omega_c)} = \frac{G(j\omega_c)}{1 + G(j\omega_c)} \tag{7.19}$$

Table 7-1 Nyquist plots for various loop transfer functions

$G(s)H(s)$	Nyquist plot (Polar plots)	Stability
$\dfrac{K}{s+a}$		Stable for all gain, $K > 0$
$\dfrac{K}{s(s+a)}$		Stable for all gain, $K > 0$
$\dfrac{K}{(s+a)(s+b)}$		Stable for all gain, $K > 0$
$\dfrac{K}{s(s+a)(s+b)}$		Unstable as shown. May become stable if gain is decreased.
$\dfrac{K(s+c)}{s^2(s+a)(s+b)}$		Stable as shown. Becomes unstable as K is increased.
$\dfrac{K(s+a)(s+b)}{s^3}$		Conditionally stable. Becomes unstable as K is decreased.

Since $G(j\omega_c) = u + jv$, where

$$u = |G(j\omega_c)| \cos \phi$$

$$v = |G(j\omega_c)| \sin \phi$$

we have

$$F(j\omega_c) = \frac{u + jv}{(1 + u) + jv} = \frac{u(1 + u) + v^2 + jv}{(1 + u)^2 + v^2}$$

If the output is represented as a magnitude M and phase θ,

$$F(j\omega_c) = Me^{-j\theta}$$

then

$$M = \left[\frac{(u(1 + u) + v^2)^2 + v^2}{((1 + u)^2 + v^2)^2}\right]^{\frac{1}{2}} \qquad \theta = \arctan \frac{v}{u(1 + u) + v^2}$$

We shall consider a more generalized form of this in a later section. It suffices to note here that we are able to obtain the steady state performance of the closed loop system from the Bode response plot.

Closed Loop Frequency Response from Nyquist Plot

The closed loop frequency response of a unity feedback system can be obtain from the Nyquist plot. Suppose the Nyquist plot for a system is as shown in Fig. 7-35. In this figure the point A on the Nyquist contour corresponds to frequency ω_c. From Eq. (7-19)

$$\left|\frac{C(j\omega_c)}{R(j\omega_c)}\right| = \frac{|G(j\omega_c)|}{|1 + G(j\omega_c)|}$$

and

$$\arg \frac{C(j\omega_c)}{R(j\omega_c)} = \arg G(j\omega) - \arg(1 + G(j\omega_c))$$

Thus we see that the closed loop amplitude and phase responses can be easily computed from the magnitudes and angles of the vectors $\overrightarrow{CA}$ and $\overrightarrow{OA}$ shown on the figure.

7.4   Root Locus Method

The previous method dwelt on the analysis of the transfer function $G(s)H(s)$ when the system was excited with a sinusoidally varying input. The various

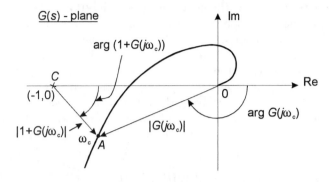

Figure 7.35: Nyquist plot of system

methods yielded the magnitude and phase of the transfer function as the frequency was varied. It was seen that although the shape of the Bode and polar plot was unchanged as the gain K was varied, the stability was greatly affected. Actually the variation of K changes the roots of the characteristic equation and hence the response of the control system. Assume that $G(s)H(s) = KP_1(s)/P_2(s)$, then the **characteristic equation** becomes

$$1 + G(s)H(s) = \frac{P_2(s) + KP_1(s)}{P_2(s)} = 0$$

Thus the zeros of the characteristic equation are given by

$$P_2(s) + KP_1(s) = 0$$

and the role of K is clearly evident. The root locus, conceived by W. R. Evans in 1948, is a graphical method that studies the roots of the characteristic equation as K varies from zero to positive infinity. Since **the roots of the characteristic equation are the poles of the closed loop transfer function** they determine the behavior of the closed loop system. Thus the root locus is a technique whereby we may study the behavior of the closed loop itself.

We consider $G(s)H(s)$ in factored **loop sensitivity form** and write the characteristic equation as

$$1 + G(s)H(s) = 1 + \frac{K \displaystyle\prod_{z=1}^{Z} (s + z_z)}{\displaystyle\prod_{p=1}^{P} (s + p_p)} = 0$$

Here K is referred to as the **static loop sensitivity**, z_z is a zero, and p_p is a pole of the open loop transfer function. There are also n poles at the origin. Unlike the factored form in the frequency plots it is not necessary to have a separate quadratic form since zeros and poles may be complex. It is essential when applying the root locus method for $G(s)H(s)$ to be written in loop sensitivity form, as defined in Eq. (4-44).

As mentioned above, studying the zeros of the characteristic equation, is equivalent to studying the values of s where $G(s)H(s) = -1$. This means that as K varies from zero to infinity, $G(s)H(s)$ must always equal -1 or equivalently

$$|G(s)H(s)| = 1$$

$$\arg G(s)H(s) = (2k+1)\pi; k = 0, \pm 1, \pm 2, \cdots$$

$$(7.20)$$

The first expression is referred to as the **magnitude condition** and the second the **angle condition**. Substituting for $G(s)H(s)$ these conditions become

$$\frac{K\,|s + z_1|\,|s + z_2|\,|s + z_3|\cdots|s + z_Z|}{|s + p_1|\,|s + p_2|\cdots|s + p_P|} = 1$$

$$\arg\left(s + z_1\right) + \arg\left(s + z_2\right) + \cdots + \arg\left(s + z_Z\right) - \arg\left(s + p_1\right)$$

$$- \arg\left(s + p_2\right) - \cdots - \arg\left(s + p_P\right) = (2k+1)\pi$$

The first of these equations may be written as

$$K = \frac{|s + p_1|\cdots|s + p_P|}{|s + z_1||s + z_2|\cdots|s + z_Z|}$$

The rest of the section will dwell extensively on satisfying the angle and magnitude condition as the static loop sensitivity K is varied.

In theory, the entire s-plane should be searched in order to find points satisfying the magnitude and angle condition. This however is unnecessary as the proper interpretation of the magnitude and angle condition yields several rules that facilitate the rapid construction of the root loci plot. Although this plot is approximate, it suffices for purposes of a preliminary study. The following rules are therefore to be considered as aids for the construction of an approximate root loci — more accurate plots can be obtained using one of the computer packages listed in Appendix D.

Root Loci Construction Rules

1. **The root loci begin at the open loop poles**. From the magnitude

condition

$$\frac{|s + z_1|\,|s + z_2|\cdots|s + z_Z|}{|s + p_1|\cdots|s + p_P|} = \frac{1}{K}$$

and since the root loci begin at $K = 0$, $G(s)H(s) \rightarrow \infty$, which can only occur at poles of $G(s)H(s)$.

2. **The root loci end at the open loop zeros.** Since the root loci end at $K = \infty$, the magnitude of $G(s)H(s)$ goes to zero and this can only occur if the root loci terminates at a zero of $G(s)H(s)$.

3. **The number of separate loci is equal to the number of open loop poles.** In accordance with the first rule, each branch of the loci begins for $K = 0$ at an open loop pole. Therefore, the number of branches is equal to the number open loop of poles. Now each branch must end at a zero and if the number of zeros Z is less than the number of poles P, then it is assumed that $P - Z$ zeros exist at infinity. This implies that Z loci begin at the poles and end at zeros, whereas $P - Z$ loci beginning at the poles end at infinity on the s-plane.

4. **The loci are symmetrical about the real axis.** The root locus is symmetrical about the real axis since the roots of $1 + G(s)H(s) = 0$ must either be real or appear as complex conjugates.

5. **Asymptotes of the root locus.** The open loop transfer function may be expressed as

$$G(s)H(s) \;=\; \frac{K\left(s^Z + a_1 s^{Z-1} + \cdots\right)}{s^P + b_1 s^{P-1} + \cdots}$$

$$=\; \frac{K}{s^n + (b_1 - a_1)s^{n-1} + \cdots} = -1$$

where $n = P - Z$. Assuming s to be large so that terms lower than s^{n-1} may be neglected

$$-K \approx s^n + (b_1 - a_1)s^{n-1} = \left(1 + \frac{b_1 - a_1}{s}\right)s^n$$

Taking the nth root and expanding via the binomial theorem gives

$$(-K)^{1/n} = \left(1 + \frac{b_1 - a_1}{ns} + \cdots\right)s$$

Substituting $s = \sigma + j\omega$, again neglecting higher-order terms, and writing $(-K)^{1/n}$ as a complex number

$$\left[\sigma + \frac{b_1 - a_1}{n}\right] + j\omega = \left|K^{1/n}\right|\left[\cos\frac{(2k+1)\pi}{n} + j\sin\frac{(2k+1)\pi}{n}\right]$$

This leads to two equations when the real and imaginary parts are equated,

$$\sigma + \frac{b_1 - a_1}{n} = \left|K^{1/n}\right|\cos\frac{(2k+1)\pi}{n}$$

$$\omega = \left|K^{1/n}\right|\sin\frac{(2k+1)\pi}{n}$$

Each of the above equations may be solved for $K^{1/n}$ and equated to each other yielding an equation for ω,

$$\omega = \tan\frac{(2k+1)\pi}{n}\left[\sigma + \frac{b_1 - a_1}{n}\right]$$

which has the form of $\omega = m(\sigma - \sigma_0)$ where m is the slope of the asymptote and σ_0 the real axis intercept. Since $n = P - Z$, we write the **slope angle of the asymptote** as

$$\frac{(2k+1)\pi}{P-Z}, \quad k = 0, \pm 1, \cdots \tag{7.21}$$

and the **real axis intercept** as

$$\sigma_0 = -\frac{b_1 - a_1}{P - Z} \qquad \begin{aligned} b_1 &= \sum_{i=1}^{P} p_i = \text{ sum of the poles} \\ a_1 &= \sum_{i=1}^{Z} z_i = \text{sum of the zeros} \end{aligned} \tag{7.22}$$

6. **Root loci on the real axis**. The angle condition established that

$$\left[\sum \text{angles from zeros}\right] - \left[\sum \text{angles from poles}\right] = (2k+1)\pi$$

Consider first the case of real poles and zeros to the left of a search point on the real axis. The angular contribution of each is zero. Also if these poles and zeros are complex, the angle contribution will be zero since the zeros and poles appear as conjugates. This leaves real poles and zeros to the right of a search point on the real axis. If the number of poles plus zeros is even, then the total angular condition is 0 or

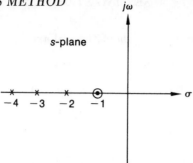

(a) Pole-zero configuration

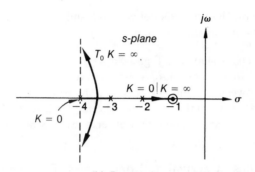

(b) Root locus

Figure 7.36: Root locus of $G(s) = \dfrac{K(s+1)}{(s+2)(s+3)(s+4)}$

$2k\pi$. However, if the number is odd, then the angular contribution will be $(2k+1)\pi$. We conclude therefore that the **root loci exist on the real axis if the total number of poles and zeros to the right is odd.**

Example 15 *Construct a root locus plot for the open loop transfer function given by*

$$G(s) = \frac{K(s+1)}{(s+2)(s+3)(s+4)}$$

The poles and zeros are shown in Fig. 7-36(a). From the six rules established so far, we may conclude:

(1) the root loci begin at $-2, -3$, and -4

(2) one branch ends at -1

(3) there are three different loci, one ending at -1 and two at infinity,

(4) the loci are symmetrical about the real axis,

(5) the asymptotes intersect the real axis at

$$\sigma_0 = -\frac{(2+3+4)-1}{3-1} = -4$$

and the angle is

$$\frac{(2k+1)\pi}{P-Z} = \frac{(2k+1)\pi}{2} = \pm\frac{\pi}{2}$$

(6) the region on the real axis between -1 and -2 as well as between -3 and -4 exists on the root loci.

The root locus is shown in Fig. 7-36(b).

In Fig. 7-36 we did not address ourselves to the question as to what the exact point, between -3 and -4, is where the root locus departs from the real axis. Also we are not sure about the angles associated with the loci as it leaves the real axis. These and other questions are considered in the next five rules.

7. **The angles of arrival and departure**. The angles of arrival and departure of the root locus may be obtained from the angle condition. Consider a search point s_0 near the pole at $-2+j$, then the angle condition yields, as shown in Fig. 7-37,

$$\theta_1 = \theta_5 - (\theta_2 + \theta_3 + \theta_4) - (2k+1)\pi \qquad (7.23)$$

and since θ_2, θ_3, θ_4, θ_5 are known, θ_1 may be computed.

8. **Breakaway point on real axis**. A breakaway point on the real axis is a point where the root locus leaves the real axis. Since the root loci are symmetrical, the number of branches leaving to go toward the positive frequencies is equal to those going toward the negative frequencies. The angles at which these loci depart may be shown to be $\pm\pi/N$, where N is the number of loci departing. The breakaway point itself is also interpreted as the point where the characteristic equation has double roots. The calculation of breakaway points depends on the pole zero configuration.

(a) *Poles and zeros on the real axis.* Let the breakaway point be at $-\alpha$. For the search point s_0 the angle condition yields

$$\theta_5 + (\pi - \theta_2) - (\theta_4 + \pi - \theta_3 + \pi - \theta_1) = (2k+1)\pi$$

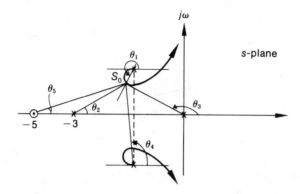

Figure 7.37: Termination angles for $G(s) = \dfrac{K(s+5)}{s(s+3)(s^2+4s+5)}$

$$\theta_5 - \theta_2 - \theta_4 + \theta_3 + \theta_1 = 0$$

From Fig. 7-38 it is seen that the angles are small, therefore they may be approximated by their tangents

$$\frac{\epsilon}{z_2 - \alpha} - \frac{\epsilon}{\alpha - z_1} - \frac{\epsilon}{p_3 - \alpha} + \frac{\epsilon}{\alpha - p_2} + \frac{\epsilon}{\alpha} = 0$$

so that

$$\frac{1}{z_1 - \alpha} + \frac{1}{z_2 - \alpha} - \frac{1}{p_2 - \alpha} - \frac{1}{p_3 - \alpha} - \frac{1}{0 - \alpha} = 0$$

The breakaway point may now be computed by trial and error. This can be generalized as follows: **the breakaway point α can be computed by solving**

$$\sum_{i=1}^{Z} \frac{1}{z_i - \alpha} - \sum_{j=1}^{P} \frac{1}{p_j - \alpha} = 0 \qquad (7.24)$$

If there are Z zeros and P poles, this equation is equivalent to solving a polynomial of order $P + Z$.

(b) *Complex poles and zeros.* Consider a conjugate set of poles as shown in Fig. 7-39. If a search point exists as s_0, then

$$\theta_1 = \arctan \frac{\omega_1}{\alpha - \sigma_1}, \qquad \theta_2 = \arctan \frac{\omega_1 - \epsilon}{\alpha - \sigma_1}$$

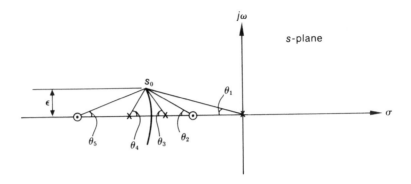

Figure 7.38: Breakaway points for poles and zeros on real axis

The angle $(\theta_1 - \theta_2)$ can be approximated by $\tan(\theta_1 - \theta_2)$ if s_0 is very close to the breakaway point. Then we have

$$(\theta_1 - \theta_2) \approx \tan(\theta_1 - \theta_2) = \frac{\tan\theta_1 - \tan\theta_2}{1 + \tan\theta_1 \tan\theta_2}$$

$$= \frac{\epsilon\,|\alpha - \sigma_1|}{(\alpha - \sigma_1)^2 + (\omega_1)^2}$$

Since the poles are conjugates, the total angle contribution is $2(\theta_1 - \theta_2)$. This contribution is positive for zeros and negative for poles. Also if the conjugate pair (pole or zero) were to the left, the sign changes. In general we can write

$$\Delta\theta = (\theta_1 - \theta_2) = \frac{\epsilon|\sigma_1 - \alpha|}{(\sigma_1 - \alpha)^2 + (\omega_1)^2} \qquad (7.25)$$

This can be generalized by computing terms like Eq. (7-25) for all the zeros and poles. **The breakaway point is then obtained by solving,**

$$\Delta\theta]_{\text{zeros}} - \Delta\theta]_{\text{poles}} = 0$$

(c) *Breakaway point not on the real axis.* In some situations the breakaway point is not situated on the real axis as shown in Fig. 7-40. The original poles are located at $0, -3, -2 + j,$ and $-2 - j$.

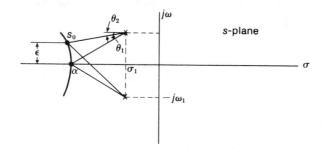

Figure 7.39: Breakaway point for complex roots and zeros

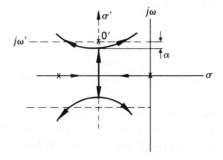

Figure 7.40: Breakaway point not on the real axis

In order to obtain the breakaway point we rotate the coordinate system with the new origin at $0'$ as shown in Fig. 7-40. The location of the new poles in the σ', $j\omega'$ system are at $0, -4, -1 - j1.5$, and $-1 + j1.5$. In the new system the breakaway point α is on the real axis and the previous equations are applicable.

(d) *Breakaway points computed analytically.* The final case considered here is the analytical method. Consider the open loop transfer function in the following form,

$$G(s)H(s) = K\frac{Q(s)}{P(s)} = K\left[G(s)H(s)\right]'$$

Suppose a breakaway point occurs for $K_0 \neq 0$. The characteristic polynomial becomes $P(s) + K_0 Q(s) = F(s)$. Since multiple roots occur at the breakaway point, $F(s)$ can be written as

$F(s) = (s - s_0)^L M(s)$, where $L \geq 2$. Thus

$$P(s) + K_0 Q(s) = (s - s_0)^L M(s)$$

$$1 + K_0 [G(s) H(s)]' = (s - s_0)^L \frac{M(s)}{P(s)}$$

Differentiating both sides we find

$$K_0 \frac{d}{ds} [G(s) H(s)]' =$$

$$L(s - s_0)^{L-1} \frac{M(s)}{P(s)} + (s - s_0)^L \frac{dM(s)/P(s)}{ds}$$

and when $s = s_0$

$$\frac{d}{ds} [G(s) H(s)]' \bigg|_{s=s_0} = 0 \qquad (7.26)$$

This requirement is independent of whether the roots are real or complex.

(e) An alternative method of finding the breakaway points on the real axis, which is often more convenient than trying to solve either Eq. (7-24) or Eq. (7-26), is obtained from the observation that the solutions of Eq. (7-26) are the stationary points of the modified open loop transfer function $[G(s)(H(s)]'$. It is in general not necessary to plot $[G(s)(H(s)]'$ to find all the turning points as it is often sufficient to bracket the breakaway points between upper and lower bounds. Also Rule 6 considerably reduces the range of values of s needing to be searched. To illustrate this point consider the transfer function given in Example 15, where we know from Rule 6 a breakaway point must lie between -3 and -4. All we need do is look for a turning point of $G(s)$ in this region.

9. **Intersection of the root loci with the imaginary axis**. In several of the previous examples, as K was varied the root locus intersected the imaginary axis and entered the other side of the complex plane. The point where the imaginary axis is intersected represents the frequency where pure oscillations occur. Had we used the Routh-Hurwitz criterion to study the characteristic equation, this condition becomes evident when one of the constants in the Routh-Hurwitz array goes to zero as was seen earlier in the chapter.

Example 16 *Obtain the root locus plot and the frequency at which it crosses the imaginary axis for the following,*

$$G(s)H(s) = \frac{K(s+5)}{s(s^2 + 4s + 5)}$$

The characteristic equation becomes

$$s^3 + 4s^2 + (5+K)s + 5K = 0$$

The Routhian array may be now formed,

$$
\begin{array}{c|cc}
s^3 & 1 & (5+K) \\
s^2 & 4 & 5K \\
\hline
s^1 & b_1 & b_2 \\
s^0 & c_1 &
\end{array}
\qquad
\begin{array}{l}
b_1 = \dfrac{20 - K}{4} \\[2mm]
b_2 = 0 \\[1mm]
c_1 = 5K
\end{array}
$$

For stability the constraint of $b_1 \geq 0$ yields that $K \leq 20$. If K reaches the critical value of 20, then $b_1 = 0$ and the auxiliary equation is formed,

$$4s^2 + 5K = 0$$

$$s = \pm j5$$

The root loci cross the imaginary axis at $\pm j5$ and the static loop sensitivity at this point is 20. Beyond this point, K increases and the system becomes unstable. In the Routh array if $b_1 < 0$, then two complex roots appear in the right half s-plane as shown in Fig. 7-41.

Once the root loci are constructed with the aid of the previous nine rules, the gain K at any specific point may be found by using the magnitude condition.

10. **The value of K on the root loci.** If $s = s_0$, then from the magnitude condition

$$K = K_0 = \left[\frac{1}{|G(s)H(s)|} \right]_{s=s_0} \tag{7.27}$$

From Example 16, when $s = j5$

$$
\begin{aligned}
K &= (A)(B)(C)/D \\
&= \frac{|(2+j) + j5|\,|(2-j) + j5|\,|j5|}{|5 + j5|} = 20
\end{aligned}
$$

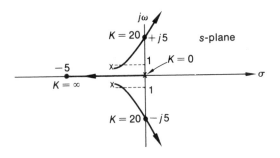

Figure 7.41: Root locus of $G(s)H(s) = \dfrac{K(s+5)}{s(s^2 + 4s + 5)}$

11. **The sum of closed loop poles is a constant.** If the order of the denominator of the open loop transfer function is greater than the numerator by at least 2, then the coefficient of s^{n-1} is independent of K. The coefficient equals the negative of the sum of the closed loop poles. This fact may be used to follow the migration of the closed loop poles as the gain K is varied. In Example 16, for the open loop transfer function, the coefficient of s^{n-1} is 4, therefore the sum of the closed loop poles is -4. At $K = 20$ we know that $s = j5$ and $-j5$. The third pole may be located by substituting $s_1 = -s_2 = j5$ into

$$s_1 + s_2 + s_3 = -4$$

to give $s_3 = -4$. Thus for $K = 20$, the three poles of the closed loop system are $j5, -j5$, and -4.

By the application of the above eleven rules the root locus may be very simply constructed. We have said previously that the root locus is very useful for studying the zeros of the characteristic equation as the gain is varied from zero to infinity. This method should be used in conjunction with other methods, previously discussed, to study the overall performance of a control system.

Once the root locus of the system is obtained, we may then obtain the transient behavior of the closed loop system. In order to show this we shall first show the **relationship between the closed loop poles and the root locus plot.** Consider a third-order closed loop transfer function,

$$\frac{C(s)}{R(s)} = \frac{G(s)}{1 + G(s)}$$

where the forward loop transfer function is

$$G(s) = \frac{K}{(s+A)(s^2+bs+c)}$$

Then

$$\frac{C(s)}{R(s)} = \frac{K}{K+(s+A)(s^2+bs+c)}$$

Now let us set the gain at $K = K_s$, then

$$\frac{C(s)}{R(s)} = \frac{K_s}{(s-p_1)(s-p_2)(s-p_3)}$$

where p_1, p_2,and p_3 are poles of the closed loop system which are equivalent to the zeros of the characteristic equation. We stated earlier that for a given value of K, the zeros of the characteristic equation are obtained whenever

$$G(s)H(s) = -1$$

is satisfied. Therefore, for a specific value of $K = K_s$, the poles p_1, p_2, and p_3 of the closed loop are the zeros of the characteristic equation as shown in Fig. 7-42. The response may now be obtained by techniques discussed in Chapter 4.

Example 17 *The open loop transfer function of a third order control system is given by*

$$G(s)H(s) = \frac{K(s^2+2s+2)}{s(s^2+1)}$$

Obtain the root locus and the values of K for which the system is stable.

The root locus plot has three branches with one going to infinity. The asymptote has an angle of π. Applying the Routh-Hurwitz criterion we can establish bounds for stability. The characteristic equation is

$$s^3 + Ks^2 + (2K+2)s + K = 0$$

The array becomes

$$
\begin{array}{c|cc}
s^3 & 1 & 2K+2 \\
s^2 & K & K \\
\hline
s^1 & b_1 & 0 \\
s^0 & c_1 & \\
\end{array}
\qquad
\begin{array}{l}
b_1 = 2K+1 \\[2ex]
c_1 = K
\end{array}
$$

The system is therefore stable for all K as is confirmed on the root locus plot shown in Fig. 7-43.

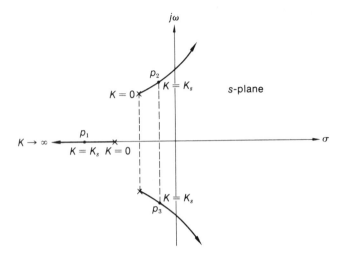

Figure 7.42: Relationship of the closed loop poles and the root locus plot

Example 18 *Construct the root locus for a unity feedback control system that has a forward loop transfer function given by*

$$G(s) = \frac{Ke^{-s\tau}}{s(s + a)}$$

(See Section 7-3 for the Nyquist plot.) Let $\tau = 1$ sec and $a = 1$ and find K and ω for marginal or neutral stability.

When the transfer function under consideration has a delay, an infinite number of branches exist, i.e. an infinite number of zeros exist at infinity. The time delay contributes an angle $\omega\tau$ to the phase. To obtain the equation for the asymptotes we note that the pole-zeros of $G(s)$ without the delay term contributes no phase to the far right of the s-plane and either 0 or π to the far left side of the s-plane depending upon whether $P - Z$ is even or odd. Since the angle condition requires a phase of $\pm(2k + 1)\pi$ for all K, we can say that the phase requirement must be

$$-\omega\tau + \sum \text{angle of } Z \text{ zeros} - \sum \text{angle of } P \text{ poles} = \pm(2k + 1)\pi$$

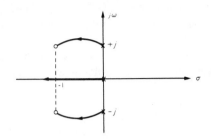

Figure 7.43: Root locus of a third-order system

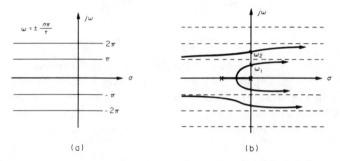

Figure 7.44: Root locus of a system with time delay

This leads to the asymptote equations:

$$\omega = \pm(2k + 1)\pi/\tau \quad \text{in right half } s\text{-plane,}$$

$$\omega = \pm(2k + 1)\pi/\tau \quad \text{in left half } s\text{-plane,} P - Z \text{ is even,}$$

$$\omega = \pm 2k\pi/\tau \quad \text{in left half } s\text{-plane,} P - Z \text{ is odd.}$$

For the case of $P - Z$ being even, the asymptotes are shown in Fig. 7-44(a).

Since the region between $s = 0$, and -1 exists on the root locus, the breakaway point must be computed. This can be obtained analytically from $d[G(s)]'/ds = 0$. We note that

$$-\frac{1}{K} = \frac{e^{-s}}{s(s + 1)}$$

so that computing $d[G(s)]'/ds = 0$ yields $s^2 + 3s + 1 = 0$ or $s = -2.618, -0.382$.

We select $s = -0.382$ as the breakaway point. The root locus is shown in Fig. 7.44(b). We note that when $\omega = \omega_1$ the magnitude of the gain is lower than at $\omega = \omega_2$ and it is the gain at ω_1 that determines stability as seen in Section 7-3.

Ordinarily a second-order system is stable for all K, but the addition of a delay can cause instability. The magnitude and phase of $G(j\omega)$ are

$$|G(j\omega)| = \frac{K}{\omega\sqrt{\omega^2 + 1}}$$

$$\arg G\left(j\omega\right) = -\omega - \frac{\pi}{2} - \arctan\omega$$

If we set the phase to be $-\pi$ then

$$-\omega - \frac{\pi}{2} - \arctan\omega = -\pi$$

$$\omega = \tan\left(\frac{\pi}{2} - \omega\right) = \cot\ \omega$$

This is satisfied at $\omega \cong 0.86$ and substituting in the magnitude condition yields $K = 1.134$. For marginal stability we have $K = 1.134$ at $\omega_1 = 0.86$ as shown in Fig. 7-44(b). We note that the same result could be obtained by substituting $s = j\omega$ in $G(s)$ and equating it to -1 and solving the two equations derived from the real and imaginary parts separately.

Root Locus from State Equations

The root locus method can be used for investigating the effect of parameters other than the static loop sensitivity on the stability of feedback systems, and it can even be used for investigating the stability of systems having no feedback. To illustrate this we apply the method to a system represented in state form.

When a system is represented in state form, then it was shown in Chapter 3 that the output $Y(s)$ is given by

$$Y(s) = \mathbf{c}[s\mathbf{I} - \mathbf{A}]^{-1}\mathbf{b}R(s)$$

where $\mathbf{A}$ is the coefficient matrix, whereas $\mathbf{c}$ and $\mathbf{b}$ are the output and input vectors. Substituting for the inverse,

$$Y(s) = \frac{\mathbf{c} \text{ adj } [s\mathbf{I} - \mathbf{A}]\mathbf{b}}{\det[s\mathbf{I} - \mathbf{A}]}R(s) \qquad (7.28)$$

so that the output is a function of the characteristic roots obtained from

$$\det[s\mathbf{I} - \mathbf{A}] = 0$$

In general this can be expanded to yield

$$s^n + a_{n-1}s^{n-1} + \cdots + a_1 s + a_0 = 0$$

or

$$1 + \frac{a_0}{s^n + a_{n-1}s^{n-1} + \cdots} = 0$$

If we set $a_0 = K$ and $G(s) = K/(s^n + a_{n-1}s^{n-1} + \cdots)$, then we obtain

$$1 + G(s) = 0$$

and we can apply the angle and magnitude conditions to $G(s)$.[4]

Example 19 *The coefficient matrix of a system is given by*

$$\mathbf{A} = \begin{bmatrix} 0 & 1 & 0 \\ 0 & 0 & 1 \\ -K & -20 & -9 \end{bmatrix}$$

Construct the root locus and set the system gain K so that the damping ratio is 0.5.

The characteristic equation is

$$\det [s\mathbf{I} - \mathbf{A}] = s^3 + 9s^2 + 20s + K = 0$$

which can be written as

$$1 + G(s) = 0$$

where

$$G(s) = \frac{K}{s^3 + 9s^2 + 20s}$$

or

$$G(s) = \frac{K}{s(s+4)(s+5)}$$

which is the function whose locus must be obtained.

The root locus plot is obtained by following the previous rules and is shown in Fig. 7-45. Since $\zeta = 0.5, \gamma = \arccos 0.5 = 60°$ and this is represented by a radial line in Fig. 7-45. Along this line $\zeta = 0.5$. This line intersects the root locus at $K = K_s$ and $s = s_s$. The gain may be obtained by using Rule 10.

$$K = \left| \frac{1}{G(s)} \right|_{s=s_s} \cong 35$$

For the closed loop response to have a damping ratio of 0.5, the gain must be set around $K = 35$.

---

[4] We can actually apply all the previous graphical methods. Since the extension is so obvious, we will not pursue it, except in this one example.

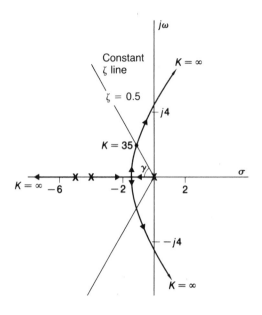

Figure 7.45: Root locus of $G(s) = \dfrac{K}{s(s+4)(s+5)}$

7.5   Dynamic Response Performance Measures

It is evident from the discussion above that feedback can be used to modify the frequency or transient response of a system. It is also clear that a imprudent choice of the gain K can actually degrade rather than improve its stability and performance. However, in most control systems it is not sufficient for the system to be just stable, for we usually want to know significantly more about its response to transient or sinusoidal inputs. We shall also subsequently see that by introducing additional dynamical elements called **compensators** or **controllers** we can often modify and hopefully improve the transient and steady state performance of a system.

We saw in Chapter 4 that the transient response of a physical system may be very complex. However, when the system has a single or complex conjugate pair of dominant poles it can often be approximated by the response of a first or second order transfer function. Because many feedback control systems exhibit these characteristics it is both helpful and

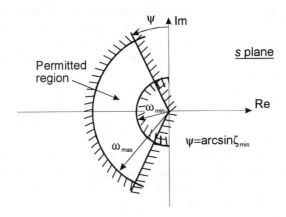

Figure 7.46: Permitted region for dominant poles of closed loop system

convenient to specify their performance under the assumption they can be modeled as first or second order systems. It is of course necessary to validate this assumption towards the end of the design cycle. Fortunately this can often be achieved by visual inspection of the closed loop poles on the s-plane. Using this approach we can derive s-plane or frequency domain performance criteria from the transient performance criteria discussed in Chapter 6.

Laplace Domain Specifications

The transient response of a second order system is completely characterized by the parameters ζ and ω_n as discussed in Section 4.2. The inter-relationship between these two parameters and time domain parameters such as T_p, T_r, M_p, etc. are given in Table 6-1. The transient response of a system can be specified in terms of time domain parameters from which bounds on ζ and ω_n can be deduced. Thus we may start, for example, with a maximum bound on the percentage overshoot M_p, and a maximum settling time T_s for a step input. Using the formulas relating M_p and T_s to ω_n and ζ given in Table 6-1 we deduce $\zeta_{\min}$ and $\omega_{\min}$. These parameters can be interpreted graphically on the s-plane as shown in Fig. 7-46 where the poles of the system must be restricted to the **permitted region**.

This diagram might suggest that dominant closed loop poles far to the left of the shaded boundary are not only acceptable but also desirable. However, in general this will not be the case for the following two reasons. Firstly, if the natural frequency of the dominant poles is large the control system will be strongly susceptible to external disturbances. Secondly, if we try to make the natural frequency of the dominant poles too large we will often experience unexpected stability problems due to unmodeled system dynamics. For these reasons a ω_{max} for the dominant poles of a system design will also often be specified. The choice of ω_{max} is often quite difficult, requiring considerable experience and judgment concerning the specific application. Obviously if the control system is to operate in an extremely noisy environment or the plant has structural resonant modal frequencies relatively close to ω_{min} then ω_{max} is likely be chosen to be close to ω_{min}. At other times there may be considerable latitude in the choice of ω_{max}.

Recalling the system discussed in Example 16 we see that these performance bounds on the s-plane will essentially fix the gain K. This is sometimes called "setting the system gain".

Example 20 *A unity feedback control system used in a numerically controlled machine tool has the plant transfer function*

$$G_p(s) = \frac{K}{s(s^2 + 30s + 306)}$$

Find the range of values of the static loop sensitivity K for which the dominant closed loop poles satisfy the transient conditions $M_p \leq 0.2$, $T_s \leq 1.77$ sec. In addition experimental modal analysis has shown that the first structural modal frequency is at 50 rad/sec, so it has been decided to take $\omega_{max} = 10$ rad/sec so as to avoid system instabilities.

Using the control system analysis package Program CC[5] , we obtain the root locus plot shown in Fig. 7-47. From Table 6-1 or from Fig. 4-9(b) we find for a peak overshoot $M_p \leq 0.2$ that $\zeta_{min} = 0.45$. Also from this same table we find $T_s = 1.77$ sec, which implies that $\omega_{min} = 5$ rad/sec. Plotting the boundaries corresponding to ω_{min}, ω_{max} and ζ_{min} on Fig. 7-47 allows us to find their intersections with the respective root loci branches. From this figure it can be seen that the dominant poles lie within the **permitted region** for $1063 < K < 2302$. When $K = 2302$ the closed loop poles are at $s = -4.718 \pm j9.47$ and $s = -20.6$. It can be seen that the complex conjugate poles are dominant so it is expected that the transient response will be close to those given in the specification. These results are confirmed in Fig. 7-48 which gives the response of the system to a unit step input. Here $M_p = 0.178$ and $T_s = 0.84$ sec.

---

[5]See Appendix D.

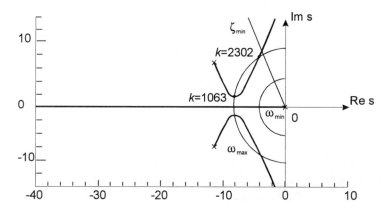

Figure 7.47: Root locus plot for $G_p(s) = \dfrac{K}{s(s^2 + 30s + 306)}$

Frequency Domain Specifications

The Nyquist plot of Fig. 7-30 indicated that the system was stable for low gain, oscillatory for some critical gain, and finally unstable as the gain was increased further. Clearly, as K varies the **degree of stability** varies, i.e. the point of closest approach of the Nyquist plot to the $(-1,0)$ point varies. If the closest point of the Nyquist plot is distant from the critical point $(-1,0)$, then the system transient response is sluggish. As the plot moves closer to $(-1,0)$ the transient response becomes less sluggish, more oscillatory, and less damped. Eventually the response becomes unstable for values of gain that force the enclosure of $(-1,0)$. The idea of performance and relative stability, or the proximity of the $GH(j\omega)$ plot to $(-1,0)$, is often correlated to an effective damping ratio of the closed loop poles.

We mentioned earlier that the Nyquist criterion was not only useful for establishing absolute stability of a control system but also its relative stability. In the following, we propose to show how this can be done.

Gain and Phase Margins Generally the **gain** and **phase margins** are used to define the performance or **relative stability** of a system in the frequency domain. Consider the Nyquist plot of a third-order control system shown in Fig. 7-49. The magnitude of the open loop gain when it crosses the real axis is $|G(j\omega_\phi) H(j\omega_\phi)|$ where ω_ϕ is the frequency at which this crossing occurs. This frequency is called the **phase crossover frequency**. The **gain margin** GM is measured at this frequency and is

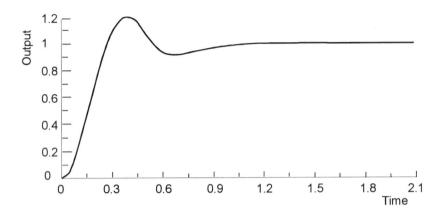

Figure 7.48: Transient response of closed loop system with $K = 2302$

defined as

GM = Additional gain needed to make the system marginally stable

This is usually measured in decibels and is

$$\text{GM} = 20 \log \left[\frac{1}{|G(s)H(s)|} \right]_{s=j\omega_\phi} \tag{7.29}$$

The gain margin is shown on the Nyquist plot and the Bode plot of Fig. 7-49. Note how easy it is to obtain this from the Bode plot. Gain margin has little meaning for first- or second-order systems since $G(j\omega)H(j\omega)$ never crosses the negative real axis.

The **phase margin** PM is a measure of relative stability in terms of the angle of $G(j\omega)H(j\omega)$. It is defined as

PM = Additional phase lag needed to make system marginally stable

This phase is measured from the point where $|G(j\omega)H(j\omega)| = 1$. The frequency where this happens is called the **gain crossover frequency** and denoted by ω_c

$$\text{PM} = \pi + \arg G\left(j\omega_c\right) H\left(j\omega_c\right) \tag{7.30}$$

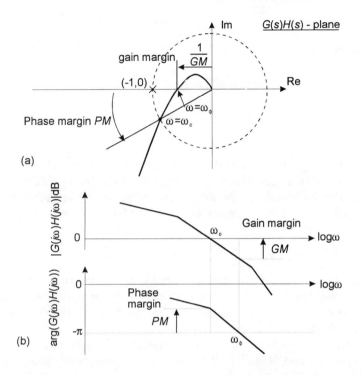

Figure 7.49: Gain and phase margin on the Nyquist and Bode plots

Example 21 *Obtain the gain and phase margins and crossover frequencies for the third-order system described by*

$$G(s)H(s) = \frac{K}{(s+1)(s+2)(s+3)}$$

It is known that $K = 22.8$.

To obtain the phase crossover frequency we separate $G(j\omega)H(j\omega)$ into real and imaginary parts and then equate the imaginary part to zero,

$$G(j\omega)H(j\omega) = \frac{22.8}{(j\omega)^3 + 6(j\omega)^2 + 11(j\omega) + 6}$$

$$= \frac{22.8}{(6 - 6\omega^2) + j(11\omega - \omega^3)}$$

Equating the imaginary part to zero yields

$$\omega = \omega_\phi = \pm\sqrt{11}$$

The magnitude at $\omega = \omega_\phi$ becomes

$$|G(j\omega_\phi)H(j\omega_\phi)| = \frac{22.8}{60}$$

so that

$$\left[\frac{1}{|GH|}\right]_{\omega=\omega_\phi} = 2.63$$

and the gain margin becomes GM $= 20\log 2.63 = 8.4$ dB which indicates that the system gain may be increased by 8.4 dB (factor of 2.63) before becoming marginally stable. The gain crossover frequency ω_c occurs when $|G(j\omega)H(j\omega)| = 1$. The frequency $\omega_c = 2.0$ and the phase margin PM becomes

$$\text{PM} = \pi + [-\arctan 2 - \arctan 1.0 - \arctan 0.67] = 39^\circ$$

which indicates that the phase can increase by 39° before the system becomes marginally stable, under the condition that the magnitude remains constant. As a general rule PM should not be less than 30° and GM about 6 dB for good transient response.

It is instructive to correlate the phase margin to system damping. An exact relationship may be derived if we consider a second-order system. For higher-order systems where there is a pair of dominant poles, the ensuing discussion applies in an approximate way only.

Consider a second-order system,

$$G(s) = \frac{\omega_n^2}{s(s + 2\zeta\omega_n)}, \quad H(s) = 1$$

The open loop frequency response is

$$G(j\omega) = \frac{\omega_n^2}{j\omega(j\omega + 2\zeta\omega_n)}$$

Defining $\nu = \omega_n/\omega$ we have

$$G(j\omega) = \frac{\nu^2}{j(2\zeta\nu + j)}$$

The phase margin occurs at the gain crossover frequency ν_c where

$$|G(j\omega)| = 1 = \frac{\nu_c^2}{(4\zeta\nu_c^2 + 1)^{1/2}}$$

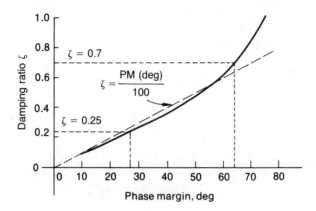

Figure 7.50: Phase margin versus damping ratio for second-order systems

Rewriting the above equation,

$$\nu_c^4 - 4\zeta\nu_c^2 - 1 = 0 \qquad (7.31)$$

which defines the necessary equation relating the damping factor ζ to the gain crossover frequency. The phase margin becomes

$$
\begin{aligned}
\text{PM} &= \pi + \arg G\left(j\nu_c\right) \\[1mm]
&= \pi + \left(-\frac{\pi}{2} - \arctan\frac{1}{2\zeta\nu_c}\right) \\[1mm]
&= \frac{\pi}{2} - \arctan\frac{1}{2\zeta\nu_c}
\end{aligned}
$$

This equation together with Eq. (7-31) is used to generate the plot shown in Fig. 7-50. For low damping ratio the phase margin versus ζ may be approximated by a straight line given by

$$\zeta = 0.01 \cdot \text{PM}$$

We can go a step further and observe that the damping ratio for a second-order system is related to the peak overshoot M_p. Therefore the peak overshoot is given by

$$M_p = e^{-\pi\zeta\sqrt{1-\zeta^2}} \qquad (7.32)$$

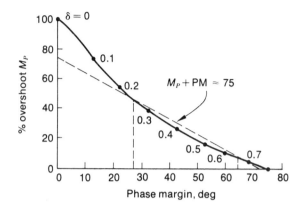

Figure 7.51: Phase margin versus percentage overshoot for second-order systems

and since ζ is related PM we may obtain a relationship of PM versus M_p. This is best approached graphically. The plot of PM versus M_p is shown in Fig. 7-51. For PM $\geq 30°$, this relationship may be approximated by a straight line

$$PM(\text{degrees}) = 75(\text{percent}) - M_p(\text{percent})$$

We note that as the value of PM increases, M_p decreases and the damping increases.

While the gain margin and the phase margin can often be useful measures of the stability margin for a system, a word of caution about their indiscriminate use needs to be given. Neither a good gain margin nor a good phase margin for a system when taken alone guarantees a satisfactory transient response to a step input. Both these situations are illustrated in the Nyquist plots shown in Fig. 7-52(a) and (b). In each case the closed loop system would have a poor transient response. Even if an open loop system has both good phase and gain margins this does not guarantee a good transient response when the feedback loop is closed. The Nyquist plot of such a system which would nevertheless have poor transient performance is illustrated in Fig. 7-52(c).

The explanation of the apparent paradox is obtained by observing that in each of the examples illustrated in Fig.7-52 the respective Nyquist plot closely approaches the critical point $(-1, 0)$. As a general rule if the Nyquist plot for a system closely approaches the critical point then it is likely to have a poor transient response. While a good gain and/or phase margin

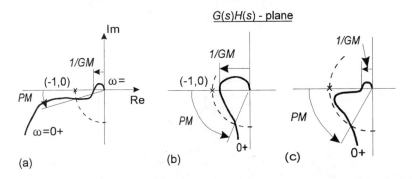

Figure 7.52: All systems having these Nyquist plots would have poor transient responses (a) Good GM, poor PM (b) Good PM, poor GM (c) Good PM and GM

ensures there is good separation of the Nyquist plot from the critical point for systems with dominant closed loop poles, this result is not guaranteed for the non-dominant case. We shall examine this situation in further detail below.

Contours of Constant Closed Loop Gain and Phase Response   Before proceeding with this discussion it is important to examine the close relationship between the transient performance and the frequency response for second order systems; just as we have done when discussing Laplace domain specifications. The transient response of a second order system is characterized by the parameters ζ and ω_n which are related to time domain parameters such as T_r, T_s, M_p, etc. by the formulas given in Table 6-1. In Section 4-4 we showed that the maximum gain M_m and the resonant frequency ω_m of the frequency response function, as shown in Fig. 7-7, are given by

$$M_{m,\max} = \frac{1}{2\zeta_{\min}\sqrt{1 - \zeta_{\min}^2}} \tag{7.33}$$

$$\omega_{m,\min} = \omega_{n,\min}\sqrt{1 - 2\zeta_{\min}^2} \tag{7.34}$$

Thus we see there is a direct link between the time domain specification and the frequency domain parameters $M_{m,\min}$ and $\omega_{m,\min}$ for second order systems.

In addition to the frequency domain specifications mentioned above, an upper bound $\omega_{m,\max}$ for the resonant frequency, will often be spec-

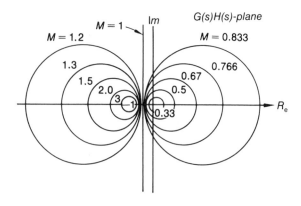

Figure 7.53: Constant M-circles on GH-plane

ified to ensure that the influences of external disturbances and the effects of unmodeled plant dynamics on the system performance are minimized. From these considerations we can conclude that if the closed loop transfer function of a feedback system has a pair of complex conjugate dominant poles, and if its frequency response satisfies the conditions $|M(\omega)| < M_{m,\text{max}}$ and $\omega_{m,\text{min}} < \omega_m < \omega_{m,\text{max}}$ then its transient response will satisfy the specified time domain parameters.

The Nyquist criterion and phase margin index are defined for the open loop transfer function $G(s)H(s)$. As we have just seen another useful index of performance is the magnitude $|M(\omega)|$ of the closed loop frequency response function which is shown in Eq. (7-32) to be related to the damping ratio ζ. The relationship between the closed loop and open loop frequency response functions was established earlier and we now return to this formulation.

Consider a unity feedback control system whose frequency response is written as

$$\frac{C(j\omega)}{R(j\omega)} = \frac{G(j\omega)}{1 + G(j\omega)} = M(\omega)e^{j\phi(\omega)}$$

If $G(j\omega)$ is written as

$$G(j\omega) = u + jv$$

then the magnitude $M(\omega)$ becomes

$$M(\omega) = \left| \frac{G(j\omega)}{1 + G(j\omega)} \right| = \frac{(u^2 + v^2)^{1/2}}{((1 + u)^2 + v^2)^{1/2}}$$

Squaring and rearranging we obtain

$$u^2 + v^2 - \frac{2M^2 u}{1 - M^2} = \frac{M^2}{1 - M^2}$$

where $M = M(\omega)$. Now completing the squares, i.e. adding $\left(M^2/\left(1 - M^2\right)\right)^2$ to both sides we obtain

$$\left(u - \frac{M^2}{1 - M^2}\right)^2 + v^2 = \frac{M^2}{(1 - M^2)^2} \qquad (7.35)$$

which is an equation of a circle with center at (U_m, V_m) and radius r_m where

$$U_m \;=\; \frac{M^2}{1 - M^2}$$

$$V_m \;=\; 0$$

$$r_m \;=\; \left| \frac{M}{1 - M^2} \right|$$

These circles are plotted in Fig. 7-53. Note that as M increases, the radii of the circles decrease until at the limit, the $(-1, 0)$ point is obtained. If enough M-circles are drawn and if $G(j\omega)$ is superimposed on the plot, then at each intersection of the $G(j\omega)$ plot with an M-circle we obtain the closed loop response at a given frequency as shown in Fig. 7-54. The smallest circle that $G(j\omega)$ is tangential to corresponds to the highest value of M, which is denoted by M_m. The frequency at which this occurs is the resonant frequency ω_m.

The phase angle of the closed loop may now be written by substituting for $G(j\omega)$ in the overall transfer function,

$$\phi = \arctan \frac{v}{u} - \arctan \frac{v}{1 + u}$$

Taking the tangent of both sides,

$$N = \tan \phi = \frac{v}{u^2 + u + v^2}$$

Rewriting and completing the squares, we obtain

$$\left(u + \frac{1}{2}\right)^2 + \left(v - \frac{1}{2N}\right)^2 = \frac{N^2 + 1}{4N^2} \qquad (7.36)$$

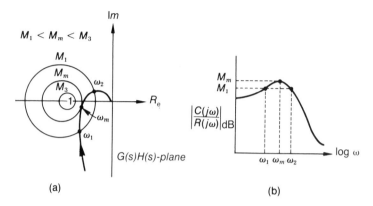

Figure 7.54: Determination of M by plotting $G(s)$ on the same plot as the M-circles

which represents the equation of a circle with center at (U_n, V_n) and radius r_n

$$U_n = -\frac{1}{2}$$

$$V_n = \frac{1}{2N}$$

$$r_n = \left|\frac{1}{2}\sqrt{1 + 1/N^2}\right|$$

These circles are plotted in Fig. 7-55 for various values of ϕ ($\tan\phi = N$). If the open loop frequency response $G(j\omega)$ were plotted on this paper, the intersections of $G(j\omega)$ with the ϕ-circles indicates the value of the closed loop phase angle.

Clearly, if M- and ϕ-circles were plotted on the same paper and if $G(j\omega)$ were superimposed on this paper, then we may obtain the magnitude and phase of the closed loop response.

When the M- and ϕ-circles are plotted on the same paper, the resulting charts are called **Nichols charts**. The plots are given in decibels versus phase shift. They are obtained in the following manner. A point on a constant M locus may be obtained by drawing a vector from the origin to a particular point on the M-circle. The vector length measured in decibels and its phase determine a point on the Nichols chart as shown in Fig. 7-56. The actual value of M is shown in parenthesis. The point $(-1, 0)$ corresponds to 0 dB and $-\pi$ phase in the Nichols chart. The above procedure is

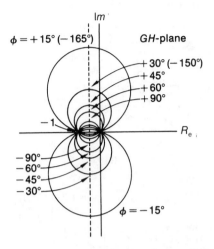

Figure 7.55: Constant ϕ-circles on GH-plane

repeated for the N circles. We therefore obtain values of M and N plotted as functions of the magnitude.

Example 22 *The open loop transfer function of a unity feedback system is given by*

$$G(s) = \frac{K}{s(s^2 + s + 1)}$$

Construct the Nichols chart for $K = 1$ and 0.5. For $K = 0.5$, obtain the peak magnitude of the closed loop response. What is the phase margin?

The Nichol chart is shown in Fig. 7-57 for the two values of K. For $K = 0.5$, $|G(j\omega)| = 1$ occurs at the point where the phase is $-130°$. The phase margin is therefore $180 - 130°$ or PM $= 50°$. The peak magnitude of the closed loop response is 3 dB at a phase of $-150°$ and at $\omega_m \approx 0.56$.

As we conclude this chapter it is important to stress a point made earlier. This pertains to the use of computers. Much of what we have discussed is amenable to a computer analysis. The time domain as well as graphical analysis may be compiled as a package of interactive programs using a time sharing computer. It is often necessary to resort to this when real systems of sufficient complexity arise. Appendix D briefly describes some of these packages. Although it is beyond the scope of this book to consider it in much detail, it is advisable that the student finish this chapter by writing

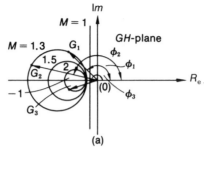

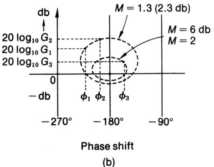

Figure 7.56: Construction of Nichol chart

some programs for obtaining the Bode plot, Nyquist plot, and the root locus for at least a few simple examples.

7.6   Summary

We have developed one analytical and two graphical techniques for analyzing the behavior of control systems.

The analytical method discussed is based upon the Routh-Hurwitz stability criterion which can be used for determining if, and how many, roots of the characteristic equation exist in the right half of the s-plane. Since this approach allows stability to be determined in terms of system parameters their range which ensures stability can be ascertained, and if more than one parameter is involved the stability regions can be plotted on a stability diagram. A disadvantage of the Routh-Hurwitz method is its failure to give

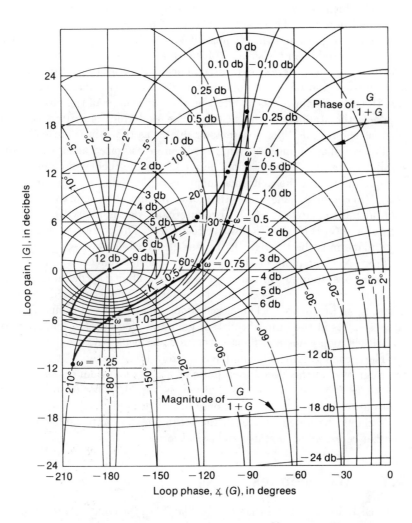

Figure 7.57: Nichol diagram for $G(s) = \dfrac{K}{s(s^2 + s + 1)}$, $K = 1, 0.5$

insight into how the system stability may be improved. As a result it has not become a popular design tool.

Of the graphical techniques discussed, the first is the frequency response method which is based upon the assumption that the input signal is a sinusoidal function. There are two types of frequency response plots, viz. Bode response plot and the Nyquist plot. The Bode response plot consists of the magnitude and phase of a transfer function plotted versus frequency. The magnitude is generally expressed in decibels. The Nyquist plot is a plot of magnitude and phase on polar paper where ω is the parameter. In each case the overall gain K has no effect on the basic shape of the plot although it affects the stability of the system. The Bode plot is shifted vertically, whereas the Nyquist plot contracts or dilates as the gain is varied.

The stability of linear systems has been investigated using the Nyquist stability criterion. This criterion, derived from the theory of complex variables and involving the conformal mapping of the right half s-plane into the $G(s)H(s)$ complex plane, states that for a system to be stable the $G(s)H(s)$ contour must encircle the $(-1,0)$ counterclockwise as many times as there are poles of $G(s)H(s)$ with positive real parts. The corresponding path on the s-plane must be the right half plane encircled clockwise.

From Nyquist plots of the open loop transfer functions, we obtained measures of system performance and relative stability by defining gain and phase margins. It was seen that the phase margin is related to system damping and overshoot. The closed loop response was obtained from the Nyquist plots in the form of constant gain and phase plots. The open loop and closed loop gains and phases were related using Nichols charts.

The second graphical technique is the root locus method. It investigates the migration of the closed loop poles of a control system on the complex s-plane as the static loop sensitivity K is varied. This technique yields information on the transient response of a system.

It should be noted that both the techniques are applicable to systems whether represented in state or classical form.

7.7   References

1. E. J. Routh, *A Treatise on the Stability of a Given State of Motion*, Macmillan, London, 1877.

2. R. R. Gantmacher, *The Theory of Matrices*, Chelsea Publishing Co., New York, 1964.

3. H. W. Bode, *Network Analysis and Feedback Amplifier Design*, D. Van Nostrand, Princeton, N. J., 1945.

4. N. Balabanian, T. A. Bickart, *Linear Network Theory*, Matrix Publishers, Inc., Beaverton, Oregon, 1981.

5. H. Chestnut, R. N. Mayer, *Servomechanisms and Regulating System Design* , John Wiley and Sons, Inc., New York, 1959.

6. R. C. Dorf, *Modern Control Systems*, Addison-Wesley Publishing Co., Reading, Mass., 1989.

7. H. Nyquist, "Regeneration Theory", *Bell Syst. Tech. J.*, Vol. 11 (1932) pp. 126-147.

8. J. J. D'Azzo, C. H. Houpis, *Linear Control System Analysis and Design — Conventional and Modern*, Mc Graw Hill Book Co., New York, 1988.

9. G. S. Brown, D. P. Campbell, *Principles of Servomechanisms*, John Wiley and Sons, Inc., New York, 1948.

10. W. R. Evans, "Graphical Analysis of Control Systems", *Trans. AIEE*, Vol. 67 (1948) pp. 547-551.

11. W. R. Evans, *Control System Dynamics*, Mc Graw Hill Book Co., New York, 1954.

12. B. C. Kuo, *Automatic Control Systems*, Prentice-Hall, Inc., Englewood Cliffs, NJ, 1987.

7.8 Problems

7-1 Apply the Routh-Hurwitz criterion to the following characteristic equations, and determine for what range of the parameter K each system is stable.

(a) $s^3 + s^2 + Ks + 2 = 0$

(b) $s^4 + 2s^3 + 2s^2 + Ks + 5 = 0$

(c) $s^5 + s^4 + Ks^3 + 8s^2 + s + K = 0$

(d) $s^3 + 22s^2 + 9s + K = 0$

(e) $s^3 + Ks^2 + 8s + 200 = 0$

7-2 Under what conditions will the systems characterized by Problem 7-1(d) exhibit pure oscillations?

7-3 Obtain the value of K for which the system of Problem 7-1(e) exhibits pure oscillations. What is the frequency of these oscillations?

7-4 Consider the unity feedback system where the open loop transfer function is

$$G_c(s)G_p(s) = K \left(\frac{s+6}{Ts+1} \right) \left(\frac{5}{s^2 + 5s + 25} \right)$$

Using the Routh-Hurwitz stability criterion find the conditions needing to be satisfied by the parameters K and T ($T \geq 0$) for the system to be stable. Using these conditions construct a stability diagram for the system.

7-5 To determine the relative stability of a system so all the roots of the characteristic equation have a minimum amount of damping substitute $\tilde{s} = s + a$ and apply the Routh-Hurwitz criterion to the modified characteristic equation. This substitution shifts all roots of the equation to the right by a distance a so the Routh criterion will determine whether any of the roots lie to the right of the line, $s = -a$.

The open loop transfer function for a servo system is

$$G(s)H(s) = \frac{K}{s(0.01s + 1)(0.1s + 1)}$$

(a) Determine the limiting value of gain K for neutral stability.

(b) It is desired to have a settling time of 0.5 second. Find the gain K satisfying this requirement.

7-6 What are the restrictions on the frequency response method of analysis?

7-7 Obtain the Bode plot for the following transfer functions:

(a) $G(s) = K(s+1)$ (f) $G(s) = \dfrac{K(s+0.02)}{s(s+0.01)}$

(b) $G(s) = \dfrac{K(s+1)}{s+2}$ (g) $G(s) = \dfrac{K}{s(s+4)(s+5)}$

(c) $G(s) = \dfrac{K}{s}$ (h) $G(s) = \dfrac{K}{s^3}$

(d) $G(s) = \dfrac{K}{s(s+1)}$ (i) $G(s) = \dfrac{K(s+100)}{s^3}$

(e) $G(s) = \dfrac{K(s+2)}{s(s+1)}$ (j) $G(s) = \dfrac{K(s+1)}{s^2+2s+10}$

7-8 Construct the Nyquist plots for the transfer functions shown in Problem 7-7.

7-9 Obtain the frequency at which the Nyquist plot intersects the real and imaginary axis, for Problem 7-7(g), and verify your results via the Routh-Hurwitz method.

7-10 Determine the stability via the Nyquist criterion for the following systems:

(a) $G(s)H(s) \;=\; K/s(s+1)(s+2)$

(b) $G(s)H(s) \;=\; K/s^2(s+1)(s+2)$

(c) $G(s)H(s) \;=\; 2.5(s+1)/(s^2+s+1)$

(d) $G(s)H(s) \;=\; 10(2-s)/s(s+10)(s-5)$

(e) $G(s)H(s) \;=\; K(s+1)/s^2(s+2)$

(f) $G(s)H(s) \;=\; 100/(s-0.5)(s+10)$

7-11 Verify the comments given for the Nyquist plots in Table 7-1.

7-12 For the following transfer functions study the system stability for:

(a) T_3, T_4 very small, and
(b) T_3, T_4 very large.

$$G(s)H(s) = \frac{K(1+sT_3)(1+sT_4)}{s^3(1+sT_1)(1+sT_2)}$$

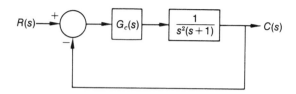

Figure P7-13

7-13 A closed loop system with a controller is shown in Fig. P7-13. The controller $G_c(s)$ is necessary to stabilize the system. If $G_c(s) = (10s + K)$, then what is the maximum value of K for which the system is stable? Obtain the phase and gain margins if $K = 5$.

7-14 The open loop transfer function of system with delay is

$$G(s)H(s) = \frac{Ke^{-sT}}{s(s+1)}$$

If $K = 5$, what is the maximum value of T for which the system is stable?

7-15 Using the Nichol chart, find the maximum overshoot and the corresponding oscillation frequency of the closed loop response if

$$G(s)H(s) = \frac{1}{s(s^2 + s + 2)}$$

7-16 Determine the stability for the following systems using the Bode plot. Compare results to those obtained by using the Nyquist plot.

(a) $G(s)H(s) = K/s(1 + Ts)$

(b) $G(s)H(s) = K(1 - T_1 s)/(1 + T_2 s)$

(c) $G(s)H(s) = K(1 - T_1 s)/s(1 + T_2 s)$

7-17 Obtain the gain necessary to achieve a damping ratio of 0.2 for

$$G(s) = \frac{K}{s(s+1)(s+5)}$$

What is the value of M_p? Obtain the gain and phase margin for this gain setting.

7-18 An idealized nuclear power plant is shown in Fig. P7-18. The controller comprises proportional plus integral control. The actuator and reactor have a time delay. It is required that the system overshoot be less than 20%. What values of K_p and K_i satisfy these criteria?

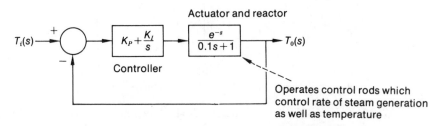

Figure P7-18

7-19 Construct the root loci for the following transfer functions:

(a) $G(s) \;=\; K/\left[s^2 + s + 1\right]$

(b) $G(s) \;=\; K(s + 9)/\left[s(s + 1)(s + 2)\right]$

(c) $G(s) \;=\; K/s^2$

(d) $G(s) \;=\; K/\left[s(s + 1)(s + 2)\right]$

(e) $G(s) \;=\; K/\left[s(s + 1)\right]$

If any of the loci cross the imaginary axis, obtain the frequency at which this happens.

7-20 Obtain the root locus plot for the transfer functions given in Problem 7-7(h) and (i). What happens to the system with the addition of an open loop zero at $s = -100$?

7-21 We would like to make sure that the closed loop response has a damping ratio of 0.5 for a unity feedback system whose forward loop $G(s)$ is given by Problem 7-19(e). What must the static loop sensitivity be set at? Also, what values must K be constrained to if $\zeta < 0.707$ and $\omega_d < 0.8$.

7-22 For the system having the following coefficient matrix:

$$\mathbf{A} = \begin{bmatrix} 0 & 1 & 0 \\ 0 & 0 & 1 \\ 0 & -8 & K \end{bmatrix}$$

show how the eigenvalues vary as K is varied.

7-23 If a damping ratio of 0.707 is required for Problem 7-22, what value of K satisfies this?

7-24 A system whose coefficient matrix is

$$A = \begin{bmatrix} 0 & 1 & 0 \\ 0 & 0 & 1 \\ -K & -12 & -7 \end{bmatrix}$$

is subjected to sinusoidal inputs. Determine the frequency and value of $K = K_c$ at which the system becomes unstable.

7-25 If $K = K_c/2$ for Problem 7-24, determine the gain margin, phase margin, and the crossover frequencies.

7-26 Sketch the root locus for a unity feedback system with a forward loop transfer function given by $G(s) = Ke^{-s}/(s+1)$.

Chapter 8

Control Strategies and Plant Sizing

8.1 Introduction

We have seen in earlier chapters how to model and analyze elementary control systems. During these considerations we have assumed the structure of the control system is known *a priori*, and for simplicity we have largely concentrated our attention on single-input single-output systems. To aid in this work we have introduced the concepts of block diagrams as well as transfer functions and state space models. These concepts ease the burden of analysis and enhance our ability to conceptualize the integrated nature of complex systems. The application of these tools using computer aided design methods has greatly extended the size and complexity of problems which can be analyzed. It has also removed much of the tedium and risks of errors which were attached to the manual methods used previously. The utility of these approaches has been greatly assisted by modern computers and workstations having extensive graphics handling capabilities.

In this chapter we consider two global problems which need to be addressed before the detailed analysis and design of a control system can proceed. These are the selection of the control strategy and topology to be employed for the control system, and the sizing of the main power handling elements for the system. For the elementary problems discussed in the text so far the control strategies which have been used are fairly straightforward. However in many practical control problems this is not the case and much careful thought and recourse to previous experience needs to be called upon in arriving at a successful solution. Also the sizing of the main power handling elements can be a particularly difficult problem due to the many imponderables in the operating environment which can affect their required size. The analysis of plant sizing will often require simplifying assumptions based upon past experience with similar applications.

While these questions are barely discussed in most control system texts the reader should not underestimate the importance of these considerations. Wrong decisions or judgments on these questions during the initial stages of a project can lead to poor system operation or even failure, and can often be very expensive to correct at a later stage; both in time and money.

8.2   Goals for Control System

Hierarchical Control

During the early period of the application of digital computers to control in the process industry, a single computer was used for controlling a large number of control loops by directly generating the input signals of the regulating elements. This approach has been referred to as **direct digital control**. At the time these machines were expensive and much less reliable than modern computers. To ensure adequate reliability many of these systems used back-up computers and also had analog controllers which could be switched into operation if the computers failed. This approach proved to be very expensive, and only practical if large numbers of control loops were present in a control system. A number of notable failures in the early period of computer control development led to it having a bad reputation at that time.

An alternative approach used the computer to determine the controller set-points for individual analog controllers which were designed and adjusted for good transient performance and disturbance rejection. This approach which regulates the controlled variables indirectly is referred to as **supervisory control**. Today many of the analog controllers are replaced by single-loop digital controllers using micro-processors. A block diagram of such a system is shown in Fig. 8-1. In these systems the computer calculates and directly controls the single-loop controller set-points using an optimization algorithm which minimizes the operating cost of the plant. In the event of computer failure, the plant continues to operate using the local controllers with their set-points remaining at the last determined values. This approach has proved very cost-effective and it has often been found that operating cost savings soon pay for the cost of the computer.

Supervisory control is an example where we have separated control at the process level from process optimization at a somewhat higher level.

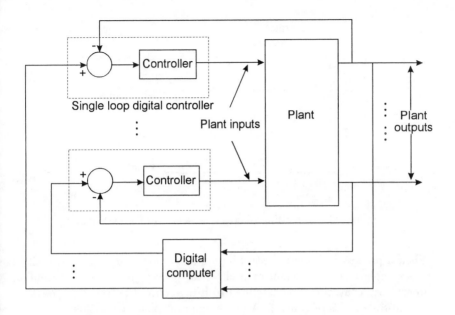

Figure 8.1: Digital computer supervisory control system

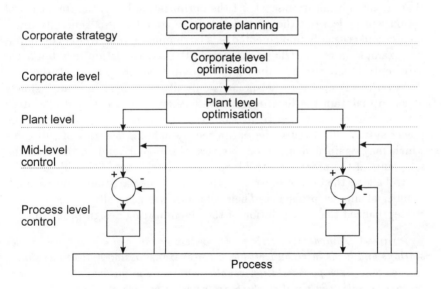

Figure 8.2: Heirarchical control system

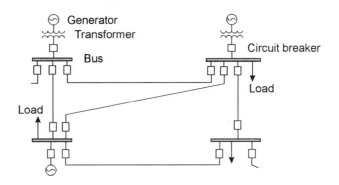

Figure 8.3: Single line diagram of section of power system

When a process has many control inputs and controlled variables, such as for example a power system or a chemical process plant, the concept of supervisory control can be extended to **hierarchical control** where there are multiple levels of control. A block diagram illustrating this concept is shown in Fig. 8-2. This approach generally works quite well since the frequency of control action at the process level usually needs to be much higher than at the optimization level. At the optimization levels the frequency of action will be less but the control algorithm complexity is generally much greater and is usually implemented on a large mini- or mainframe-computer. For example in an electric power station the control action in a temperature control loop may be quite rapid to compensate for plant disturbances. However the temperature set-point is only likely to need to change slowly so as to maintain the best operating efficiency as the external operating environment of the power plant changes. Hierarchical control allows the power system efficiency to be optimized by defining a series of sub-goals which may again define a further series of sub-goals and so on until the set-points of the process control loops are defined. In addition hierarchical control allows the task of controlling the plant to be partitioned into a small number of simpler problems. Their solutions will generally be more easily conceptualized than the solution of the all-embracing design problem.

Example 1 *A modern electric power system consists of a number of power stations and loads interconnected by a power transmission network as shown in Fig. 8-3. Each power station usually contains a number of electrical generators operating in parallel which are powered by steam or water turbines. The loads connected to the power system vary over daily, weekly and seasonal cycles. Since energy cannot be stored in an electric power system, an*

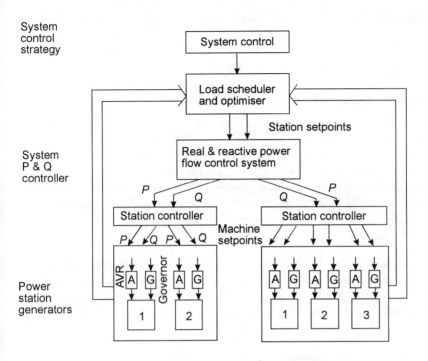

System control strategy

System P & Q controller

Power station generators

Figure 8.4: Hierarchical control of power system

energy balance must be maintained within the system at all times.

Control of a modern power system is usually arranged hierarchically. At the lowest level of control each generator is equipped with an automatic voltage regulator (AVR) and a turbine speed governor, so that if the generator was to operate on a stand-alone basis the generator terminal voltage and output frequency would be regulated, and held at the AVR and governor set-points respectively. When a generator is connected to the power system it is useful to introduce the concepts of real power P and reactive power Q. In this case its real power is controlled by the governor setting and its reactive power flow is controlled by the AVR setting.

In a practical power system the flow of real and reactive power is controlled by the higher level control systems as shown in Fig. 8-4. At the lowest level the set-points, of the governor and AVR for each machine, are determined by the power station controller on the basis of the demanded output from the system controller. At the next higher level the overall operation of the power system is determined by the load scheduler which attempts to optimize the operation of the power system while taking into account all system constraints and plant operating efficiencies. At the highest level the system control strategies and their interaction with management and corporate strategies are determined.

Control and Controlled Variable Identification

As can be seen from the above discussion the design of a control system must proceed using a top-down approach by establishing a series of hierarchical control levels. At each level the **control** and **controlled variables** must be identified, and in some cases new concepts, such as real and reactive power, must be introduced to aid this process. Thus initially we need to identify the overall goal or goals of the control system, and then by moving down the hierarchy we are able to identify the control and controlled variables at each level.

Example 2 *Consider the fluid storage system shown in Fig. 8-5(a) which is part of a chemical processing plant. Suppose the goal of this control system is to maintain the flow q_o at some constant value in spite of variations of the inlet pressure p and the uncontrolled inlet flow q_d. Determine suitable control and controlled variables for this system.*

The dynamical equation for the tank fluid height h is given by

$$A\frac{dh}{dt} = q_i + q_d - q_o \tag{8.1}$$

where A is the tank cross-sectional area, q_i is the fluid inflow, q_d is the disturbance fluid inflow, and q_o is the outflow to the mixing tank in the next stage of the process.

There are a number of possible means of controlling q_o. For instance, suppose the flow characteristics of the outlet pipe is defined by the well known relationship

$$q_o = C_p\sqrt{h} \tag{8.2}$$

where C_p is the pipe discharge coefficient whose value depends upon the pipe diameter, length and surface roughness, and h is the tank fluid height defined above. In this case we might control h which will in turn determine the outlet flow rate q_o. Since q_d is variable and unpredictable we need to control q_i by placing a control valve at A. The angular rotation of the control valve stem θ_A thus becomes the control variable for the system. Since there is usually some uncertainty in the value of C_p it would be difficult to achieve precise control of q_o by regulating h which is sensed by the level transducer LT.

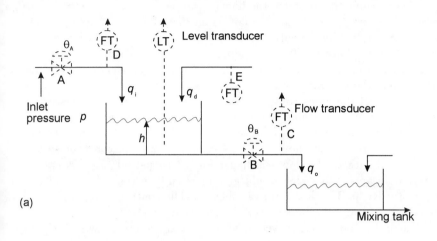

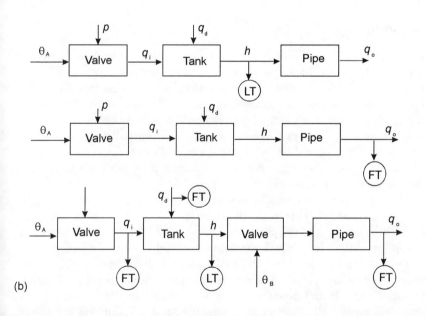

Figure 8.5: (a) Fluid storage system used in chemical processing plant (b) Block diagrams of system showing control and controlled variables

To overcome this problem the flow q_o can be made the controlled variable, and it can be sensed by the flow transducer FT at C. The flow could still be controlled by the valve A which would effect the regulation of q_o by controlling h. Since the flow transducer accurately measures q_o it can be anticipated that a control system using this as the feedback variable would also ensure the accurate control of q_o.

In both of the cases considered above we have ignored the disturbing effect of the unpredictable inflow q_d. This can be quite severe because the typically large volume of the storage tank will result in the feedback system, controlling the flow rate q_o, having a long time constant so that any disturbance to the flow will linger for a long period. In this case an alternative approach which would give more precise control is to place a control valve at B with control variable θ_B. The control valve at A is still required to ensure the tank neither overflows nor runs dry.

Block diagrams for the systems described above are shown in Fig. 8-5(b).

Control Strategy

For the comparatively simple system to be controlled, which was discussed in Example 2, we have identified three controlled variables q_i, q_o, and h, and two control variables θ_A and θ_B. In addition, for some control strategies it can be useful to measure q_d even though it is assumed that it cannot be controlled. Once these variables have been identified the next step is to devise a **control strategy**. While control strategies will be discussed in more detail in Section 8-4 let us examine a feasible strategy for the control of the system shown in Fig. 8-5. A schematic diagram for the control system is shown in Fig. 8-6(a).

A flow control loop controlling q_i is shown using the control valve A and the flow transducer D. This loop will attenuate the effect of the inflow pressure disturbance p. The set-point $q_{i,sp}$ for this control loop is determined by an outer control loop which regulates the fluid level h. In this control arrangement the inflow q_d and the outflow q_o are disturbances within the level control system. Finally a flow control loop is used to regulate q_o using the control valve B and flow transducer C. If correctly tuned this control system will ensure the flow q_o is insensitive to the disturbing effects of fluctuations in the pressure p and the flow q_d.

The block diagram representation of this control system, shown in Fig. 8-6(b), indicates how the various sub-systems interact. If models of the elements are included then the time response of the system to transient disturbances can be found using the methods discussed in earlier chapters. It will be noted that the level control system has multiple feedback loops.

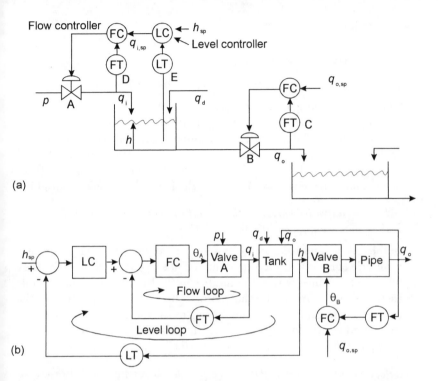

Figure 8.6: (a) Control system for fluid storage system (b) Block diagram of fluid storage control system

This technique is referred to as **cascade control** in the process control field.

8.3 Control and Controlled Variables

Even for the relatively simple fluid storage system considered above we have found six variables which are important in its operation. For this example we have identified the **controlled variables** to be q_i, q_o and h, the **control variables** to be θ_A and θ_B, and the variable q_d to be an uncontrollable **disturbance variable**. This situation where we can identify a multitude of variables related to the operation of the system is not uncommon. It becomes our task once they have been identified to classify them into **control**, **controlled**, and **disturbance variables**. Whilst in many instances this step is often not difficult some simple guidelines listed below are useful.

Controlled Variables

1. *Identify the principal controlled variables for the system or subsystem.*

   The top-down design philosophy identifies these variables readily. For example in a servo drive the output shaft position or speed may be the principal controlled variable. In a numerically controlled machining center the position of the cutter tip may be the principal controlled variable. To simplify analysis we can resolve this position into three orthogonal coordinates X, Y and Z which are the principal variables.

2. *Select as controlled variables those that are constrained by either physical or operational conditions.*

   In the case of a position servo for a numerically controlled machine the armature current and speed are important secondary variables as they must be restrained within the limits specified by the manufacturer so as to prevent damage to the motor. In a gas turbine engine the turbine blade temperature and speed must not exceed limits set by the manufacturer for safety reasons. In this case these variables must be regulated by the engine control system.

3. *Select as controlled variables those that are strongly coupled to a control variable.*

   If the controlled variable does not have a strong, direct and responsive coupling to a control variable then considerable control effort will need to be expended to cause it to change.

   A satellite in circular orbit, which has been studied by Brockett, has the equations of motion

$$\dot{r}(t) = r(t)\dot{\theta}(t)^2 - \frac{k}{r(t)^2} + u_1(t) \tag{8.3}$$

$$\ddot{\theta}(t) = -\frac{2\dot{\theta}(t)\dot{r}(t)}{r(t)} + \frac{1}{r(t)}u_2(t) \tag{8.4}$$

where $r(t)$ and $\theta(t)$ are the orbit radius and angle respectively. He has shown that even if the radial force $u_1(t)$ is zero, due to a defective radial thruster, the radial orbit can still be controlled by the tangential thrust force $u_2(t)$. However, examination of the equations shows that $r(t)$ is only affected indirectly through cross-coupling from the second equation. As the coupling is indirect the required control effort would be much larger than that required if the radial thruster is operative so that $u_1(t)$ can be non-zero.

4. *Choose output variables that are strongly cross-coupled to other controlled variables.*

It is often highly advantageous to control variables which if uncontrolled would act as disturbances to sub-systems involving other controlled variables. For the fluid storage system discussed above it can be seen that outflow q_o is sensitive to the fluid level h through Eq. 8-1. Thus large or rapid fluctuations in h will act as disturbances to the flow control loop regulating q_o. This effect can be minimized by taking h as a controlled variable which is regulated as shown in Fig. 8-6.

Control Variables

1. *Select as control variables those input quantities which have high sensitivity to controlled variables.*

To ensure adequate controllability it is important that the controlled variables be highly sensitive to the selected control variables, and if possible insensitive to disturbance variables. This characteristic is usually achieved by careful design of the plant. For example in an aircraft the elevators strongly affect its pitch angle. Alternatively the ailerons which are strongly coupled to the aircraft roll and yaw angles only have a weak effect upon the pitch angle. Consequently it would make little sense to control the longitudinal motion of the aircraft using the ailerons except in emergencies.

2. *Take as control variables those input variables which rapidly affect the controlled variables.*

In the fluid storage system considered above we have seen that the outflow q_o can be controlled by regulating the fluid level h. Since the speed of response of h is generally quite slow this approach is not recommended. As an alternative the control valve B can be controlled quite rapidly from fully-opened to fully-closed, so that use of θ_B as a control variable will result in rapid control of q_o.

For an aircraft travelling at high speed the ailerons are preferred to the rudder for controlling its heading. This arises because the aircraft roll moment of inertia is generally much less than its yaw moment of inertia so that the aircraft will respond much more rapidly to a changed control command using the ailerons.

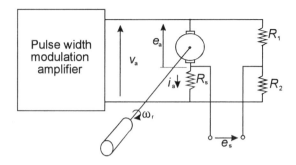

Figure 8.7: Position servo mechanism with speed sensing circuit

3. *Choose as control variables those input variables which directly affect the controlled variables.*

For an aircraft operating at low speed the rudder not only generates a yaw moment, but also a roll moment about the aircraft center of mass. However this roll moment, which is caused by rudder asymmetry about the longitudinal axis, is only a secondary effect compared with the roll moment directly generated by the ailerons. Therefore in a landing approach it is much preferred to used the ailerons for direct control of aircraft roll, and the rudder for controlling its yaw and heading.

In addition to the considerations discussed above for selecting control and controlled variables other factors such as the ability to measure these variables need to be considered. For instance to be able to achieve good control of a variable it is important to have an accurate, noise-free measurement of its value. In addition it is important that these measurements not introduce large time delays which can have an adverse effect upon system stability. It will be seen in Chapter 9 that time delays can severely degrade the response of the system to transient command inputs and disturbances.

At times it may not be practical, economic, or convenient to directly measure controlled variables. This is especially the case when it is not necessary to control the variable with high accuracy, such as the fluid level h in the system discussed above. In such cases it is often convenient to compute the value of the variable from other variables which can be more easily measured or whose measurement is required as part of the overall control system operation.

Example 3 *Consider the position servomechanism driven by a d.c. servo-motor as shown schematically in Fig. 8-7. Since the armature voltage and current are readily measured it is proposed to use these variables to estimate the armature speed. The circuit shown in the above figure can be used for this purpose in place of a tachometer. Analyze this circuit and determine what conditions must be satisfied so that the output signal is a function of the rotor speed.*

From Eq. (2-31) the back e.m.f. is given by

$$e_a(t) = K_B \omega_r(t) \tag{8.5}$$

Also from Eq. (2-32) and Fig. 8-7, when the motor is operating under steady load, the amplifier output voltage $v_a(t)$ is related to the back e.m.f. by

$$v_a(t) = i_a(t)(R_a + R_s) + e_a(t) \tag{8.6}$$

Thus the output $e_s(t)$ of the speed sensing circuit shown in Fig. 8-7 is

$$
\begin{aligned}
e_s(t) &= \frac{R_2}{R_1 + R_2} v_a(t) - R_s i_a(t) \\[2mm]
&= v_a(t)\left[\frac{R_2}{R_1 + R_2} - \frac{R_s}{R_a + R_s}\right] + \frac{R_s}{R_a + R_s} e_a(t)
\end{aligned}
\tag{8.7}
$$

Now if $R_1/R_2 = R_a/R_s$ then Eq. (8-7) becomes

$$e_s(t) = \frac{R_s}{R_a + R_s} e_a(t) = \frac{R_s}{R_a + R_s} K_B \omega_r(t) \tag{8.8}$$

This shows that the output $e_s(t)$ of the speed sensing circuit is directly proportional to the motor angular velocity $\omega_r(t)$.

This circuit or some variant of it is commonly used as part of an internal speed feedback loop with position servo systems where a high accuracy measurement of the motor speed is not required. In practice this method can only be used to measure speed to an accuracy of 2 or 3 percent.

Multiple-Input Multiple-Output (MIMO) Processes

While to date we have concentrated our attention on single-input single-output (SISO) systems we have seen earlier in this chapter that many physical processes have multiple-inputs and multiple-outputs, so we briefly divert our attention to methods of characterizing the interactions in these processes. In the next section we will examine how these models can be used to develop control strategies for MIMO systems by treating them as a series of SISO processes.

Two approaches will be discussed. The first is a natural extension of the transfer function concept introduced in Chapter 3. Suppose we have a system with r inputs which we denote by the r-vector $\mathbf{U}(s)$, and m outputs denoted by the m-vector $\mathbf{Y}(s)$. The relationship between the inputs and outputs can be characterized by the $m \times r$ transfer function matrix

$$\mathbf{G}(s) = \begin{bmatrix} G_{11}(s) & G_{12}(s) & \cdots & G_{1r}(s) \\ G_{21}(s) & G_{22}(s) & \cdots & G_{2r}(s) \\ \vdots & \vdots & & \vdots \\ G_{m1}(s) & G_{m2}(s) & \cdots & G_{mr}(s) \end{bmatrix} \tag{8.9}$$

Suppose the numbers of inputs and outputs are equal i.e. $r = m$, and the components of $\mathbf{U}(s)$ and $\mathbf{Y}(s)$ are ordered so that the strongest couplings are between $U_i(s)$ and $Y_i(s)$, for $i = 1, \cdots, m$. In this case we can view $G_{ii}(s)$ as being the principal transfer function between the input $U_i(s)$ and output $Y_i(s)$ and the cross coupling terms $G_{ij}(s)$, $j \neq i$ can be viewed as being secondary transfer functions or alternatively as disturbance inputs affecting $Y_i(s)$. When the transfer functions $G_{ii}(s)$ are dominant we can often devise useful control strategies based upon the idea that the system consists of m SISO subsystems as will be discussed in the next section.

The second approach is to extend the concept of state equations as discussed in Section 3-6. We suppose the system has n state variables which we denote by the n-vector $\mathbf{x}(t)$ in addition to r inputs and m outputs, which are denoted by the vectors $\mathbf{u}(t)$ and $\mathbf{y}(t)$. In this case we replace the $(n \times 1)$ and $(1 \times n)$ matrices $\mathbf{b}$ and $\mathbf{c}$, by the $(n \times r)$ and $(m \times n)$ matrices $\mathbf{B}$ and $\mathbf{C}$ respectively in Eqs. (3-26) and (3-27). These become

$$\begin{aligned} \dot{\mathbf{x}}(t) &= \mathbf{A}\mathbf{x}(t) + \mathbf{B}\mathbf{u}(t) \\ \mathbf{y}(t) &= \mathbf{C}\mathbf{x}(t) \end{aligned} \tag{8.10}$$

where $\mathbf{A}$ is the **coefficient matrix** as defined in Section 3-6. The **input matrix** $\mathbf{B}$ and **output matrix** $\mathbf{C}$ are given by

$$\mathbf{B} = \begin{bmatrix} b_{11} & b_{12} & \cdots & b_{1r} \\ b_{21} & b_{22} & \cdots & b_{2r} \\ \vdots & \vdots & & \vdots \\ b_{n1} & b_{n2} & \cdots & b_{nr} \end{bmatrix} \quad \mathbf{C} = \begin{bmatrix} c_{11} & c_{12} & \cdots & c_{1n} \\ c_{21} & c_{22} & \cdots & c_{2n} \\ \vdots & \vdots & & \vdots \\ c_{m1} & c_{m2} & \cdots & c_{mn} \end{bmatrix}$$

While both methods of describing MIMO systems are extensively used the transfer function approach clearly exhibits much of the topological structure of the system being considered. This is often lost with the state variable approach.

8.4 Reducing Goals to Control Strategies

Once the control system variables have been identified the next step in a control system design is to develop the control strategy. By this we mean how the control variables and controlled variables are to be inter-connected with the system controllers so as to realize the overall operating goals of the system. Theory is not of immediate use at this stage as it does not tell the designer what parameters to control and how to control them. Instead, at these initial stages, the designer needs to have:

(a) A detailed knowledge of the operation and performance of the system or process to be controlled.

(b) A detailed knowledge of the control elements.

(c) A quantitative and qualitative understanding of the system performance requirements. (It is often difficult to characterize all system performance requirements quantitatively.)

Control strategies are usually evolved by a mixture of the following:

1. Seeing what has been done before on similar plants or systems.

2. Examining where previous strategies have failed on similar plants.

3. Devising methods of remedying previous problems with the control of similar processes.

4. Mulling over the performance requirements of the new system to find feasible control strategies, and then closely examining the advantages and disadvantages of each strategy.

In many cases the control method is quite straight-forward, and does not require any careful research or experimental investigation for its implementation. For example in the control of the temperature of a cooker in a food processing plant it usually suffices to use a single-loop temperature controller to regulate the steam flow which controls its temperature. Fortunately many industrial applications are of this type so that the controller can be installed with the process and then **tuned** so as to give satisfactory operation.

It should also be noted that for many types of plants the control systems have evolved with the development of the process or plant. In these circumstances the control system design for a new plant does not start from scratch. To see this let us examine a typical situation such as the evolution of the control of electric power generating systems:

(a)   In the early period one generator fed the load with the engine speed controlled by a governor. The field excitation was initially controlled manually, but very rapidly automatic voltage regulators were developed.

(b)   With increased loads parallel operation of generators was instituted and to simplify their control the governors and automatic voltage regulators were designed to have speed and voltage droops so as to achieve satisfactory load sharing.

(c)   As the systems became larger with geographically separated power stations, new types of transmission and equipment protection systems were developed, but real and reactive power flows were controlled manually.

(d)   With the development of improved communication systems automatic techniques for load/frequency and volt/reactive power flow control were developed.

(e)   As the boilers for the plants increased in size integrated computer control and fast-valving techniques have been introduced to improve their operation.

This evolutionary process continues to this day. Similar stories apply in many other fields, such as in petrochemical plants where the evolution of control systems has taken place over the last 100 years. In aircraft over the last 70 years complex flight control systems have gradually been introduced beginning with the work of Sperry.

We have seen in earlier sections that the **goals** have to be reduced through a series of hierarchical levels to a set of control tasks. For example in the case of a robot the goal may be to specify the location and orientation of the robot gripper in space which might be specified by six coordinates. The control strategy will determine the sequence of operations of the operating servo mechanisms which are required to achieve this goal. We now discuss some control strategies which have been found useful in various fields.

Cascade (or Minor Loop) Control

If the output of one controller is used as the input to a second controller in a control system then when the system is represented in block diagram form it will be seen to have **nested** feedback loops. This type of control system is variously known as having **cascade** or **minor-loop** feedback control.

To illustrate this concept we refer to the system shown in Fig. 8-6(a). It will be noted that the flow q_i is controlled by the flow controller FC which regulates the setting of the valve A. The set-point for this controller is in this case the output of the level controller LC, which has an independent set-point and a measured variable input given by the output of the level transducer LT. Referring to the block diagram for this system shown in Fig. 8-6(b) it can be seen that the *flow* loop is nested inside *level* loop. In this case the flow loop is considered to be a **minor** or **secondary loop** in relation to the level or **primary loop**.

A key reason for using this type of control is to improve the performance of systems when they are subjected to external disturbances. In cascade control some secondary controlled variable is selected which responds to the disturbance much more rapidly than the primary controlled variable. If this variable is controlled then the effect of the disturbance on the primary controlled variable can be minimized.

In the example discussed above, the disturbance input is the pressure p, which directly and rapidly affects the inflow q_i. Changes in q_i affect the primary controlled variable h, but with a long time constant. The most straightforward control strategy is to directly control the level h by regulating the control valve A. This approach will give unsatisfactory performance because of the slow response of this loop and the fact that corrective action in a feedback system only begins after the disturbance has occurred. By using cascade control q_i is regulated with a fast inner flow loop which results in only small deviations from its set-point value due to the pressure disturbances. This set-point is controlled by the level controller LC. Because of the fast response of the inner loop the effect of pressure disturbance on the performance of the outer loop is largely eliminated.

This type of control is also extensively used with servo mechanisms and with many other types of electro-mechanical control systems. As a second example let us examine a servo mechanism containing a DC motor, coupled to an inertial load. In this case it is common practice to use a control system as shown in Fig. 8-8, where there are three control loops. The inner most is the current loop. This loop is required to limit the current flowing through the armature circuit. If uncontrolled, excessive current could pass through the motor with the risk of damaging the power amplifier as well as causing mechanical damage to the system. This is usually achieved by designing

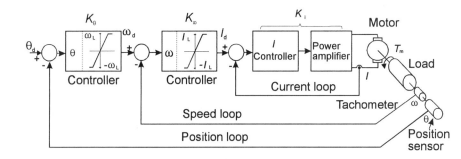

Figure 8.8: Position servo mechanism showing control loops

the current loop control system to be fast acting and to limit the command signal at the output of the speed controller. It is also common to have a speed loop in addition to the position loop so as to improve the transient performance of the system.

Loop Tuning

In systems with minor-loop feedback the question arises of how to adjust the controller parameters of the internal loops. In the case of the servo system described above it is usually aimed to make the speed of response of the control system as fast as possible with minimal overshoot to transient disturbances, and with zero steady-state error for a step input change. For this purpose it is common to use a proportional plus integral controller with the adjustable current limits set in the speed controller so the maximum current will be limited to $\pm I_L$. The design of minor-loop control systems will be discussed further in Section 9-5.

Feedforward Control

Feedforward control can be used quite beneficially when a system experiences external disturbances, where the disturbances or some function of them can be measured. It can be of particular use when feedback alone gives unsatisfactory disturbance rejection. Since feedforward control alone does not introduce any feedback signals its use does *not* lead to any stability problems.

In practice feedforward control is often used in conjunction with feedback control to achieve superior system performance to what can be achieved by either approach alone. In this case the feedforward control is used to

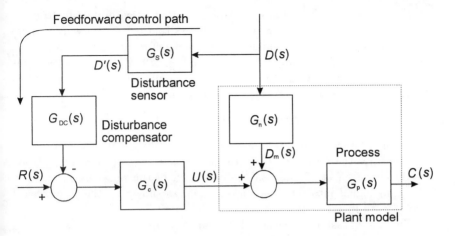

Figure 8.9: System with disturbance input $D(s)$ and feedforward control

minimize the effects of measurable disturbances, while the feedback control is used to allow for small uncertainties in the plant model, measurement noise, and other unmeasured disturbances.

To understand the benefits of feedforward control let us first examine some of the characteristics of feedback control. Its main advantages are:

(a) Feedback action compensates for small errors in the plant model.

(a) Feedback action tends to nullify disturbances of unknown origin.

(c) Corrective action begins as soon as the output deviates from the command input.

However along with these advantages, feedback control presents certain difficulties:

(a) Since the system response relies on the error between the controlled variable at the output and the input command, perfect control is unachievable.

(b) When plants have long time constants external disturbances may cause large output errors for long periods of time.

(c) To maintain small system errors, large gains and considerable control activity is required which can lead to system stability problems.

For these reasons feedback control can often prove unsatisfactory for reducing the effects of disturbances. In these situations feedforward control can be quite beneficial. To understand how feedforward control functions suppose a plant can be modeled as shown in Fig. 8-9. Here the plant output $C(s)$ is determined by the sum of the input signal $U(s)$ and the modified disturbance input signal $D_m(s)$, and is given by

$$C(s) = G_p(s)[U(s) + G_n(s)D(s)] \tag{8.11}$$

If the disturbance $D(s)$ can be measured by the disturbance sensor $G_S(s)$ then we have the estimate $D'(s)$. In feedforward control we can use $D'(s)$ to cancel the effect of $D(s)$ on the operation of the plant by injecting a cancelling signal into the plant input signal as shown by the feedforward control path in Fig. 8-9.

The design task is to select a disturbance compensator $G_{DC}(s)$ so as to minimize the effect of the disturbance $D(s)$ upon the output $C(s)$. From Fig. 8-9 we can see if a feedforward compensator is used then

$$C(s) = G_p(s)G_c(s)R(s) + G_p(s)[G_n(s) - G_c(s)G_{DC}(s)G_S(s)]D(s) \tag{8.12}$$

Thus the disturbance compensator $G_{DC}(s) = G_n(s)/G_c(s)G_S(s)$ will yield perfect disturbance cancellation. Uncertainties in the transfer functions $G_c(s), G_S(s)$, and particularly $G_n(s)$, make this ideal situation unachievable in practice. In addition $G_{DC}(s)$ may be physically unrealizable or unstable. Still, great improvements can be achieved even with imperfect cancellation. This approach is widely used for boiler control and in the process industries.

Both **static** and **dynamic** methods of feedforward control are used. With the first method only the steady-state relationship between the disturbance $D(s)$ and the plant output $C(s)$ is considered, while with the latter method the dynamic effects are also considered.

As mentioned above feedback is often combined with feedforward control to compensate for uncertainties in the plant model, and disturbances which may be unmeasurable for practical reasons. There are many control system arrangements which can be used in combining feedforward and feedback control. Two possibilities are shown in Fig. 8-10. In the first the feedforward controller lies outside the feedback path and so does not affect system stability, while in Fig. 8-10(b) the feedforward controller lies within the feedback loop. It is preferable to arrange dynamic feedforward controllers to lie outside the feedback path so as to ensure they do not affect system stability.

Example 4 *Consider the fluid storage system discussed in Example 2. Use feedforward and feedback control to minimize the effect of the disturbance flow rate q_d and pressure disturbance p upon the fluid level h.*

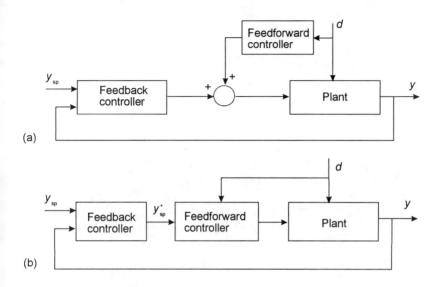

Figure 8.10: Combined feedback and feedforward controllers

From Eq. 8-1 we can see that the fluid level h will remain constant if the material balance

$$q_i + q_d - q_o = 0 \qquad (8.13)$$

is maintained, where the flow rates are as indicated in Fig. 8-11. In addition the flow rate q_i can be shown to be given by

$$q_i = C_v(\theta)\sqrt{p} \qquad (8.14)$$

where $C_v(\theta)$ is the flow coefficient of valve A. As a consequence q_i is functionally dependent upon the inlet pressure.

To minimize the effects of variations in q_o and q_d Eq. (8-13) shows material balance will be maintained if $q_i = q_o - q_d$. The material balance feedforward loop FF1 is combined with the level feedback loop to give the inlet flow rate set-point q_{id}. If $h = h_{sp}$ it will be noted that q_{id} equals the output flow rate q_o, less the disturbance flow rate q_d, thus maintaining material balance. If q_d or q_o experiences a transient change then q_{id} changes by the required amount almost instantaneously, whereas if the feedforward loop was absent then q_{id} would change slowly because of the slow dynamics of the level feedback control loop. The level feedback loop will correct for small errors in the material balance which would otherwise cause the level to drift away from the set-point h_{sp}.

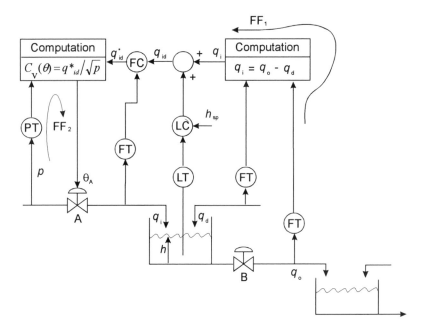

Figure 8.11: Level control system using combined feedforward and feedback control

To compensate for changes in pressure p the feedforward loop FF2 computes the required valve setting θ for the flow set-point q_{id}^* which is the output of the flow feedback controller. The flow feedback control system is included to compensate for small errors in the computation of θ which would lead to q_i differing from q_{id}.

It will be note that the feedforward control systems FF1 and FF2 correspond to those illustrated in Figs. 8-10(a) and (b) respectively.

Variable Structure Control Systems

For applications with electrical motion control systems a need can arise for the system to have a variable control structure. Such controllers can be easily implemented these days using a digital micro processor. An example is the dual-mode controller shown in Fig. 8-12. In this system the non-linear controller could be a relay controller of the type to be discussed in Chapter 11. This controller would be designed to give rapid response to large transient disturbances, so that the error $e(t)$ is minimized quickly. Non-

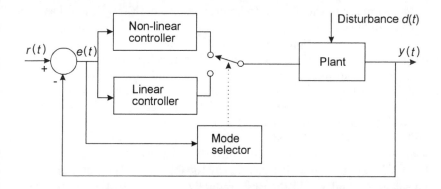

Figure 8.12: Dual-mode control system

linear controllers of this type often exhibit erratic behavior when required to follow slowly varying inputs or disturbances due to the finite switching time of the relay and the sampling time of the control logic. For this reason when $|e(t)|$ and $|de(t)/dt|$ are small, linear controllers often give better performance. Thus the control system has two modes of operation, and the mode controller must select which mode is most appropriate for the given operating conditions.

Another type of variable structure control system is illustrated in Fig. 8-13. Here the control system must operate in two modes where the controlled variable changes when various operating constraints are encountered. This situation can arise in servo applications such as with robot grippers.

In the design of variable structure systems careful consideration needs to be given to the way they transit from one mode to another so as to prevent instabilities or limit cycles from occurring.

Example 5 *A variable structure control system of the type shown in Fig. 8-13 is to be used as an actuator for a robot gripper. The control system is to operate in two modes, these being position control, and force control. Investigate the control strategy required for this system to operate with smooth transfer between modes.*

The output $y_1(t)$ is the displacement of the fingers which engage the object to be gripped, while $y_2(t)$ is the thrust force experienced by the fingers when the object is engaged. Initially the fingers will be disengaged from the object so the registered force will be zero. In this state the fingers will be operated in the position control mode. Once the fingers engage the object the operation of the control system must switch to the force control mode, as many of the objects to be gripped are both fragile and have very low compliance.

During the transition from one mode to another it is necessary for it to occur in a "bump-less" manner. Since PID controllers, of the type discussed in Chapter 6 are to be used for the position and force control loops, care needs to be taken to ensure that the integrators are reset at the correct instants. One approach is to reset the PID2 integrator when the measured force $y_2(t)$ is zero and to operate PID1 as a position servo. In this case the fingers which are disengaged from the object would be gradually closed until a small force $+\varepsilon$ is registered, at which time the system would transfer to a force servo controlled by PID2. Initially the force control input $y_{2d}(t)$ would be set at $+\varepsilon$ and would then be gradually increased to the desired value. While the object is gripped the control input $y_{1d}(t)$ for the position loop would be set equal to the measured value of $y_1(t)$ and the PID1 integrator would be reset.

The process of releasing the object would reverse the above procedures. While the general operating sequence is as described above, the detailed operation of the system as it switches from one mode to another needs to be carefully considered to ensure smooth operation.

Computed Variable Control

It is not always convenient or practical to directly measure a variable which is to be controlled. In this case we may estimate its value using other measurable variables, and then use the computed value as the feedback signal in the control system.

In general this type of control system will not be as accurate as one which uses a directly measured variable, but can be quite useful with a feedforward system which functions as part of an overall feedback system. An illustration of this approach is given in Example 3.

Interacting Control Systems

As we have seen MIMO processes occur frequently. When there are non-zero off-diagonal transfer functions in Eq. 8-9 there are cross couplings between the various control and controlled variables. In this case the development of control strategies becomes more difficult. While many approaches have been developed over the last 30 years, no complete solution for the design of MIMO systems has been achieved.

Criteria for selecting control and controlled variables have been discussed earlier. Once these variables have been selected the task of pairing

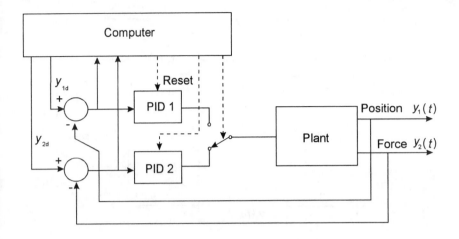

Figure 8.13: Variable structure control system

them for control purposes arises. Some guidelines developed by Houghen are valuable for this purpose:

1. Use control loops which isolate the effects of disturbances upon the system.

   In the fluid flow system discussed earlier, regulating the inflow rate q_i minimizes the disturbing effect of the pressure fluctuations p upon the fluid level h.

2. Feedforward control can reduce the effects of disturbances.

3. Use control elements which respond considerably more rapidly than the system to be controlled.

4. Reduce interactions and couplings between parts of system.

   In the fluid flow system discussed regulating the fluid level h minimizes its disturbing effect upon downstream operation. This together with proper pairing of control and controlled variables has the effect of diagonalizing the transfer function matrix given by Eq. (8-9).

5. Pair control and controlled variables having greatest sensitivity.

6. If a controlled variable cannot be measured use computed variable control for its regulation.

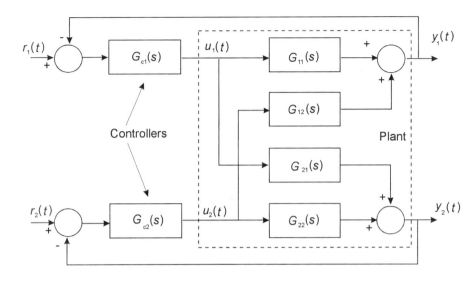

Figure 8.14: Multiple loop feedback control system

Using these guidelines control strategies based upon single-input single-output control system design methods can be developed for MIMO systems. It is important that these considerations only be taken as a guide because interactions between the various loops can at times lead to system instabilities.

Other methods of control and controlled variable pairing such as the relative gain array and singular gain array methods have been developed. Also various design techniques for these systems, such as the diagonal dominance, sequential loop closing, and Nyquist array methods have been intensively developed.

To illustrate how the SISO approach may be used consider the 2-input, 2-output, example shown in Fig. 8-14. Suppose that the cross-couplings are small compared with the main couplings between the control and controlled variables. According to the above guidelines we can design the two control loops independently. In this approach when designing one particular control loop we disregard the influence of the other. This is equivalent to assuming $G_{12}(s)$ and $G_{21}(s)$ both equal zero. However since there is cross-coupling there is a risk that the interactions may degrade the operation of the total system, or even cause instability. Experience has shown that often the single-loop design approach works quite well when the interactions are not severe, so that many practical systems have been designed this way.

Ignoring the possibility of instability for the present, it can be seen that the disturbance rejection capability of feedback systems assists us to understand why the single-input single-output approach works. This can be understood by observing that the signal introduced into the upper loop due to the cross-coupling transfer function $G_{12}(s)$ can be visualized as a disturbance signal. If the upper loop is well designed this disturbance will be strongly rejected. A similar situation will exist for the lower loop, so it can be expected that each control loop will function as if the other is absent. These observations are often borne out in practice when the cross-couplings are weak.

When the single-loop approach fails to given satisfactory performance other techniques, which take the cross-coupling into account need to be used. One approach is the use of **decoupling controllers**, where the cross-coupling is either fully or partially cancelled by the inclusion of additional elements in the controller. Various approaches have been considered including **static** and **dynamic decoupling**. For the former case, only the static effects of the cross-coupling are removed, while in the latter case the dynamic effects are also cancelled. Difficulties can arise with dynamic decoupling because some of the introduced elements may be physically unrealizable so that the scheme cannot be implemented. Also when the decoupling is computed for a nominal plant, uncertainties in the plant model will lead to a partially decoupled plant. Single-loop control system designs based on the decoupled plant can yield poor overall system designs as often the success of these designs depends upon having an accurate plant model. A block diagram for a 2-input, 2-output system showing the decoupling elements and the single loop feedback paths is shown in Fig. 8-15.

Numerous other techniques including the method of **diagonal dominance**, pioneered by Rosenbrock, have proved useful because they can take the uncertainties in the plant model into account during the design process.

8.5 Examples from Applications

Our goal in this section is to illustrate how the concepts introduced in the earlier parts of the chapter can guide us in devising control strategies for a system. Examples from the two disparate fields of chemical engineering and electrical engineering have been chosen so as to illustrate the broad applicability of these principles. In the first example we will examine the control of a distillation column, while the second will consider the speed control of an alternating current induction motor. In our treatment neither of these topics will be discussed exhaustively. The interested reader should refer to the existing extensive bodies of literature for detailed discussions

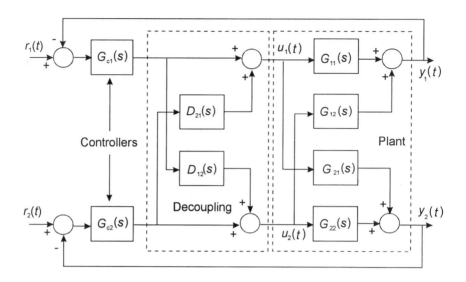

Figure 8.15: Multiple loop feedback control system with decoupling

of these applications. Some useful references are listed at the end of the chapter.

Control of a Distillation Column

Distillation columns are extensively used in the chemical industry for purifying a fluid stream containing a mixture of various products. In this example we will assume the feed F moles/hr consists of a binary mixture of light and heavy fractions. The column separates the feed into D moles/hr of distillate and B moles/hr of bottom-product. We investigate a suitable control scheme for this column so as to assure high quality distillate.

In a distillation column, as illustrated in Fig. 8-16, the feed F is introduced part way up the column. The vapor is generated by the re-boiler and as it rises in the columns it becomes richer in the light fraction and leaner in the heavy fraction. When the vapor reaches the top of the column it is condensed in a heat exchanger and collected in the reflux drum. By careful control of the column operating conditions the distillate will principally contain light fractions, while the bottom-product will mainly contain heavy fractions.

In such a column the controlled variables are identified to be the distillate and bottom-product qualities, the re-boiler and reflux drum liquid

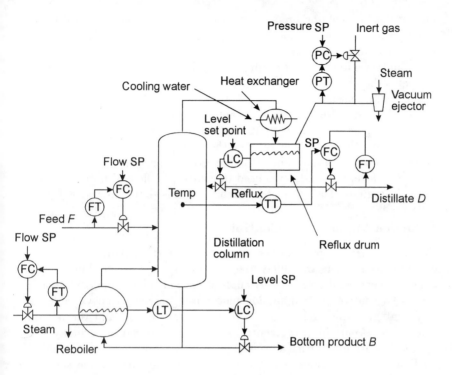

Figure 8.16: Control system for distillation column

levels, and the column pressure. The control variables on the other hand
are identified as the product flow rates, the reflux flow rates, the heat input,
and heat removal. The product composition is inferred by the temperatures
at the top and bottom of the column, as it is difficult to directly measure
product quality.

Since material balance must be maintained, the product outputs must
satisfy $F = B + D$ at all times. Also suppose the fraction of the distillate
in the feed F is known. As the goal is to control the quality of the distillate
D we need to set D to be just below this fraction if high quality distillate
is to be obtained. Thus the control scheme must regulate the distillate flow
rate as shown in Fig. 8-16. The bottom-product B is thus determined from
the above relationship.

Since we wish to control the distillate composition and it is known the
temperature at the top of the column is a good measure of this quality,
it is proposed to use this temperature to regulate the flow D. It is also

known that the column pressure must be accurately maintained if column temperature is to be a good measure of product composition. To achieve this a vacuum ejector and inert gas bleed is used for pressure control. Also the liquid level in the reflux drum is controlled by a level control loop which regulates the flow in the reflux circuit. The control system for the top-end of the column is shown in the upper part of Fig. 8-16.

In addition, to maintain a constant vapor flow in the column it is necessary to regulate the heat input into the re-boiler. This can be achieved by regulating the steam flow through the re-boiler using a flow control loop on the steam line as shown in the figure. Also to prevent the re-boiler from flooding a level control loop is used to control the flow of bottom-product from the system. The complete control system is shown in Fig. 8-16.

Induction Motor Speed Control

Electrical variable speed drives using a.c. induction motors with power electronic controllers are widely used in industry, and are gradually replacing other types of drives in many applications. In this example we examine the speed control of an induction motor using a computed variable control strategy[1] which regulates the motor stator current and the machine load angle. Apart from the control strategy considered here, many other approaches are used or have been suggested by researchers in the field.

Examining the block diagram for the variable speed drive shown in Fig. 8-17 we see the power-electronic circuitry consists of two major subsystems. The first is an a.c. to d.c. converter where the magnitude of the output voltage is determined by the control signals at point A. The other sub-system is an inverter which converts d.c. to a.c. The output frequency of the inverter is controlled by the signals at point B. Thus we can envisage the plant as consisting of the d.c. converter, the a.c. inverter and the induction motor, and as having two control variables e_A and e_B.

The measured variables for this system are the converter output current, measured at C, the stator currents and voltages measured at D, and the motor speed measured at E. From these measured variables the controlled variables, motor magnetizing current $|i_{mr}|$ and load angle δ, are computed by the feedback acquisition block shown in Fig. 8-17.

For successful operation of this type of variable speed drive the load angle δ and magnetizing current i_{mr} should be controlled. However direct control of $|i_{mr}|$ is not possible. The control strategy adopted is shown Fig. 8-17. The set point ω_r^* is compared with the measured speed ω_r, and the speed error is used to compute the required electrical torque τ_e^*

---

[1] This description is based upon the work of Mr. T.J. Bergin which was submitted as a requirement for the Master of Engineering Degree.

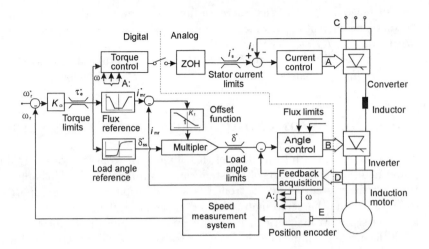

Figure 8.17: Variable speed drive control system block diagram

which is limited by the torque limits. These limits are generally set to values determined from the motor design. From a theoretical analysis it can be shown that the optimum stator current i_s^* can be computed from the electrical torque τ_e^*, the motor load angle δ, and the rotor current i_{mr}. To prevent damage to the power electronic systems and the motor, it is necessary to limit the stator current as shown in the figure. After limiting, i_s^* is used as the command input for the current feedback control loop which regulates the stator current i_s.

For the induction motor it is also necessary to control the machine load angle δ. Analysis shows that under steady state conditions this angle should have a constant value with its sign determined by the sign of τ_e^*. In the figure the load angle command δ^* is computed, and after being limited by the load angle limits is fed to the load angle controller which regulates the machine load angle δ.

Under transient conditions it is not feasible to control the load angle to the value determined for steady-state conditions. Instead the load angle must be varied by the offset function K_f, and only allowed to approach its steady state value when the measured value of the magnetizing current $|i_{mr}|$ approaches its computed steady state value $|i_{mr}^*|$. The rate at which this can occur is limited by the rate at which the airgap magnetic flux can change to assume its new steady state value. This rate is slow when compared with the time response of other parts of the electrical system. To understand how this part of the system operates imagine there is a step increase in τ_e^*. The flux reference function computes the steady state magnetizing current $|i_{mr}^*|$, which is compared with $|i_{mr}|$ and then fed to the offset function K_f. Under these conditions $(|i_{mr}^*| - |i_{mr}|)$ will be positive, K_f will be less than unity and $\delta^* = K_f \delta_{ss}^*$ will be less than its steady state value. As $|i_{mr}|$ increases to its steady state value δ^* approaches δ_{ss}^*.

8.6   Sizing of Components and Subsystems

As mentioned in the introduction of this chapter the selection of the main power handling components of a system can be a difficult task. This applies whether the system being controlled is a distillation column in a petro-chemical plant, the servo-motor driving an equatorial mount for a large radio telescope, or magnetic bearings in a micro-machine drive motor which might fit inside a 1 mm cube. In practice this arises because there are many factors affecting their sizing which cannot be easily quantified.

In our present discussion it is impossible to consider questions relating to the sizing of every type of plant. We restrict our consideration, in the following, to the selection of motors with or without geared couplings for servo mechanisms. Many of the principles used in the selection of the motors and geared couplings can be translated with obvious modifications to other fields. It will be left to the reader to make these connections.

Estimating the Range of Controlled Variables

It is usually possible to arrive at the performance specification of a control system by considering the operation of the system to which it is coupled. However it is often only possible to characterize the system performance in very general terms where the types of operations considered are idealized to simplify analysis. As so many influences are neglected it is common to use fairly generous factors of safety when selecting the power elements.

Example 6 *A numerically controlled machining center has a servo-drive for longitudinal control of the compound table as shown in Fig. 8-18(a). Suppose the cutter radius $r_c = 10$ mm and is absorbing a power of 30 kW when rotating at 3000 rpm. The table can be assumed to have a mass $M = 200$ kg, and the slides to have a coefficient of friction $\mu = 0.1$. The recirculating ball screw which is used to couple the rotary drive motor to the linear motion of the table has an efficiency $\eta = 0.8$ and a pitch $p = 25$ mm. Estimate the required size of the servo drive motor assuming the table is travelling at a constant speed of 10 mm/sec.*

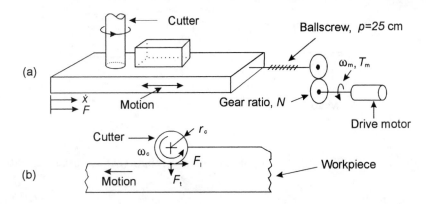

Figure 8.18: Longitudinal servo-system for numerically controlled machine

While it is difficult to precisely estimate the magnitude and direction of the cutting forces, the worst case will occur when the transverse force F_t, defined in Fig. 8-18(b), is assumed to be zero, and the longitudinal force F_l is assumed to equal the cutter tip force. The cutter torque $T = F_l r_c$, so that the spindle input power is given by

$$P_s = \omega_c F_l r_c$$

From the data given above

$$F_l = \frac{30 \times 10^3}{\dfrac{2\pi \times 3000}{60} \times 0.01} = 9549 \text{ N}$$

The table friction force

$$F_f = \mu M g = 0.1 \times 9.81 \times 200 = 196 \text{ N}$$

Since the table is travelling at 10 mm/sec and the screw efficiency $\eta = 0.8$ we find the servo motor power is

$$P_m = \frac{(9549 + 196) \times 10 \times 10^{-3}}{0.8} = 121.8 \text{ watt}$$

Since many factors have been neglected in this analysis a safety factor of 2 is advisable, so that the motor power would be 244 W.

It will also be noted that the speed of the ball screw is 24 rpm when the table is travelling at 10 mm/sec. Since most electric servo-motors operate at speeds up to about 3000 rpm it can be seen that the geared coupling should have a ratio of at least 20:1.

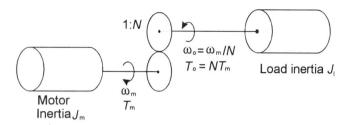

Figure 8.19: Servo motor coupled to load through gear-train having ratio $1 : N$

Factors of Importance in Component Selection

Some of the factors which need to be examined when selecting a d.c. drive motor for a servo system are now considered. The relative importance of these factors will depend upon the particular application. Also, similar considerations apply to the selection of other types of plant.

Maximum Slew Rate   Mechanical strength and other design considerations limit the maximum rotational speed $\omega_{m,\max}$ of electrical servo-motors. This speed is usually stated by the motor manufacturer in their specification information. For other types of actuators the maximum speed of operation is also usually specified by the manufacturer. For example flow limitations due to turbulence in hydraulic motors can limit their maximum speed.

Since it is usually desirable for a servo system to operate from zero to some maximum output speed $\omega_{o,\max}$, it is often necessary to use a gearbox or other method of coupling the motor to the load as shown in Fig. 8-19. In this case we find the gear ratio N must satisfy

$$N \leq \frac{\omega_{m,\max}}{\omega_{o,\max}} = N_{vl} \tag{8.15}$$

Friction Loading   Ignoring inertial loading, which is equivalent to assuming the system is operating at constant speed, the main form of load on the drive motor is due to the friction torque T_f. In this case the output power $P_o = \omega_o T_o = \omega_o T_f$. Supposing the motor is coupled to the load by a gearbox having an efficiency η we find

$$P_o = \omega_o T_o = \eta \omega_m T_m$$

The torque-speed characteristic of a typical d.c. servo-motor is shown in Fig. 8-20. The motor output torque T_m is related to the motor speed ω_m

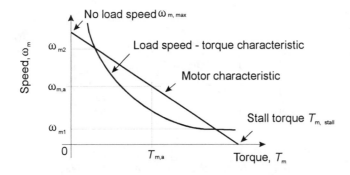

Figure 8.20: Torque-speed characteristic for d.c. motor

by

$$\omega_m = P_o/\eta T_m \tag{8.16}$$

This relationship is plotted in Fig. 8-20 and it can be seen for speeds between ω_{m1} and ω_{m2} the motor output torque exceeds the load torque reflected through the gearbox. Thus the motor is capable of driving the load if the gearbox ratio N lies in the range

$$N_{fl1} = \frac{\omega_{m1}}{\omega_o} \leq N \leq \frac{\omega_{m2}}{\omega_o} = N_{fl2}$$

Torque to Friction Ratio Suppose the friction torque acting on the gearbox output shaft is T_f. Practical experience has shown that if a servo is to smoothly track a slowly varying reference input then the maximum motor output torque referred to the gearbox output shaft should be at least an order of magnitude greater than T_f. Thus the gearbox ratio must satisfy

$$N \geq 10\frac{T_f}{T_{m,\text{stall}}} = N_f \tag{8.17}$$

Initial Acceleration The initial output acceleration $\ddot{\theta}_o$ of the servo-system can be found from the equation

$$(J_\ell + \eta N^2 J_m)\ddot{\theta}_o = \eta N T_{m,\text{stall}} - T_f \tag{8.18}$$

It is useful to plot the initial acceleration $\ddot{\theta}_o$ as a function of the gear ratio N as shown in Fig. 8-21. Also indicated on this diagram are the gear ratio

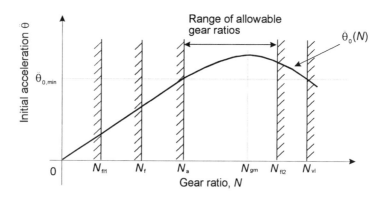

Figure 8.21: Initial output acceleration of geared d.c. motor as a function of gear ratio N

limits, N_{vl}, N_{fl}, N_{fl2}, and N_f. If a minimum initial acceleration $\ddot{\theta}_{o,\min}$ is specified for the servo-system then the gear ratio N_a can be read from this diagram. Also from this diagram we can determine the range of gear ratios which meet all of the constraints discussed in this section.

The gear ratio for maximum initial acceleration occurs at $N_{g,\max}$ which can be computed from Eq. (8-18). For the case shown in Fig. 8-21 $N_{g,\max} > N_f$, and it can easily be demonstrated that

$$N_{g,\max} \cong \sqrt{\frac{J_\ell}{\eta J_m}} \tag{8.19}$$

Range of Output Speeds for Minimum Acceleration $\ddot{\theta}_{o,\min}$ For the case illustrated in Fig. 8-21 the gear ratio N can be selected to lie in the range $N_a \le N \le N_{fl2}$. If the selected $N > N_a$ then there will be a range of speeds for which the acceleration $\ddot{\theta}_{o,\min}$ can be maintained. From Eq. (8-18)

$$T_{m,a} = \frac{1}{\eta N}\left[T_f + \ddot{\theta}_{o,\min}(J_\ell + \eta N^2 J_m)\right] \tag{8.20}$$

Using the result obtained in Eq. (8-20) the speed $\omega_{m,a}$ can be read from Fig. 8-20. The output acceleration will exceed $\ddot{\theta}_{o,\min}$, for all output speeds from zero to $\omega_{o,a} = \omega_{m,a}/N$.

Static Resolution   When **coulomb friction** is present in a servo system having proportional control the reference input needs to change by a small increment before the system responds. The magnitude of this small change which is referred to as the **static resolution** ultimately determines the system accuracy irrespective of the accuracy of any transducer used for measuring the servo output variable. While the subject of **error budgeting** is beyond the scope of this book it suffices to observe that for digitally controlled systems if the servo static resolution is less than the resolution of the digital-to-analog converter (DAC) at the computer output, then the static resolution is determined by the computer.

Suppose the coulomb friction torque at the output of the position servo shown in Fig. 8-8 is T_f, and suppose the proportional gains of the position and speed controllers are K_θ and K_ω respectively. In addition the sensitivity of the current control system is denoted by K_i. To compute the static resolution imagine that the position reference input signal θ_d is gradually increased until the motor output torque $T_m = T_f$, and the shaft is just on the verge of rotating. In this case $T_f = K_i K_\omega K_\theta \theta_d$, so the

$$\text{Static Resolution} = \frac{T_f}{K_i K_\omega K_\theta} \tag{8.21}$$

From this result we see that the static resolution is determined by the gains of the elements in the forward path and is unaffected by the elements in the feedback paths.

Example 7 *A servo-motor having $T_{m,\text{stall}} = 4.55 \times 10^{-3}\ Nm$ and $J_m = 0.89 \times 10^{-7}\ kg\ m^2$ is to be coupled to a load of $150 \times 10^{-7}\ kg\ m^2$ by a gearbox. Find the gear ratio which maximizes the initial acceleration, and find this acceleration if the gearbox efficiency $\eta = 0.8$.*

From Eq. (8-19)

$$N_{g,\text{max}} = \sqrt{\frac{150 \times 10^{-7}}{0.89 \times 10^{-7} \times 0.8}} = 14.5$$

From Eq. (8-18)

$$\ddot{\theta}_o = \frac{0.8 \times 14.5 \times 4.55 \times 10^{-3}}{300 \times 10^{-7}} = 1759\ \text{rad/sec}$$

Example 8 *An instrument servo is to be designed to drive an inertial load $J_\ell = 8.4 \times 10^{-6}\ kg\ m^2$ which is subjected to a friction torque $T_f = 1.4 \times 10^{-3}\ Nm$. The load output speed ω_o should exceed 200 rpm, and its initial acceleration should be greater than 1200 rad/sec². Select a servo-motor and gearbox ratio for this application assuming that the maximum motor speed is 3000 rpm. It may be assumed that the class of motors has a stall torque to inertia ratio $T_{m,\text{stall}}/J_m = 30,000$, and the gearbox efficiency $\eta = 0.8$.*

Since $\omega_o = 200$ rpm we find from Eq. (8-15) that the velocity limit gear ratio is

$$N_{vl} = \frac{3000}{200} = 15$$

If the load inertia is operating at a constant speed of 200 rpm (20.94 rad/sec) then the output power $P_o = 1.4 \times 10^{-3} \times 20.94 = 29.32$ mW. Since the gearbox efficiency $\eta = 0.8$, from Eq. (8-16)

$$\omega_m T_m = 29.32 \times 10^{-3}/0.8 = 36.65 \times 10^{-3} \text{ W}$$

Taking a trial value for the motor stall torque $T_{m,\text{stall}} = 0.005$ N m the motor performance and load speed-torque characteristic can be plotted as shown in Fig 8-20. From this plot the speeds $\omega_{m1} = 7.47$ rad/sec and $\omega_{m2} = 306.5$ rad/sec are obtained, and since $\omega_o = 20.94$ rad/sec we find $N_{fl1} = 0.357$ and $N_{fl2} = 14.6$.

From the torque-to-friction ratio condition we find

$$N_f = \frac{10 \times 1.4 \times 10^{-3}}{0.005} = 2.8$$

The above limiting ratios for the gearbox ratios are plotted on Fig. 8-22. From Eq.(8-18) we find

$$\ddot{\theta}_o = \frac{4000N - 1400}{8.4 + 0.1667N^2}$$

which is also plotted as a function of N on Fig. 8-22.

For the selected motor all constraint conditions, including an initial acceleration of 1200 rad/sec^2, are achieved for gear ratios $3.7 < N < 14.6$. Practical considerations such as the availability of gearboxes with certain ratios may dictate the final choice of gear ratio.

Suppose the selected gear ratio is $N = 10$. The range of output speeds for which $\ddot{\theta}_{o,\text{min}}$ can be maintained can now be determined from Eq. (8-20)

$$\begin{aligned} T_{m,a} &= \left[1.4 \times 10^{-3} + 1200 \left(8.4 \times 10^{-6}\right.\right. \\ &\quad \left.\left. + 0.8 \times 10^2 \times 0.1667 \times 10^{-6}\right)\right] / (0.8 \times 10) \\[6pt] &= 3.44 \times 10^{-3} \text{ N m} \end{aligned}$$

From the speed-torque characteristic of the motor the above acceleration can be maintained for motor speeds from zero to 938.7 rpm, which corresponds to an output speed of 93.87 rpm.

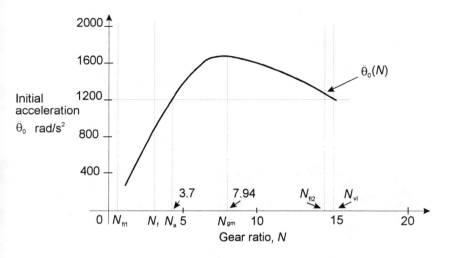

Figure 8.22: Initial acceleration of instrument servo as a function of gear ratio N

Thermal Constraints of Motor and Controller Most d.c. servo-motors have a speed-torque characteristic similar to the one shown in Fig. 8-20. However, if the motor is loaded with a constant torque then the maximum operating torque will be limited to a value considerably less than $T_{m,\text{stall}}$ because of heating effects due to the current in the armature winding. For this reason the RMS output torque and RMS armature current should always be less than the RMS ratings of the motor or controller.

Example 9 *Consider the numerically controlled machining center discussed in Example 6, and suppose the velocity profile of the longitudinal motion is as shown in Fig. 8-23. The characteristics of some typical servo-motors are shown in Table 8-1. Find the most suitable gear ratio and servo-motor for driving this machine, and determine the peak and RMS current ratings for the motor and controller.*

From the frictional and cutting forces computed in Example 6, and the inertial forces required to accelerate the table, which are computed for the intervals ab, cd, de, fg, the force generated by the lead-screw in the various intervals is shown in Fig. 8-23(b). These forces are referred to the motor output shaft by taking into account the gear-ratio N, the screw-pitch p,

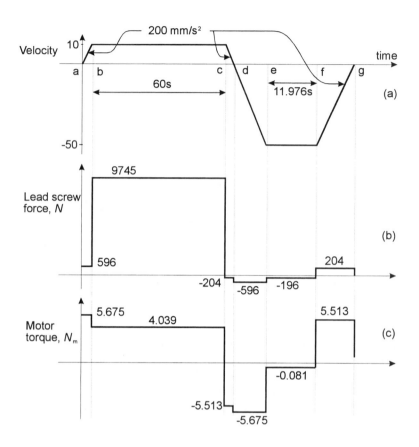

Figure 8.23: Load characteristics of servo-system for numerically controlled machine

Table 8-1 Typical servo motor characteristics

Motor Type No.	1	2	3	4
$T_{m,\text{stall}}[\text{N m}]$	35.5	50	50	50
$T_{m,\text{rms}}[\text{N m}]$	3.8	3.8	3.8	3.8
n, no load [rpm]	2096	2962	4192	5388
n, max. oper. [rpm]	1500	2000	3000	4000
J_m [kg m^2]	0.0009	0.0009	0.0009	0.0009
K_T [N m/A]	0.82	0.58	0.41	0.29

and the screw efficiency η. Referring to Fig. 8-18 it can be shown that

$$\dot{x} = \frac{p}{2\pi N}\omega_m, \quad \text{and } T_m = \frac{p}{2\pi N \eta}F_l$$

We take as a trial $N = 12$. Using these relationships, and including the motor armature inertial load, the motor torque in the various intervals is as given in Fig. 8-23(c). The RMS torque is computed from this plot and found to be $T_{m,rms} = 3.69$ N m which shows that the Motor Type No. 1 is suited for this application. Using the armature torque constant $K_T = 0.82$ N m/A, we find $I_{arm,max} = 5.675/0.82 = 6.93$ A, and $I_{arm,rms} = 3.69/0.82 = 4.5$ A. The motor controller must be capable of safely handling this RMS and peak current.

8.7   A Design Example

To further illustrate the application of the concepts discussed above we examine an approach to the design of a road load simulating dynamometer for use as part of a system for measuring the pollution emission and the fuel economy of motor vehicles. These systems are used in many countries for determining whether new vehicle designs meet their strict environmental and fuel economy standards.

In these machines the driven wheels of the vehicle to be tested are positioned on rollers which are loaded by an electrical variable speed (VS) drive so as to simulate the loading on the vehicle as it is driven according to a specified driving cycle. This arrangement is illustrated in Fig. 8-24. At the same time the exhaust emissions and fuel consumption are monitored.

To simulate the road operation the road load force F_R will be

$$F_R = A + Bv + Cv^2 + Mg\sin\theta + M\frac{dv}{dt} \qquad (8.22)$$

where the coefficients A, B, C represent the rolling friction and wind drag, v is the vehicle speed, M is the vehicle mass, $g = 9.81$ m/sec^2, and θ is the road grade angle. The dynamometer must simulate the operation of a vehicle according to the law given in Eq. (8-22) with an appropriate set of coefficients.

One approach to the simulation of the road load law is to measure the road speed v and acceleration dv/dt and then calculate the road load force F_R. This load force can then be used as the input command for the VS drive which is arranged as a feedback system controlling its output torque (or force F_R) as shown in Fig. 8-24(a). As the VS drive is coupled to the machine rollers this arrangement controls the vehicle tractive effort F_R. A

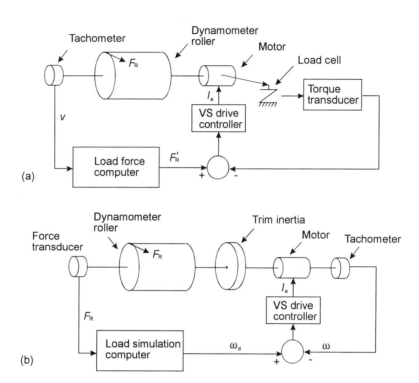

Figure 8.24: Road load simulating dynamometer (a) Load force control strategy (b) Vehicle speed control strategy

disadvantage of this approach is that the inertia of the motor armature and rollers, as well as the various friction forces, appear outside the control loops. As a consequence these effects must be independently measured by a complicated calibration procedure and then compensated for in computing the force F_R. Thus if the residual effects are A_r, B_r, C_r and M_r then the command signal F_R' to the force control system is

$$F_R' = (A - A_r) + (B - B_r)v + (C - C_v)v^2 + Mg\sin\theta + (M - M_r)\frac{dv}{dt}$$

The troublesome problem with this approach is that the vehicle mass M must exceed M_r for the system to be stable. This means the effective roller mass must be lighter than the lightest vehicle to be tested which poses severe mechanical design difficulties. Another problem is the measurement of the acceleration dv/dt, which must be computed from the tachometer output signal, and will thus be a very noisy signal. This will result in significant errors in F_R'.

To overcome these difficulties an alternative strategy is considered where the tractive effort F_R is directly measured and the speed corresponding to this force is computed by integrating the differential equation Eq. (8-22) in the simulation computer. This computed speed is used as the control variable for a high performance VS drive which will accurately simulate the load law given by Eq. (8-22). The key advantages of this approach are that all residual effects are contained within the speed control loops, and no derivatives need to be calculated. As a consequence no calibration procedure is required to compute the residual values of the parameters A, B, C and M. An additional benefit of this approach is that vehicle masses smaller than the roller mass can be easily simulated. This control scheme is illustrated in Fig. 8-24(b) and is the one which is considered in the remainder of this section.

We now consider the selection of the motor and drive controller which are the key power handling elements. Mechanical considerations show that when the vehicle is travelling at 100 kph the roller speed is 435 rpm for 1.22 m rollers. Also it is assumed that the maximum motor speed is 2500 rpm. The resulting gear ratio between the roller and drive motor is 5.75. It has been computed that the roller inertia $J_r = 300$ kg m^2 and the vehicles to be simulated have masses lying in the range 450 kg to 2700 kg. It has also been specified that the maximum vehicle acceleration is 3.6 m/sec^2, and the maximum drag force, which is due to the first three terms in Eq. (8-22), is 580 N.

From the above data, when the vehicle is accelerating the maximum tractive effort

$$F_R = 3.6 \times 2700 + 580 = 10,300 \text{ N}$$

which translates to a roller torque of 6283 N m. Taking into account the roller inertia, and the gearing ratio, the peak motor output torque is

$$T_m = \frac{6283 - 300(3.6/0.61)}{5.75} = 785 \text{ Nm}$$

It will be observed in the above equation that by increasing the roller inertia we can reduce the peak motor output torque. Thus by operating the drive motor in all four quadrants (i.e. forward and reverse, and motoring and generating) and increasing the roller inertia by adding a trim flywheel, we can reduce the motor size. The optimum inertia for the trim flywheel is $J_f = 286$ kg m^2. In this case the base load inertial, including the roller inertia, is 586 kg m^2. Computing the motor torque for the extreme vehicle masses and for maximum acceleration and deceleration we obtain the results shown in Table 8-2. These results show that by operating in all four

Table 8-2 Motor torque for extreme operating conditions

Vehicle Mass kg	Acceleration $+3.6$ m/sec^2	Deceleration -3.6 m/sec^2
2700	-490.0 N m	-366.5 N m
450	370.0 N m	492.0 N m

quadrants the peak motor torque is considerably reduced from 785 N m to 490 N m.

In the absence of further information about the load cycle of the motor a conservative approach in selecting the motor and drive is to assume it continuously accelerates and decelerates at ± 3.6 m/sec^2. This indicates the dynamometer needs a 130 kW VS drive with a maximum speed of 2500 rpm, and that the motor should be geared to the roller with a ratio of 1 : 5.75.

8.8   Summary

We have examined two facets of control systems which are of utmost importance in their development and design. These are control system strategies and power element sizing. Poor decisions about either of these aspects can often lead to difficult and costly remedies at a later stage, so it is important to give careful consideration to these questions during the initial stages of a project.

The development of control strategies often involves an amalgam of careful analysis, inspired insight, and depth of experience with similar systems. Although the procedures for evolving control strategies are not subject to mathematical precision we have given some guidelines which have proven useful in a number of fields such as process control and servo-mechanisms.

The selection of power handling elements also poses difficulties because of many uncertainties in the system operating environment which can affect their required size. It has been shown, by identifying the range of the key parameters to be controlled, and by idealizing the types of operations required, that useful information about the sizing of these elements can be obtained. Once the operating envelope of the controlled variables is determined analytical techniques can be used for the sizing of the power elements. However, because so many influences are often neglected or overlooked it is common to use fairly large factors of safety to minimize the risk of a poor choice.

We have used examples from various fields to illustrate these problems.

However theory is no substitute for careful thought, experience, and cautious experimentation in the various fields of application.

8.9   References

1. J. O. Houghen, *Measurement and Control Applications*, Instrument Society of America, Pittsburgh, Pennsylvania, 1979.

2. F. G. Shinskey, *Process Control Systems — Application, Design and Adjustment*, Mc Graw Hill Book Co., N. Y., 1988.

3. D. G. Seborg, J. F. Edgar, D. A. Mellichamp, *Process Dynamics and Control*, John Wiley and Sons, Inc., N. Y., 1989.

4. E. O. Doebelin, *Control System Principles and Design*, John Wiley and Sons, Inc., N. Y., 1985.

5. O. J. Elgerd, *Energy System Theory*, Mc Graw Hill Book Co., N. Y., 1982.

6. R. W. Brockett, *Finite Dimensional Liner Systems*, John Wiley and Sons, Inc., N. Y., 1970.

7. W. H. Rosenbrock, *Computer Aided Control System Design*, Academic Press, Inc., London, 1974.

8. K. J. Astrom, B. Wittenmark, *Computer-Controlled Systems*, Prentice-Hall International, Inc., Englewood Cliffs, N. J., 1990.

9. P. Harriott, *Process Control*, Mc Graw Hill Book Co., N. Y., 1964.

10. W. L. Luyben, *Process Modeling, Simulation, and Control for Chemical Engineers*, Mc Graw Hill Book Co., N. Y. 1990.

11. T. J. Bergin, *Active Torque Four Quadrant Operation of a Squirrel Cage Induction Motor Fed from a Current Source Inverter*, M. Eng. dissertation, Royal Melbourne Institute of Technology, Melbourne, 1989.

12. J. G. Gibson, F. B. Tuteur, *Control System Components*, Mc Graw Hill Book Co., N. Y., 1958.

13. W. R. Ahrendt, C. J. Savant, *Servomechanism Practice*, Mc Graw Hill Book Co., N. Y., 1960.

14. E. F. Kohn, "The Problem in Selecting the Proper DC Servo-Drive System", Proc. 3rd Internat. Motor Conf., Sept. 28-30, 1982, Geneva, Switzerland.

15. S. Jenkins, "Selection of Servo Motors for Best Performance", Muirhead Ltd., England, May 1973.

8.10   Problems

8-1 A paper making machine is an example of a complex multi-variable system where the settings of a number of control and controlled variables must be coordinated to ensure satisfactory product quality. In these machines pulp feedstock (finely pulped wood suspended in water), is pumped into the head-box, which is pressurized by air to control the flow of stock through the slice onto a wire mesh belt travelling at high speed. The paper forms on this belt as the water drains away to form a damp thin continuous strip which then passes through a row of drying and pressing rollers before being spooled at the end of the machine line. Such a machine which can produce paper at over 1000 m/min is illustrated in Fig. P8-1. Identify the goals, and key variables for this system, and investigate the control strategies used for such machines. Explain how the various control problems are addressed.

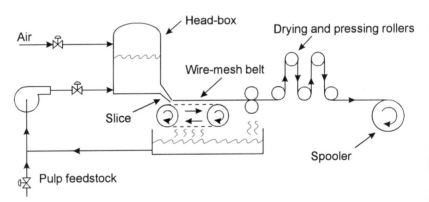

Figure P8-1

8-2 Modern fan-jet gas turbine engines are highly complex and require multi-variable control systems. In many of these engines with high bypass ratios there are usually two or even three power turbines coupled to fans or compressors whose speeds must be separately controlled.

In addition it is necessary to control other parameters such as turbine blade temperature, compressor flow and stall characteristics, fuel flow and so forth. Investigate the control strategies used for these systems giving particular attention to the control techniques used to ensure these systems have high operating integrity and defined performance degradation in the event of non-critical subsystem failures. These questions can be of paramount concern with such critical elements in aircraft, and can heavily influence the control system design.

8-3 Modern interconnected electrical power systems often contain many power stations coupled by high voltage transmission lines to widely distributed loads. In addition modern trends are to link separate power systems operating over wide geographical areas so as to exploit the effect of time zone differences on load patterns. Thus the control of these systems is becoming highly complex. The control of these systems is usually effected by the control of the real and reactive power flows using the excitation and governor controls of the various generators in the power stations. Investigate the methods currently being used for effecting these control strategies. To ensure security of supply to consumers, various strategies have been developed for shedding load and partitioning power systems when unanticipated plant failure and shutdown occurs, which is incapable of being supplied by spinning reserve. Investigate the various underlying philosophies used in developing these control strategies.

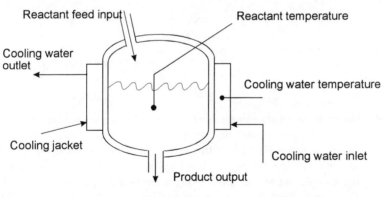

Figure P8-4

8-4 Tank reactors are commonly used in chemical plants. For exothermic chemical processes these tanks must be cooled to prevent thermal runaway, and to keep a constant reactor temperature so as to maintain uniform product quality. Such a tank is shown in Fig. P8-4.

Since there can be considerable thermal lags in such a system, investigate various control strategies for maintaining the reactor temperature constant in the presence of disturbances which arise from cooling water temperature changes, and reactant composition changes.

8-5 Steam generation is an important process in power stations as well as many types of process plants. It is principally used in steam turbines for converting heat energy into mechanical energy and in process plants for heating. These boilers and steam generators are complex multivariable systems which must be closely controlled to ensure safe and efficient operation. In power system operation quite advanced control strategies have been developed over many years. Investigate various control strategies that have been developed for combustion and steam flow control in large power station boilers and determine the main features of each approach.

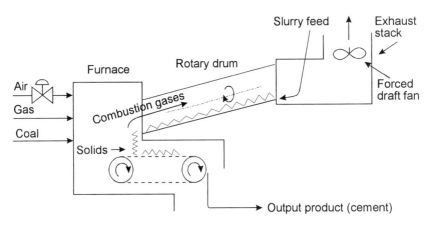

Figure P8-6

8-6 Direct fired rotary kilns are widely used for drying and manufacturing various materials such as cement. In these plants a fuel such as coal, or oil, or gas is burnt in a furnace and the hot combustion gases pass along a sloping refractory lined cylindrical rotating drum and then up an exhaust stack, after being scrubbed and filtered, as shown in Fig. P8-6. In a cement kiln a slurry is introduced at the stack end of the drum which gradually works its way down in a counter flow direction to the combustion gases towards the furnace where it is collected. As the slurry travels along the drum it is gradually dried and converted to cement. The goal of the control system for this process is to produce a cement of high and consistent quality at the lowest possible cost. Investigate control strategies for these systems which need to operate in the presence of changes in calorific value of the fuel and liquid content of the slurry.

8-7 High speed laser profile cutting machines commonly use two axis computer numerical control for the positioning of the laser cutter relative to the work piece. For high speed cutting it is not sufficient to simply have two position servos controlling each axis separately as significant position errors can occur during high speed operation. Various methods of addressing this problem have been and are still being developed. Suppose the laser cutting machine is arranged as shown in Fig. P8-7 with two position servo mechanisms which position the cutting head in the x and y directions. Also suppose that the velocity error coefficients of these servos are K_x and K_y respectively, which in general differ in value. Investigate control strategies which can be used to control such a machine so as to ensure high cutting accuracy. For this purpose you should investigate the path errors when cutting a commanded circular path of radius R when the laser has a cutting speed of V.

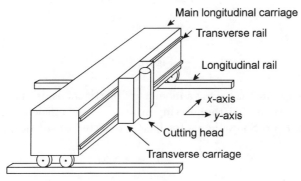

Figure P8-7

8-8 Diabetes is a disease affecting a very large number of people around the world which can be fatal if untreated. It is due to the inability of the person suffering the disease to make normal use of the ingested sugars, because of insufficient production of insulin, so that their blood sugar level increases to dangerous levels. Physician have found that the amount of insulin required depends upon the individual patients diet and exercise regime, and that both excessive or insufficient insulin can lead to dangerous conditions for the patient. In recent years attempts have been made to model the dynamics of the blood sugar and insulin levels in the body and to develop control systems for automatically regulating the flow of insulin so as to keep the blood sugar level constant. Investigate the control strategies which can be used for automatically metering insulin into the blood stream of a diabetes sufferer to control their blood sugar level.

8-9 A radar antenna is tracking an aircraft travelling at a minimum range $R_o = 2000$ m, a height $Z_0 = 1000$ m and a speed of 800 kph as shown in Fig. P8-9. Find the azimuth and elevation angles, as well as their velocity and acceleration time histories. From this information determine a set of performance criteria for the servos controlling the azimuth and elevation of the antenna mount.

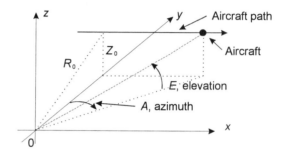

Figure P8-9

8-10 A servo-motor having a moment of inertia J_m is driving a load inertia J_ℓ through a gearbox having ratio $1 : N$ and efficiency η as shown in Fig. 8-19. Show the motor inertia reflected to the output is

$$J_m' = \eta N^2 J_m$$

and the output acceleration

$$\left(J_\ell + \eta N^2 J_m\right) \ddot{\theta}_o = \eta N T_m - T_f$$

where T_m is the motor output torque, and T_f is the load friction torque. From this result show that the maximum acceleration occurs for a gear ratio

$$N_{g,\max} = \sqrt{\frac{J_\ell}{\eta J_m}}$$

if T_f can be neglected.

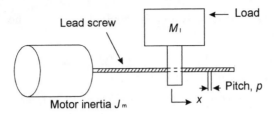

Figure P8-11

8-11 A servo motor having inertia J_m is driving a load M_ℓ through a lead screw having pitch p, and efficiency η, as shown in Fig. P8-11. Show the load inertia reflected to the motor is

$$J'_\ell = \frac{1}{\eta} M_\ell \left(\frac{p}{2\pi}\right)^2$$

8-12 Consider the following requirements for a servo to drive the table of the numerically controlled milling machine shown in Fig. P8-12. Suppose that the performance requirements are:

Static resolution $\epsilon_s = 0.01$ mm

Slew rate $\dot{x}_{max} = 50$ mm/sec

Initial acceleration $\ddot{x}_{max} = 1.0$ m/sec^2

The table of the milling machine weighs 50 kg and the slide has coefficient of friction of 0.3. Suppose the maximum motor speed $\omega_{m,max} = 314$ rad/sec (3000 rpm), and for this class of motors the relationship between the inertia J_m and the stall torque $T_{m,stall}$ is given by

$$\frac{J_m \omega_{m,max}}{T_{m,stall}} = 0.6$$

Select a motor and gear ratio to satisfy these requirements and specify the motor size and characteristics.

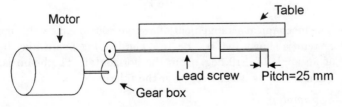

Figure P8-12

8-13 A position servo system is required to satisfy the following requirements:

Maximum output torque $= \pm 3$ N m

Static resolution $= 3 \times 10^{-3}$ radian

Maximum slew rate for 3 N m load $= 12$ rad/sec

Select a suitable motor and gearbox ratio to satisfy these requirements, which at the same time maximizes the initial shaft output angular acceleration. Determine the value of this acceleration.

8-14 Consider the design of the y-axis drive servo for the laser profile cutting machine discussed in Problem 8-7. Assume the y-axis carriage weighs 400 kg, and there is a maximum frictional force of 25 N resisting its motion. It should also be assumed that the guide rails can only be leveled to an accuracy of ± 2 parts in 1000. The maximum traverse speed should be 3000 mm/min, and performance requirements indicate that the carriage should be able to accelerate to this speed in a travel distance of 1 mm. The drive motor is coupled by gearing to a friction wheel driving onto the rail as shown in Fig. P8-14. Determine a set of performance specifications for the y-axis drive servo, and select a suitably sized servo-motor and gear box ratio assuming the gear train efficiency $\eta = 0.55$.

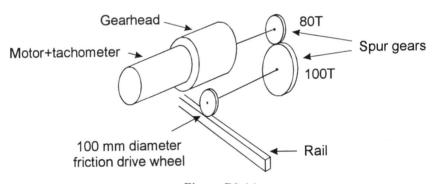

Figure P8-14

8-15 The mechanical arrangement of a servo controlled deadweight loading system is shown in Fig. P8-15. The loading beam may be inclined at an angle of ± 10 deg. from the horizontal. The beam is balanced when the mass is centered over the fulcrum.

Servomotor

The parameters describing the family of servo motors are:

Torque to inertia ratio $T_m/J_m = 2 \times 10^4$ rad/sec^2

Maximum motor speed $\omega_m = 314$ rad/sec

Motor friction torque $T_c = 1.4 \times 10^{-2}$ N m

Screw

Permissible screw pitch $p = 5, 10, 15$ or 25 mm

Screw efficiency $\eta = 0.9$

Load Mass

The load mass experiences a friction of $F_f = 10$ N as it moves along the beam.

Select a servo motor and appropriate lead screw for this system to meet the following requirements:

(i) Maximum slew rate $\dot{x} \geq 0.6$ m/sec.

(ii) Maximum acceleration of mass $\ddot{x} = 1.2$ m/sec^2. This acceleration is to be able to be sustained for $0 < \dot{x} \leq 0.6$ m/sec.

The inclination of the loading beam should be taken into account in the analysis.

You should also find the steady state motor torque when the mass is stationary and the beam is inclined at an angle of 10 degrees. Comment on why the magnitude of this steady state torque may be an important consideration in the selection of the motor.

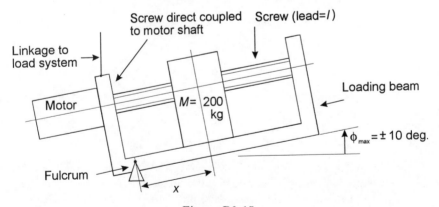

Figure P8-15

Chapter 9

System Compensation

9.1   Introduction

It was mentioned earlier that the performance of a control system is measured by its stability, accuracy, and speed of response. In general these items are specified when a system is being designed to satisfy a specific task. Quite often the simultaneous satisfaction of all these requirements cannot be achieved by using the basic elements in the control system. Even after introducing controllers and feedback as shown in Chapter 6, we are limited as to the choice we may exercise in selecting a certain transient response while requiring a small steady state error. We will show how the desired transient as well as the steady state behavior of a system may be obtained by introducing **compensatory elements** (also called **equalizer networks**) into the control system loop. These compensation elements are designed so that they help achieve system performance, i.e. bandwidth, phase margin, peak overshoot, steady state error, etc. without modifying the entire system in a major way.

From our experience so far we recognize that any changes in system performance can be achieved only through varying the forward loop gain. Consider the third-order unity feedback system with the following forward loop transfer function

$$G(s) = \frac{K}{s(s+a)(s+b)}$$

From the Routh-Hurwitz criterion we know that stability requires

$$K \leq ab(a+b)$$

We also know that the steady state error to a ramp input is

$$e(\infty) = \lim_{s \to 0} s \left[\frac{1}{s^2} \cdot \frac{1}{1 + G(s)} \right] = \frac{ab}{K}$$

Obviously if it is necessary to minimize the steady state error, the gain K should be increased. Since K is constrained to a maximum value of $ab(a + b)$, the minimum steady state error becomes

$$e(\infty)|_{min} = \frac{1}{a + b}$$

A further decrease in the error requires an increase in K which in turn has a destabilizing effect on the system. It is therefore clear that the forward "gain game" is rather limited.

9.2   Stabilization of Unstable Systems

Since increasing the forward loop gain K tends to destabilize a system, we must find ways to **compensate** it in such a way as to stabilize it again. It was established in Chapter 7 that the addition of a pole in $G(s)H(s)$ tends to have a destabilizing influence on system response. Can we then reverse the argument and say that the addition of a zero tends to have a stabilizing influence on system response? Let us answer this by considering an example. Consider the control system with its transfer function given in Chapter 7, Example 13. This system is unstable if $K > K_c$.

Now consider the same system but with the addition of a zero

$$G(s)H(s) = \frac{K(s\tau_3 + 1)}{s(s\tau_1 + 1)(s\tau_2 + 1)} \tag{9.1}$$

This is the type of function we obtain if we were to add derivative and proportional control to a third-order servomechanism. The characteristic equation becomes

$$\frac{s^3 \tau_1 \tau_2 + s^2(\tau_1 + \tau_2) + (K\tau_3 + 1)s + K}{s(s\tau_1 + 1)(s\tau_2 + 1)} = 0$$

and the zeros of the characteristic equation are determined from

$$s^3 \tau_1 \tau_2 + s^2(\tau_1 + \tau_2) + (K\tau_3 + 1)s + K = 0$$

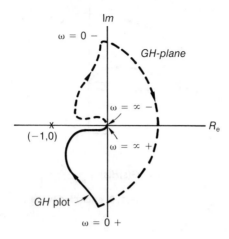

Figure 9.1: Nyquist plot for $G(s)H(s) = \dfrac{K(s\tau_3 + 1)}{s(s\tau_1 + 1)(s\tau_2 + 1)}$

The Routh array becomes

s^3	$\tau_1 \tau_2$	$K\tau_3 + 1$
s^2	$\tau_1 + \tau_2$	K
s^1	b_1	b_2
s^0	c_1	

$$b_1 = \frac{(K\tau_3 + 1)(\tau_1 + \tau_2) - K\tau_1\tau_2}{\tau_1 + \tau_2}$$

$$b_2 = 0$$

$$c_1 = K$$

For stability $b_1 \geq 0$, and therefore

$$K(\tau_1\tau_3 + \tau_2\tau_3 - \tau_1\tau_2) + (\tau_1 + \tau_2) > 0$$

Clearly, with a proper selection of the time constants, this may be satisfied for all $K > 0$. The Nyquist plot for this is shown in Fig. 9-1.

As another example, let us begin with a closed loop unstable system whose open loop transfer function is given by

$$G(s)H(s) = \frac{K}{s^2(s\tau_1 + 1)}$$

and whose Nyquist plot is shown in Fig. 9-2(a). We now add a zero to the transfer function so that

$$G(s)H(s) = \frac{K(s\tau_2 + 1)}{s^2(s\tau_1 + 1)} \tag{9.2}$$

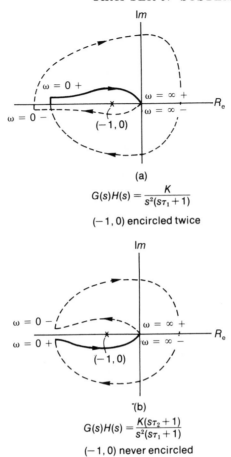

(a)

$$G(s)H(s) = \frac{K}{s^2(s\tau_1 + 1)}$$

$(-1, 0)$ encircled twice

(b)

$$G(s)H(s) = \frac{K(s\tau_2 + 1)}{s^2(s\tau_1 + 1)}$$

$(-1, 0)$ never encircled

Figure 9.2: Effect of adding a zero

and the new Nyquist plot is shown in Fig. 9-2(b). We again note that the
addition of a zero has stabilized the system for all gains.

   In both of these examples the effect of the zero is to add leading phase to
the lagging phase of each plant in the region near the critical point resulting
in the Nyquist contours being shifted so they no longer encircle the point
$(-1, 0)$. The positioning of the zero break frequency is fairly critical, for
the further its position is to the left of the origin of the s-plane the smaller
is its stabilizing effect. Methods of positioning the zero will be discussed
when we consider phase-lead compensators in Section 9-4.

   It is instructive to also examine the effect of the introduction of the zeros
upon the root locus diagrams for these systems. The root locus diagram

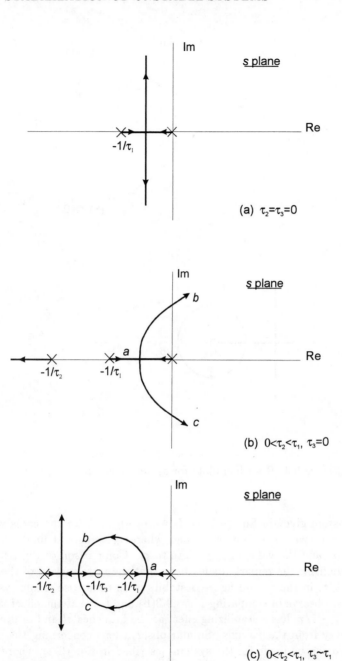

Figure 9.3: Root loci plots for system given in Eq. (9-1)

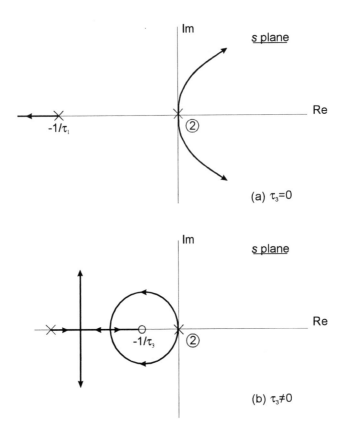

Figure 9.4: Root loci plots for system given in Eq. (9-2)

for the system given in Eq. (9-1) is shown in Fig. 9-3 for the cases where $\tau_2 = \tau_3 = 0$, where $0 < \tau_2 < \tau_1$ and where $\tau_3 = 0$, and finally where $0 < \tau_2 < \tau_1$ and the value of τ_3 is close to τ_1. Comparison of Figs. 9-3(a) and (b) confirms the remark made above that the introduction of the pole at $s = -1/\tau_2$ in the second figure destabilizes an otherwise stable system. We further observe in comparing Figs. 9-3(b) and (c) that the effect of the zero at $s = -1/\tau_3$ has a stabilizing effect as the branches ab and ac are now curved away from the $j\omega$ axis. Similar observations concerning the effect of the zero can be made for the system described in Eq. (9-2), where root locus plots are shown in Fig. 9-4.

The above observations can be summarized in the following two **Golden**

Rules for Root Loci:

1. **Zeros attract root loci.** The addition of zeros to a system in the neighborhood of the branches of the unmodified system attract these branches towards themselves. Thus zeros added in the left half of the s-plane tend to pull the root loci to the left and to stabilize the system.

2. **Poles repel root loci.** The addition of poles to a system in the left half of the s-plane tend to push the root loci to the right.

Since the addition of poles and zeros not only affect the system stability but also its performance characteristics, the location of these **compensating poles** and **zeros** must be carefully determined. Whilst the above two rules play no quantitative role their importance is in guiding the designer in correctly locating the compensating poles and zeros. They will thus be valuable in compensator design which will be discussed next.

9.3   Types of Compensation

The performance of a control system may be modified by adding compensation elements in the control loop. The types of compensation we shall discuss fall into the following categories:

(a) Cascade or series compensation

(b) Feedback compensation

(c) Feedforward compensation.

These schemes are illustrated in Fig. 9-5.

The study and design of compensated control systems can be most readily carried out in the frequency or complex domain or via the root locus. The use of the frequency domain is straightforward and quite simple. When the characteristics of the closed loop response are desired, use is made of Nichols charts. The root locus is generally used to study the poles and zeros directly.

The consideration of compensation elements in this chapter shall be restricted to passive elements consisting of resistors and capacitors. We shall assume that they are physically realizable. This generally implies that the transfer function of compensation networks must be rational algebraic functions and all with real coefficients in which the number of poles equals or exceeds the number of zeros.

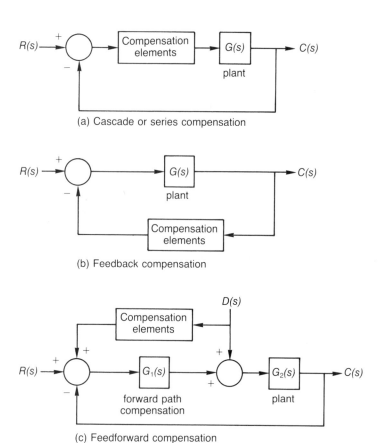

(a) Cascade or series compensation

(b) Feedback compensation

(c) Feedforward compensation

Figure 9.5: Types of compensation

9.4   Cascade Compensation

As indicated in Fig. 9-5(a) cascade compensation consists of placing elements in series with the forward loop transfer function. Such compensation may be classified into the following categories:

(a) Phase-lag compensation

(b) Phase-lead compensation

(c) Lag-lead compensation

(d) Compensation by cancellation.

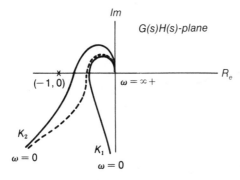

Figure 9.6: Nyquist plot of a third-order system for different gains

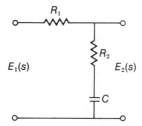

Figure 9.7: A phase-lag network

The details of these methods is the subject of this section.

Phase-Lag Compensation

Consider a unity feedback control system whose forward loop transfer function represents a third-order system with its Nyquist plot shown in Fig. 9-6. It is required that the gain be K_1 to satisfy the margin of stability whereas it should be K_2 to satisfy the steady state performance. This seemingly contradictory requirement may be satisfied if we were to reshape the plot to the one indicated by the dotted lines. The reshaped plot may be obtained if the low-frequency part of K_1 is rotated clockwise while the high-frequency part remains unaltered. Since the phase of the low-frequency part of K_1 must lag, the type of compensation used to achieve this is phase-lag compensation. Such compensation is obtained by a phase-lag element.

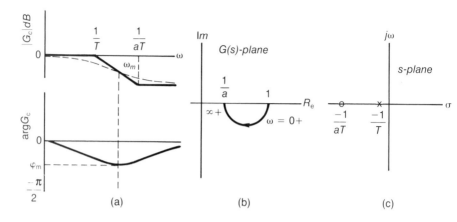

Figure 9.8: Behavior of a lag network

When the output of an element lags the input in phase and the magnitude decreases as a function of frequency, the element is called a **phase-lag element**. Consider the lag network of Fig. 9-7. The transfer function for this is

$$\frac{E_2(s)}{E_1(s)} = G_c(s) = \frac{1 + aTs}{1 + Ts} \tag{9.3}$$

where

$$T = C(R_1 + R_2), \quad \text{and } a = \frac{R_2}{R_1 + R_2} < 1$$

The Bode, Nyquist, and pole-zero plots for Eq. (9-3) are shown in Fig. 9-8. We observe that the magnitude decreases with increasing frequency and lagging phase angle. The value of a determines the separation of the pole and the zero. The minimum phase ϕ_m occurs at ω_m which is the geometric average of the corner frequencies

$$\log \omega_m = \frac{1}{2}\left(\log \frac{1}{T} + \log \frac{1}{aT}\right)$$

$$\omega_m = \frac{1}{T\sqrt{a}} \tag{9.4}$$

The phase angle becomes

$$\phi_m = + \arctan aT\omega_m - \arctan T\omega_m$$

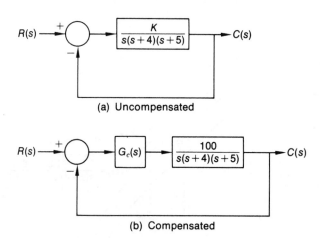

(a) Uncompensated

(b) Compensated

Figure 9.9: Phase-lag compensation

or

$$\sin \phi_m = (a - 1)/(a + 1)$$

The maximum phase lag is strictly a function of a. Let us investigate how such a network alters the performance of a feedback control system.

Example 1 *For the system shown in Fig. 9-9(a), it is desired that the velocity error constant be at least 5 sec^{-1} and that the phase margin be 40° or more. Design an appropriate lag compensator for this system. Construct the Bode and Nichol plots with and without compensation. Compute the maximum gain and minimum steady state error with and without compensation.*

Since

$$K_v = K_1 = \lim_{s \to 0} sG(s) = \frac{K}{20} = 5$$

the gain becomes

$$K = 20K_1 = 100$$

We therefore construct the Bode plot for

$$G(s) = \frac{100}{s(s + 4)(s + 5)}$$

as shown in Fig. 9-10. We note that the gain crossover frequency is $\omega_c = 3.25$ rad/sec and the phase margin is 18°, which is not satisfactory. Since

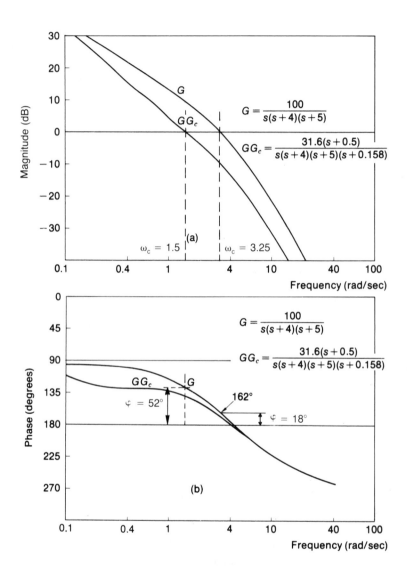

Figure 9.10: Phase-lag compensation for system shown in Fig. 9.9

a lag network will contribute additional phase lag (generally between 5° and 15°) the new crossover frequency will occur where the phase margin of $G(j\omega)$ is 52°. We arrive at this figure by taking the required phase margin of 40° and add the expected 12° phase lag of the compensator $G_c(j\omega)$. This occurs at $\omega_c = 1.5$ rad/sec. Since $G(j\omega)$ has a gain of 10 dB at $\omega_c = 1.5$ rad/sec the compensator must have a gain of -10 dB to achieve an overall gain of 0 dB at the crossover frequency. To achieve this we employ a phase-lag network which provides the additional attenuation. The magnitude is approximated by

$$|G_c(j\omega)| = 20\log a$$

so that

$$a = 10^{|G_c(j\omega)|/20}$$

Since the lag network must provide the -10 dB gain

$$a = 0.316$$

This then provides the separation between the corner frequencies.

The zero corner frequency for the lag network is generally selected between an octave and decade below the new crossover frequency which is 1.5 rad/sec. We select $\omega = 0.5$ so that

$$aT = \frac{1}{\omega} = \frac{1}{0.5} = 2$$

The other corner frequency will be determined from

$$T = \frac{2}{0.316} = 6.42$$

and the transfer function of the lag network becomes

$$G_c(s) = \frac{1 + 2s}{1 + 6.42s}$$

The Bode plot of the compensated system is shown in Fig. 9-10 and the Nichols chart is shown in Fig. 9-11. From these plots it is seen that the phase crossover requirement is now satisfied, and from the Nichol plot we find the uncompensated system and compensated system goes unstable for $K = 180$ and 620 respectively.

The steady state error of the uncompensated system for a ramp input is given by

$$e(\infty) = \frac{1}{K_v} = 0.2$$

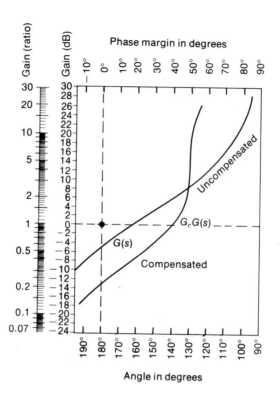

Figure 9.11: Nichols chart for uncompensated system and compensated system using phase-lag compensation

However, if K is increased to its maximum value, and the system is still to be stable, then

$$e(\infty)|_{min} = \frac{1}{K_{1,max}}$$

where

$$K_{1,max} = \lim_{s \to 0} s\frac{K_{max}}{s(s+4)(s+5)} = \frac{K_{max}}{20}$$

Since $K_{max} = 180$, $K_{1,max} = 9$, and $e(\infty)_{min} = 0.11$. When the system is compensated however, $K_{max} = 620$. Hence the maximum error constant is $K_{1,max} = 620/20 = 31$, and $e(\infty)_{min} = 1/31 = 0.0323$.

Example 2 *For the same system shown in Fig. 9-9 it is desired that the velocity error constant be at least 5 sec^{-1} and the maximum overshoot to*

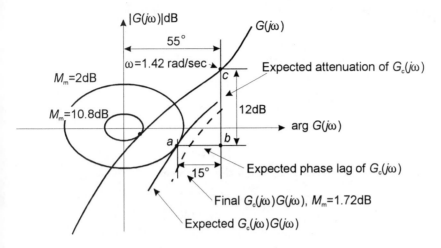

Figure 9.12: Nichol plot for system of Example 2, with $M_m = 2$ dB contour.

a step input is not to exceed 20 percent. Design an appropriate lag compensator using the Nichol plot and root locus methods and find the settling
time for the system.

Assuming the closed loop system has dominant complex conjugate poles,
the step response specification implies $\zeta_{min} = 0.45$. Also from Eq. (7-33)
$M_{m,max} = 2$ dB. We have also seen in the last example that if $K_v = 5$ sec^{-1}
then $K = 100$. From the Nichol plot shown in Fig. 9-11 we find for the
uncompensated system that $M_m = 10.8$ dB and $\omega_m = 3.48$ rad/sec.

In Fig. 9-12 we show the $M_m = 2$ dB contour with the frequency
response of $G(j\omega)$ shown for $K = 100$ on the Nichol plot. It is expected
that the compensated frequency response $G(j\omega)G_c(j\omega)$ will be tangential
to the 2 dB contour at a which corresponds to a phase lag of $-140°$. The
horizontal line ab corresponds to the residual phase lag of 15° which we
have allowed for the compensator, and the projected line bc which intersects
the frequency response contour of $G(j\omega)$ at c is the assumed attenuation
of the lag compensator. The frequency corresponding to c will be the
resonant frequency $\omega_m = 1.42$ rad/sec of the compensated system if the
compensator has the assumed phase and attenuation given by the lines ab
and bc respectively at this same frequency. From the above figure we find
the compensator attenuation corresponding to bc is 12 dB. Since the lag

network must provide an attenuation of 12 dB we find

$$a = 10^{-12/20} = 0.251$$

Selecting the lag compensator zero corner frequency as $\omega = 1.42/5 = 0.284$ rad/sec we obtain

$$aT = \frac{1}{0.284} = 3.52 \text{ sec}$$

and

$$T = 14.0 \text{ sec}$$

The compensator transfer function thus becomes

$$G_c(s) = \frac{1 + 3.52s}{1 + 14.0s}$$

The Nichol plot of the compensated system shows that it will have $M_m = 1.72$ dB and a resonant frequency $\omega_m = 0.89$ rad/sec.

The root locus diagram for the uncompensated system is shown in Fig. 9-13(a) with the $\zeta = 0.45$ boundary also being shown. The loop sensitivity at the intersection of the root locus branch and this boundary line is $K = 38$, which is unacceptable since from the specification requirement $K_v = 5$ sec^{-1} we must have $K = 100$. We thus need to move the point corresponding to $K = 100$ back along the root locus contour toward the real axis. To achieve this we find the coefficient a of the compensator

$$a = \frac{K \text{ at intersection with } \zeta = 0.45 \text{ boundary}}{K \text{ required from steady state error specification}} = \frac{38}{100} = 0.38$$

Since a defines the separation between the compensator pole and zero, and thus its effectiveness, we generally find a needs to be somewhat less than the figure calculated above, so we take $a = 0.33$. If the crossover point on the $\zeta = 0.45$ boundary is not to change significantly, we see from the angle condition given in Section 7-4 that the net angle contributed by the compensator pole and zero should be small, say 5° to 10°. To achieve this and yet at the same time have the separation defined by a, the compensator pole and zero must be close to the origin. From the construction line ab on Fig. 9-13(a) which is positioned at an angle of 10° from line Oa, corresponding to $\zeta = 0.45$, the zero will be located at b if

$$aT = \frac{1}{0.3} = 3.33, \quad \text{so that } T = \frac{3.33}{0.33} = 10$$

The compensator transfer function thus becomes

$$G_c(s) = \frac{1 + 3.33s}{1 + 10s}$$

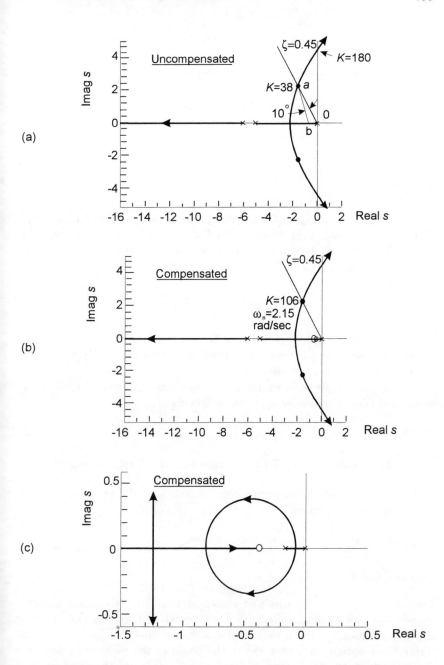

Figure 9.13: Root locus plot for uncompensated and compensated system

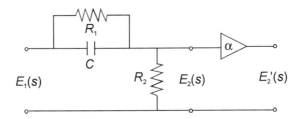

Figure 9.14: A phase-lead network

The root locus diagram for the compensated system given in Figs. 9-13(b) and (c) shows that $K = 106$ and $\omega_n = 2.15$ rad/sec at the point where the dominant pole branch intersects the $\zeta = 0.45$ line. Since the dominant poles are at $s = 0.98 \pm j1.90$ the settling time T_s is

$$T_s = \frac{4}{0.45 \times 2.15} = 4.13 \text{ sec}$$

In the above examples it is shown the effect of lag compensation is to increase the velocity error constant which decreases the steady state error for a given transient performance.

Also from the previous examples we may conclude that the phase-lag method of compensation achieves the following:

1. Reduces high-frequency gain and improves the phase margin.

2. Increases the velocity error constant for a fixed relative stability.

3. The gain crossover frequency is decreased. This also reduces the bandwidth of the system.

4. The time response usually gets slower.

Phase-Lead Compensation

Let us return to the Nyquist plot shown in Fig. 9-6. We could have reshaped the plot by beginning with the Nyquist plot for K_2 and rotating the high-frequency part in the counterclockwise direction but without altering the low-frequency part. Since the phase of the high-frequency part must now lead, the type of compensation used to achieve this is phase-lead compensation. Such compensation is achieved by a phase-lead element.

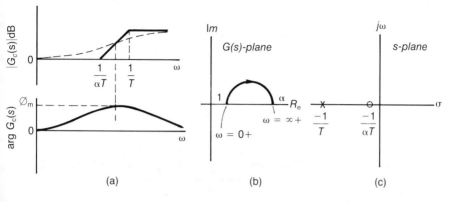

Figure 9.15: Behavior of a lead network

When the output of an element leads the input in phase and the magnitude increases as a function of frequency, the element is called a **phase-lead element**. Consider the lead element shown in Fig. 9-14. The transfer function is

$$\frac{E_2(s)}{E_1(s)} = \frac{1}{\alpha}\frac{1+\alpha Ts}{1+Ts} \tag{9.5}$$

where

$$T = \frac{R_1 R_2 C}{R_1 + R_2}, \quad \text{and} \quad \alpha = \frac{R_1 + R_2}{R_2} > 1$$

The phase-lead compensator transfer function is

$$\frac{E_2'(s)}{E_1(s)} = G_c(s) = \frac{1+\alpha Ts}{1+Ts}$$

where the lead network is followed by an amplifier having a gain α.

The Bode, Nyquist, and pole-zero plots are shown in Fig. 9-15. We note that the magnitude increases with increasing frequency. The value of α determines the separation of the pole and the zero. The maximum phase lead ϕ_m occurs at ω_m. Using the previous method, it may be shown that

$$\omega_m = \frac{1}{T\sqrt{\alpha}}, \quad \sin\phi_m = \frac{\alpha-1}{\alpha+1}, \quad |G(j\omega_m)| = \sqrt{\alpha}$$

Example 3 *Obtain a lead compensation network for the servo system shown in Fig. 9-16(a). It is required that the phase margin be 45° while the velocity error must be less than 2 percent.*

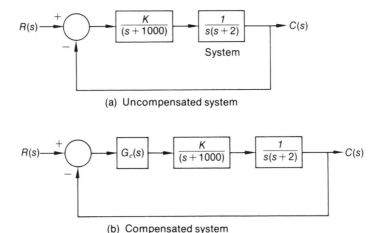

(a) Uncompensated system

(b) Compensated system

Figure 9.16: System compensated by lead network

Since the steady state error to a ramp input is

$$e(\infty) = \frac{2000}{K} = 0.02$$

the necessary gain becomes $K = 100,000$ and the uncompensated forward loop transfer function becomes

$$G(s) = \frac{100,000}{s(s+1000)(s+2)}$$

The Bode plot of $G(s)$ is given in Fig. 9-17. The gain crossover frequency is close to 10 rad/sec and the phase margin is about 11° which is 34° less than desired. The phase-lead network must provide this phase. We select a phase margin addition of 38° in order to offset any shift in gain crossover frequency caused by the addition of the compensating network. Since $\phi_m = 38°$

$$\sin \phi_m = \frac{\alpha - 1}{\alpha + 1} = 0.616 \quad \text{so that } \alpha = 4.21$$

This establishes the separation of the lead network corner frequencies. Recall that the maximum phase-lead occurs at the geometric mean of $1/\alpha T$ and $1/T$. This maximum phase lead should occur at the new crossover frequency ω_c. Since the gain of the compensator at frequency ω_m is

$$|G_c(j\omega_m)|_{\text{dB}} = 20 \ \log \sqrt{4.21} = +6.25 \text{ dB}$$

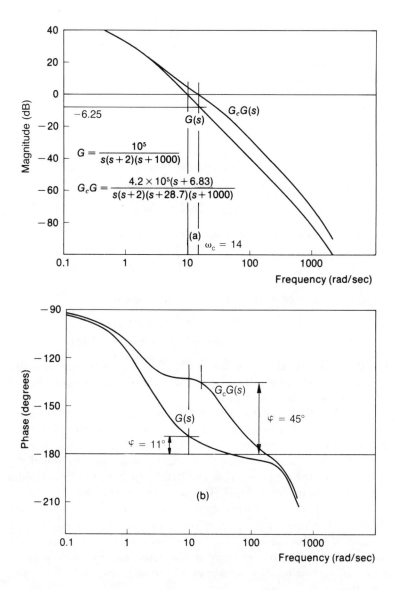

Figure 9.17: Lead compensation of system shown in Fig. 9-16

the gain crossover frequency is determined from the relation

$$|G(j\omega_c)|_{dB} = -6.25 \text{ dB}$$

From Fig. 9-17(a) the new gain crossover frequency $\omega_c = 14$ rad/sec, and since this should also equal the frequency ω_m at which the maximum compensator phase lead occurs we have

$$\omega_c = \omega_m = \frac{1}{\sqrt{\alpha T}} = 14$$

Thus

$$\frac{1}{\alpha T} = 6.83 \text{ rad/sec}$$

and

$$\frac{1}{T} = 28.7 \text{ rad/sec}$$

The transfer function of the lead network becomes

$$G_c(s) = 4.2 \left(\frac{s + 6.83}{s + 28.7} \right)$$

so the compensated open loop transfer function is

$$G_c(s)\,G(s) = \frac{420,000(s + 6.83)}{s(s + 28.7)(s + 2)(s + 1000)}$$

The Bode plot for this is shown in Fig. 9-17 and it is seen that the phase margin is in the satisfactory range. The magnitude versus phase shift is shown on the Nichol chart in Fig. 9-18.

From the Nichol plot given in Fig. 9-18 the 3 dB bandwidth for the uncompensated and compensated systems are found to be 15.5 and 23.7 rad/sec. Generally lead compensation is used to improve the transient response of a system by moving the dominant pole root loci branches further to the left in the s-plane. This result follows from the first Golden Rule of root loci, enunciated above, if the compensator zero is close to the pole at $s = -2$. The pole of the compensator is generally chosen to be far to the left of the zero on the pole-zero diagram so that the effect of the zero is enhanced. Moving it too far to the left however will make the system susceptible to noise disturbances; so the determination of its position must always be a compromise. As a general rule α is chosen to be in the range 5 to 20.

Example 4 *Obtain a lead compensator for the control system shown in Fig. 9-16(a) so that the velocity error is less than 2 percent, and the dominant poles satisfy $\zeta = 0.45$ and $\omega_n = 35$ rad/sec.*

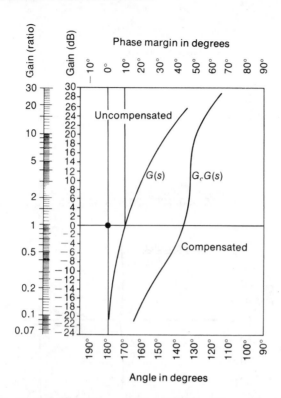

Figure 9.18: Nichol chart for uncompensated system and compensated system using phase-lead compensation

As discussed in Example 3, for the steady state error to be less than 2 percent we must have $K = 10^5$. The root locus plot of the uncompensated system is shown in Fig. 9-19(a), where it is seen the dominant poles for $K = 10^5$ have $\omega_n = 10$ and $\zeta = 0.096$. If the loop sensitivity is selected so $\zeta = 0.45$ for the dominant poles we find $K = 4800$. Since neither ω_n nor ζ meet the requirements listed above we introduce a lead compensator having its zero in the neighborhood of $s = -2$, and adjust the location of the compensator pole so relevant branches of the root locus plot pass through the point on the s-plane corresponding to $\zeta = 0.45$ and $\omega_n = 35$ rad/sec. In this example there is considerable latitude in the choice of the zero position which allows the dominant pole specification to be met, but it will be found that the required loop sensitivity varies as the zero position is changed.

As shown in Fig. 9-20, for the branch of the compensated system to pass through the point a it is necessary, from the angle condition given in

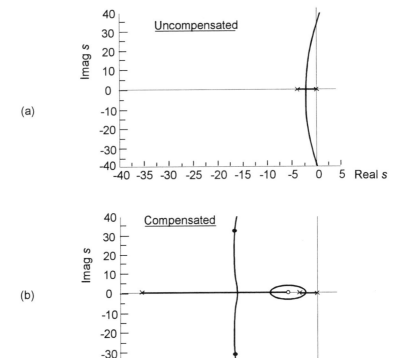

Figure 9.19: Root locus of uncompensated and compensated system for Example 4 in neighborhood of origin. Note in each case there is a pole at $s = -1000$

Eq. (7-20) for

$$\beta_1 - \alpha_4 - \alpha_3 - \alpha_2 - \alpha_1 = -180$$

Thus the compensator pole and zero must be positioned so that the angle ϕ subtended at point a is given by

$$\phi = \beta_1 - \alpha_4 = \alpha_1 + \alpha_2 + \alpha_3 - 180$$

Since the zero must be close to the pole at $s = -2$ let us take $1/\alpha T = 4$. By a succession of trials we find the pole should be positioned so $1/T = 35$. Consequently $\alpha = 8.75$ and

$$G_c(s) = \frac{1 + s/4}{1 + s/35}$$

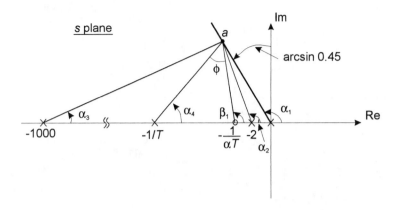

Figure 9.20: Requirement for satisfaction of angle condition at point a

The root locus plot for the compensated system is shown in Fig. 9-19(b). Utilizing the gain condition given in Eq. (7-20) we find the loop sensitivity for the point a is $K = 1.485 \times 10^5$. It will be noted that the introduction of the lead compensator has not only led to an increase in ω_n for the dominant poles but also the velocity error coefficient has increased. In our example we see that the velocity error coefficient has increased from $K_v = 2.4$ sec^{-1} for the uncompensated case to $K_v = 74.3$ sec^{-1} for the compensated case.

For systems compensated by phase-lead networks the following is concluded:

1. The velocity constant is increased and therefore the steady state error to a ramp input is decreased for a given relative stability.

2. The damping ratio is increased and the overshoot is reduced while the phase margin is increased.

3. The gain crossover frequency is increased and the bandwidth is usually increased.

4. The rise time is faster.

Phase Lag-Lead Compensation

Phase-lag compensation was seen to improve the steady state response although the rise time became slower. The phase-lead compensation, on the other hand, decreases the rise time and can also substantially decrease the

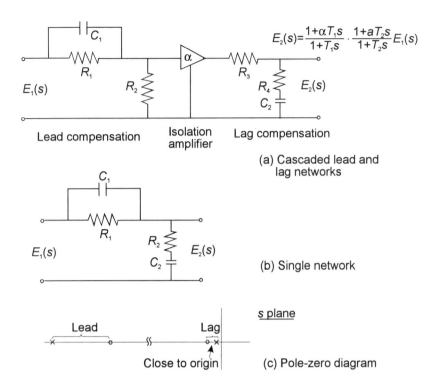

Figure 9.21: Lag-lead compensators

overshoot. It is often necessary to combine these different properties so as to simultaneously satisfy the steady state as well as the transient performance specifications of control systems. Compensation elements that combine these properties are called **lag-lead networks**.

In order to embody the characteristics of lag and lead networks, it is possible to cascade two independent networks as shown in Fig. 9-21(a). However, there are individual networks that possess the combined property of lead and lag. Consider the lag-lead network shown in Fig. 9-21(b). The transfer function for this network is

$$\frac{E_2(s)}{E_1(s)} = \frac{(1 + R_1 C_1 s)(1 + R_2 C_2 s)}{1 + (R_1 C_1 + R_1 C_2 + R_2 C_2)s + R_1 R_2 C_1 C_2 s^2}$$

This may be cast in the familiar lead and lag notation by defining

$$T_1 a^{-1} = R_1 C_1, \quad T_2 a = R_2 C_2, \quad T_1 T_2 = R_1 R_2 C_1 C_2$$

Substituting

$$\frac{E_2(s)}{E_1(s)} = \overbrace{\left[\frac{1 + sT_1a^{-1}}{1 + sT_1}\right]}^{\text{lead}} \overbrace{\left[\frac{1 + sT_2a}{1 + sT_2}\right]}^{\text{lag}} \tag{9.6}$$

The computation procedure of obtaining a, T_1, and T_2 is as before. It is appropriate to offer a word of caution here. If two independent networks are cascaded to realize a lag-lead compensator it is advisable to first design the lead component and to follow it by the lag compensator design. This order is preferred because the examples above show that a lead compensator not only improves the transient performance, but also decreases the steady state error, while the lag compensator can be designed to have only a minor influence on the transient performance.

The previous examples seemed to yield the necessary results in a straight-forward manner. In practice this does not occur. Instead the application of compensation elements requires, in general, an iterative procedure.

Compensation by Cancellation

Although we have tried to categorize and generalize the compensation and design of control systems, it is important to realize that the real way to design a system is to have some "feel" for the problem. It is possible that none of the techniques we have considered really help, or possibly a very simple method exists. One such simple technique is the method of direct cancellation.

Consider a second-order system whose forward loop transfer function is given by

$$G_c(s) = \frac{K_0}{s(s + a)}$$

Let us assume that the system performance is not satisfactory. We therefore introduce a compensation network

$$G_c(s) = \frac{s + \alpha_1}{s + \alpha_2}$$

so that the new forward loop transfer function becomes

$$G_c(s)G(s) = \frac{K_0(s + \alpha_1)}{s(s + a)(s + \alpha_2)}$$

If we now select $\alpha_1 = a$, then

$$G_c(s)G(s) = \frac{K_0}{s(s + \alpha_2)}$$

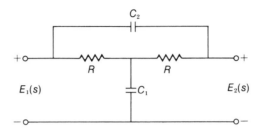

Figure 9.22: A bridged-T network

Clearly, the appropriate selection of α_2 allows us to obtain whatever response we desire.

The technique of cancellation is particularly attractive when the open loop poles are complex conjugates. If the location of the complex conjugates is not satisfactory, then new zeros may be introduced to cancel their effect. A compensation network that achieves this is shown in Fig. 9-22. This network is known as a Bridged-T network. The transfer function[1] is

$$G_c(s) = \frac{E_2(s)}{E_1(s)} = \frac{R^2C_1C_2s^2 + 2RC_2s + 1}{R^2C_1C_2s^2 + R(C_1 + 2C_2)s + 1} \qquad (9.7)$$

Consider the system shown in Fig. 9-23(a). The location of the poles is not satisfactory. We therefore introduce, in cascade, the Bridged-T network as shown in Fig. 9-23(b). The new open loop transfer function becomes

$$G_cG(s) = \frac{[s^2 + (2/RC_1)\,s + 1/R^2C_1C_2]K}{[s^2 + (1/RC_2 + 2/RC_1)s + 1/R^2C_1C_2]s(s^2 + bs + c)}$$

If we now select $2/RC_1 = b$, and $1/R^2C_1C_2 = c$, then

$$G_cG(s) = \frac{K}{s(s^2 + (b' + b)s + c)}$$

where $b' = 1/RC_2$. The value of b' is selected on the basis of desired response. The introduction of such a cancellation compensator is shown on the root locus in Fig. 9-24. This method is very useful when the open loop poles are close to the imaginary axis.

---

[1]The poles of $G_c(s)$ are $s = \left[(C_1 + 2C_2) \pm \sqrt{C_1^2 + 4C_2^2}\right]/2RC_1C_2$, and are always real.

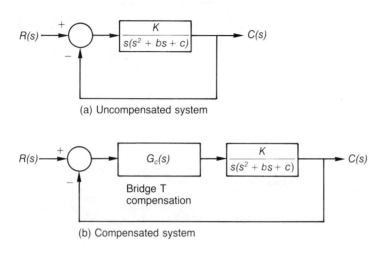

(a) Uncompensated system

(b) Compensated system

Figure 9.23: Compensation by a bridged-T network

Example 5 *A simplified block diagram of the control system for one axis of a magnetic bearing is shown in Fig. 9-25 where it will be noted that the actuator transfer function is open loop unstable. Obtain a compensator which stabilizes the bearing control system and find the loop sensitivity so that $\zeta = 0.45$. For this loop sensitivity find the settling time for a step transient input.*

The root locus plot of the uncompensated system is shown in Fig. 9-26(a) where it is seen the system is unstable for all values of loop sensitivity K. From the root locus plot it can be seen that a compensator zero must be introduced to attract the unstable branch of the root locus into the left-half of the s-plane. A lead compensator whose transfer function is

$$G_c(s) = \frac{1 + s/300}{1 + s/3000}$$

was chosen, and the root locus plots for the compensated system are shown in Figs. 9-26(b) and 9-26(c). It can be seen in Fig. 9-26(c) that the dominant poles at s_1 and s_2 have $\zeta = 0.45$ and $\omega_n = 88$ rad/sec, and the loop sensitivity $K = 2750$. The settling time of the system for a step input will be

$$T_s = \frac{4}{\zeta \omega_n} = \frac{4}{0.45 \times 88} = 0.1 \, \text{sec}$$

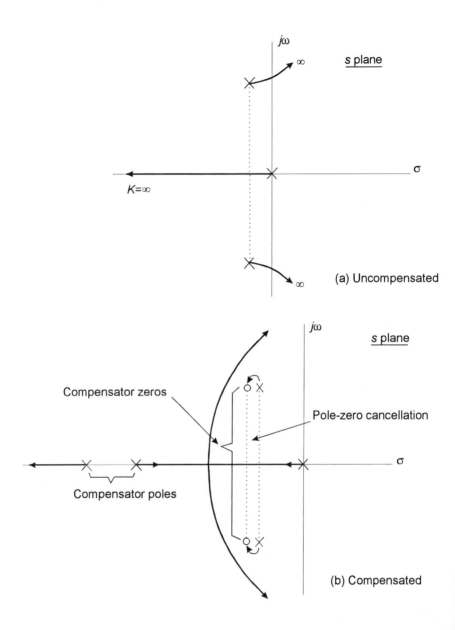

Figure 9.24: Root locus showing the effect of compensation using a bridged-T network

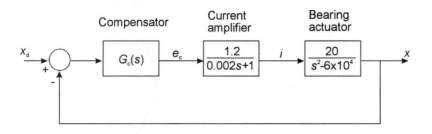

Figure 9.25: Block diagram of magnetic bearing control system

9.5   Feedback Compensation

Although the passive networks discussed so far were for cascade compensation, they may be incorporated into the feedback path of a control system. Here we shall investigate the effect of introducing compensation in the feedback loop.

Consider the control system of Fig. 9-27 where $H_c(s)$ is the transfer function of a compensation network. The overall transfer function is

$$\frac{C(s)}{R(s)} = \frac{G(s)}{1 + G(s)H_c(s)}$$

The characteristic equation becomes

$$1 + G(s)H_c(s) = 0$$

and is identical to that obtained if the compensation elements are added in cascade. The difference is in the addition of zeros in the closed loop function. Let $H_c(s) = P_1(s)/P_2(s)$, then the closed loop transfer function becomes

Feedback compensation: $\quad \dfrac{C(s)}{R(s)} = \dfrac{G(s)P_2(s)}{P_2(s) + G(s)P_1(s)}$

Cascade compensation: $\quad \dfrac{C(s)}{R(s)} = \dfrac{G(s)P_1(s)}{P_2(s) + G(s)P_1(s)}$

Thus it will be observed that the zeros of the closed loop transfer function for a system with feedback compensator $H_c(s)$ are the zeros of $G(s)$ and the poles of $H_c(s)$, whereas for the same system with the cascade compensator $H_c(s)$ the zeros are the zeros of $G(s)$ and $H_c(s)$. However in both cases

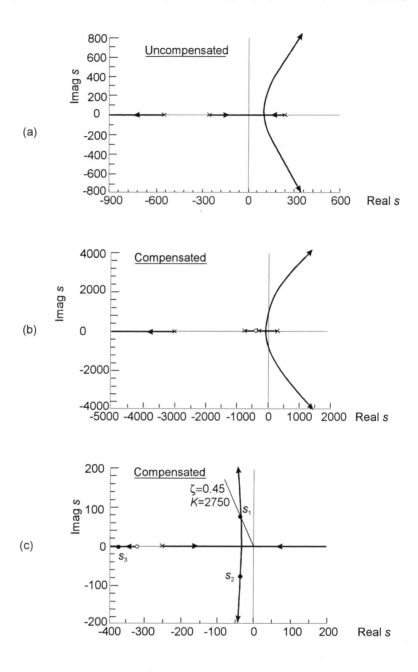

Figure 9.26: Root locus diagram for magnetic bearing control system

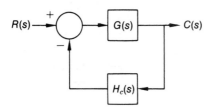

Figure 9.27: Compensation in the feedback loop

the closed loop poles are the same. These differences can on occasion be advantageously exploited.

Rewriting the equation for feedback compensation, we have

$$\frac{C(s)}{R(s)} = \left[\frac{G(s)H_c(s)}{1 + G(s)H_c(s)}\right] \frac{1}{H_c(s)}$$

Since $|GH| \gg 1$ in the low frequency range, we have

$$\frac{C(s)}{R(s)} \approx \frac{1}{H_c(s)} \quad \omega \to \text{small} \tag{9.8}$$

In the high-frequency range, $|GH| \ll 1$ and

$$\frac{C(s)}{R(s)} \approx G(s) \quad \omega \to \text{high} \tag{9.9}$$

We conclude, then, that the feedback compensated control system behaves as the inverse of the feedback characteristics for low frequency and as the system itself at high frequencies.

Conditions for Retention of System Type

Consider the system shown in Fig. 9-27 where $G(s)$ is of type n. For the closed loop system to retain the System Type of $G(s)$ the feedback transfer function $H_c(s)$ must have a zero at the origin of order $m \geq n$. To see this suppose $G(s) = \widetilde{G}(s)/s^n$ and $H_c(s) = \widetilde{H}(s)s^m$. The closed loop transfer function is

$$\frac{C(s)}{R(s)} = \frac{\widetilde{G}(s)}{s^n \left[1 + \widetilde{G}(s)\widetilde{H}(s)s^{m-n}\right]}$$

which shows it is Type n.

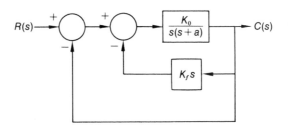

Figure 9.28: Tachometric feedback

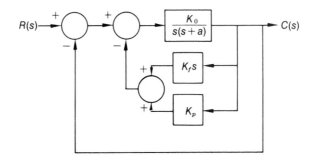

Figure 9.29: Tachometric feedback with proportional control

Tachometer Feedback

Besides the use of passive networks in feedback control, the use of **tachometric control** is very common. Here a tachometer is used to feed back a signal proportional to the derivative of the output variables. Consider a second-order system with tachometric control as shown in Fig. 9-28. The overall transfer function becomes

$$\frac{C(s)}{R(s)} = \frac{K_o}{s^2 + (a + K_o K_f)s + K_o} \tag{9.10}$$

We observe that the addition of tachometric compensation has increased system damping.

Consider tachometric feedback with proportional control as shown in Fig. 9-29. The overall transfer function becomes

$$\frac{C(s)}{R(s)} = \frac{K_o}{s^2 + (a + K_f K_o)s + (1 + K_p)K_o}$$

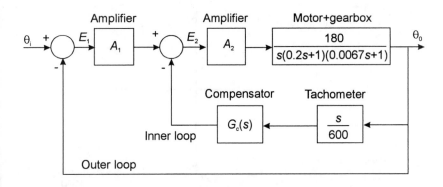

Figure 9.30: Block diagram of tachometer feedback servomechanism

Here we are able to affect the natural frequency as well as the damping ratio of the system.

Example 6 *The block diagram for a servomechanism utilizing rate feedback is illustrated in Fig. 9-30. This control system is required to be able to follow a velocity input of 12 deg/sec with the steady state error being less than 0.1 deg. In addition it is required that the dominant poles must have $\zeta \geq 0.45$ and $5 \leq \omega_n \leq 10$ rad/sec.*

We first investigate the case where the compensator $G_c(s) = 1$, which is the case of **tachometer feedback**. Examining the inner loop, the open loop transfer function is

$$G(s)H(s) = \frac{K}{(s+5)(s+150)}$$

The root locus plot for this system is given in Fig. 9-31(a). Since it will be shortly seen the outer loop root locus branches migrate toward the imaginary axis we need to have $\zeta = 0.7$, say, for the inner loop dominant poles. From Fig. 9-31(a) we find the loop sensitivity $K = 11,550$, so that $A_2 = 51.3$. The closed loop transfer function for the inner loop becomes

$$\frac{\theta_0(s)}{E_2(s)} = \frac{6.93 \times 10^6}{s[(s+77.5)^2 + 79.3^2]}$$

The root locus plot for the outer loop is shown in Fig. 9-31(b), where it will be noted the poles s_1 and s_2 have the same locations as s_1 and s_2 in Fig. 9-31(a). Since the outer loop dominant poles must have natural

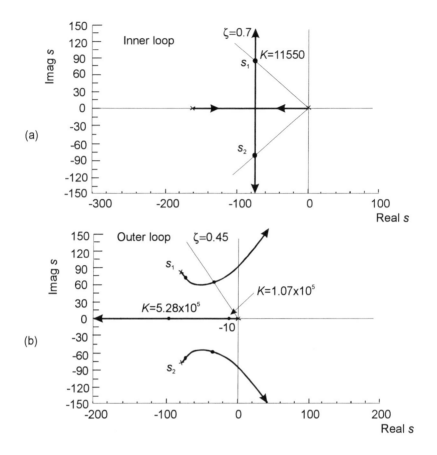

Figure 9.31: Root locus for tachometer feedback $G_c(s)$

frequencies $\omega_n \leq 10$ rad/sec, we find the loop sensitivity $K = 1.07 \times 10^5$, so that $A_1 = 0.015$. The steady state velocity error coefficient

$$K_v = \lim_{s \to 0} s A_1 \frac{6.93 \times 10^6}{s[(s+77.5)^2 + 79.3^2]}$$

$$= \frac{0.015 \times 6.93 \times 10^6}{77.5^2 + 79.3^2} = 8.45 \, \text{sec}^{-1}$$

Thus

$$e_1(\infty) = \frac{12}{8.45} = 1.42 \, \text{deg}$$

It will be noted in the above analysis it is the root on the real axis which is the dominant pole for the system.

It is seen that the magnitude of the steady state error is excessively large. We now see that replacing $G_c(s)$ by a lead network will reduce the steady state error by a significant factor. This is known as **tachometer plus lead compensation**. Experimenting with the position of the lead compensator pole suggests

$$G_c(s) = \frac{s}{s + 2.4}$$

Repeating the above analysis for the inner loop gives the root locus plot shown in Fig. 9-32(a). Taking $\zeta = 0.7$ for the dominant poles we find for a loop sensitivity $K = 1.156 \times 10^4$, or $A_2 = 51.4$, the inner closed loop transfer function is

$$\frac{\theta_0(s)}{E_2(s)} = \frac{6.936 \times 10^6(s + 2.4)}{s[(s + 78.6)^2 + 80.5^2](s + 0.142)}$$

The root locus plots for the outer loop are shown in Figs. 9-32(b) and 9-32(c). Again since the outer loop dominant poles must have $\omega_n \leq 10$ rad/sec, we determine that the loop sensitivity $K = 1.45 \times 10^5$, so that $A_1 = 0.021$. It will be noted that the introduction of the lead compensator $G_c(s)$ in the inner loop effectively introduces a lag compensator in the open loop transfer function of the outer loop. By careful positioning of the pole of $G_c(s)$ we can determine the position of the effective lag compensation pole, which in this case is at $s = 0.142$.

The steady state velocity error coefficient

$$K_v = \lim_{s \to 0} sA_1 \frac{6.936 \times 10^6(s + 2.4)}{s[(s + 78.6)^2 + 80.5^2](s + 0.142)}$$

$$= 194.5 \, \text{sec}^{-1}$$

The steady error for a ramp input becomes

$$e_1(\infty) = \frac{12}{194.5} = 0.062 \, \text{deg}$$

We thus obtain a twenty two-fold reduction in the steady state error and consequently satisfy the specification.

9.6 Feedforward Compensation

The concept of feedforward compensation has been found useful for reducing the effect of external disturbances upon system response, and also for

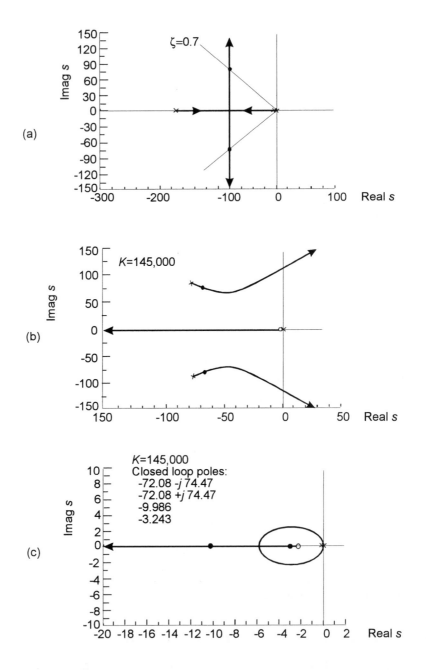

Figure 9.32: Root locus for tachometer plus lead network feedback

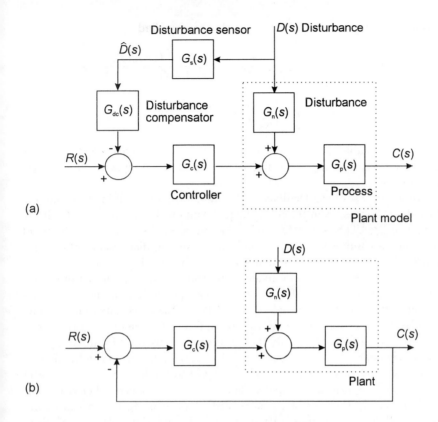

Figure 9.33: (a) Disturbance compensation by feedforward control (b) Feedback system with disturbance input

compensating command inputs.

To illustrate the application of **disturbance feedforward compensation** suppose the process has an input $R(s)$ and disturbance $D(s)$ and is modeled as shown in Fig. 9-33(a). We can often measure some function of the disturbance. Assuming this function $G_s(s)$ together with the functions $G_n(s)$ and $G_c(s)$ are precisely known then, in principle, a disturbance compensator $G_{dc}(s)$ can be found which will completely nullify the influence of $D(s)$ upon the output $C(s)$. In practice things are not so simple because even if $G_{dc}(s)$ is stable and physically realizable unknown errors in $G_c(s)$, $G_n(s)$ and $G_s(s)$ will make the ideal situation unachievable.

Feedback can also be used to minimize the effect of disturbances upon the output of a system. This is illustrated in Fig. 9-33(b) where it is assumed that $|G_c(s)G_p(s)| \gg 1$. In this case we find

$$
C(s) = \frac{G_p(s)G_c(s)}{1 + G_p(s)G_c(s)} R(s) + \frac{G_n(s)G_p(s)}{1 + G_p(s)G_c(s)} D(s)
$$

$$
\approx R(s) + \frac{G_n(s)}{G_c(s)} D(s)
$$

If $|G_c(s)|$ is sufficiently large the effect of the disturbance $D(s)$ upon the output $C(s)$ can be minimized irrespective of the accuracy with which $G_n(s)$ is known.

In many cases both feedback and feedforward are combined to compensate disturbances, even though it might seem on the surface that feedback would suffice. The reason for their joint use arises because often the plant dynamics are quite slow so that significant transient disturbance of the output $C(s)$ can occur before feedback action can come into play. In this case the feedforward path can be made to be fast acting. As an example, this type of control is quite often used in such applications as power station steam boilers, as the thermal and fluid flow time constants can be quite long, so that if a step change in the steam flow rate is demanded the boiler would respond quite sluggishly if only feedback is used. If feedforward control is used in conjunction with feedback the steam flow rate can be directly measured, and this signal can be fed forward to the heating and feedwater control loops much more rapidly than could be achieved otherwise. The performance in the face of transient disturbances using feedforward combined with feedback compensation will often be superior to the use of either alone.

To see the effect of feedforward disturbance compensation on the operation of a feedback system consider the system shown in Fig. 9-34 with and without feedforward control. The overall transfer functions for the two cases become

No feedforward: $C(s) = \dfrac{R(s)G_1(s)G_2(s)}{1 + G_1(s)G_2(s)} + \dfrac{D(s)G_2(s)}{1 + G_1(s)G_2(s)}$

Feedforward: $C(s) = \dfrac{R(s)G_1(s)G_2(s)}{1 + G_1(s)G_2(s)} + \dfrac{D(s)G_2(s)[1 - G_0(s)G_1(s)]}{1 + G_1(s)G_2(s)}$

We notice that the inclusions of feedforward control has reduced the effect of the disturbance $D(s)$ on system response by a factor $(1 - G_0(s)G_1(s))$.

Example 7 *Consider the feedback system shown in Fig. 9-35, with command and disturbance inputs $R(s)$ and $D(s)$ respectively. Assuming the*

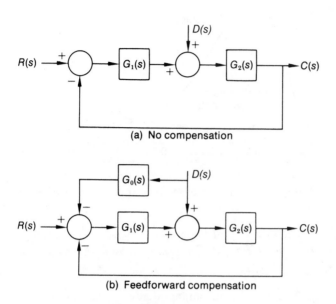

(a) No compensation

(b) Feedforward compensation

Figure 9.34: Using feedforward compensation for decreasing disturbance effects in feedback system

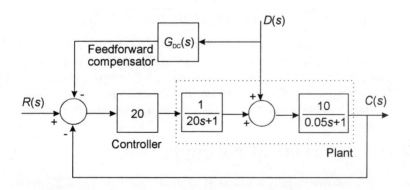

Figure 9.35: Feedback system with feedforward compensation

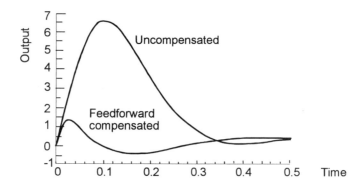

Figure 9.36: System time response with unit step disturbance input

command input $R(s) = 0$, find the output response of the system to a unit step disturbance input for (a) $G_{DS}(s) = 0$, and (b) $G_{DS}(s) = 1/20 + s/(0.01s + 1)$.

From Fig. 9-35, for the uncompensated case the output $C(s)$ is given by

$$C(s) = \frac{10(20s + 1)}{s^2 + 20.05s + 201} D(s)$$

while for the feedforward compensated case we have

$$C(s) = \frac{2s^2}{(s^2 + 20.05s + 201)(0.01s + 1)} D(s)$$

The time response to a unit step disturbance for the uncompensated and feedforward compensated cases are plotted in Fig. 9-36. It will be observed in this figure that not only does the system with feedforward compensation settle much more rapidly than the uncompensated case but it will also be noted that the peak magnitude of the output is considerably reduced.

9.7   A Practical Example

In this section we employ a practical example, discussed in Chapter 1, to illustrate the procedure of design, analysis, and compensation of a control system.

Let us assume that we wish to design a platform for use as a calibration test stand for a digital sun sensor. This requires the platform to remain

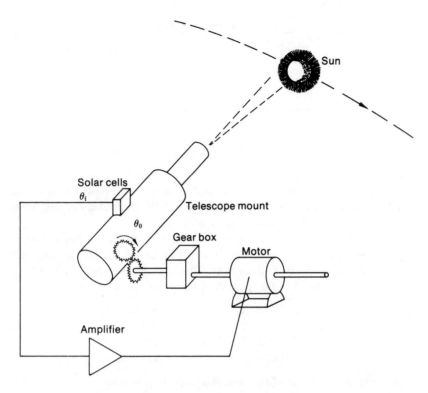

Figure 9.37: Schematic of Sun Tracker

in a fixed orientation with respect to the sun, i.e. the platform should automatically[2] track the sun. Since the precision of this sun tracker would limit the precision of the calibration of the digital sun sensor, it is necessary to limit the tracking error to no more than $0.0004°$.

The postulated system, that we thought could do the necessary job, is pictorially illustrated in Fig. 9-37. Briefly, it consists of an astronomical telescope mount, two silicon solar cells, an amplifier, a motor, and gears. The solar cells are attached to the polar axis of the telescope so that if the pointing direction is in error, more of the sun's image falls on one cell than the other. This pair of cells, when connected in parallel opposition, appear as a current source and act as a positional error sensing device. A simple differential input transistor amplifier can provide sufficient gain so that the

---

[2]Manual tracking using a rheostat is ruled out since the precision desired would require tremendous continuous concentration.

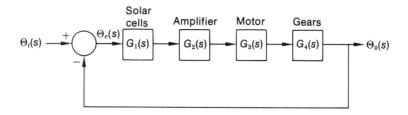

Figure 9.38: Block diagram of SunTracker

small error signals produce an amplifier output sufficient for running the motor. This motor sets the rotation rate of the polar axis of the telescope mount to match the apparent motion of the sun.

Having postulated one possible system, we now draw a block diagram of the sun tracker as shown in Fig. 9-38. We idealize the system and assume that it shall behave in a linear way. The various transfer functions can be obtained as outlined in Chapter 2. These transfer functions are

$$
\begin{aligned}
G_1(s) &= K_1 && \text{Solar cells} \\
G_2(s) &= K_2 && \text{Amplifier} \\
G_3(s) &= K_3/s(As + B) && \text{Motor} \\
G_4(s) &= K_4 && \text{Gear ratio}
\end{aligned}
$$

The output and error of the postulated system become

$$
\frac{\Theta_o(s)}{\Theta_i(s)} = \frac{K}{s(As + B) + K}
$$

$$
\frac{\Theta_e(s)}{\Theta_i(s)} = \frac{s(As + B)}{s(As + B) + K}
$$

where $K = K_1 K_2 K_3 K_4$.

In obtaining these relationships we have assumed that the time constant of the amplifier is small and can be therefore neglected; there is no friction or backlash, etc., in the gears; the gear ratio is so high that the reflected inertia of the telescope mount is small and therefore can be neglected; and finally the system is performing in a linear manner.

The sun is assumed to move at a constant rate over the region of interest so that $\theta_i(t) = mt$ where m is 2π radians per day or about 4.14×10^{-3} deg/sec. The output becomes

$$
\Theta_o(s) = \frac{4.14 \times 10^{-3} K}{s^2[s(As + B) + K]}
$$

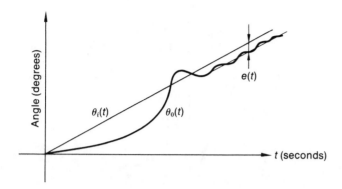

Figure 9.39: The input and output of the Sun Tracker

Assuming that the system is under damped, the error and output can be obtained in the time domain from the techniques developed in Chapter 4,

$$\theta_0(t) = C_0 + C_1 t + C_2 e^{-\sigma t} \sin(\omega t + \phi)$$

where C_0 is the constant error and $C_1 = 4.14 \times 10^{-3}$. The last term goes to zero in the steady state. The error is

$$C_0 = \frac{B(4.14 \times 10^{-3})}{K}$$

If this is constrained to 0.0004, then the overall gain must be set at $K = 10.35B$. The predicted output and error of the linear system are shown in Fig. 9-39.

Satisfied that the idealized linear system meets the specifications, the actual hardware was built and tested[3]. The hardware used were off-the-shelf items with the following characteristics

$$\begin{aligned} A &= 20 \text{ g cm}^2 \\ B &= 0.45 \text{ g cm/rad} \\ K_1 &= 80 \text{ mA/deg} \end{aligned}$$

Using the previous analysis of the idealized system, the overall gain was set at

$$K = (10.35)(0.45) = 4.65$$

---

[3] This was built and tested by George Bush at the Applied Physics Laboratory of John Hopkins University.

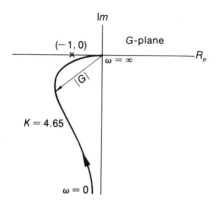

Figure 9.40: Nyquist plot of the Sun Tracker

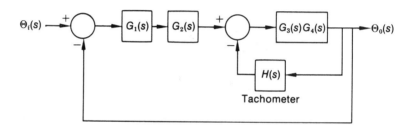

Figure 9.41: Sun Tracker compensated by tachometric feedback

Initial tests showed that the performance was erratic with the error suddenly changing and not being well damped. After careful analysis, it was concluded that the erratic behavior was due to noise and varying friction in the gear train which had frequency components commensurate with the system itself. Since the amplifier gain was set very high, the Nyquist plot of the open loop system looks like that shown in Fig. 9-40. We therefore concluded that the system was being excited at and operating near the resonant peak. It was apparent that a compensation network would be necessary to increase the damping and thereby suppress the resonant peak.

A tachometric generator[4] as shown in Fig. 9-41, was employed for compensation.

$$H(s) = K_5 s$$

---

[4] This was readily available.

The effect of this compensator is to add system damping and therefore move the Nyquist plot away from the resonant area. The output now becomes

$$\frac{\Theta_o(s)}{\Theta_i(s)} = \frac{K}{s(As + B_e) + K}$$

where

$$B_e = K_3 K_4 K_5 + B$$

This allows us some control over the system damping. The above scheme was found adequate to correct the problem and meet the needs of the calibration facility. Notice that the amplifier gain must now be increased if the same steady state positional error is desired.

9.8 Pole-Placement Design

The approach to compensator design described in this chapter so far can involve a considerable amount of trial and error where compensators are introduced in an ad hoc manner to meet the required system specifications. If the proposed compensator fails to yield a satisfactory system design then the designer, guided by his previous experience, must propose and then evaluate alternative compensators to find if they are suitable. In the discussion below we show **state variable feedback** allows the system closed loop poles to be located at any desired positions. Thus state feedback can be used to develop a useful design method provided all state variables are directly measurable. We also show that even when some variables cannot be measured they can often be estimated using a state observer and its state variables can be used instead of the actual variables in the feedback scheme.

State Variable Feedback

We start our investigation by assuming all state variables are directly measurable. Consider a single-input single-output system whose transfer function has a state equation realization

$$\dot{\mathbf{x}}(t) = \mathbf{A}\mathbf{x}(t) + \mathbf{b}u(t) \tag{9.11}$$

$$y(t) = \mathbf{c}\mathbf{x}(t) \tag{9.12}$$

The roots of the characteristic polynomial $\det[s\mathbf{I} - \mathbf{A}]$ are shown in Chapter 4 to be the poles of $G(s)$. State variable feedback is obtained by letting the system input

$$u(t) = r(t) - \mathbf{k}\mathbf{x}(t) \tag{9.13}$$

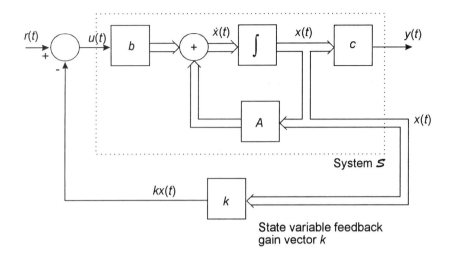

Figure 9.42: Control system with state variable feedback

where $r(t)$ is the command input and

$$\mathbf{k} = [\ k_1 \quad k_2 \quad \cdots \quad k_n\ ]$$

is the state feedback gain vector. A block diagram for a system with state variable feedback is shown in Fig. 9-42. It is immediately evident from Eqs. (9-11) and (9-13) that the state equation becomes

$$\dot{\mathbf{x}}(t) = (\mathbf{A} - \mathbf{b}\mathbf{k})\mathbf{x}(t) + \mathbf{b}r(t) \tag{9.14}$$

It can be seen from Eq. (9-14) that the modified characteristic polynomial for this system is $\det[s\mathbf{I} - \mathbf{A} + \mathbf{b}\mathbf{k}]$. In general, the term $\mathbf{b}\mathbf{k}$ will lead to some or all of the system closed loop poles being relocated from their open loop positions, as given by the poles of $G(s)$.

Ackermann's Formula

Assuming the system given in Eq. (9-11) is completely state controllable, it can be shown there exists a matrix $\mathbf{T}$ such that Eq. (9-11) is transformed to the controller canonical form (refer Section 3-7)

$$\dot{\mathbf{z}}(t) = \widetilde{\mathbf{A}}\mathbf{z}(t) + \widetilde{\mathbf{b}}u(t) \tag{9.15}$$

and from Eq. (9-13)

$$u(t) = r(t) - \widetilde{\mathbf{k}}\mathbf{z}(t) \tag{9.16}$$

where

$$
\tilde{\mathbf{A}} =
\begin{bmatrix}
0 & 1 & 0 & \cdots & 0 \\
0 & 0 & 1 & \cdots & 0 \\
\vdots & \vdots & \vdots & & \vdots \\
0 & 0 & 0 & \cdots & 1 \\
-a_0 & -a_1 & -a_2 & \cdots & -a_{n-1}
\end{bmatrix}
\qquad
\tilde{\mathbf{b}} =
\begin{bmatrix}
0 \\
0 \\
\vdots \\
0 \\
1
\end{bmatrix}
$$

and $\mathbf{x}(t) = \mathbf{T}\mathbf{z}(t)$. In this case it can be easily seen the characteristic polynomial is $s^n + a_{n-1}s^{n-1} + \cdots + a_1 s + a_0$. Supposing it is desired the characteristic polynomial for the closed loop system is

$$\Delta(s) = s^n + p_{n-1}s^{n-1} + \cdots + p_1 s + p_0 \qquad (9.17)$$

then we find the feedback gain vector is

$$\tilde{\mathbf{k}} = \begin{bmatrix} (p_0 - a_0) & (p_1 - a_1) & \cdots & (p_{n-1} - a_{n-1}) \end{bmatrix}$$

We now seek to find the feedback gain vector $\mathbf{k}$ for the system given by Eq. (9-11). From Eq. (9-16)

$$u(t) = r(t) - \tilde{\mathbf{k}}\mathbf{T}^{-1}\mathbf{x}$$

so that $\mathbf{k} = \tilde{\mathbf{k}}\mathbf{T}^{-1}$. From Eq. (3-59) we know

$$\mathbf{T} = \mathcal{C}\left(\mathbf{A}, \mathbf{b}\right) \mathcal{C}\left(\tilde{\mathbf{A}}, \tilde{\mathbf{b}}\right)^{-1} \qquad (9.18)$$

We can avoid calculating $\mathbf{T}$ explicitly by noting the special forms of the matrices $\tilde{\mathbf{A}}$ and $\tilde{\mathbf{b}}$ give

$$
\mathcal{C}\left(\tilde{\mathbf{A}}, \tilde{\mathbf{b}}\right)^{-1} =
\begin{bmatrix}
a_1 & a_2 & \cdots & a_{n-1} & 1 \\
a_2 & a_3 & \cdots & 1 & 0 \\
\vdots & \vdots & & \vdots & \vdots \\
1 & 0 & \cdots & 0 & 0
\end{bmatrix}
$$

so that

$$\begin{bmatrix} 1 & 0 & \cdots & 0 \end{bmatrix} \mathcal{C}\left(\tilde{\mathbf{A}}, \tilde{\mathbf{b}}\right) = \begin{bmatrix} 0 & 0 & \cdots & 1 \end{bmatrix} \qquad (9.19)$$

From Eq. (9-17) the matrix polynomial

$$\Delta\left(\tilde{\mathbf{A}}\right) = \tilde{\mathbf{A}}^n + p_{n-1}\tilde{\mathbf{A}}^{n-1} + \cdots + p_0 \mathbf{I} \qquad (9.20)$$

and from the Cayley-Hamilton theorem we have

$$\tilde{\mathbf{A}}^n + a_{n-1}\tilde{\mathbf{A}}^{n-1} + \cdots + a_0 \mathbf{I} = 0$$

Solving for $\widetilde{\mathbf{A}}^n$ and substituting into Eq. (9-20) gives

$$\Delta\left(\widetilde{\mathbf{A}}\right) = (p_{n-1} - a_{n-1})\widetilde{\mathbf{A}}^{n-1} + \cdots + (p_0 - a_0)\mathbf{I}$$

The first row of the matrix $\widetilde{\mathbf{A}}^k$ contains all zeros excepting for the $(k+1)$th element which is unity. Therefore we can write

$$\widetilde{\mathbf{k}} = \begin{bmatrix} 1 & 0 & \cdots & 0 \end{bmatrix} \Delta\left(\widetilde{\mathbf{A}}\right)$$

so that

$$\mathbf{k} = \widetilde{\mathbf{k}}\mathbf{T}^{-1} = \begin{bmatrix} 1 & 0 & \cdots & 0 \end{bmatrix} \Delta\left(\widetilde{\mathbf{A}}\right)\mathbf{T}^{-1}$$

$$= \begin{bmatrix} 1 & 0 & \cdots & 0 \end{bmatrix} \mathbf{T}^{-1}\Delta(\mathbf{A})$$

Replacing $\mathbf{T}^{-1}$ by Eq. (9-18) and using the identity given in Eq. (9-19) gives

$$\mathbf{k} = \begin{bmatrix} 0 & 0 & \cdots & 0 & 1 \end{bmatrix} \mathcal{C}(\mathbf{A}, \mathbf{b})^{-1} \Delta(\mathbf{A}) \qquad (9.21)$$

This is known as **Ackermann's formula** and it gives the state feedback gain vector $\mathbf{k}$ for a general state variable system as given in Eq. (9-11). It will be noted that the system needs to be state controllable for the formula to be applicable.

It is important to note state variable feedback only re-positions the system pole locations. The zeros of the closed loop system remain at the positions of the open loop zeros of the plant. This observation can be seen by applying the transformation $\mathbf{x}(t) = \mathbf{T}\mathbf{z}(t)$ to Eq. (9-12). In this case Eq. (9-15) and the output equation

$$y(t) = \mathbf{c}\mathbf{T}\mathbf{z}(t) = \widetilde{\mathbf{c}}\mathbf{z}(t) \qquad (9.22)$$

represent the system given by Eqs. (9-11) and (9-12) in controller canonical form. It can be seen from Section 3-7 the elements of the output matrix $\widetilde{\mathbf{c}}$ equal the coefficients of the numerator polynomial of $G(s)$ which determines its zeros. From the discussion given in the development of the Ackermann formula state variable feedback modifies the elements of the companion matrix $\widetilde{\mathbf{A}}$, yet $\widetilde{\mathbf{c}}$ remains unchanged. Thus the zeros are unaffected by state variable feedback. While the positions of the zeros are unaffected by feedback any number of them may be canceled by positioning closed loop poles at their positions. Thus there exist the relationship

$$(P - Z)_{\text{closed}} = (P - Z)_{G(s)} \qquad (9.23)$$

for the open loop and closed loop systems, where P and Z are the number of poles and zeros respectively.

Example 8 *Consider the magnetic bearing discussed in Example 5, where the transfer function of the current amplifier and bearing actuator*

$$G(s) = \frac{1.2}{0.002s + 1} \cdot \frac{20}{s^2 - 6 \times 10^4}$$

Using the actuator position, velocity and input current as state variables find the state variable feedback gain vector k so that the closed loop pole positions are at $\zeta = 0.45$ and $\omega_n = 90$ rad/sec for the complex poles and $s = -400$ for the real poles. The gain vector k should be found by the method of equating coefficients and by using Ackermann's formula.

Defining $x_1(t) = i(t)$, $x_2(t) = x(t)$, $x_3(t) = \dot{x}(t)$ and $u(t) = e_c(t)$, where $x(t)$, $i(t)$ and $e_c(t)$ are the bearing rotor position, coil current and the amplifier input voltage respectively. We obtain the state equation

$$\begin{bmatrix} \dot{x}_1(t) \\ \dot{x}_2(t) \\ \dot{x}_3(t) \end{bmatrix} = \begin{bmatrix} -500 & 0 & 0 \\ 0 & 0 & 1 \\ 20 & 6 \times 10^4 & 0 \end{bmatrix} \begin{bmatrix} x_1(t) \\ x_2(t) \\ x_3(t) \end{bmatrix} + \begin{bmatrix} 600 \\ 0 \\ 0 \end{bmatrix} u(t) \quad (9.24)$$

The general linear state variable feedback is given by

$$u(t) = r(t) - k_1 x_1(t) - k_2 x_2(t) - k_3 x_3(t)$$

Substituting this relationship into Eq. (9-14) gives

$$\dot{x}(t) = \begin{bmatrix} -500 - 600k_1 & -600k_2 & -600k_3 \\ 0 & 0 & 1 \\ 20 & 6 \times 10^4 & 0 \end{bmatrix} x(t) + \begin{bmatrix} 600 \\ 0 \\ 0 \end{bmatrix} r(t)$$

The characteristic equation for this state equation is

$$s^3 + (500 + 600k_1)s^2 + (1.2k_3 - 6) \times 10^4 s + \left(1.2 \times 10^4 k_2 \right.$$
$$\left. -36 \times 10^6 k_1 - 30 \times 10^6\right) = 0$$

From above the desired characteristic equation is

$$\Delta(s) = (s^2 + 2 \times 0.45 \times 90s + 90^2)(s + 400)$$

$$= s^3 + 481s^2 + 40500s + 3.24 \times 10^6 = 0$$

Equating coefficients gives the equations

$$500 + 600k_1 = 481$$

$$12000k_3 - 60000 = 40500$$

$$12000k_2 - 36 \times 10^6 k_1 - 30 \times 10^6 = 3.24 \times 10^6$$

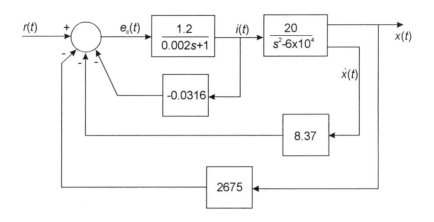

Figure 9.43: Magnetic bearing with state variable feedback

Solving these equations gives

$$\mathbf{k} = \begin{bmatrix} -0.0316 & 2675 & 8.375 \end{bmatrix}$$

We now apply Ackermann's formula to find $\mathbf{k}$. The inverse of the controllability matrix

$$\mathcal{C}(\mathbf{A}, \mathbf{b})^{-1} = \begin{bmatrix} 1.67 \times 10^{-3} & 0 & 4.167 \times 10^{-2} \\ 0 & 4.167 \times 10^{-2} & 8.33 \times 10^{-5} \\ 0 & 8.33 \times 10^{-5} & 0 \end{bmatrix}$$

and substituting the matrix $\mathbf{A}$ into the characteristic equation $\Delta(s)$ gives

$$\Delta(\mathbf{A}) = \begin{bmatrix} -2.176 \times 10^{7} & 0 & 0 \\ -380 & 3.21 \times 10^{7} & 1.005 \times 10^{5} \\ 2.2 \times 10^{6} & 6.03 \times 10^{9} & 3.21 \times 10^{7} \end{bmatrix}$$

Thus the state variable feedback gain matrix becomes

$$\mathbf{k} = \begin{bmatrix} 0 & 0 & 1 \end{bmatrix} \mathcal{C}(\mathbf{A}, \mathbf{b})^{-1} \Delta(\mathbf{A})$$

$$= \begin{bmatrix} -0.0316 & 2670 & 8.37 \end{bmatrix}$$

which equals the result given by equating coefficients. The block diagram for the magnetic bearing with state variable feedback is shown in Fig. 9-43.

Servo System Design

In the above discussion we have only considered the problem of positioning the system closed loop poles using state variable feedback so as to optimize its response to plant transient disturbances. In this case, where the command input $r(t)$ is assumed to be identically zero, the poles are usually positioned so that the states are rapidly driven to zero. This is referred to as the **regulator problem**. The **servo problem** is one where the input $r(t) \neq 0$ and where the objective is for the output, and in some cases the states, to follow some desired trajectory. In general for servo design where the command signal $r(t)$ is varying it is necessary to feed-forward compensate the signal $r(t)$.

One common type of servo problem is to ensure that small errors in the plant model do not lead to undesired errors in the steady-state error. We have seen when using block diagram design methods that these types of errors can be eliminated by using integral control. This approach can also be introduced with state variable feedback design.

Supposing $r_c(t)$ is the command input and $y(t)$ is the system output, integral control can be introduced as shown in Fig. 9-44(a). Rearranging this diagram as shown in Fig. 9-44(b), it can be seen that the integrator augments the number of state variables so that the closed loop pole locations can be determined by the choice of the augmented state variable feedback gains $k_1, \ldots, k_{n+1}$. In this case if the closed loop poles for the augmented system are stable then the output $y(\infty) = r_c(\infty)$, which is the steady state value of $r_c(t)$.

Once the number of state variables has been augmented by the integrator as shown in Fig. 9-44(b) the design procedure followed for servo design follows along similar lines to that discussed in Example 8. To see this we write the state equations for the system shown in Fig. 9-44(b) as

$$
\begin{aligned}
\underline{\dot{\mathbf{x}}}(t) &= \underline{\mathbf{A}}\,\underline{\mathbf{x}}(t) + \underline{\mathbf{b}}_1 u(t) + \underline{\mathbf{b}}_2 r_c(t) \\
y(t) &= \underline{\mathbf{c}}\,\underline{\mathbf{x}}(t)
\end{aligned} \tag{9.25}
$$

where

$$
\underline{\mathbf{x}}(t) = \begin{bmatrix} \mathbf{x}(t) \\ x_{n+1}(t) \end{bmatrix}, \quad \underline{\mathbf{A}} = \begin{bmatrix} \mathbf{A} & 0 \\ -\mathbf{c} & 0 \end{bmatrix}, \quad \underline{\mathbf{b}}_1 = \begin{bmatrix} \mathbf{b} \\ 0 \end{bmatrix}, \quad \underline{\mathbf{b}}_2 = \begin{bmatrix} 0 \\ 1 \end{bmatrix} \tag{9.26}
$$

and

$$
\underline{\mathbf{c}} = \begin{bmatrix} \mathbf{c} & 0 \end{bmatrix}
$$

The state feedback is given by

$$
u(t) = -\mathbf{k}\mathbf{x}(t) - k_{n+1}x_{n+1}(t) = -\underline{\mathbf{k}}\,\underline{\mathbf{x}}(t) \tag{9.27}
$$

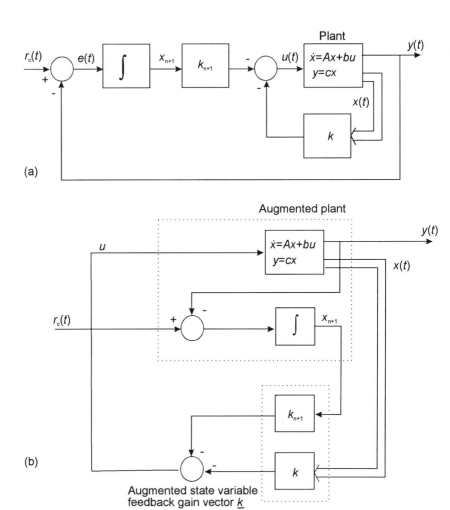

Figure 9.44: Servo system with state feedback and integral control

where $\underline{\mathbf{k}} = \begin{bmatrix} \mathbf{k} & k_{n+1} \end{bmatrix}$, so that the state equation for the complete system with state feedback is

$$\dot{\mathbf{x}}(t) = (\underline{\mathbf{A}} - \underline{\mathbf{b}}_1\underline{\mathbf{k}})\mathbf{x}(t) + \underline{\mathbf{b}}_2 r_c(t) \tag{9.28}$$

It will be noted the matrix

$$\underline{\mathbf{A}} - \underline{\mathbf{b}}_1\underline{\mathbf{k}} = \begin{bmatrix} \mathbf{A} - \mathbf{bk} & -\mathbf{b}k_{n+1} \\ -\mathbf{c} & 0 \end{bmatrix} \tag{9.29}$$

Example 9 *Consider the magnetic bearing discussed in Examples 5 and 8, where the state equation is given in Eq. (9-24), and the variables $x_1(t)$, $x_2(t)$, $x_3(t)$, and $e_c(t)$ are the coil current, rotor position and velocity, and amplifier input voltage respectively. To ensure the system under steady state conditions tracks the input $r_c(t)$ accurately we propose to use integral control as shown in Fig. 9-44. Find the augmented state variable feedback gain vector $\mathbf{k}$ for the system using Ackermann's formula, and find the position response of the system when $r_c(t)$ is a unit step input.*

The augmented state equation is

$$\begin{bmatrix} \dot{x}_1(t) \\ \dot{x}_2(t) \\ \dot{x}_3(t) \\ \dot{x}_4(t) \end{bmatrix} = \begin{bmatrix} -500 & 0 & 0 & 0 \\ 0 & 0 & 1 & 0 \\ 20 & 6\times10^4 & 0 & 0 \\ 0 & -1 & 0 & 0 \end{bmatrix} \begin{bmatrix} x_1(t) \\ x_2(t) \\ x_3(t) \\ x_4(t) \end{bmatrix} +$$

$$\begin{bmatrix} 600 \\ 0 \\ 0 \\ 0 \end{bmatrix} u(t) + \begin{bmatrix} 0 \\ 0 \\ 0 \\ 1 \end{bmatrix} r_c(t)$$

$$y(t) = \begin{bmatrix} 0 & 1 & 0 & 0 \end{bmatrix} \mathbf{x}(t)$$

From Fig. 9-44 we see the state feedback is

$$u(t) = - \begin{bmatrix} k_1 & k_2 & k_3 & k_4 \end{bmatrix} \mathbf{x}(t)$$

From Example 8 we find the desired characteristic equation is

$$\begin{aligned} \Delta(s) &= (s^2 + 2\times0.45\times90s + 90^2)(s+400)^2 \\ &= s^4 + 881s^3 + 232900s^2 + 1.944\times10^7 s + 1.296\times10^9 = 0 \end{aligned}$$

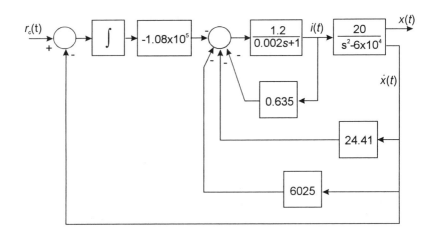

Figure 9.45: Magnetic bearing control system using state feedback with integral control

We now find the inverse of the controllability matrix

$$\mathcal{C}\left(\mathbf{A}, \mathbf{b}_1\right)^{-1}$$

$$
= \begin{bmatrix}
1.67 \times 10^{-3} & 0 & 4.167 \times 10^{-2} & 2500 \\
0 & 4.167 \times 10^{-2} & 8.33 \times 10^{-5} & 5 \\
0 & 8.33 \times 10^{-5} & 0 & -4.166 \times 10^{-2} \\
0 & 0 & 0 & -8.33 \times 10^{-5}
\end{bmatrix}
$$

and substituting the matrix $\mathbf{A}$ into the characteristic equation $\Delta(s)$ gives

$$
\Delta\left(\mathbf{A}\right) = \det \begin{bmatrix}
2.176 \times 10^9 & 0 & 0 & 0 \\
2.048 \times 10^6 & 1.887 \times 10^{10} & 7.23 \times 10^7 & 0 \\
4.22 \times 10^8 & 4.338 \times 10^{12} & 1.887 \times 10^{10} & 0 \\
-7620 & -7.23 \times 10^7 & -2.929 \times 10^5 & 1.296 \times 10^9
\end{bmatrix}
$$

Thus the state variable feedback gain matrix becomes

$$\mathbf{k} = \begin{bmatrix} 0 & 0 & 0 & 1 \end{bmatrix} \mathcal{C}\left(\mathbf{A}, \mathbf{b}_1\right)^{-1} \Delta(\mathbf{A})$$

$$= \begin{bmatrix} 0.635 & 6025 & 24.41 & -1.08 \times 10^5 \end{bmatrix}$$

The block diagram for the magnetic bearing control system using state variable feedback with integral control is shown in Fig. 9-45. The output

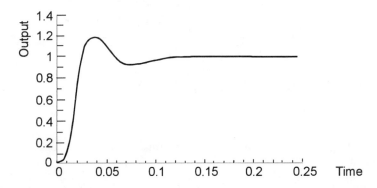

Figure 9.46: Output response of magnetic bearing control system for $r_c(t)$ being a unit step

response of this magnetic bearing for a unit step applied to the input $r_c(t)$ is given in Fig. 9-46. It can be seen that the settling time is approximately 0.1 second and the peak overshoot is 20 percent.

Linear Quadratic Regulator Problem

We have shown in this section that state variable feedback enables us to place the closed loop poles for a system at any desired location in the complex plane thus solving the **pole-placement regulator problem**. It is natural to enquire if there is a "best solution", in some sense, to this problem. Extending the ideas developed in Chapter 6, Section 3 we pose the linear quadratic optimal regulator problem: Given a system $\mathcal{S}$ described by the state equation

$$\dot{\mathbf{x}}(t) = \mathbf{A}\mathbf{x}(t) + \mathbf{b}u(t) \tag{9.30}$$

find the state variable gain matrix $\mathbf{k}$ for the feedback law

$$u(t) = -\mathbf{k}\mathbf{x}(t) \tag{9.31}$$

which minimizes the performance index

$$J = \int_0^\infty \mathbf{x}(t)^T \mathbf{x}(t)\, dt \tag{9.32}$$

This is not the most general problem of this class but it will suffice to illustrate the approach to their solution. A block diagram for this system is shown in Fig. 9-47. Substituting Eq. (9-31) into Eq. (9-30) gives

$$\dot{\mathbf{x}}(t) = (\mathbf{A} - \mathbf{b}\mathbf{k})\,\mathbf{x}(t) = \mathbf{N}\mathbf{x}(t) \tag{9.33}$$

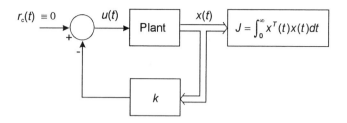

Figure 9.47: Linear quadratic optimal regulator system

where $\mathbf{N} = \mathbf{A} - \mathbf{b}\mathbf{k}$.

In order to minimize the function J given in Eq. (9-32), we assume the existence of a differential of the form

$$\frac{d}{dt}\left(\mathbf{x}^T \mathbf{Q} \mathbf{x}\right) = -\mathbf{x}^T \mathbf{x} \tag{9.34}$$

where the constant symmetric matrix $\mathbf{Q}$ must be determined. Substituting this into Eq. (9-32),

$$\begin{aligned} J &= \int_0^\infty -\frac{d}{dt}\left(\mathbf{x}^T \mathbf{Q} \mathbf{x}\right) dt \\ &= \mathbf{x}^T(0)\mathbf{Q}\mathbf{x}(0) - \left[\mathbf{x}^T(t)\,\mathbf{Q}\mathbf{x}(t)\right]_{t=\infty} \end{aligned} \tag{9.35}$$

The last term of the above equation is assumed to be zero, thereby insuring system stability. We have therefore

$$J = \mathbf{x}^T(0)\mathbf{Q}\mathbf{x}(0) \tag{9.36}$$

The minimization of this satisfies the objective of our design. The matrix $\mathbf{Q}$ can be related to the system by carrying out the differentiation of Eq. (9-34), and substituting Eq. (9-33)

$$\begin{aligned} \frac{d}{dt}\left(\mathbf{x}^T \mathbf{Q} \mathbf{x}\right) &= \frac{d\mathbf{x}}{dt}^T \mathbf{Q}\mathbf{x} + \mathbf{x}^T \mathbf{Q}\frac{d\mathbf{x}}{dt} \\ &= \mathbf{x}^T \left[\mathbf{N}^T \mathbf{Q} + \mathbf{Q}\mathbf{N}\right] \mathbf{x} \end{aligned}$$

We can satisfy Eq. (9-34) only if

$$\mathbf{N}^T \mathbf{Q} + \mathbf{Q}\mathbf{N} = -\mathbf{I} \tag{9.37}$$

The use of Eqs. (9-36) and (9-37) together satisfy our design requirements. When the size of the matrices are large, which is often the case in modern systems, the use of a digital computer is necessary.

Example 10 *A system is characterized by* $\dot{\mathbf{x}} = \mathbf{A}\mathbf{x} + \mathbf{b}u$ *where*

$$\mathbf{A} = \begin{bmatrix} 0 & 1 \\ -\alpha & -\beta \end{bmatrix}, \quad \mathbf{b} = \begin{bmatrix} 0 \\ 1 \end{bmatrix}$$

It is desired to minimize the performance index of Eq. (9-32) and to use state feedback where $u = -x_1 - ax_2$. *Obtain the coefficient* a *so as to minimize the value of* J. *Let* $x^T(0) = \begin{bmatrix} 1 & 0 \end{bmatrix}$.

From $u = -x_1 - ax_2$, we have

$$\mathbf{k} = \begin{bmatrix} 1 & a \end{bmatrix}$$

The **N** matrix becomes

$$\mathbf{N} = \begin{bmatrix} 0 & 1 \\ -\alpha & -\beta \end{bmatrix} - \begin{bmatrix} 0 \\ 1 \end{bmatrix} \begin{bmatrix} 1 & a \end{bmatrix}$$

$$= \begin{bmatrix} 0 & 1 \\ -\alpha - 1 & -\beta - a \end{bmatrix}$$

We now satisfy Eq. (9-37)

$$\begin{bmatrix} 0 & -\alpha - 1 \\ 1 & -\beta - a \end{bmatrix} \begin{bmatrix} q_{11} & q_{12} \\ q_{12} & q_{22} \end{bmatrix} + \begin{bmatrix} q_{11} & q_{12} \\ q_{12} & q_{22} \end{bmatrix} \begin{bmatrix} 0 & 1 \\ -\alpha - 1 & -\beta - a \end{bmatrix}$$

$$= \begin{bmatrix} -1 & 0 \\ 0 & -1 \end{bmatrix}$$

which leads to the following equations

$$-q_{12}(1 + \alpha) - q_{12}(1 + \alpha) = -1$$

$$-q_{22}(1 + \alpha) + q_{11} - q_{12}(\alpha + \beta) = 0$$

$$2q_{12} - 2q_{22}(\alpha + \beta) = -1$$

Solving these equations gives

$$q_{11} = \frac{(2 + \alpha)(1 + \alpha) + (a + \beta)^2}{2(a + \beta)(1 + \alpha)}, \quad q_{12} = \frac{1}{2(1 + \alpha)},$$

$$q_{22} = \frac{2 + \alpha}{2(1 + \alpha)(a + \beta)}$$

The performance index is

$$J \;=\; \mathbf{x}^T(0)\mathbf{Q}\mathbf{x}(0)$$

$$=\; \begin{bmatrix} 1 & 0 \end{bmatrix} \begin{bmatrix} q_{11} & q_{12} \\ q_{12} & q_{22} \end{bmatrix} \begin{bmatrix} 1 \\ 0 \end{bmatrix} = q_{11}$$

Minimizing J with respect to a

$$\frac{dJ}{da} = \frac{4(a+\beta)(1+\alpha)(a+\beta) - 2(1+\alpha)[(2+\alpha)(1+\alpha) + (a+\beta)^2]}{4(a+\beta)^2(1+\alpha)^2} = 0$$

Solving this, we obtain

$$a = \sqrt{(2+\alpha)(1+\alpha)} - \beta$$

Gain Considerations for State Variable Feedback

It will be observed in Example 8 that as the magnitude of the real parts of the desired closed loop poles become larger the coefficients of $\Delta(s)$ increase. This results in increased values for the feedback gain coefficients and increased chance of non-linear system operation, and degraded system performance. In general it is desirable to keep system gains as low as practicable to prevent non-linear operation and excessive sensitivity to external disturbances and noise.

It is extremely difficult to reconcile these conflicting demands so that they can be absorbed into a general theory beyond the linear analysis given above. In practical systems great reliance often needs to be placed upon the system designers judgment, and also upon extensive simulation. A limited theoretical understanding of non-linear control system operations can sometimes be obtained, and some useful techniques for system analysis are given in Chapter 11.

9.9   State Observers

In our study of the pole-placement design method we assumed that all the state variables are directly measurable. While this assumption might be true for low order systems it is likely to be manifestly untrue for high order systems, either because the states are not directly accessible or because it is impractical to measure them. In order to be able to apply pole-placement design methods we need a means of estimating the unmeasurable states. In this section we shall examine the design of so-called **full order Luenberger state observers**. In these observers the states are reconstructed

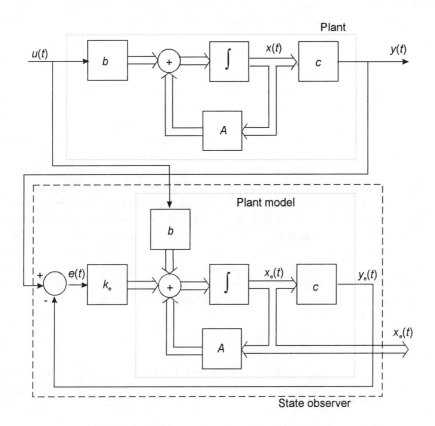

Figure 9.48: Operation of state observer using negative feedback

using a dynamic system model of the plant whose states are being estimated.

Full-Order Luenberger State Observer

To understand operation of an observer suppose the plant is a single-input
single-output system whose state equations are

$$\dot{\mathbf{x}}(t) = \mathbf{A}\mathbf{x}(t) + \mathbf{b}u(t) \tag{9.38}$$

$$y(t) = \mathbf{c}\mathbf{x}(t) \tag{9.39}$$

Assume the state $\mathbf{x}(t)$ is to be estimated by the state $\mathbf{x}_e(t)$ of the system model

$$\dot{\mathbf{x}}_e(t) = \mathbf{A}\mathbf{x}_e(t) + \mathbf{b}u(t) \tag{9.40}$$

where both systems have the same input, and the model coefficients $\mathbf{A}, \mathbf{b}$, and $\mathbf{c}$ are identically the same as those of the plant. If both the plant and the model start operation with the same initial conditions then it is clear from Eqs. (9-38) and (9-40) that $\mathbf{x}_e(t) = \mathbf{x}(t)$ for $t > t_0$.

If their initial conditions differ however, then $\mathbf{x}_e(t)$ and $\mathbf{x}(t)$ will differ for all time. We will now show that by using negative feedback, where the plant and model outputs $y(t)$ and $y_e(t)$ respectively, are compared, we can ensure $\mathbf{x}_e(t)$ approaches $\mathbf{x}(t)$ asymptotically provided the feedback system is stable. To see this we examine the models for the plant and the observer shown in Fig. 9-48. It will be noted the model is identical to the plant except for the injected signal $\mathbf{k}_e e(t)$. If the states $\mathbf{x}_e(t)$ and $\mathbf{x}(t)$ differ then it will be seen that the error signal $e(t)$ is non-zero, but we expect the negative feedback to act to reduce the magnitude of $e(t)$. In these circumstances it also seems likely that the difference between $\mathbf{x}_e(t)$ and $\mathbf{x}(t)$ will approach zero. To show this we find from Fig. 9-48

$$\dot{\mathbf{x}}(t) = \mathbf{A}\mathbf{x}(t) + \mathbf{b}u(t)$$

and

$$\dot{\mathbf{x}}_e(t) = \mathbf{A}\mathbf{x}_e(t) + \mathbf{b}u(t) + \mathbf{k}_e e(t) \tag{9.41}$$

Thus

$$\begin{aligned} \frac{d}{dt}\left(\mathbf{x}(t) - \mathbf{x}_e(t)\right) &= \mathbf{A}\left(\mathbf{x}(t) - \mathbf{x}_e(t)\right) - \mathbf{k}_e\left(y(t) - y_e(t)\right) \\ &= (\mathbf{A} - \mathbf{k}_e\mathbf{c})\left(\mathbf{x}(t) - \mathbf{x}_e(t)\right) \end{aligned} \tag{9.42}$$

Now if the eigenvalues of $(\mathbf{A} - \mathbf{k}_e\mathbf{c})$ all lie in the left half of the complex plane then $\mathbf{x}(t) - \mathbf{x}_e(t) \to 0$ asymptotically, so that $\mathbf{x}_e(t)$ tracks the plant state $\mathbf{x}(t)$.

By analogy with the analysis of state variable feedback it can be shown that the eigenvalues of the matrix $(\mathbf{A} - \mathbf{k}_e\mathbf{c})$ can be positioned anywhere in the complex plane if the plant described by Eqs. (9-38) and (9-39) is completely observable. Suppose $\Delta(s)$ is the desired characteristic polynomial for the state observer, where

$$\Delta(s) = (s - s_1)(s - s_2)\cdots(s - s_n) \tag{9.43}$$

and $s_1, s_2, \cdots, s_n$ are the desired observer eigenvalues. Using arguments similar to those developed for state variable feedback **Ackermann's for-**

mula gives the **state observer gain matrix**

$$\mathbf{k}_e = \Delta(\mathbf{A})\mathcal{O}(\mathbf{c}, \mathbf{A})^{-1} \begin{bmatrix} 0 \\ 0 \\ \vdots \\ 0 \\ 1 \end{bmatrix} \tag{9.44}$$

where $\mathcal{O}(\mathbf{c}, \mathbf{A}) = \begin{bmatrix} \mathbf{c}^T & \mathbf{A}^T\mathbf{c}^T & (\mathbf{A}^T)^2\mathbf{c}^T & \cdots & (\mathbf{A}^T)^{n-1}\mathbf{c}^T \end{bmatrix}^T$ is the observability matrix as defined in Chapter 5.

Example 11 *Consider the system whose state equation is*

$$\dot{\mathbf{x}}(t) = \begin{bmatrix} 0 & 5 \\ 10 & 0 \end{bmatrix}\mathbf{x}(t) + \begin{bmatrix} 0 \\ 2 \end{bmatrix}u(t)$$

$$y(t) = \begin{bmatrix} 1 & 1 \end{bmatrix}\mathbf{x}(t)$$

Design a full-order state observer with the observer poles at $\omega_n = 10$ rad/sec and $\zeta = 0.707$, using both the method of equating coefficients, and Ackermann's formula. If the initial states of the plant are at $x_1(0) = 0.5$, $x_2(0) = 0$, and those of the observer are $x_{1e}(0) = x_{2e}(0) = 0$ find the observer state errors as a function of time.

It is seen from Eq. (9-41) that the observer state equation for the system is

$$\dot{\mathbf{x}}_e(t) = \begin{bmatrix} -k_{1e} & 5 - k_{1e} \\ 10 - k_{2e} & -k_{2e} \end{bmatrix}\mathbf{x}_e(t) + \begin{bmatrix} 0 \\ 2 \end{bmatrix}u(t) + \begin{bmatrix} k_{1e} \\ k_{2e} \end{bmatrix}y(t)$$

From this equation we obtain the state observer characteristic polynomial

$$\Delta(s) = \det\begin{bmatrix} s + k_{1e} & -5 + k_{1e} \\ -10 + k_{2e} & s + k_{2e} \end{bmatrix}$$

$$= s^2 + (k_{1e} + k_{2e})s - 50 + 10k_{1e} + 5k_{2e}$$

Since the observer poles must have $\omega_n = 10$ rad/sec and $\zeta = 0.707$ its characteristic polynomial must be

$$\Delta(s) = s^2 + 14.14s + 100$$

Equating coefficients in these two expressions gives

$$k_{1e} + k_{2e} = 14.14$$

$$10k_{1e} + 5k_{2e} = 150$$

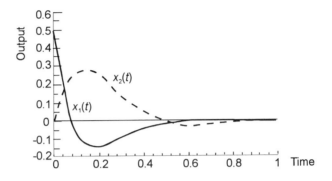

Figure 9.49: Observer errors for system given in Example 11

and the solution is

$$\mathbf{k}_e = \begin{bmatrix} 15.86 \\ -1.72 \end{bmatrix}$$

Let us now apply Ackermann's formula to find $\mathbf{k}_e$. From the characteristic equation we find

$$\Delta(\mathbf{A}) = \mathbf{A}^2 + 14.14\mathbf{A} + 100\mathbf{I}$$

$$= \begin{bmatrix} 150 & 70.7 \\ 141.4 & 150 \end{bmatrix}$$

The inverse of the observability matrix

$$\mathcal{O}(\mathbf{c}, \mathbf{A})^{-1} = \begin{bmatrix} -1 & 0.2 \\ 2 & -0.2 \end{bmatrix}$$

so the observer gain matrix becomes

$$\mathbf{k}_e = \Delta(\mathbf{A})\mathcal{O}(\mathbf{c}, \mathbf{A})^{-1} \begin{bmatrix} 0 \\ 1 \end{bmatrix} = \begin{bmatrix} 15.86 \\ -1.72 \end{bmatrix}$$

The observer error which is described by Eq. (9-42) is given by

$$\frac{d}{dt}(\mathbf{x}(t) - \mathbf{x}_e(t)) = \begin{bmatrix} 15.86 & -10.86 \\ 11.72 & 1.72 \end{bmatrix} (\mathbf{x}(t) - \mathbf{x}_e(t))$$

A simulation of the error response for $\mathbf{x}(0) - \mathbf{x}_e(0) = \begin{bmatrix} 0.5 & 0 \end{bmatrix}$ is shown in Fig. 9-49, where it will be observed the error $(\mathbf{x}(t) - \mathbf{x}_e(t))$ has all but disappeared in approximately 0.5 second. A block diagram for the observer system is shown in Fig. 9-50.

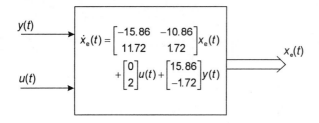

Figure 9.50: Block diagram of observer

Selection of Observer Gain k_e

The effect of measurement noise upon the operation of state observers is determined by the observer gain $\mathbf{k}_e$. In general if the magnitude of this gain is large then the state observer error settling time will be short. A corollary to this is that the observer will have a wide frequency bandwidth. Thus the measurement and disturbance noise superimposed on $u(t)$ and $y(t)$ will only be slightly filtered, and so may even be large enough to totally mask the undisturbed estimated state $\mathbf{x}_e(t)$. If however $\mathbf{k}_e$ is small the dynamics of the observer will be relatively slow, and if it is used to estimate the plant states as part of a state feedback control system design the observer dynamics will interfere with the operation of the system. This means the value selected for the observer gain $\mathbf{k}_e$ often will be a compromise between fast response on the one hand and insensitivity to noise on the other. A systematic method for making this choice is the subject of Kalman-Bucy filter theory.

Reduced-Order Observers

The state observers discussed above are said to be of full order because the plant model they use is of the same order as the plant. Luenberger has shown that it is not necessary for the observer to be of full order, and in fact he has shown it suffices for the observer to be of order $(n-1)$, where the plant is of order n. It can be shown that the states of systems with multiple outputs can also be observed with observers having orders less than $(n-1)$. The reader is referred to references at the end of the chapter for further discussion of reduced-order observers.

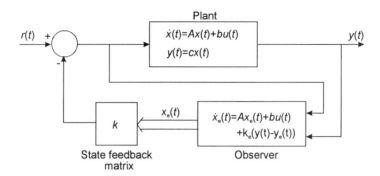

Figure 9.51: Combined state variable feedback and state observer control system

9.10   Pole-Placement Design with State Observer

We have observed in the discussion above that while not all states of a plant are measurable, we can use a state observer to estimate the unmeasurable states, provided the plant is completely observable. In this case a pole-placement design can use the estimated states rather than actual plant states.

The steps to be followed in any pole-placement design are as follows:

1. Design the state variable feedback system on the assumption that all states are available and measurable.

2. Design a state observer to determine the inaccessible or unmeasurable states.

3. The estimated states $\mathbf{x}_e(t)$ of the observer are used rather than the plant states $\mathbf{x}(t)$ as the inputs to the state variable gain matrix $\mathbf{k}$.

A block diagram of a control system using a combination of state variable feedback and an observer is shown in Fig. 9-51.

It is useful to examine the effect of the integration of the state observer with the pole-placement state feedback design. To see this consider the adjoined state equations of both systems, where for simplicity we assume $r(t) \equiv 0$. Introducing the new variable $\mathbf{z}(t) = \mathbf{x}(t) - \mathbf{x}_e(t)$ and noting from

Fig. 9-51 that $u(t) = -\mathbf{k}\mathbf{x}_e(t)$ we see the plant equation becomes

$$
\begin{aligned}
\dot{\mathbf{x}}(t) &= \mathbf{A}\mathbf{x}(t) - \mathbf{b}\mathbf{k}\mathbf{x}(t) + \mathbf{b}\mathbf{k}\mathbf{z}(t) \\
&= (\mathbf{A} - \mathbf{b}\mathbf{k})\mathbf{x}(t) + \mathbf{b}\mathbf{k}\mathbf{z}(t)
\end{aligned}
\tag{9.45}
$$

Also from the plant and observer equations given in Fig. 9-51 we have

$$
\dot{\mathbf{z}}(t) = \dot{\mathbf{x}}(t) - \dot{\mathbf{x}}_e(t) = (\mathbf{A} - \mathbf{k}_e\mathbf{c})\mathbf{z}(t)
\tag{9.46}
$$

Adjoining these two state equations we obtain

$$
\begin{bmatrix} \dot{\mathbf{x}}(t) \\ \dot{\mathbf{z}}(t) \end{bmatrix} = \begin{bmatrix} \mathbf{A} - \mathbf{b}\mathbf{k} & \mathbf{b}\mathbf{k} \\ 0 & \mathbf{A} - \mathbf{k}_e\mathbf{c} \end{bmatrix} \begin{bmatrix} \mathbf{x}(t) \\ \mathbf{z}(t) \end{bmatrix}
\tag{9.47}
$$

and $\mathbf{x}_e(t) = \mathbf{x}(t) - \mathbf{z}(t)$.

The characteristic equation for this augmented state equation is

$$
\det(s\mathbf{I} - \mathbf{A} + \mathbf{b}\mathbf{k})\det(s\mathbf{I} - \mathbf{A} + \mathbf{k}_e\mathbf{c}) = 0
\tag{9.48}
$$

Thus we find the roots of the characteristic equation for the combined state variable feedback and observer control system are the same as the poles of the state variable feedback system, and of the observer, when each is designed separately. This result justifies the design procedure given above since the interconnection does not cause any shift in their pole positions.

Example 12 *Consider the magnetic bearing discussed in Example 8, where the state equation is given in Eq. (9-24). Suppose that the pole placement state variable feedback system designed in the above Example is to be used for the control of the magnetic bearing, but the state variables are unmeasurable. Design a full state observer for this system with its poles at $s = -150$, and compare the pole locations of the closed loop system with those of the separate elements.*

From Example 8 the plant state equations are

$$
\dot{\mathbf{x}}(t) = \begin{bmatrix} -500 & 0 & 0 \\ 0 & 0 & 1 \\ 20 & 6 \times 10^4 & 0 \end{bmatrix} \mathbf{x}(t) + \begin{bmatrix} 600 \\ 0 \\ 0 \end{bmatrix} u(t)
$$

$$
y(t) = \begin{bmatrix} 0 & 1 & 0 \end{bmatrix} \mathbf{x}(t)
$$

Applying Ackermann's formula we find the observer gain matrix

$$
\mathbf{k}_e = \begin{bmatrix} -2.14 \times 10^6 & -50 & 1.53 \times 10^5 \end{bmatrix}^T
$$

The adjoined **A** matrix given in Eq. (9-47) is

$$
\begin{bmatrix} \mathbf{A} - \mathbf{bk} & \mathbf{bk} \\ 0 & \mathbf{A} - \mathbf{k}_e\mathbf{c} \end{bmatrix} =
$$

$$
\begin{bmatrix}
-481 & -1.61 \times 10^6 & -5.03 \times 10^6 & -19 & 1.61 \times 10^6 & 5020 \\
0 & 0 & 1 & 0 & 0 & 0 \\
20 & 60000 & 0 & 0 & 0 & 0 \\
0 & 0 & 0 & -500 & 2.14 \times 10^6 & 0 \\
0 & 0 & 0 & 0 & 50 & 1 \\
0 & 0 & 0 & 20 & -92500 & 0
\end{bmatrix}
$$

Numerical analysis shows that the adjoined matrix has eigenvalues at -400, $-40.5 \pm j80.37$, -150, -150, -150 which corresponds to the pole locations for the state variable feedback controller, and the state observer, when each is treated independently.

9.11   Summary

We have been concerned with different compensating techniques useful for changing system performance in order to meet specifications. The compensating elements considered here were passive elements.

Compensation of control systems may be achieved by cascading, feedback, feedforward, or cancellation techniques.

The method of compensation was most easily understood using frequency plots. The effect of compensation on closed loop performance was obtained using Nichols charts.

It was shown how pole-placement design using state variable feedback could be applied when all state variables are measurable. Also an elementary example of the optimal linear quadratic problem was analyzed. Full order state observers for completely observable systems were introduced and it was shown how it could be designed using Ackermann's formula. It was shown that state observers could be combined with pole-placement state variable feedback for control system design, thus relaxing the requirement for all state variables to be measurable.

9.12   References

1. K. Ogata, *Modern Control Engineering*, Prentice-Hall International, Inc., Englewood Cliffs, N. J., 1990.

2. B. C. Kuo, *Automatic Control Systems*, Prentice-Hall International, Inc., Englewood Cliffs, N. J., 1987.

3. R. C. Dorf, *Modern Control Systems*, Addison Wesley Publishing Co., Reading, Mass., 1989.

4. J. J. D'Azzo and C. H. Houpis, *Linear Control System Analysis and Design - Conventional and Modern*, Mc Graw Hill Book Co., New York, 1988.

5. E. B. Canfield, *Electromechanical Control Systems and Devices*, John Wiley and Sons, Inc., New York, 1965.

6. E. O. Doebelin, *Control System Principles and Design*, John Wiley and Sons, Inc., New York, 1985.

7. K. J. Astrom and B. Wittenmark, *Computer Controlled Systems*, Prentice-Hall International, Inc., Englewood Cliffs, N. J., 1990.

8. J. M. Maciejowski, *Multivariable Feedback Design*, Addison Wesley Publishing Co., Reading, Mass., 1989.

9. R. Brockett, "Poles, Zeros and Feedback: State Space Interpretation", *IEEE Trans. Autom. Control*, **AC-10**, No. 2 (1965) pp. 129-135.

10. G. F. Franklin, J. D. Powell and M. L. Workman, *Digital Control of Dynamic Systems*, Addison Wesley Publishing Co., Reading, Mass., 1990.

11. T. Kailath, *Linear Systems*, Prentice-Hall International, Inc., Englewood Cliffs, N. J., 1980.

12. B. D. O. Anderson and J. Moore, *Optimal Control - Linear Quadratic Methods*, Prentice-Hall International, Inc., Englewood Cliffs, N. J., 1989.

13. D. G. Luenberger, "Observers for Multivariable Systems", *IEEE Trans. Autom. Control*, **AC-11**, No. 2 (1966) pp. 190-97.

14. D. G. Luenberger, "An Introduction to Observers", *IEEE Trans. Autom. Control*, **AC-16**, No. 6 (1971) pp. 596-602.

9.13   Problems

9-1 A second-order control system has an open loop transfer function given by

$$G(s) = \frac{K}{s(s+2)}$$

It is required that the phase margin be 50° and the velocity error constant be 20 sec^{-1}. Design compensating networks using:

(a) A phase-lead network

(b) Cancellation techniques

Discuss the relative advantages of each method.

9-2 The closed loop system having the plant transfer function

$$G(s) = \frac{2}{s^2(1+0.1s)}$$

and a simple proportional controller is unstable for all values of loop sensitivity. It is to be stabilized using a lead circuit cascade compensator whose transfer function is

$$G_c(s) = \frac{1+1.585s}{1+0.158s}$$

Show the compensated and uncompensated frequency response functions on the Bode plot. Indicate all the changes brought about by the compensator.

9-3 A third-order servomechanism is shown in Fig. P9-3. It is required that the steady state error be 0.1 of the final output velocity and the phase margin be 50°. What suitable values of R_1, R_2, and C would satisfy the above conditions?

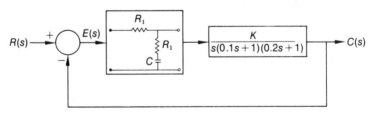

Figure P9-3

9-4 The block diagram of a servomechanism with feedback compensation is shown in Fig. P9-4. Study the effect of varying T on system response.

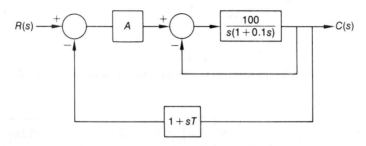

Figure P9-4

9-5 It is known that tachometric feedback affects the relative damping of a system. What value of K_f will provide a damping ratio of 0.707, for the system shown in Fig. P9-5?

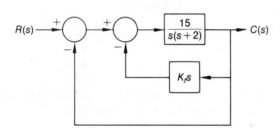

Figure P9-5

9-6 A feedforward compensation scheme is shown in Fig. P9-6. Assume that $D(s)$ is a step disturbance whose magnitude is 0.5. Let $R(s)$ also be a unit step input. Study the response as a function of K.

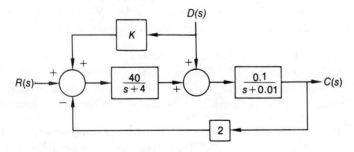

Figure P9-6

9-7 The block diagrams of two compensating systems are shown in Fig. P9-7, where

$$G_1 = K/s(s + 1000)(s + 20)$$

$$G_2 = \frac{\tau_1 \tau_2 s + \tau_2}{\tau_1 \tau_2 s + \tau_1}, \quad \tau_1 = \frac{1}{120}, \quad \tau_2 = \frac{1}{1200}$$

K is to be selected so that the velocity constant is 10 sec^{-1}. Compare the two schemes of compensation and discuss the behavior of each in the high-frequency and low-frequency range.

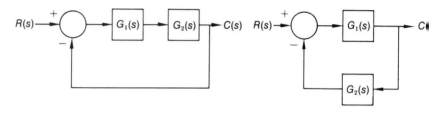

Figure P9-7

9-8 Select a cascade compensator for a unity feedback system with a plant

$$G(s) = K/(s + 10)(s + 30)$$

It is required that the velocity error constant be greater than 90 sec^{-1}, and the maximum overshoot be less than 40 percent.

9-9 A unity feedback control system has an open loop transfer function

$$G(s) = \frac{K}{s^2(s + 1)}$$

We would like to use a cascaded phase-lead or phase-lag network to stabilize it with a phase margin of 12 deg. Which network would you suggest? Obtain the transfer function for the one you select.

9-10 Could the control system of Problem 9-9 be stabilized using a lead or lag network in the feedback loop? Would the transfer function change? What is the principal difference of having a compensation element in the feedback loop as opposed to cascaded elements?

9-11 What is the effect of placing the compensating network of Problem 9-2 in the feedback path? How does the behavior differ at the high-frequency from the low-frequency region?

9-12 A hydraulic servo-motor is described by the following state equations

$$\dot{\mathbf{x}}(t) = \begin{bmatrix} -2 & 12 \\ 0 & 0 \end{bmatrix} \mathbf{x}(t) + \begin{bmatrix} 0 \\ 8 \end{bmatrix} u(t)$$

$$y(t) = \begin{bmatrix} 1 & 0 \end{bmatrix} \mathbf{x}(t)$$

(a) Find the controller gain matrix $\mathbf{k}$ for the state feedback control law $u(t) = -\mathbf{k}\mathbf{x}(t) + r_c(t)$ so that the closed loop poles have $\zeta = 0.7$ and $\omega_n = 30$ rad/sec.

(b) Investigate the steady state error when a unit step is applied at the input $r_c(t)$.

9-13 A control system is described by the state equations

$$\dot{\mathbf{x}}(t) = \begin{bmatrix} 0 & 1 & 0 \\ 0 & -5 & 8 \\ 0 & -5 & -20 \end{bmatrix} \mathbf{x}(t) + \begin{bmatrix} 0 \\ 0 \\ 100 \end{bmatrix} u(t)$$

$$y(t) = \begin{bmatrix} 1 & 0 & 0 \end{bmatrix} \mathbf{x}(t)$$

Find the gain matrix $\mathbf{k}$ for the state feedback control law $u(t) = -\mathbf{k}\mathbf{x}(t)$ so that the roots of the characteristic equation for the closed-loop system are located at $-10, -3 \pm j3$.

9-14 Given the system described by

$$\dot{\mathbf{x}}(t) = \begin{bmatrix} 1 & 3 \\ 8 & 9 \end{bmatrix} \mathbf{x}(t) \begin{bmatrix} 1 \\ 8 \end{bmatrix} u(t)$$

Find a linear, state variable feedback controller

$$u(t) = -\mathbf{k}\mathbf{x}(t)$$

so that the closed loop poles are at $s = -10$ and -20.

9-15 Suppose the plant transfer function for a control system is

$$G_p(s) = \frac{100}{s\left(\dfrac{s}{25} + 1\right)\left(\dfrac{s^2}{2600} + \dfrac{s}{26} + 1\right)}$$

(a) Taking the phase variables $y(t), y^{(1)}(t), y^{(2)}(t)$ and $y^{(3)}(t)$ as state variables, design a regulator state variable feedback system (i.e. $r_c(t) \equiv 0$) as shown in Fig. P9-15. A pair of complex closed loop poles should be positioned at $\zeta = 0.425$ and $\omega_n = 75$ rad/sec, and the remaining poles should be real and at $s = -200$ rad/sec.

(b) Suppose the system is to be used as a servo-system. Investigate by simulation the error signal $e(t) = r_c(t) - y(t)$ when the input $r_c(t)$ is a unit step.

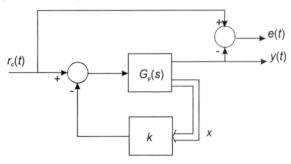

Figure P9-15

9-16 Consider the system discussed in Problem 9-15.

(a) Design an integral control state variable feedback system so that under steady state conditions the system output $y(t)$ tracks the input $r_c(t)$. One pair of closed loop poles should be complex with $\zeta = 0.425$ and $\omega_n = 75$ rad/sec, and the remaining poles should be real and located at $s = -200$ rad/sec.

(b) Simulate the error signal $e(t)$ time response for a unit ramp input and determine its peak and steady state values.

9-17 Design an observer for the hydraulic servo discussed in Problem 9-12, where the observer poles are at $\zeta = 0.7$ and $\omega_n = 120$ rad/sec.

9-18 Design an observer for the system discussed in Problem 9-13, where the observer poles are at $s = -40$ rad/sec.

9-19 Suppose the observer designed in Problem 9-18 is combined with the state variable controller designed in Problem 9-13. Using simulation investigate the relative time response performance of the system designed in Problem 9-13, and the combined system discussed in this problem. Repeat this analysis, but with the observer poles at $s = -7$.

Compare the results for these three cases and draw any relevant conclusions.

9-20 (Astrom and Wittenmark) A system of two tanks is shown in Fig.
P9-20, where the control variable $u(t)$ is the input flow rate, and
the state variables are $x_1(t)$ and $x_2(t)$. The state equations for this
system are

$$\dot{\mathbf{x}}(t) \;=\; \begin{bmatrix} -0.05 & 0 \\ 0.015 & -0.02 \end{bmatrix} \mathbf{x}(t) + \begin{bmatrix} 0.07 \\ 0 \end{bmatrix} u(t)$$

$$y(t) \;=\; [\,0\ \ 1\,]\,\mathbf{x}(t)$$

(a) Design a state variable feedback controller so that the closed loop
poles have $\zeta = 0.5$ and $\omega_n = 0.05$ rad/sec.

(b) Design an integral control state variable feedback system which
ensures the steady state output error is zero. Take two closed
loop poles to be complex with $\zeta = 0.5$ and $\omega_n = 0.05$ rad/sec,
and the remaining poles to be real with $s = -1$.

(c) Supposing only the variable $y(t)$ is measured. Design a Luenberger observer to estimate $x_1(t)$ and $x_2(t)$ with real poles at
$s = -1$, and then design a system with combined state variable
feedback and state observer.

(d) Simulate the systems discussed above and compare their relative
performances.

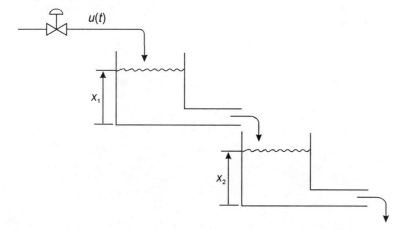

Figure P9-20

9-21 If the initial state vector in Example 10 is $\mathbf{x}^T(0) = \begin{bmatrix} 0 & 1 \end{bmatrix}$ and $r = -ax_1 - ax_2$, obtain the value of a necessary to minimize the performance index shown in Eq. (9-32).

9-22 A system is characterized by $\dot{\mathbf{x}} = \mathbf{A}\mathbf{x} + \mathbf{b}u$ where

$$\mathbf{A} = \begin{bmatrix} 0 & 1 \\ -1 & -1 \end{bmatrix}, \quad \mathbf{b} = \begin{bmatrix} 0 \\ 1 \end{bmatrix}$$

and $u = -x_1 - ax_2$. If the initial condition is $\mathbf{x}^T(0) = \begin{bmatrix} 0 & 1 \end{bmatrix}$, obtain the value of a that minimizes the performance index of Eq. (9-32).

Chapter 10

Discrete Time Control Systems

10.1   Introduction

The study of linear systems has so far been restricted to the consideration of signals that are continuous with respect to time. Since digital computers have become such valuable aids to modern control systems, instances of systems operating with intermittent data are many. For example, a computer enables us to solve, in real time, navigation and guidance problems. As another example, the increased plant efficiency and product quality in chemical processes is the direct result of computer control. In many instances the existence of telemetry links, time sharing due to the multiplicity of signals, or high precision computations, leads to the use of digitized information. It is therefore necessary that we modify our previous approach to allow for the inclusion of discrete signals in control systems.

This chapter is concerned with control systems receiving data at intermittent intervals. In these systems some of the signals are either in the form of pulse trains or data streams. The digital computer is an important example of a system which accepts and generates data streams. Such systems are know as **sampled data systems** or **digital control systems** although the former implies a more general system. We shall assume that although the signals are intermittent, they are received at regular intervals, i.e. they are periodic. A fairly general sampled-data system is shown in Fig. 10-1. Let us take a close look at it. The signals $r(t)$, $c(t)$, $b(t)$, $e(t)$ and $m(t)$ are continuous, whereas $e^*(t)$ and $m^*(t)$ are intermittent. The signal $e(t)$, after it passes through the sampling device (called a **sampler**),

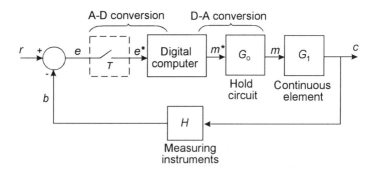

Figure 10.1: Representation of a general sample-data control system

becomes intermittent in nature. As a matter of fact, the amplitude of $e^*(t)$ is equal to $e(t)$ when the sampler closes every T seconds. As digital computers can only process data streams, the signal $e^*(t)$ is transformed into a **data sequence**, which we will denote by the symbol $\{e_n\}$, within the block identified as the digital computer. We will often refer to this process of sampling and conversion to a data stream as **analog-to-digital** conversion, and the converse process as **digital-to-analog** conversion. The discrete signal $m^*(t)$ is the input to G_0 which reconstructs[1] it to a continuous signal $m(t)$. (G_0 is called a **hold circuit**.)

To illustrate these ideas consider the example of a position servo system for a numerically controlled machine tool. In the example shown in Fig. 10-2 the servo system is separated from the digital process so that the procedures discussed in earlier chapters can be used for its design. The computer is simply used to store and control the flow of position coordinate data to the servo system. This arrangement was used in the early development of numerically controlled machines. The second example shown in Fig. 10-3 is an example of computer numerical control, where digital signals flow in certain parts of the closed loop, as shown in Fig. 10-1. Conventional control system design methods cannot be used for this type of system.

The simultaneous existence of continuous and discrete signals requires a unified theory that may be applied to both type of signals. The operational technique best suited for this is the $\mathcal{Z}$-transform. We shall first introduce the mathematics of the sampler and then develop the techniques used to obtain the response of sampled systems. It is not intended here to intro-

---

[1] It will be seen in a later section, that sampling introduces higher harmonics into the signal which must be suppressed before input to a continuous element. Data reconstruction by a hold circuit minimizes the effect of these harmonics.

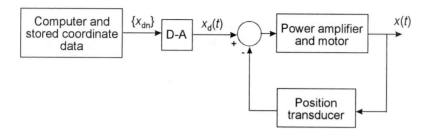

Figure 10.2: Position axis servo for numerically controlled machine tool

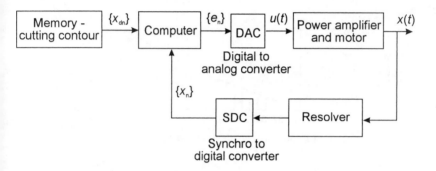

Figure 10.3: Position axis servo for computer numerically controlled machine tool

duce mathematics for sampled systems with rigor or include all its subtle nuances. We take several liberties in order to introduce the subject matter in as straightforward a way as possible.

10.2   The Sampling Process

A sampler, or sampling switch, is a device that converts a continuous signal to a discrete signal. A schematic representation of a sampler is shown in Fig. 10-4. The input signal is a continuous signal, whereas the output signal is in the form of pulses. In a **practical sampler** these pulses have a nonzero width. In an **ideal sampler** the pulses are assumed to have negligible width. This assumption greatly simplifies the analysis. The pulse amplitude of the output of the ideal sampler, then, is equal to the input at $t = 0, T, 2T, \cdots$. We assume that the output of the ideal sampler is defined

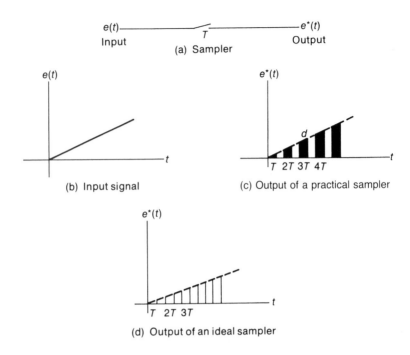

Figure 10.4: Schematic representation of the sampling process

and finite at the sampled instants, i.e. there are no jump discontinuities in the system.

In addition to the ideal sampler we introduce the **zero-order-hold** (abbreviated as ZOH). Here we imagine the output of an ideal sampler is fed to the zero-order-hold whose input and output are as shown in Fig. 10-5. The output signal

$$\hat{e}(t) = e(0)\left[H(t) - H(t - T)\right] + e(1)\left[H(t - T) - H(t - 2T)\right] + \cdots$$

and its Laplace transform is

$$\hat{E}(s) = \left[\frac{1 - e^{-sT}}{s}\right]\left[e(0) + e(1)e^{-sT} + e(2)e^{-2sT} + \cdots\right]$$

Since the first term in brackets is independent of the input signal $e(t)$ it can be imagined to be the transfer function of the zero-order-hold circuit. The second term can be seen to be a signal dependant upon $e(t)$ which we

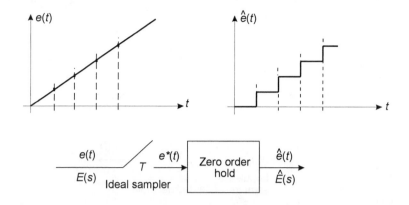

Figure 10.5: Input and output signals of combined ideal sampler and zero order hold circuit (ZOH)

denote by

$$E^*(s) = e(0) + e(1)e^{-sT} + e(2)e^{-2sT} + \cdots$$

$$= \sum_{n=0}^{\infty} e(nT)e^{-nsT} \qquad (10.1)$$

By taking the inverse Laplace transform of $E^*(s)$ we find

$$e^*(t) = e(0)\delta(t) + e(1)\delta(t-T) + e(2)\delta(t-T) + \cdots \qquad (10.2)$$

This signal can be visualized to be the output of an **impulse sampler** so that the system shown in Fig.10-6(b) is equivalent to the ideal sampler and zero-order-hold circuit shown in Fig. 10-5(c).

Although the Laplace transform of $e^*(t)$ is a series, it does converge for cases when $E(s)$ is a rational algebraic function. Defining the unit impulse train $I(t)$ as

$$I(t) = \sum_{n=0}^{\infty} \delta(t - nT) \qquad (10.3)$$

an alternate form of $E^*(s)$ may be obtained from the theory of complex convolution

$$\mathcal{L}\left[e^*(t)\right] = \mathcal{L}\left[e(t) \cdot I(t)\right]$$

or

$$E^*(s) = \mathcal{L}\left[e(t)\right] * \mathcal{L}\left[I(t)\right] \qquad (10.4)$$

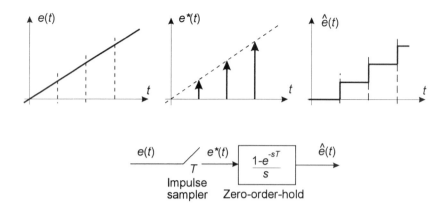

Figure 10.6: Equivalent block diagram of sampler and zero order hold circuit using impulse sampler

where $*$ represents the complex convolution operation. Since $\mathcal{L}\left[I(t)\right]$ is known, Eq. (10-4) becomes

$$E^*(s) = E(s) * \frac{1}{1 - e^{-sT}} = \sum_{\substack{\text{poles} \\ \text{of } E(p)}} \text{Res}\left[E(p)\frac{1}{1 - e^{-(s-p)T}}\right] \qquad (10.5)$$

where

$$E(p) = \frac{N(p)}{D(p)}$$

Applying the residue theorem and assuming k simple poles,

$$E^*(s) = \sum_{n=1}^{k} \frac{N(s_n)}{D'(s_n)} \frac{1}{1 - e^{-(s-s_n)T}} \qquad (10.6)$$

where

$$D'(s_n) = \left[\frac{dD(p)}{dp}\right]_{p=s_n}$$

The restriction on the use of this equation is that

$$\lim_{s \to \infty} sE(s) = 0$$

Similarly it may also be shown that

$$E^*(s) = \frac{1}{T} \sum_{n=-\infty}^{\infty} E\left(s + jn\omega_s\right)$$

provided that $\lim_{s \to \infty} s E(s) = 0$.

Example 1 *Obtain the Laplace transform of the output if the input to a sampler is $e(t) = \sin \omega t$.*

From Eq. (10-1)

$$e(nT) = \sin n\omega T$$

$$E^*(s) = \sum_{n=0}^{\infty} \sin n\omega T \, e^{-nsT}$$

Since

$$\sin n\omega T = \frac{e^{jn\omega T} - e^{-jn\omega T}}{2j}$$

then

$$E^*(s) \quad = \quad \sum_{n=0}^{\infty} \frac{e^{-n(sT - j\omega T)}}{2j} - \sum_{n=0}^{\infty} \frac{e^{-n(j\omega T + sT)}}{2j}$$

$$= \quad \frac{1}{2j\left(1 - e^{-(sT - j\omega T)}\right)} - \frac{1}{2j\left(1 - e^{-(j\omega T + sT)}\right)}$$

which is combined to yield

$$E^*(s) = \frac{e^{-sT} \sin \omega T}{1 + e^{-2sT} - 2e^{-sT} \cos \omega T}$$

This may also be obtained via the residue theorem. Using Eq. (10-6) and noting that there are 2 poles,

$$E^*(s) = \sum_{n=1}^{2} \frac{N(s_n)}{D'(s_n)} \frac{1}{1 - e^{-(s - s_n)T}} \qquad \begin{array}{l} N(s) = \omega \\[2ex] D'(s) = 2s \end{array}$$

Substituting this we obtain

$$E^*(s) \quad = \quad \frac{\omega}{j2\omega} \left[\frac{1}{1 - e^{-(s - j\omega)T}} - \frac{1}{1 - e^{-(s + j\omega)T}} \right]$$

$$= \quad \frac{e^{-sT} \sin \omega T}{1 + e^{-2sT} - 2e^{-sT} \cos \omega T}$$

which is the same as obtained previously.

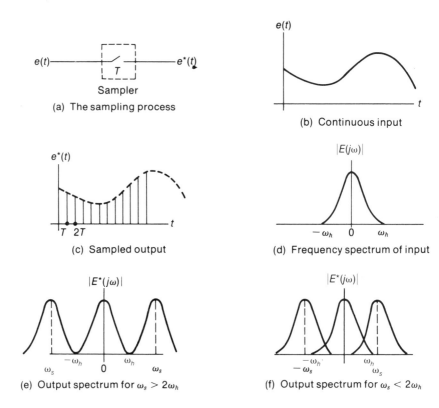

Figure 10.7: Characteristics of a sampler

Having obtained the Laplace transform of the sampled signal, let us see what are its characteristics in the frequency domain. Substituting[2] $s + jm\omega_s$ for s in Eq. (10-1)

$$E^*(s + jm\omega_s) = \sum_{n=0}^{\infty} e(nT)e^{-n(s+jm\omega_s)T}$$

$$= \sum_{n=0}^{\infty} e(nT)e^{-jnm2\pi}e^{-nsT}$$

---

[2]By doing this we are stating that $E^*(s)$ is a periodic function. The ensuing development confirms and therefore justifies this statement.

Since the term $e^{-jnm2\pi} = 1$ for all n and m, the above simplifies to

$$E^*(s + jm\omega_s) = E^*(s) \tag{10.7}$$

This states that $E^*(s)$ is a periodic function with frequency ω_s, which is the sampling frequency. In other words, sampling has introduced an infinite number of higher frequency components into the signal as shown in Fig. 10-7. In the frequency domain Eq. (10-7) becomes

$$E^*(j\omega + jm\omega_s) = E^*(j\omega)$$

so that if the sampler input frequency spectrum is as shown in Fig.10-7(d) then the output spectrum $E^*(j\omega)$ will be as shown in Figs. 10-7(e) or (f). If the bandwidth frequency of the input signal is ω_h and $\omega_s > 2\omega_h$, then the output of the sampler exhibits no overlapping of its complementary components. However, if $\omega_s < 2\omega_h$ then overlapping does indeed occur. Introducing the concept of an ideal low pass filter as one having a pass-band and stop-band with bandwidth $\omega_s/2$, as shown in Fig. 10-8(a), it is clear that the spectrum of the signal $\hat{e}(t)$ at the output of the filter shown in Fig. 10-8(b) will be the same as the spectrum of signal $e(t)$ when $\omega_h < \omega_s/2$. Thus provided $\omega_h < \omega_s/2$ then the signal can be fully recovered from its sampled version $e^*(t)$. It is evident that if $\omega_h > \omega_s/2$, then even an ideal filter will be unable to recover the input signal completely. If the input signal is to be completely recovered, then ω_s must be *at least* equal to $2\omega_h$. These observations were first made by C. E. Shannon and form the basis of Shannon's sampling theorem.

Sampling Theorem

A signal $f(t)$ is fully represented by its samples $\{f(nT)\}$ if and only if it is band-limited with bandwidth ω_h less than $\omega_s/2$, where ω_s is the sampling frequency.

A corollary of this result, as observed above, is that if a signal has a bandwidth ω_h it can be fully recovered if the sample frequency $\omega_s > 2\omega_h$. The frequency $w_N = \omega_s/2$ is referred to as the **Nyquist frequency**.

10.3   Data Reconstruction

We saw in the sampling theorem that an ideal low pass filter can completely remove the higher harmonies thus giving perfect recovery of the input signal. However this filter is physically unrealizable so that other methods of **data reconstruction** must be sought. Two classes of data reconstruction

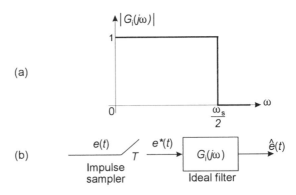

Figure 10.8: Filtering impulse sampled signal $e(t)$ with an ideal low pass filter

or hold circuits based upon either **interpolation** or **extrapolation** have been used, although extrapolation is more commonly used as it usually introduces less phase lag which is important for feedback control systems. The ideal low pass filter is an example of an interpolator while the zero-order and first-order holds are examples of extrapolators.

A signal $e(t)$ between $t = nT$ and $t = (n + 1)T$ may be expressed as a power series,

$$e(t) = e(nT) + \dot{e}(nT)(t - nT) + \frac{\ddot{e}(nT)}{2!}(t - nT)^2 + \cdots \qquad (10.8)$$

In order to construct $e(t)$, we must have the derivatives $\dot{e}(nT)$, $\ddot{e}(nT)$, etc., which, in general, are unavailable. These derivatives can however be estimated from the sampled data itself. The use of sampled data for $t < nT$ requires time delays and this generally has a destabilizing effect on system stability. Most hold circuits, therefore, do not employ higher-order derivatives.

The simplest form of a data reconstruction device *holds* the amplitude of the sample from one sampling instant to the next, i.e.

$$e(t) = e(nT) \quad nT \leq t \leq (n + 1)T \qquad (10.9)$$

Such a device is called a **zero-order-hold**[3] . The output of a zero-order-hold, shown in Fig. 10-9, corresponds exactly to the sampled signal at

---

[3] It is also referred to as a clamper, data hold, or staircase generator.

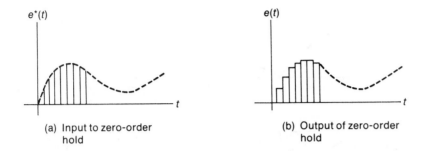

(a) Input to zero-order hold

(b) Output of zero-order hold

Figure 10.9: A zero order hold

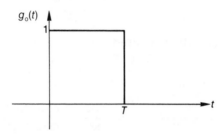

Figure 10.10: Impulse response of a zero order hold

$t = 0, T, 2T, \cdots$. The impulse response of a zero-order-hold is shown in Fig. 10-10. The Laplace transform of the impulse response is

$$G_0(s) = \mathcal{L}[g_0(t)] = \frac{1}{s} - \frac{e^{-sT}}{s} = \frac{1 - e^{-sT}}{s} \qquad (10.10)$$

The frequency response is obtained by forming $G_0(j\omega)$ and is

$$G_0(j\omega) = \frac{1 - e^{-j\omega T}}{j\omega} = \frac{2\sin(\omega T/2)}{\omega} e^{-j\omega T/2}$$

or

$$G_0(j\omega) = T\frac{\sin(\pi\omega/\omega_s)}{\pi(\omega/\omega_s)} e_s^{-j(\pi\omega/\omega_s)}$$

The amplitude of $G_0(j\omega)$ is given by

$$|G_0(j\omega)| = T\left|\frac{\sin(\pi\omega/\omega_s)}{\pi(\omega/\omega_s)}\right|$$

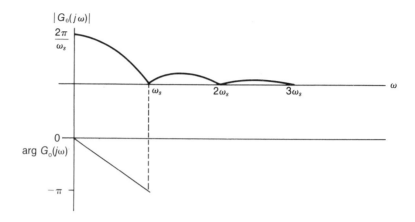

Figure 10.11: Characteristics of a zero order hold

and the phase is given by

$$\arg G_0(j\omega) = -\frac{\pi\omega}{\omega_s} \cdot \frac{\sin(\pi\omega/\omega_s)}{|\sin(\pi\omega/\omega_s)|}$$

and is shown in Fig. 10-11. We note that the behavior is quite similar to that of a low pass filter. As a matter of fact it is possible to approximate a ZOH by passive elements. Let us show this by expanding $G_0(s)$ in a power series of e^{sT}

$$G_0(s) = \frac{1 - e^{-sT}}{s} = \frac{1}{s}\left[1 - \frac{1}{1 + sT + \cdots}\right] \approx \frac{T}{1 + sT} \tag{10.11}$$

This is recognized as the transfer function of an RC network as shown in Fig. 10-12. This approximation is sometimes referred to as an exponential hold device. As we seek better approximations to a ZOH device we need to include additional terms of the power series expansion. This, however, gets complicated; therefore, networks employing operational amplifiers are used to implement ZOH circuits more accurately. Let us now return to the power series expansion of $e(t)$ between $t = nT$ and $t = (n + 1)T$. We have discussed the ZOH approximation, i.e. $e(t) = e(nT)$. Now if we take one additional term in $e(t)$, then,

$$e(t) \approx e(nT) + \dot{e}(nT)(t - nT)$$

This is called a **first-order hold**. Notice that we are now using information

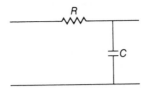

Figure 10.12: RC low pass filter as a hold device

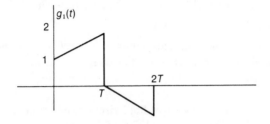

Figure 10.13: Impulse response of a first-order hold

about the derivative of $e(t)$. The first derivative may be represented using finite difference as,

$$\dot{e}(nT) = \frac{1}{T}[e(nT) - e((n-1)T)]$$

Substituting this, we obtain

$$e(t) = e(nT) + \frac{e(nT) - e((n-1)T)}{T}(t - nT) \qquad (10.12)$$

which is recognized as a ramp function. The impulse response of a first-order hold is shown in Fig. 10-13. This response is obtained by applying an impulse of unit strength at $t = 0$ as input to a first-order hold device. The transfer function $G_{01}(s)$ of the first-order hold may be obtained by taking the Laplace transform of the impulse response and is

$$G_{01}(s) = \mathcal{L}[g_0(t)] = \frac{1+sT}{T}\frac{(1-e^{-sT})^2}{s} \qquad (10.13)$$

The analysis of $G_{01}(s)$ in the frequency domain is left as an exercise. The output of a first-order hold is shown in Fig. 10-14. Although a first-order

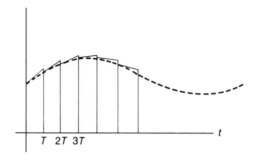

Figure 10.14: Output of a first-order hold

hold generally gives more accurate signal recovery it does indeed compli-
cate the problem of system stability due to increased phase lag compared
with the ZOH. Additionally, the complexity of construction makes it rather
unattractive.

There are several other types of hold circuits, however their representation is complex. Also the additional time delay introduced by higher-order
approximations has an adverse effect on system stability. For these reasons the zero-order-hold is widely used for signal recovery in sampled data
systems.

10.4   The $\mathcal{Z}$-Transform

When the signal in a given system is represented in a discrete form, then
the Laplace transform is seen to possess terms like e^{-nsT}

$$E^*(s) = \sum_{0}^{\infty} e(nT)e^{-nsT} \tag{10.14}$$

The term e^{-nsT} renders $E^*(s)$ a non algebraic function of s thereby adding
greatly to the complexity of analysis necessary to study discrete systems.
This can be overcome however if we define a change of variables,

$$z = e^{sT} \tag{10.15}$$

where s is the Laplace operator and T is the sampling period. Substituting
this into Eq. (10–14) we obtain

$$E(z) = E^*\left(s = \frac{1}{T}\ln z\right) = \sum_{0}^{\infty} e(nT)z^{-n} \tag{10.16}$$

Here $E(z)$ is called the Z-**transform** of $e^*(t)$, and is often denoted by

$$E(z) = \mathcal{Z}\left[e^*(t)\right] = \mathcal{Z}\left[e\left(nT\right)\right] = \sum_0^\infty e(nT)z^{-n}$$

Expanding Eq. (10-16) we have

$$E(z) = e(0) + e(T)z^{-1} + e(2T)z^{-2} + \cdots$$

It is interesting to note the role z plays in this equation. Clearly, it indicates the time at which the signal is defined. For example, if we have

$$E(z) = 1 + 2z^{-1} + 2.1z^{-2} + 2.3z^{-3} + \cdots$$

then $e(t) = 1$ at $t = 0$, $e(t) = 2$ at $t = T$, $e(t) = 2.1$ at $t = 2T$, $e = 2.3$ at $t = 3T$ and so on. This is to be expected since z^{-n} replaces e^{-nsT} which represents a time delay of nT.

In general, any continuous function possessing a Laplace transform also has a Z-transform. If the Laplace transform is convergent on the s-plane, then generally the Z-transform is convergent on the z-plane.

Mapping Between the s-Plane and z-Plane

It is perhaps instructive to explore the relationship between the s-plane and the z-plane. To establish this we shall map the s-plane onto the z-plane. However, since we have established that $E^*(s)$ is a periodic function in s, we need map only the primary strip shown in Fig. 10-15. Representing $s = \sigma + j\omega$ we have

$$z = e^{\sigma T}e^{j\omega T} \tag{10.17}$$

Therefore, for any point $s = \sigma + j\omega$ on the s-plane the corresponding point on the complex z-plane has a magnitude $e^{\sigma T}$ and angle ωT. When $s = 0, z = 1$ with zero phase. As the imaginary part of s increases, the point z changes phase but not its magnitude. Finally, when $s = j\omega_s/2, z$ has a phase of

$$\omega T = \left(\frac{\omega_s}{2}\right)\left(\frac{2\pi}{\omega_s}\right) = \pi$$

Now as the imaginary part of s stays constant while the real part goes to $-\infty$, the magnitude of z goes to zero but the phase stays at π. Note that the strips from $j\omega_s/2$ to $j3\omega_s/2, -j\omega_s/2$ to $-j3\omega_s/2$, etc. plot into the same region on the z-plane as did the primary strip. We conclude then that the region inside the unit circle on the z-plane correspond to the left half of the s-plane. The right half of the s-plane corresponds to the region outside the unit circle on the z-plane. The imaginary axis in the s-plane corresponds to the circumference of the unit circle on the z-plane.

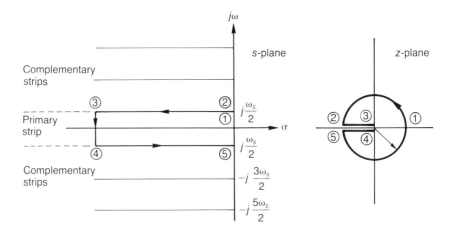

Figure 10.15: Mapping of the s-plane onto the z-plane

Relationship Between $E(s)$, $E^*(s)$, and $E(z)$

We have seen that if the signal $e(t)$ is applied to the input of a sampler then its output is

$$
e^*(t) = \begin{cases} e(nT) & t = nT \\ \\ 0 & t \neq nT \end{cases}
$$

We can thus define the sequence $\{e_n\}$ by taking $e_n = e(nT)$, and note that it is directly linked to the time function $e^*(t)$.

The Laplace transforms of $e(t)$ and $e^*(t)$, which have been denoted by $E(s)$ and $E^*(s)$ respectively, have been shown in Eq. (10-5) to be related, in that given $E(s)$ we can always find $E^*(s)$. The sampling theorem shows that provided $e(t)$ is band limited with bandwidth ω_h, and the Nyquist frequency $\omega_N = \omega_s/2$ exceeds ω_h, then in principle we can exactly derive $E(s)$ from $E^*(s)$.

The $\mathcal{Z}$-transform of $e^*(t)$ was defined by using the substitution $z = e^{sT}$ in $E^*(s)$. Thus there is a one-to-one correspondence between the $\mathcal{Z}$-transform of $e^*(t)$, denoted by $E(z)$, and $E^*(s)$. With these definitions the commutative diagram given in Fig. 10-16 shows the relationship between the signals $e(t)$, $e^*(t)$ and $\{e_n\}$, and their respective Laplace transforms and $\mathcal{Z}$-transform. It will be seen from this figure that the $\mathcal{Z}$-transform of the sequence $\{e_n\}$ can be directly related to $E(s)$ by substituting Eq. (10-15)

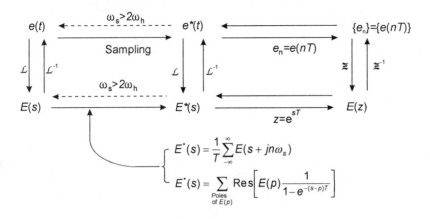

Figure 10.16: Commutative diagram showing the relationship between $E(s)$, $E^*(s)$, and $E(z)$

into Eq. (10-5) to give

$$E(z) = E^*(s)|_{s=\frac{1}{T}\ln z} = \sum_{\substack{\text{poles} \\ \text{of } E(s)}} \text{Res} \left[E(s) \frac{z}{z - e^{sT}} \right]_{s=s_i} \qquad (10.18)$$

If $E(s)$ has k simple poles at $s_1, \cdots, s_k$ then the residues may be evaluated from

$$\text{Res} \left[E(s) \frac{z}{z - e^{sT}} \right]_{s=s_i} = \lim_{s \to s_i} (s - s_i) E(s) \frac{z}{z - e^{sT}} \qquad (10.19)$$

while if there is a multiple pole s_i of multiplicity m

$$\text{Res} \left[E(s) \frac{z}{z - e^{sT}} \right]_{s=s_i} = \lim_{s \to s_i} \frac{d^{m-1}}{ds^{m-1}} \left[(s - s_i)^m E(s) \frac{z}{z - e^{sT}} \right] \qquad (10.20)$$

Having established the basic concepts of the Z-transform, let us consider a few examples.

Example 2 *Obtain the Z-transform of $e(t) = H(t)$.*

We have already established that

$$E^*(s) = \frac{1}{1 - e^{-sT}}$$

Substituting Eq. (10-15),

$$E(z) = \frac{1}{1 - z^{-1}} = \frac{z}{z - 1}$$

Example 3 *Obtain the Z-transform of* $e(t) = t$ *using the convolution method.*

Since

$$E(s) = \frac{1}{s^2}$$

there is a double pole. In this case,

$$E(z) = \sum \text{Res} \left[\frac{E(s)}{1 - e^{sT} z^{-1}} \right]_{s=s_i}$$

Defining

$$R(s) = \frac{E(s)}{1 - e^{sT} z^{-1}}$$

the residue at $s = 0$ is

$$\text{Res } [R(s)]_{s=0} = \frac{d}{ds} \left[s^2 R(s) \right]_{s=0} = \frac{Tz^{-1}}{\left(1 - z^{-1}\right)^2}$$

Therefore,

$$E(z) = \frac{Tz}{(z - 1)^2}$$

Properties of Z-Transform

Some of the basic properties, given without proof, of Z-transforms are listed below. A table of Z-transforms incorporating some of these ideas appears in Appendix A.

Linearity   If $e_1 (nT)$ and $e_2 (nT)$ have Z-transforms $E_1 (z)$ and $E_2 (z)$, then

$$E (z) = Z \left[a_1 e_1 (nT) + a_2 e_2 (nT) \right] = a_1 E_1 (z) + a_2 E_2 (z)$$

where a_1 and a_2 are constants.

Real translation   When a function $e(nT)$ is delayed by mT, then the Z-transform is given by

$$Z[e(nT \pm mT)] = z^{\pm m} E(z)$$

Multiplication by an exponential   When $e(nT)$ is multiplied by $e^{n\alpha T}$ in the time domain, then

$$\mathcal{Z}\left[e^{n\alpha T}e\left(nT\right)\right] = E\left(ze^{-\alpha T}\right)$$

Initial value theorem   If $e(nT)$ has the $\mathcal{Z}$-transform $E(z)$, then

$$\lim_{n\to 0} e(nT) = \lim_{z\to\infty} E(z)$$

provided the limit exists.

Final value theorem   Let the function $e(nT)$ have the $\mathcal{Z}$-transform $E(z)$ and further let $(1 - z^{-1})E(z)$ have no poles on or outside the unit circle centered at the origin on the z-plane[4], then

$$\lim_{n\to\infty} e(nT) = \lim_{z\to 1} \left(\frac{z-1}{z}\right) E(z)$$

As an example, consider

$$E(z) = \frac{0.4z^2}{(z-1)(z^2 - 0.5z + 0.3)}$$

The initial value of $e(nT)$ is zero while the steady state or final value is

$$\lim_{n\to\infty} e(nT) = \lim_{z\to 1} \frac{z-1}{z} E(z)$$

Substituting for $E(z)$ we have

$$\lim_{n\to\infty} e(nT) = \lim_{z\to 1} \frac{0.4z}{(z^2 - 0.5z + 0.3)} = 0.5$$

This can be verified by expanding $E(z)$,

$$E(z) = 0.4z^1 + 0.6z^{-2} + 0.58z^{-3} + 0.51z^{-4} + \cdots$$

We note that $e(nT)$ is rapidly converging to 0.5.

---

[4] This condition is identical to the requirements of the Final Value Theorem for the Laplace transform.

10.5   The Inverse $\mathcal{Z}$-Transform

In order to obtain the response in the time domain, we need to obtain the inverse $\mathcal{Z}$-transform of functions expressed in the z-domain. It should be emphasized that the inverse transform yields information at $t = 0, T, 2T, \cdots$, i.e. at discrete time steps. It gives no information about the variable in between the time steps.

The inverse $\mathcal{Z}$-transform can be obtained from tables only for the most elementary functions. For complex functions, other methods for obtaining the inverse are used. We shall consider the following four methods:

1. The partial fraction method

2. The power series expansion method

3. The difference method

4. The inversion integral method.

In each case we shall assume that the function in the z-domain appears as

$$E(z) = \frac{N(z)}{D(z)}$$

where $N(z)$ and $D(z)$ are polynomials in z.

Partial Fraction Method

Analogous to the Laplace transform we expand the z-transform function in partial fractions and take the inverse of each term. Consider the function $E(z)$ expressed as

$$E(z) = \frac{N(z)}{D(z)} = \frac{N(z)}{\left(z - e^{-a_1 T}\right)\left(z - e^{-a_2 T}\right) \cdots \left(z - e^{-a_m T}\right)} \qquad (10.21)$$

We expand $E(z)/z$ in partial fractions,

$$\frac{E(z)}{z} = \frac{K_1}{z - e^{-a_1 T}} + \frac{K_2}{z - e^{-a_2 T}} + \cdots \qquad (10.22)$$

where the coefficients are obtained as in the Laplace transform method. The inverse $\mathcal{Z}$-transform is taken term by term,

$$\mathcal{Z}^{-1}[E(z)] = \mathcal{Z}^{-1}\left[\frac{K_1 z}{z - e^{-a_1 T}}\right] + \cdots + \mathcal{Z}^{-1}\left[\frac{K_m z}{z - e^{-a_m T}}\right]$$

or

$$e(nT) = K_1 e^{-a_1 nT} + \cdots + K_m e^{-a_m nT} \qquad (10.23)$$

As an example, consider

$$E(z) = \frac{z}{(z-1)(z-e^{-T})}$$

Expanding this,

$$\frac{E(z)}{z} = \frac{K_1}{z-1} + \frac{K_2}{z-e^{-T}}$$

where

$$K_1 = \lim_{z \to 1} \left(\frac{z-1}{z}\right) E(z) = \frac{1}{1-e^{-T}}$$

$$K_2 = \lim_{z \to e^{-T}} \left(\frac{z-e^{-T}}{z}\right) E(z) = -\frac{1}{1-e^{-T}}$$

Substituting

$$E(z) = \frac{1}{1-e^{-T}} \left[\frac{z}{z-1} - \frac{z}{z-e^{-T}}\right]$$

and the inverse is

$$e(nT) = \frac{1}{1-e^{-T}} \left[1 - e^{-nT}\right]$$

When the function $E(z)$ has multiple poles we form the coefficients in a manner analogous to the Laplace transform method.

Power Series Expansion Method

This method involves the representation of $E(z)$ in the form of a series. We represent $E(z)$ as

$$E(z) = e_0 + e_1 z^{-1} + e_2 z^{-2} + \cdots \qquad (10.24)$$

Since multiplication by z^{-m} means that the function is shifted in time by mT time units, we may rewrite Eq. (10-24) in the time domain as

$$e^*(t) = e_0 \delta(t) + e_1 \delta(t-T) + e_2 \delta(t-2T) + \cdots \qquad (10.25)$$

As an example, consider

$$E(z) = \frac{z^2 + 2z - 1}{z^2 + 3z - 3}$$

Using long division we obtain the following series

$$E(z) = 1 - z^{-1} + 5z^{-2} - 18z^{-3} + \cdots$$

and the inverse is

$$e^*(t) = \delta(t) - \delta(t-T) + 5\delta(t-2T) - 18\delta(t-3T) + \cdots$$

Difference Method

This method is iterative and can be programmed very easily on a computer. Assume that we know

$$\frac{C(z)}{R(z)} = \frac{z^2}{z^2 - \alpha z + \beta}$$

Dividing through by z^2 and multiplying we obtain

$$C(z)[1 - \alpha z^{-1} + \beta z^{-2}] = R(z)$$

Recognizing z^{-1} as a time shift and letting $t = nT$, we have

$$c(nT) - \alpha c\,[(n-1)\,T] + \beta c\,[(n-2)\,T] = r(nT)$$

or

$$c(nT) = r(nT) + \alpha c[(n-1)T] - \beta c[(n-2)T]$$

An example using this method is considered in a later section.

Inversion Integral Method

A less laborious method of finding the inverse transform follows from the inversion integral, which is obtained from the theory of complex variables. In this method the time domain response is given by

$$e(nT) = \frac{1}{2\pi j} \int_{\mathbf{C}} z^{n-1} E(z)\, dz \qquad (10.26)$$

where $\mathbf{C}$ is a closed contour about the origin in the complex z-plane. This integral can be easily evaluated using the method of residues, thus

$$e(nT) = \sum_{\substack{\text{poles of} \\ z^{n-1}E(z)}} \text{Res}\,\left[z^{n-1}E(z)\right] \qquad (10.27)$$

Example 4 *Find the inverse Z-transform of*

$$E(z) = \frac{z}{(z+1)(z+2)}$$

From Eq. (10-27) we obtain

$$\begin{aligned}
e(nT) &= \lim_{z \to -1}(z+1)z^{n-1}E(z) + \lim_{z \to -2}(z+2)z^{n-1}E(z) \\[2mm]
&= \lim_{z \to -1}\frac{z^n}{z+2} + \lim_{z \to -2}\frac{z^n}{z+1} \\[2mm]
&= (-1)^n - (-2)^n = 0, 1, -3, 7, \cdots
\end{aligned}$$

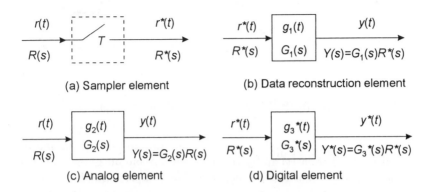

Figure 10.17: Digital control system transfer function elements

The inverse is

$$e^*(t) = \delta(t - T) - 3\delta(t - 2T) + 7\delta(t - 3T) + \cdots$$

10.6 Digital Transfer Functions

Four types of transfer function elements are used in the study of digital control systems. As shown in Fig. 10-17 these are the:

(a) Sampler — analog input and sampled output signals

(b) Data Reconstruction — sampled input and analog output signals

(c) Analog Element — analog input and output signals

(d) Digital Element — sampled input and output signals.

The first two elements which have been introduced in earlier sections of this chapter are shown in block diagram form in Figs. 10-17(a) and (b), where $G_1(s)$ is the transfer function of the data reconstruction element. The analog element shown in Fig. 10-17(c) has the transfer function $G_2(s)$. The digital and data reconstruction elements will now be examined further.

Digital Element

This element is used to perform linear arithmetic operations on a data sequence to transform it into a new data sequence, such as might occur

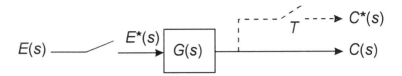

Figure 10.18: Block diagram with sampled signal

in a digital computer. To this end consider the digital system shown in Fig. 10-17(d) where $r^*(t)$ and $y^*(t)$ are the input and output sampled time functions or data sequences respectively. The sampled time function $g_3^*(t)$ is the response of the element to a single sample period pulse applied at time $t = 0$. Its Laplace transform is denoted by $G_3^*(s)$. It can be easily shown, because the system is assumed to be linear, that

$$y(nT) = \sum_{m=0}^{\infty} g_3(nT - mT)r(mT) \qquad (10.28)$$

and taking the $\mathcal{Z}$-transform of this equation gives

$$Y(z) = G_3(z)R(z) \qquad (10.29)$$

or

$$Y^*(s) = G_3^*(s)R^*(s) \qquad (10.30)$$

The function $G_3(z)$ will be called the **digital transfer function** of the element, while Eq. (10-28) will be called the **digital convolution operator**.

Data Reconstruction Element

When a signal is not continuous but consists of periodic pulses, then the output of the transfer function shown in Fig. 10-18 becomes

$$C(s) = G(s)E^*(s) \qquad (10.31)$$

where $E^*(s)$ is the pulsed transform of $E(s)$ and can be written as

$$E^*(s) = \frac{1}{T} \sum_{n=-\infty}^{\infty} E(s + jn\omega_s)$$

Similarly, the output $C(s)$ may be denoted by the pulsed transform as

$$C^*(s) = \frac{1}{T} \sum_{-\infty}^{\infty} C(s + jn\omega_s) = \frac{1}{T} \sum_{-\infty}^{\infty} G(s + jn\omega_s)E^*(s + jn\omega_s)$$

The pulsed function $E^*(s)$ was shown to be a periodic function with period ω_s, and as a consequence

$$E^*(s) = E^*(s + jn\omega_s)$$

Therefore the above may be modified to

$$C^*(s) = E^*(s)\frac{1}{T}\sum_{-\infty}^{\infty} G(s + jn\omega_s)$$

Since

$$G^*(s) = \frac{1}{T}\sum_{-\infty}^{\infty} G(s + jn\omega_s)$$

we obtain

$$C^*(s) = E^*(s)G^*(s) \qquad\qquad (10.32)$$

Now taking the $\mathcal{Z}$-transform

$$C(z) = G(z)E(z)$$

or

$$\frac{C(z)}{E(z)} = G(z) \qquad\qquad (10.33)$$

We use this consequence to develop the transfer function of systems having sampled signals. If there is more than one sampler, we shall assume that they are all synchronized.

The sampler at the output of the block denoted $G(s)$ in Fig. 10-18 is often referred to as a **fictitious sampler** as it may not always be physically present in a system. The main purpose for its introduction is to facilitate the representation of a data reconstruction element as an equivalent digital element, as shown in Eqs. (10-32) and (10-33).

Combinations of Pulsed Transfer Functions

Consider the cascaded elements shown in Fig. 10-19(a). Using the notation developed previously the output $C(s)$ is,

$$C(s) = G_1(s)G_2(s)E^*(s) \qquad\qquad (10.34)$$

Defining $G(s) = G_1(s)G_2(s)$ we have

$$C(s) = G(s)E^*(s)$$

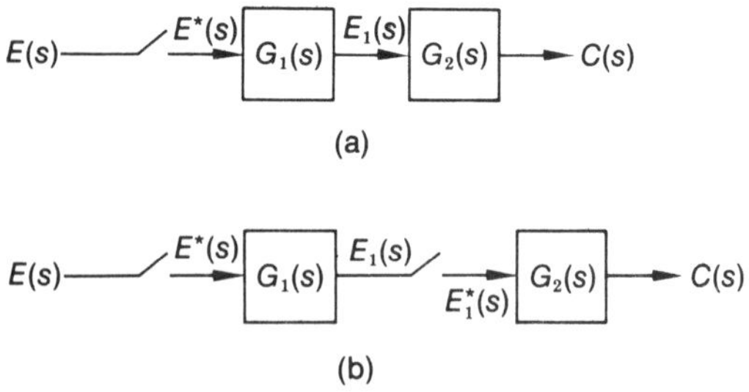

Figure 10.19: Cascaded elements with sampler

and using the result developed in the previous section

$$C^*(s) = G^*(s)E^*(s)$$

Taking the $\mathcal{Z}$-transform

$$C(z) = G(z)E(z) \qquad (10.35)$$

where $G(z) = \mathcal{Z}[G_1(s)G_2(s)]$, i.e. the $\mathcal{Z}$-transform is taken *after* $G_1(s)G_2(s)$ is formed. Now consider the same cascaded elements but with a sampler between each transfer function as shown in Fig. 10-19(b). By inspection we write

$$
\begin{aligned}
E_1(s) &= G_1(s)E^*(s) \\
C(s) &= G_2(s)E_1^*(s)
\end{aligned}
\qquad (10.36)
$$

Taking the pulsed transform we obtain

$$E_1^*(s) = G_1^*(s)E^*(s)$$

and substituting in Eq. (10-36) and again taking the pulsed transform

$$C^*(s) = G_1^*(s)G_2^*(s)E^*(s)$$

Taking the $\mathcal{Z}$-transform we obtain

$$C(z) = G_1(z)G_2(z)E(z) \qquad (10.37)$$

and clearly this is different from Eq. (10-35). We conclude that in general

$$G_1(z)G_2(z) \neq \overline{G_1 G_2}(z) \qquad (10.38)$$

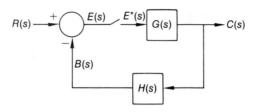

Figure 10.20: Error sampled closed loop system

As an example, assume

$$G_1(s) = \frac{1}{s^2} \quad \text{and} \quad G_2(s) = \frac{a}{s+a}$$

then for Fig. 10-19(a) we have

$$\frac{C(z)}{E(z)} = \overline{G_1 G_2}(z) = \mathcal{Z}\left[\frac{a}{s^2(s+a)}\right]$$

$$= \frac{Tz}{(z-1)^2} - \frac{(1-e^{-aT})z}{a(z-1)(z-e^{-aT})}$$

whereas for Fig. 10-19(b) we have

$$\frac{C(z)}{E(z)} = G_1(z)G_2(z) = \mathcal{Z}\left[\frac{1}{s^2}\right]\mathcal{Z}\left[\frac{a}{s+a}\right]$$

$$= \frac{aTz^2}{(z-1)^2(z-e^{-aT})}$$

Closed Loop Pulsed Systems

The previous techniques may be extended to include closed loop systems with sampled data. Consider the closed loop system of Fig. 10-20. We can write the following equations by inspection

$$E(s) = R(s) - B(s)$$

$$C(s) = G(s)E^*(s)$$

$$B(s) = H(s)C(s)$$

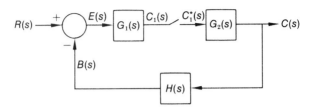

Figure 10.21: Sampled data system with sampler between cascaded element

Substituting the second and third into the first equation,

$$E(s) = R(s) - \overline{GH}(s)E^*(s)$$

Taking the pulsed transform of this and rearranging, we obtain

$$E^*(s) = R^*(s)/\left(1 + \overline{GH}^*(s)\right)$$

Now substituting for $E^*(s)$ we obtain the overall transfer function,

$$C^*(s) = G^*(s)R^*(s)/\left(1 + \overline{GH}^*(s)\right)$$

and taking the $\mathcal{Z}$-transform

$$\frac{C(z)}{R(z)} = \frac{G(z)}{1 + \overline{GH}(z)} \tag{10.39}$$

The overall transfer function given by Eq. (10-39) may be applied only to the closed loop system shown in Fig. 10-20 where the error is being sampled. If the sampler were moved to a different location, a different transfer function would result. Consider the sampled data system shown in Fig. 10-21. The following equations may be written

$$E(s) \;\;=\;\; R(s) - B(s)$$

$$C_1(s) \;\;=\;\; G_1(s)E(s)$$

$$C(s) \;\;=\;\; G_2(s)C_1^*(s)$$

$$B(s) \;\;=\;\; H(s)C(s)$$

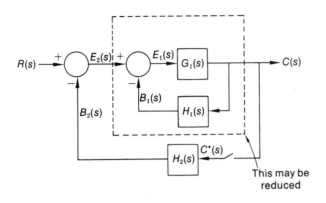

Figure 10.22: Multi-loop sampled data system

Substituting these equations, rearranging, taking the pulsed transform, and then taking the $\mathcal{Z}$-transform yields

$$C(z) = \frac{\overline{RG_1}(z)G_2(z)}{1 + \overline{G_1G_2H}(z)} \tag{10.40}$$

Since generally $R(z)G(z) \neq \overline{RG}(z)$, we are unable to obtain the transfer function $C(z)/R(z)$ but can only obtain the output in terms of $\overline{RG}(z)$ as shown in Eq. (10-40). This is not uncommon in sampled data control systems.

Example 5 *Consider the multi-loop system shown in Fig. 10-22. Find the output $C^*(s)$ of the system.*

We may write down the following equations by inspection

$$E_2(s) = R(s) - B_2(s)$$

$$B_2(s) = H_2(s)C^*(s)$$

$$C(s) = E_2(s)\frac{G_1(s)}{1 + G_1(s)H_1(s)} = G_2(s)E_2(s)$$

We note that the last equation is the result of reducing the block diagram within the dotted lines indicated in Fig. 10-22. These equations can be shown to yield

$$C^*(s) = \frac{\overline{G_2R}^*(s)}{1 + \overline{G_2H}_2^*(s)}$$

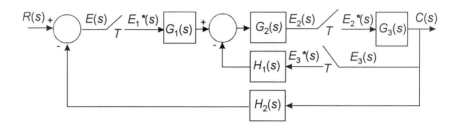

Figure 10.23: Multi-sampler digital control system

and again we obtain the output but not the "usual" transfer function.

Digital control systems may have more than one sampler. The next example illustrates a general method for analyzing these systems.

Example 6 *Consider the multi-loop multi-sampler digital control system shown in Fig. 10-23. Find the closed loop transfer function for this system.*

To analyze this system we determine the signals at the input to each sampler by inspection

$$E_1(s) = R(s) - H_2(s)G_3(s)E_2^*(s)$$

$$E_2(s) = G_2(s)G_1(s)E_1^*(s) - G_2(s)H_1(s)E_3^*(s)$$

$$C(s) = G_3(s)E_2^*(s) = E_3(s)$$

Taking the pulsed transform of each equation gives

$$E_1^*(s) = R^*(s) - \overline{H_2G_3}^*(s)E_2^*(s)$$

$$E_2^*(s) = \overline{G_2G_1}^*(s)E_1^*(s) - \overline{G_2H_1}^*(s)E_3^*(s)$$

$$E_3^*(s) = G_3^*(s)E_2^*(s)$$

Taking the $\mathcal{Z}$-transform and applying simple algebraic manipulation gives

$$\frac{C(z)}{R(z)} = \frac{G_3(z)\overline{G_2G_1}(z)}{1 + \overline{G_2G_1}(z)\overline{H_2G_3}(z) + \overline{G_2H_1}(z)G_3(z)}$$

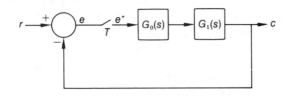

Figure 10.24: Error sampled system

10.7   System Response

Consider the unity feedback system with sampled error signal shown in Fig. 10-24 where $G_0(s)$ is a zero-order-hold and $G_1(s)$ represents a continuous element. The pulsed transfer function is

$$\frac{C(z)}{R(z)} = \frac{\overline{G_0 G_1}(z)}{1 + \overline{G_0 G_1}(z)}$$

For any given $R(z)$, the response may be obtained using the classical techniques developed in the previous sections.

Second Order System

For the system shown in Fig. 10-24, let $G_0(s)G_1(s)$ be a second order system

$$G_0(s)G_1(s) = \frac{K}{s^2 + 2\zeta\omega_n s + \omega_n{}^2}$$

Then

$$\frac{C(z)}{R(z)} = \frac{K'z}{z^2 + (K' - \alpha)z + \beta} \qquad (10.41)$$

where

$$K' = Ke^{-\zeta\omega_n T}\sin\omega_d T$$

$$\alpha = 2e^{-\zeta\omega_n T}\cos\omega_d T$$

$$\beta = e^{-2\zeta\omega_n T}$$

and

$$\omega_d = \omega_n\sqrt{1 - \zeta^2}$$

Thus, just as is the case for continuous time control systems, the location of the closed loop poles in the z-plane determines the transient performance of second order digital control systems.

To see this clearly it is appropriate at this time to recall how the s-plane and z-plane are related. As indicated in the Section 10-4, the left half s-plane is mapped inside the region $|z| < 1$ in the z-plane, the right half s-plane is mapped into the region $|z| > 1$. Points at $-\infty$ on the s-plane are mapped into the z-plane origin, whereas the s-plane origin is mapped into the point $z = 1$ on the z-plane. To directly relate the real and imaginary parts on the s- and z-planes we observe that if

$$s = \sigma + j\omega, \quad \text{and } z = \alpha + j\beta$$

then

$$z = e^{sT} = e^{\sigma T} e^{j\omega T} = e^{\sigma T} \left[\cos \omega T + j \sin \omega T \right]$$

which yields

$$\alpha = e^{\sigma T} \cos \omega T$$

$$\beta = e^{\sigma T} \sin \omega T$$

Now since we have already established the relationship between the transient response of continuous closed loop systems and the positions of their closed loop poles, we may take advantage of this, as well as the observations made above, to ascertain the discrete response as a function of closed loop poles on the z-plane. This is shown in Fig. 10-25, and you are asked to verify the various responses as an exercise. In Fig. 10-26 the z-plane contours corresponding to constant ζ and ω_n contours in the s-plane are plotted. This figure is useful when using the root locus method for digital control system design.

Higher Order Systems

The roots of the characteristic equation and consequently the partial fractions are easy to obtain when the order of the polynomial is small as in Eq. (10-41). However, when the order of the characteristic equation is large, then it is not always easy to obtain the closed loop poles. In such situations, we may use techniques other than the partial fraction method to obtain the system response. The techniques popularly used are the contour integration method, the division or power series method, and finally the difference equation method. We favor the last two methods and shall consider examples using them.

Example 7 *Consider a sampled data system whose output is given by*

$$C(z) = \frac{\alpha_4 z^4 + \alpha_3 z^3 + \alpha_2 z^2 + \alpha_1 z + \alpha_0}{\beta_5 z^5 + \beta_4 z^4 + \beta_3 z^3 + \beta_2 z^2 + \beta_1 z + \beta_0}$$

Obtain $c(nT)$ by the power series method.

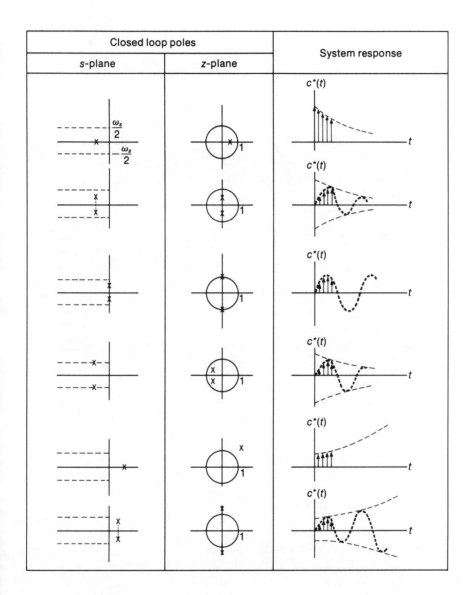

Figure 10.25: Characteristics of discrete system response

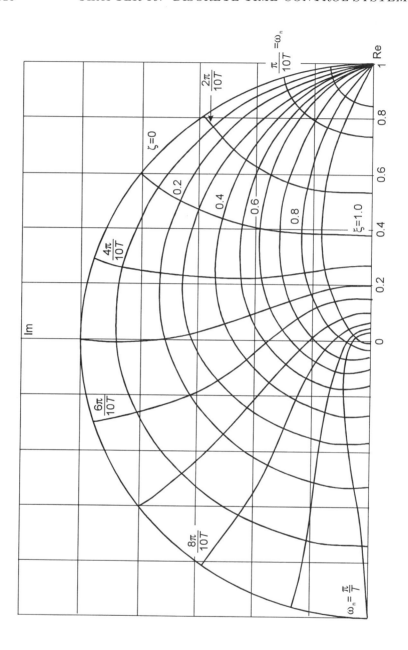

Figure 10.26: Contours of constant ω_n and ζ in the z-plane

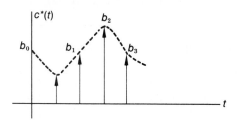

Figure 10.27: Output sequence of Example 10-7

Long division yields a power series

$$C(z) = b_0 z^{-1} + b_1 z^2 + b_2 z^{-3} + \cdots$$

where

$$b_0 = \frac{\alpha_4}{\beta_5}, \quad b_1 = \frac{\alpha_3 \beta_5 - \beta_4 \alpha_4}{\beta_5^2}$$

Substituting $z = e^{sT}$,

$$C^*(s) = b_0 e^{-sT} + b_1 e^{-2sT} + \cdots$$

and the inverse yields

$$c^*(t) = b_0 + b_1 \delta(t - T) + b_2 \delta(t - 2T) + \cdots$$

which is a series of delayed impulses of magnitude b_0, b_1, b_2 occurring at $t = 0, T, 2T, \cdots$ as shown in Fig. 10-27.

Example 8 *Using the difference equation method and assuming a unit impulse input at $t = 0$, obtain $c(nT)$ for the system having an overall transfer function given by*

$$\frac{C(z)}{R(z)} = \frac{z^2}{z^2 - 1.5z + 0.5}$$

Dividing through by z^2 we obtain

$$C(z)[1 - 1.5z^{-1} + 0.5z^{-2}] = R(z)$$

Recognizing z^{-1} as a time shift and letting $t = nT$, we can write this as

$$c(nT) - 1.5c[(n-1)T] + 0.5c[(n-2)T] = r(nT) \qquad (10.42)$$

If we assume $r(t)$ to be a unit impulse, then we can form a table as shown in Table 10-1. The output $c^*(t)$ is in the form of a series of delayed impulses at $t = 0, T, 2T, \ldots$ given in the last column.

Table 10-1 Difference table for Eq. (10-42)

n	$1.5c\left[(n-1)T\right]$	$-0.5c\left[(n-2)T\right]$	$r(nT)$	$c(nT)$
0	0.0	0.0	1	1.0
1	1.5	0.0	0	1.5
2	2.25	-0.5	0	1.75
3	2.63	-0.75	0	1.88
4	2.83	-0.875	0	1.945
5	2.917	-0.94	0	1.963
6	2.945	-0.972	0	1.973

Steady State Response

In order to fully understand the behavior of sampled systems, we must not only be concerned with the transient behavior but must also know its steady state response. Analogous to continuous systems, we examine the steady state response of discrete systems by defining error constants. These error constants are defined with the assumption that the function under consideration converges to a finite value.

The error constants for a discrete system are defined as:

$$K_0 = \lim_{z \to 1}[G(z)]$$

$$K_1 = \frac{1}{T}\lim_{z \to 1}[(z-1)G(z)] \qquad (10.43)$$

$$K_2 = \frac{1}{T^2}\lim_{z \to 1}\left[(z-1)^2 G(z)\right]$$

where K_0 is the positional error constant, K_1 is the velocity error constant, and K_2 is the acceleration error constant. For a step input to a unity feedback sampled system, the steady state error $e(\infty)$ becomes

$$e(\infty) = \lim_{n \to \infty} e(nT) = \lim_{z \to 1}\left[\frac{1}{1+G(z)}\right] = \frac{1}{1+K_0}$$

For a ramp input, $R(z) = Tz/(z-1)^2$, the error becomes

$$
\begin{aligned}
e(\infty) &= \lim_{z \to 1} \left[\frac{1}{1+G(z)} \cdot \frac{T}{(z-1)} \right] \\
&= T \lim_{z \to 1} \left[\frac{1}{(z-1)(1+G(z))} \right] \\
&= \frac{1}{K_1}
\end{aligned}
$$

Finally, for a parabolic input, $R(z) = T^2 z(z+1)/2(z-1)^3$, the error becomes

$$
e(\infty) = \lim_{z \to 1} \left[\frac{1}{1+G(z)} \cdot \frac{T^2}{(z-1)^2} \right] = \frac{1}{K_2}
$$

Having defined the error constants of discrete systems, we may now classify them as to Type. In discrete systems the number of open loop poles at $z = 1$ determines the System Type. If $G(z)$ has 0, 1, or 2 poles at $z = 1$, then the discrete system is referred to as Type 0, 1, or 2. Let us consider an example clarifying some of the points made so far.

Example 9 *The open loop transfer function of a system is given by*

$$
G(z) = \frac{0.98z + 0.66}{(z-1)(z-0.368)}
$$

Compute the error constants and relate them to the steady state error.

The error constants are

$$
K_0 = \lim_{z \to 1} \left[\frac{0.98z + 0.66}{(z-1)(z-0.368)} \right] = \infty
$$

$$
K_1 = \frac{1}{T} \lim_{z \to 1} [(z-1)G(z)] = \frac{2.59}{T}
$$

$$
K_2 = \frac{1}{T^2} \lim_{z \to 1} [(z-1)^2 G(z)] = 0
$$

The steady state error for a step input is

$$
e(\infty) = 0
$$

If the input is a ramp,

$$
e(\infty) = \frac{T}{2.59}
$$

Table 10-2 Error constants and steady state error for discrete systems

Open Loop Poles at $z = 1$	System Type	Error Constants			Steady State Error, $e(\infty)$		
		K_0	K_1	K_2	Position	Ramp	Parabola
0	0	F	0	0	F	∞	∞
1	1	∞	F	0	0	F	∞
2	2	∞	∞	F	0	0	F
3	3	∞	∞	∞	0	0	0

F – Nonzero and finite

and finally, for a parabolic input

$$e(\infty) = \infty$$

The steady state error indicates that this system can follow a step input with no error and a ramp input with a finite error of $T/2.59$. However, for a parabolic input the error pulses continue to diverge from the input pulses as time becomes large.

If it was desired that the error be finite for a parabolic input in the previous example, the System Type would have to be 2, i.e. another open loop pole at $z = 1$ must be introduced. A table showing the error constants and System Type for discrete systems is shown in Table 10-2.

10.8   Stability Tests

The previous section has shown us how we may study the transient as well as the steady state behavior of sampled data system. No mention was made of the stability of the control systems, i.e. we had no *a priori* knowledge about the roots of the characteristic equation. In a practical situation, we would like to extract some information from the characteristic equation *before* going down the tortuous path of obtaining time responses. In this section we shall consider two popular methods for studying the characteristic equation, viz. **Jury's test** and the **Routh-Hurwitz** test.

We begin by assuming that the pulse transfer function can be represented[5]

---

[5] The pulse transfer function can sometimes not be derived owing to the location of

by $P(z)$

$$P(z) = \frac{C(z)}{R(z)} = \frac{\sum\limits_{j=0}^{k} b_j z^j}{\sum\limits_{i=0}^{n} a_i z^i} \qquad (10.44)$$

The denominator of $P(z)$, when equated to zero, yields the characteristic roots. For stability it is desired that there be no characteristic roots outside the unit circle on the z-plane.

The stability criterion called **Jury's test** provides stability information for discrete-time systems just as the Routh-Hurwitz test provides stability information for continuous time systems. Let the characteristic equation of Eq. (10-44) be represented as

$$F(z) = \sum\limits_{i=0}^{n} a_i z^i = 0 \qquad (10.45)$$

We assume that all a_i's are real coefficients and $F(z)$ is written so that $a_n > 0$. We now form an array as follows.

Row				Array				
1	a_0	a_1	a_2	$\cdots$	$\cdots$	$\cdots$	a_{n-1}	a_n
2	a_n	a_{n-1}	a_{n-2}	$\cdots$	$\cdots$	$\cdots$	a_1	a_0
3	b_0	b_1	b_2	$\cdots$	$\cdots$	$\cdots$	b_{n-1}	
4	b_{n-1}	b_{n-2}	b_{n-3}	$\cdots$	$\cdots$	b_1	b_0	
5	c_0	c_1	c_2	$\cdots$	$\cdots$	c_{n-2}		
6	c_{n-2}	c_{n-3}	c_{n-4}	$\cdots$	$\cdots$	c_0		
$\vdots$	$\vdots$	$\vdots$	$\vdots$	$\vdots$	$\vdots$			
$m-2$	x_0	x_1	x_2	x_3				
$m-1$	x_3	x_2	x_1	x_0				
m	y_0	y_1	y_2					

The first row consists of the original coefficients from a_0 to a_n while the second row consists of the same coefficients but in *reverse* order. Each row is computed using Eq. (10-46) and then re-entered in *reverse* order. We notice that the number of coefficients in each row is one less than the previous row. The array is terminated when a row of three numbers is

---

the sampler. In such cases the output may be written as the ratio of two polynomials in z. The denominator polynomial when equated to zero yields the characteristic roots.

obtained. The various coefficients are evaluated as follows

$$b_i = \begin{vmatrix} a_0 & a_{n-i} \\ a_n & a_i \end{vmatrix} \qquad c_i = \begin{vmatrix} b_0 & b_{n-1-i} \\ b_{n-1} & b_i \end{vmatrix}$$

$$d_i = \begin{vmatrix} c_0 & c_{n-2-i} \\ c_{n-2} & c_i \end{vmatrix} \qquad y_i = \begin{vmatrix} x_0 & x_{3-i} \\ x_3 & x_i \end{vmatrix}$$

(10.46)

Having obtained the array, we may now apply Jury's test. This test states that the roots of $F(z) = 0$ will be inside of the unit circle on the z-plane if and only if

$$F(1) > 0$$
$$(-1)^n F(-1) > 0$$

and

$$|a_0| < a_n$$
$$|b_0| > |b_{n-1}|$$
$$|c_0| > |c_{n-2}|$$
$$|d_0| > |d_{n-3}|$$
$$\vdots$$
$$|y_0| > |y_2|$$

If any of the above requirements are not satisfied, then $F(z) = 0$ has roots outside the unit circle on the z-plane and the system response will be unstable. The test yields no information as to the number of roots outside the unit circle on the z-plane.

Example 10 *Determine the stability of a system governed by a cubic characteristic equation given by*

$$F(z) = z^3 + 2z^2 + 1.9z + 0.8$$

Before the array is formed we check $F(1)$ and $F(-1)$

$$F(1) > 0 : \quad F(1) = 1 + 2 + 1.9 + 0.8 = 5.7$$

$$F(-1) < 0 : \quad F(-1) = -1 + 2 - 1.9 + 0.8 = -0.1$$

Having satisfied these tests, we form the array

Row	Array			
1	0.8	1.9	2.0	1.0
2	1.0	2.0	1.9	0.8
3	-0.36	-0.48	-0.3	

We now check the constraints

$$|a_0| < a_3 : \quad 0.8 < 1$$

$$|b_0| > |b_2| : \quad |-0.36| > |-0.3|$$

Since all the requirements are met, Jury's test states that the characteristic equation has no roots outside the unit circle on the z-plane.

If the characteristic equation is in the form

$$F(z) = a_3 z^3 + a_2 z^2 + a_1 z + a_0 = 0$$

then this test may be used to ascertain the relationship between the coefficients such that $F(z) = 0$ has roots inside the unit circle on the z-plane.

Although Jury's test tells us whether roots exist inside or outside the unit circle on the z-plane, it gives no information as to how many roots exist outside the unit circle as mentioned previously. In this respect it is a weaker test than the Routh-Hurwitz test although it is much easier to apply.

In order to use the Routh-Hurwitz criterion we need to work on a plane similar to the s-plane. This is achieved by using the bi-linear transformation[6] which essentially maps the interior of a unit circle on the z-plane into the left half of the r-plane. The Routh-Hurwitz criterion may be applied to the r-plane where stability requires that the roots of the characteristic equation be in the left half r-plane. All the notions developed about the Routh-Hurwitz criterion on the s-plane may be applied on the r-plane. Let us take an example to elucidate this further.

Example 11 *Consider the characteristic equation given in Example 10*

$$F(z) = z^3 + 2z^2 + 1.9z + 0.8$$

Use the bi-linear transformation

$$z = \frac{1+r}{1-r} \tag{10.47}$$

and the Routh-Hurwitz criterion to determine stability.

From Eq. (10-47) we obtain

$$F\left(z = \frac{1+r}{1-r}\right) = F(r) = \left(\frac{1+r}{1-r}\right)^3 + 2\left(\frac{1+r}{1-r}\right)^2 + 1.9\left(\frac{1+r}{1-r}\right) + 0.8 = 0$$

---

[6] The relation between the frequency on the r-plane and the s-plane to the sampling frequency is discussed in the next section.

After a considerable amount of algebra we obtain the equation which is the characteristic equation representation on the r-plane. The Routh array becomes

$$
\begin{array}{c|cc}
r^3 & 0.1 & 0.7 \\
r^2 & 1.5 & 5.7 \qquad b_0 = 0.32 \\
\hline
r^1 & b_0 & b_1 \qquad b_1 = 0 \\
r^0 & c_0 & \qquad c_0 = 5.7
\end{array}
$$

which shows that there are no sign changes in the first column and therefore the system is stable as we expected. If however there had been sign reversals in the first column, then the number of reversals is equal to the number of zeros in the right half r-plane which is identical to the region exterior to the unit circle on the z-plane.

10.9   Graphical Analysis Methods

Before we may apply the graphical techniques developed in Chapter 7 to sampled data systems, certain modifications are in order. Since the pulse transfer function is obtained in the $\mathcal{Z}$-transform domain where the exterior of the unit circle corresponds to the right half of the s-plane, it is necessary to avoid having poles outside the unit circle for the system to be stable. While it is possible to study graphical methods applied directly to the z-plane it has not been found convenient; instead a method will be developed where a bi-linear transformation to the r-plane is used. This procedure takes the interior and exterior of the unit circle on the z-plane and conformally maps it into the complex left and right half of the r-plane respectively. The frequency response may now be studied as $-\infty \leq \omega_r \leq +\infty$ or $0 \leq \omega_r \leq \infty$ since the polar frequency plot is symmetrical about the real axis. When the s-plane was mapped onto the z-plane, we saw that it is sufficient to consider the range $-\omega_s/2 \leq \omega \leq \omega_s/2$ where ω_s is the sampling frequency. This is true since the complementary strips on the s-plane, shown in Fig. 10-15, also map into the same region on the z-plane as the primary strip. Thus all points in the primary strip of the s-plane map into the complete region of the r-plane.

The frequency ω_r on the r-plane may now be written as

$$
r|_{r=j\omega_r} = \left.\frac{z-1}{z+1}\right|_{z=e^{j\omega T}}, \quad \text{or } j\omega_r = \frac{e^{j\omega T}-1}{e^{j\omega T}+1} = j\tan\left(\frac{\omega T}{2}\right)
$$

and since $T = 2\pi/\omega_s$ we obtain $\omega_r = \tan \pi\omega/\omega_s$. This yields the following

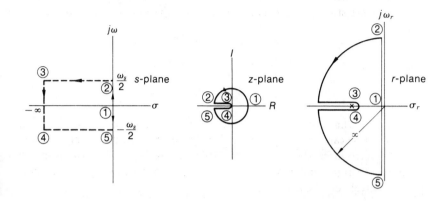

Figure 10.28: Relationships of s-, z-, and r-planes

limiting values

$$\omega = 0 \qquad \omega_r = 0$$

$$\omega = \frac{\omega_s}{2} \qquad \omega_r = \infty$$

$$\omega = -\frac{\omega_s}{2} \qquad \omega_r = -\infty$$

The mapping of the unit circle onto the r-plane is shown in Fig. 10-28 and so is the relationship with the s-plane. When ω varies from $-\omega_s/2$ to $\omega_s/2$ on the s-plane, the corresponding frequency on the r-plane varies from $-\infty$ to $+\infty$. It may be easily deduced that the entire left half s-plane, while mapping into the unit circle on the z-plane, also maps into the left half r-plane.

The Bode and Nyquist plots, discussed previously for continuous systems, may now be modified for sampled systems by focusing attention on the r-plane. The root locus technique can, however, be studied on the z-plane without any difficulty.

The Bode Plot

In making frequency plots for continuous time systems we assumed that the output frequency was the same as the input frequency, although the output had a different magnitude and phase shift. For sampled systems we saw earlier that the output sequence has an envelope that is also sinusoidal having the same frequency as the input. Therefore the frequency concepts

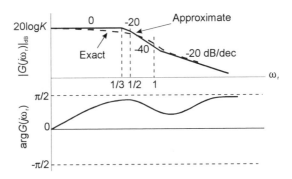

Figure 10.29: Bode plot of $G(z) = K(z+1)/(z-2)(z-3)$ constructed on the r-plane

developed for continuous systems may be applied to sampled systems without modification.

The Bode plot is a plot of the log magnitude and phase shift as a function of frequency. We obtain the frequency function by substituting $r = j\omega_r$ and then use the method described earlier in Chapter 7 to construct the plot.

Example 12 *Consider the pulsed transfer function*

$$G(z) = \frac{K(z+1)}{(z-2)(z-3)}$$

Obtain the Bode plot of $G(z)$ with z replaced by the bi-linear transformation.

Substituting the bi-linear transformation we obtain

$$G(r) = \frac{K(1-r)}{(3r-1)(2r-1)}$$

Writing the frequency function in terms of magnitude and phase shift gives

$$G(j\omega_r) = 20 \log K + 20 \log \sqrt{1 + \omega_r^2} - 20 \log \sqrt{1 + 9\omega_r^2} - 20 \log \sqrt{1 + 4\omega_r^2}$$

$$\arg G(j\omega_r) = -\arctan \omega_r + \arctan 3\omega_r + \arctan 2\omega_r$$

The three corner frequencies are at $1, 1/3$, and $1/2$. The Bode plot is shown in Fig. 10-29. The dotted line in the figure is the exact plot. As before the maximum error occurs at the corner frequencies.

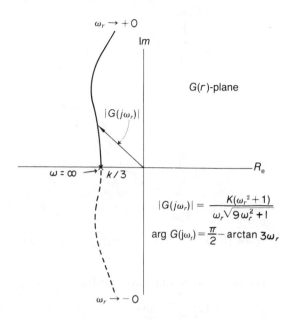

Figure 10.30: Nyquist plot of $G(z) = Kz/(z-1)(z-0.5)$ constructed on the r-plane

The Nyquist Plot

The Nyquist plot may be directly constructed from the Bode plot or independently from the transfer function. To illustrate its application consider the open loop transfer function

$$G(z) = \frac{Kz}{(z-1)(z-0.5)} \tag{10.48}$$

for a unity feedback control system. Representing this in the r-domain,

$$G(r) = \frac{K(1-r^2)}{r(1+3r)}$$

The Nyquist plot for this transfer function is shown in Fig. 10-30 where it will be observed that for $\omega_r = \infty$ the above function has a magnitude of $K/3$ and phase of $-\pi$.

The concept of the Nyquist criterion may be readily extended to sampled

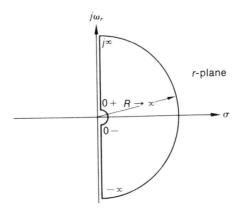

Figure 10.31: The Nyquist path on the r-plane

data systems. The closed loop pulse transfer function is

$$\frac{C(z)}{R(z)} = \frac{G(z)}{1 + \overline{GH}(z)} = M(z)$$

The characteristic equation is

$$1 + \overline{GH}(z) = 0$$

and the roots of the characteristic equation must lie within the unit circle for stability.

Now we have seen that a bi-linear transformation plots the exterior of the unit circle onto the right half r-plane. Therefore, if the function $\overline{GH}(z)$ is transformed using a bi-linear transformation, then the Nyquist path shown in Fig. 10-31 may be employed. We now look for the encirclement of the $(-1, 0)$ point on the $\overline{GH}(r)$-plane. All the techniques developed before may be applied without modification. Let us examine the application of this technique to the above example.

Applying the bi-linear transformation to the characteristic equation

$$F(z) = (z - 1)(z - 0.5) + Kz = 0$$

yields

$$F(r) = (3 - K)r^2 + r + K = 0$$

Applying either Jury's test to $F(z)$ or the Routh-Hurwitz criteria to $F(r)$ shows that $K < 3$ for stability.

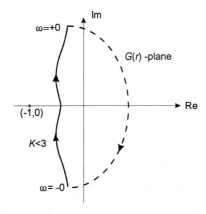

Figure 10.32: Nyquist plot of $G(z) = Kz/(z-1)(z-0.5)$ on the $G(r)$-plane

For the open loop transfer function given in Eq. (10-48) the complete Nyquist plot is shown in Fig. 10-32. The system is stable since $(-1,0)$ is not encircled. If $K > 3$, then the system becomes unstable.

We note that although a second-order system is always stable for a continuous control system having negative feedback, the inclusion of sampling which converts it into a discrete time control system tends to introduce instability.

The Root Locus

The root locus technique may be directly adapted to the analysis of sampled data systems on the z-plane. The principle difference between the continuous and sampled case is that whereas in the former the loci is on the s-plane and the crossover of the imaginary axis is important, with sampled systems the loci is on the z-plane and the unit circle is significant for investigating stability. All the other rules for constructing the root locus on the s-plane still apply on the z-plane.

Consider a second-order system having an open loop transfer function,

$$\overline{G_0G}(z) = \frac{Ke^{-T}(1 - e^{-T})z}{(z - e^{-T})(z - e^{-2T})}$$

The transfer function is not only a function of K but also of the sampling period T. For a sampling period $T = 1$ sec,

$$\overline{G_0 G}(z) = \frac{0.233 K z}{(z - 0.368)(z - 0.135)}$$

Applying the previously enunciated rules for root loci construction, we obtain:

1. The root loci begin at $z = 0.368$ and $z = 0.135$

2. The root loci end at $z = 0$ and $z = \infty$

3. There are two branches of root loci

4. The loci are symmetrical about the real axis

5. Asymptotes have an angle π. [The real axis intercept is no applicable. Why?]

6. The region between $z = 0.368$ and $z = 0.135$ on the real axis is part of the root loci and so is the region between $z = 0$ and $z = -\infty$

7. Not applicable. [Why?]

8. Breakaway points due to real poles and zeros are determined from

$$-\frac{1}{\alpha} + \frac{1}{(0.368 + \alpha)} + \frac{1}{(0.135 + \alpha)} = 0$$

Solving the quadratic, $\alpha = \pm 0.229$. [The same result may be obtained by setting the derivative of the function with respect to z equal to zero.]

9. Intersection with the unit circle is obtained by first forming the characteristic equation $1 + \overline{G_0 G}(r) = 0$

$$(0.542 + 0.233 K) r^2 + 1.897 r + (1.562 - 0.233 K) = 0$$

and from the Routh-Hurwitz method, $K \leq 6.67$. [We could have used Jury's test and worked on the z-plane and obtained the same result.]

The root locus plot may now be constructed and is shown in Fig. 10-33. Since $G(z)$ is a function of the sampling time constant T, let us observe the effect of varying T. The transfer functions for two different values of T are

$$T = 0.1 \, \text{sec} : \quad \overline{G_0 G}(z) = \frac{0.086 \, K z}{(z - 0.905)(z - 0.819)}$$

$$T = 10 \, \text{sec} : \quad \overline{G_0 G}(z) = \frac{0.454 \times 10^{-4} K z}{(z - 0.206 \times 10^{-8})(z - 0.454 \times 10^{-4})}$$

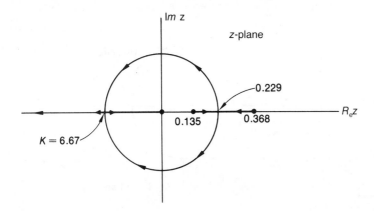

Figure 10.33: Root locus for $G(z) = 0.233Kz/(z-0.368)(z-0.135)$

The root loci plots are shown in Figs. 10-34(a) and (b). The characteristic equation for these two cases are

$$T = 0.1\,\text{sec}:\quad (0.017 + 0.086K)r^2 + 0.518r + (3.465 - 0.086K) = 0$$

$$T = 10\,\text{sec}:\quad (1 + 0.454 \times 10^{-4}K)r^2 + 2r + (1 - 0.454 \times 10^{-4}K) = 0$$

and the unit circle is crossed at $K = 40$ and $K = 22,000$ respectively.

Example 13 *The forward loop transfer function of an error-sampled control system is given by*

$$G(z) = \overline{G_0 G_1}(z) = \frac{K\left(1 - e^{-4T}\right)z}{4\,(z-1)\,(z - e^{-4T})}$$

Obtain the output for a unit impulse input when $K = 1$. Sketch the root locus and determine the value of K for marginal stability. Assume a one-second sampling rate.

For $T = 1$ sec,

$$G(z) = \frac{0.245Kz}{(z-1)(z-0.0183)}$$

The overall transfer function becomes

$$\frac{C(z)}{R(z)} = \frac{G(z)}{1 + G(z)}$$

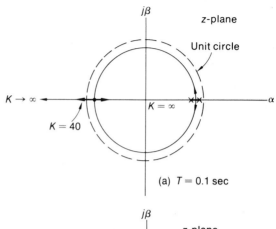

(a) $T = 0.1$ sec

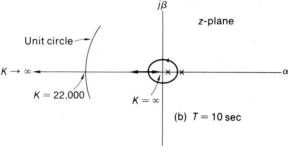

(b) $T = 10$ sec

Figure 10.34: Root loci for $\overline{G_0 G}(z) = \dfrac{K e^{-T}\left(1 - e^{-T}\right) z}{\left(z - e^{-T}\right)\left(z - e^{-2T}\right)}$

Substituting $R(z) = 1$, $K = 1$ and $G(z)$ from above we obtain

$$C(z) = \frac{0.245z}{z^2 - 0.528z + 0.0183} = \frac{0.245z}{(z - 0.491)(z - 0.037)}$$

The partial fraction expansion yields,

$$C(z) = \frac{0.54z}{z - 0.491} - \frac{0.54z}{z - 0.037}$$

which yields the output as

$$c(nT) = 0.54(e^{-0.7nT} - e^{-3.3nT})$$

where it is recalled that $T = 1$ sec.

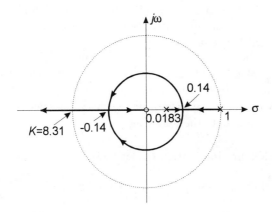

Figure 10.35: Root locus for $\overline{G_0G}(z) = \dfrac{K\left(1 - e^{-4T}\right)z}{4\left(z - 1\right)\left(z - e^{-4T}\right)}$

From the open loop transfer function we note that the root locus loci start at $z = 1$ and $z = 0.0183$. The loci end at $z = 0$ and $z = -\infty$. A plot of the root locus is shown in Fig. 10-35. The break away point can be calculated from

$$\frac{1}{1 - \alpha} + \frac{1}{\alpha} + \frac{1}{\alpha - 0.0183} = 0$$

and is around $\alpha = 0.14$.

For determining the value of K for neutral stability we write the characteristic equation,

$$z^2 + (0.245K - 1.0183)z + 0.0183 = 0$$

Substituting $z = (1 + r)/(1 - r)$ and simplifying yields

$$(2.0366 - 0.245K)r^2 + 1.964r + 0.245K = 0$$

For stability the first term must be positive which requires $K \le 8.31$.

10.10   Digital Compensators

With the widespread use of micro-computers and micro-controllers in discrete time control systems **digital compensators** are extensively used, because they are usually cheaper, more reliable and more accurate than their continuous time counterparts. To this end digital equivalents of analog lag, lead, and lag/lead, compensators are often used. Digital implementations of proportional, integral, and derivative (PID) compensators are more easily implemented than their analog realizations. These implementations which can accurately model ideal PID compensators can easily include algorithms to prevent the controller output from exceeding a pre-defined limit and at the same time prevent integrator wind-up.

While there are many ways of obtaining these equivalent digital transfer functions we examine only the following:

(a) Forward difference integration (Euler method)

(b) Backward difference integration

(c) Trapezoidal rule (Tustin or bi-linear)

(d) Zero-order-hold (ZOH) equivalence.

To illustrate these various methods consider a system represented by an integrator where

$$y(t) = \int_0^t x(t)dt \qquad (10.49)$$

Hence

$$y(kT) = y(kT - T) + \int_{(k-1)T}^{kT} x(t)dt \qquad (10.50)$$

Forward Difference

Examining Fig. 10-36(a) it can be seen the integral term in Eq. (10-50) can be approximated by the area $x(kT - T)T$ of the rectangle $abcd$, so that

$$y(kT) \approx y(kT - T) + x(kT - T)T$$

Taking the $\mathcal{Z}$-transform of this expression gives

$$\frac{Y(z)}{X(z)} = G_F(z) = \frac{Tz^{-1}}{1 - z^{-1}} = \frac{T}{z - 1} \qquad (10.51)$$

Thus the integral operation given in Eq. (10-49) which is represented in the Laplace domain by $G(s) = 1/s$ can be represented in the $\mathcal{Z}$-domain by the **forward difference transfer function** given in Eq. (10-51).

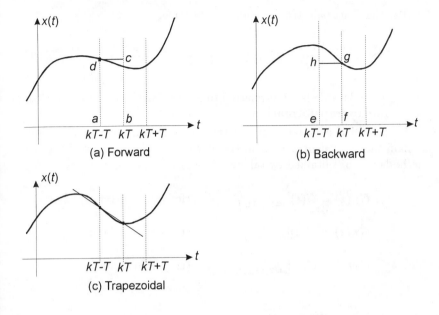

Figure 10.36: Integral equivalence between continous time and discrete time integrators

Backward Difference

The integral term in Eq. (10-50) can also be approximated by the area of the rectangle $efgh$ in Fig. 10-36(b), so that

$$y(kT) \approx y(kT - T) + x(kT)T$$

In this case the integral operation given in Eq. (10-50) has the **backward difference transfer function**

$$\frac{Y(z)}{X(z)} = G_B(z) = \frac{T}{1 - z^{-1}} = \frac{zT}{z - 1} \qquad (10.52)$$

Trapezoidal Rule

A further approximation to the integral in the Eq. (10-49) can be obtained by using a triangular approximation for the function $x(t)$ in the interval $(k - 1)T \leq t \leq kT$, as shown in Fig. 10-36(c). In this case

$$y(kT) \approx y(kT - T) + \frac{T}{2}[x(kT) + x(kT - T)]$$

so that the $\mathcal{Z}$-domain transfer function becomes

$$\frac{Y(z)}{X(z)} = G_T(z) = \frac{T}{2}\frac{1+z^{-1}}{1-z^{-1}} = \frac{T}{2}\frac{z+1}{z-1} \qquad (10.53)$$

The trapezoidal rule is often referred to as the **Tustin approximation** or the **bi-linear transformation**.

In general if a system has the transfer function $G(s)$ in the Laplace domain the equivalent $\mathcal{Z}$-domain transfer function using each of the above methods can be obtained by taking

$$G_F(z) = G(s)|_{s=(z-1)/T} \qquad \text{(forward difference)}$$

$$G_B(z) = G(s)|_{s=(z-1)/zT} \qquad \text{(backward difference)}$$

$$G_T(z) = G(s)|_{s=2(z-1)/(z+1)T} \quad \text{(trapezoidal rule)}$$

Frequency Pre-Warping

The trapezoidal rule has the advantage that the transformed system $G_T(z)$ in the $\mathcal{Z}$-domain is stable if and only if the transfer function $G(s)$ in the s-domain is stable. Neither of the other transformations guarantee this situation. However the trapezoidal rule leads to considerable **frequency warping** because the complete $j\omega$-axis in the s-plane is mapped into the boundary of the unit circle in the z-plane. The trapezoidal rule will now be modified to correct this effect over a limited frequency range. To see this let us examine the effect of frequency warping on the first order transfer function

$$G(s) = \frac{a}{s+a} \qquad (10.54)$$

From above we see that the equivalent digital transfer function is

$$G_T(z) = \frac{a}{\dfrac{2}{T}\left[\dfrac{z-1}{z+1}\right] + a} \qquad (10.55)$$

It will be noted the corner frequency for the transfer function given in Eq. (10-54) is $\omega_c = a$. Suppose we apply the data sequence $\{e^{j\omega nT}\}$ to the transfer function $G_T(z)$. From Eq. (10-16) this is equivalent to taking

$z = e^{j\omega T}$ in Eq. (10-55) so that

$$G_T^*(j\omega) = G_T\left(e^{j\omega T}\right) = \cfrac{a}{\cfrac{2}{T}\left[\cfrac{e^{j\omega T}-1}{e^{j\omega T}+1}\right]+a}$$

$$= \cfrac{a}{j\cfrac{2}{T}\tan\left(\cfrac{\omega T}{2}\right)+a}$$

The corner frequency for this transfer function is

$$\omega_c = \frac{2}{T}\arctan\left(\frac{aT}{2}\right)$$

which differs from $\omega_c = a$ for $G(s)$ if $aT/2 \gg 1$. Suppose we wish $G_T^*(j\omega)$ and $G(j\omega)$ to have the same corner frequency then we must modify the trapezoidal transformation rule to

$$s = \frac{a}{\tan\dfrac{aT}{2}} \cdot \frac{z-1}{z+1}$$

so that the **pre-warped trapezoidal transformation** is

$$G_P(z) = G(s)|_{s=a(z-1)/(z+1)\tan(aT/2)} \tag{10.56}$$

where a is the frequency at which the frequency warping is to be minimized. For Eq. (10-54) we find

$$G_P^*(j\omega) = a/\left(\frac{a}{\tan\left(aT/2\right)}\left(\frac{z-1}{z+1}\right)+a\right)\Bigg|_{z=e^{j\omega T}}$$

$$= \cfrac{\tan\left(\cfrac{aT}{2}\right)}{j\tan\left(\cfrac{\omega T}{2}\right)+\tan\left(\cfrac{aT}{2}\right)}$$

Thus the corner frequency now occurs when $\tan\left(\omega_c T/2\right) = \tan\left(aT/2\right)$, that is when $\omega_c = a$.

Zero-Order-Hold Equivalence

Consider the continuous time compensation $G_C(s)$ shown in Fig. 10-37(a). Here we propose to obtain an equivalent digital transfer function by using

(a)

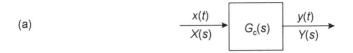

(b)

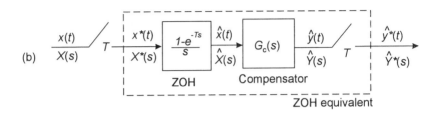

Figure 10.37: Zero-order equivalent digital transfer function

a zero-order hold in conjunction with $G_C(s)$ so that the response of the digital compensator approximates that of the analog compensator $G_C(s)$. The arrangement we will consider is shown in Fig. 10-37(b). Here we suppose the input signal $x(t)$ is assumed to be sampled and then fed to a zero-order hold before being applied to the compensator $G_C(s)$. To obtain the digital equivalent of this arrangement we assume $\hat{y}(t)$ is sampled as shown in the figure. The **zero-order-hold equivalent digital transfer function** is taken as the elements shown in the box surrounded by the broken line. It will be observed that if the sample period T is extremely short then to a good approximation $\hat{x}(t) \approx x(t)$ so that $\hat{y}(t) \approx y(t)$. In this case the response of the ZOH equivalent transfer function is a good approximation to the continuous time compensator.

From Fig. 10-37(b) we see

$$\hat{X}(s) = \frac{1 - e^{-Ts}}{s} X^*(s)$$

so that

$$\hat{Y}(s) = \left[\frac{1 - e^{-Ts}}{s} G_C(s) \right] X^*(s) \qquad (10.57)$$

Defining $G_{ZC}(s) = \left[(1 - e^{-Ts}) G_C(s) \right] / s$ and taking the $\mathcal{Z}$-transform of Eq. (10-57) gives

$$\hat{Y}(z) = G_{ZC}(z) X(z)$$

The $\mathcal{Z}$-transform of $G_{ZC}(s)$ is seen to be

$$G_{ZC}(z) = \mathcal{Z}\left[\frac{(1 - e^{-Ts})G_C(s)}{s}\right]$$

Thus the **zero order hold equivalent digital transfer function** is

$$G_{ZC}(z) = (1 - z^{-1})\mathcal{Z}\left[\frac{G_C(s)}{s}\right] \qquad (10.58)$$

Example 14 *Consider the lag compensator*

$$G_C(s) = \frac{1 + \alpha T_1 s}{1 + T_1 s}$$

where $\alpha < 1$. Find the ZOH equivalent transfer function.

From Eq. (10-58) we find

$$G_{ZC}(z) = (1 - z^{-1})\mathcal{Z}\left[\frac{1 + \alpha T_1 s}{s(1 + T_1 s)}\right]$$

$$= \frac{z - 1}{z}\mathcal{Z}\left[\frac{\frac{1}{T_1}}{s\left(s + \frac{1}{T_1}\right)} + \frac{\alpha}{s + \frac{1}{T_1}}\right]$$

From the table of $\mathcal{Z}$-transforms we have

$$G_{ZC}(z) = \frac{z - 1}{z}\left[\frac{z\left(1 - e^{-T/T_1}\right)}{(z - 1)\left(z - e^{-T/T_1}\right)} + \frac{\alpha z}{z - e^{-T/T_1}}\right]$$

This simplifies to

$$G_{ZC}(z) = \alpha\,\frac{z - 1 + \left(1 - e^{-T/T_1}\right)/\alpha}{z - e^{-T/T_1}}$$

PID Controllers

The **proportional, integral, and derivative (PID) controller** is widely used, especially in process industries and in aerospace systems. We examine the digital implementation of these controllers.

The transfer function of the analog PID controller is given by

$$u(t) = K_P\left[e(t) + C_D\frac{de(t)}{dt} + C_I\int_0^t e(t)dt\right] \qquad (10.59)$$

In this equation we can use finite differences to approximately evaluate the differential and integral terms. Thus if we define $D(kT)$ as below, we have

$$D(kT) = \frac{de(kT)}{dt} \approx \frac{e(kT) - e(kT - T)}{T} \tag{10.60}$$

and defining $I(kT)$ as

$$I(kT) = \int_0^{kT} e(t)dt$$

gives

$$I(kT) = I(kT - T) + \int_{(K-1)T}^{KT} e(t)dt \tag{10.61}$$

$$\approx I(kT - T) + e(kT)T$$

Thus

$$u(kT) = K_P\left[e(kT) + C_D D(kT) + C_I I(kT)\right] \tag{10.62}$$

and if we take the $\mathcal{Z}$-transform of Eqs. (10-59) to (10-61) we obtain

$$U(z) = \left\{K_P\left[1 + C_D\frac{z-1}{Tz} + C_I\frac{Tz}{z-1}\right]\right\}E(z)$$

The term in braces which we denote by

$$G_C(z) = K_P\left[1 + C_D\frac{z-1}{Tz} + C_I\frac{Tz}{z-1}\right]$$

is the **PID controller digital transfer function.**

10.11   Summary

In this chapter we have been concerned with control systems receiving data at intermittent intervals as might occur in a many modern digital control systems. A sampler was employed to convert continuous signals to intermittent data. It was assumed that the sampler was periodic, i.e. data was received at regular intervals. Also the width of the periodic pulse was neglected. We observed that sampling introduces higher frequency components in the sampled signal and before applying the signals to a continuous element these undesirable components must be removed. This was why it was necessary to employ data reconstruction devices. The simplest one was the zero-order-hold.

The existence of sampling also necessitated that we shift our analysis from the s-plane to the z-plane. Given a transfer function in the z-domain, we showed how the transient response may be obtained using partial fractions, division, and finite difference methods. For better understanding of the steady state performance, we defined error coefficients and the steady state error in a manner analogous to continuous systems.

The stability of sampled systems was investigated using Jury's test and the Routh-Hurwitz criterion. The former could be applied on the z-plane, whereas the latter required a bi-linear transformation to the r-plane.

Next, we modified our graphical techniques for application to sampled systems. The Bode and Nyquist plots required a bi-linear transformation, whereas the root locus was constructed on the z-plane. The example considered indicated how the stability of a sampled data control systems was a function of the sampling period.

Finally we examined how the digital equivalents of common compensators may be obtained using a number of different finite difference approximations.

10.12 References

1. E. G. Gilbert, *Notes on Sampled-Data Systems,* Department of Aero-Space Engineering, University of Michigan, Ann Arbor, 1965.

2. B. C. Kuo, *Digital Control Systems,* Holt, Rinehart and Winston, New York, 1980.

3. B. C. Kuo, *Analysis and Synthesis of Sampled-Data Control Systems,* Prentice-Hall Inc., N. J., 1963.

4. J. R. Ragazzini, G. F. Franklin, *Sampled-Data Control Systems,* Mc Graw Hill Book Co., New York, 1958.

5. J. T. Tou, *Digital and Sampled-Data Control Systems,* Mc Graw Hill Book Co., New York, 1959.

6. E. I. Jury, *Theory and Applications of Z-Transform Method,* John Wiley and Sons, Inc., New York, 1964.

7. H. Freeman, *Discrete-Time Systems,* John Wiley and Sons, Inc., New York, 1964.

8. R. Iserman, *Digital Control Systems,* Springer-Verlag, Berlin, 1981.

9. K. J. Astrom, B. Wittenmark, *Computer-Controlled Systems — Theory and Design,* Prentice-Hall International, Inc., N. J., 1990.

10. K. Ogata, *Discrete-Time Control Systems,* Prentice-Hall International, Inc., N. J., 1987.

11. G. F. Franklin, J. D. Powell, M. L. Workman, *Digital Control of Dynamic Systems,* Addison Wesley Publishing Co., Reading, Massachusetts, 1990.

12. R. J. Mayhan, *Discrete-Time and Continuous-Time Linear Systems,* Addison Wesley Publishing Co., Reading, Massachusetts, 1984.

13. C. H. Houpis, G. B. Lamont, *Digital Control Systems — Theory, Hardware, Software,* Mc Graw Hill, Inc., New York, 1992.

14. C. L. Phillips, H. T. Nagle, *Digital Control System Analysis and Design,* Prentice-Hall, Inc., N. J., 1984.

10.13   Problems

10-1 Applying Shannon's sampling theorem, determine appropriate values of the sampling period T for the signals: (a) $e(t) = \sin 2\pi t$, (b) $e(t) = e^{-5t}\sin 10\pi t$.

10-2 Consider the Laplace transform of the signals:

(a) $E(s) = \dfrac{100}{s(s^2 + 25s + 100)}$

(b) $E(s) = \dfrac{100(s + 10)}{s(s^2 + 25s + 100)}$

Applying Shannon's sampling theorem, determine suitable values of the sampling period T.

10-3 Suppose the spectrum of a signal $e(t)$, given in Fig. P10-3(a) is sampled by an ideal sampler as shown in Fig. P10-3(b).

(a) Find the spectrum of the sampled signal, and plot it graphically.

(b) Suppose the signal $e^*(t)$ is passed through a low pass filter $G_1(j\omega)$ whose frequency response function is shown in Fig. P10-3(c). Determine the time function $\hat{e}(t)$ at the output of the filter.

(c) Suppose the signal $e^*(t)$ is passed through a band pass filter $G_2(j\omega)$ whose frequency response function is shown in Fig. P10-3(d). Determine the time function $\tilde{e}(t)$ at the output of the filter.

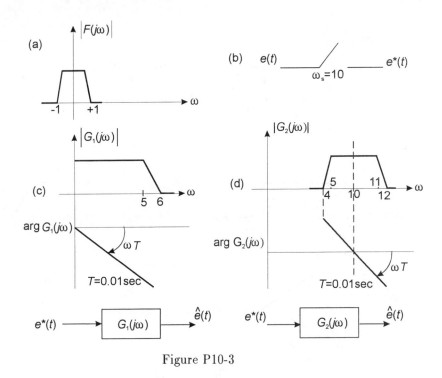

Figure P10-3

10-4 Suppose the spectrum of a signal $f(t)$ is as given in Fig. P10-4(a).

(a) Determine the spectrum of the signal $f^*(t)$ at the output of the ideal sampler shown in Fig. P10-4(b), assuming $\omega_s = 24$ rad/sec.

(b) Suppose the signal is passed through the low pass filter $G_1(j\omega)$ whose frequency response function is shown in Fig. P10-4(c). Determine the time function $\hat{f}(t)$ at the output of the filter.

(c) Taking $\omega_s = 4$ rad/sec repeat (a) above, and suppose the signal $f^*(t)$ is passed through the band pass filter shown in Fig. P10-4(d). Determine the time function $\tilde{f}(t)$ at the output of the filter.

(d) Discuss your results and determine the minimum sampling frequency ω_s that can be used, so as to fully recover the original signal $f(t)$.

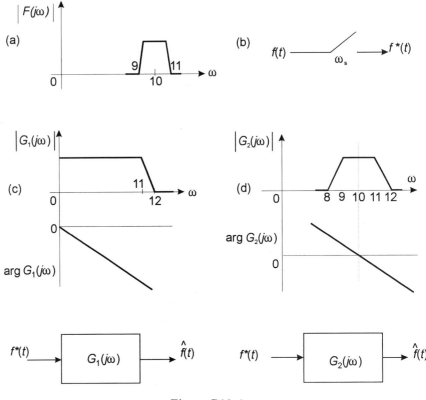

Figure P10-4

10-5 Derive the $\mathcal{Z}$-transform of the following time functions: (a) e^{-at} (b) $\cos \omega t$ (c) At (d) At^2.

10-6 Find the $\mathcal{Z}$-transform of (a) $e^{-at} \cos bt$ (b) te^{-at}.

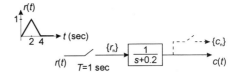

Figure P10-7

10-7 A pulse sequence $\{r_n\}$ obtained from an ideal sampler is applied to a low pass filter as shown in Fig. P10-7: (a) Find the $\mathcal{Z}$-transform of $\{r_n\}$ and $\{c_n\}$ (b) Find the pulse train $\{c_n\}$ for $n = 0, 1, \cdots, 8$. (c) Find c_0 and c_∞ using the initial and final value theorems.

10-8 Find the $\mathcal{Z}$-transforms of:

(a) $G(s) = \dfrac{1}{s(s+1)^2}$

(b) $G(s) = \dfrac{1}{s^3(s+2)(s+10)}$

using partial fractions and the residue formula.

10-9 Find the inverse $\mathcal{Z}$-transform of each $F(z)$ using the methods presented above. For each case compute the first four terms in the data sequence and compare their values.

(a) $F(z) = \dfrac{0.5z}{(z-1)(z-0.3)}$

(b) $F(z) = \dfrac{3+z^{-2}}{(1-0.4z^{-1})^2(1-z^{-1})}$

(c) $F(z) = \dfrac{z(z-0.8)}{(z-1)(z-0.3)}$

(d) $F(z) = \dfrac{z^3+5z+1}{z(z-1)(z-0.2)}$

10-10 Find the control ratio $C(z)/R(z)$ for the digital control systems shown in Fig. P10-10, assuming that all samplers are synchronized.

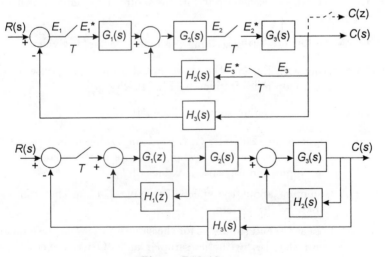

Figure P10-10

10-11 Obtain $C(s)$ by generating the composite block diagram for the control system shown in Fig. P10-11.

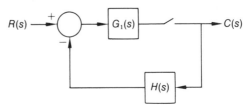

Figure P10-11

10-12 Obtain the modified block diagram and then the pulsed output for Fig. P10-12. Was it necessary to generate the modified block diagram? How is this control system fundamentally different from all the others discussed so far?

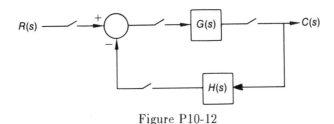

Figure P10-12

10-13 Obtain the pulsed output for the sampled systems shown in Fig. P10-13.

10-14 Verify Eq. (10-13) describing the first-order hold device.

10-15 Starting with Eq. (10-41), show that the output response for different closed loop poles is given by Fig. 10-25.

10-16 Consider the system shown in Fig. P10-16.

   (a) Suppose the controller $G_c(z) = 1$ and T has values in the range 0.5 to 4 sec. Plot the system response of the system for a unit step input and for various values of T.

   (b) Comment about the effect of varying T upon the transient response of the system.

   (c) Repeat (a) and (b) above for the same system, but with an analog controller, i.e. with the sampler and ZOH removed, and with $G_c(z)$ replaced by $G_c(s) = 1$.

(d) Compare the transient responses with those obtained in (a) and (b) above.

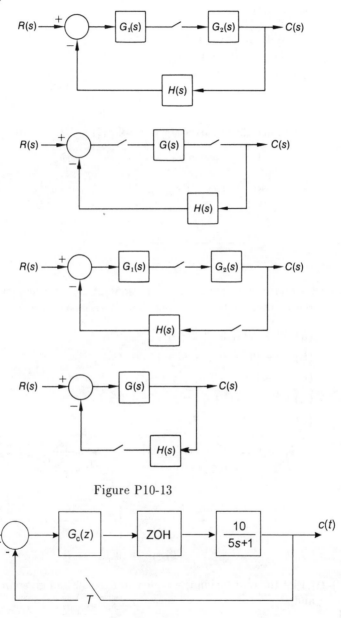

Figure P10-13

Figure P10-16

10-17 A regulator control system is shown in Fig. P10-17, in which the output $c(t)$ is to be maintained constant in the presence of a disturbance $d(t)$.

(a) With $G_c(z) = 1$, find the steady state value of $c(t)$ if the disturbance $d(t)$ is a unit step function.

(b) Let
$$G_c(z) = 1 + \frac{0.1z}{z - 1}$$

Determine the steady state value of $c(t)$ if $d(t)$ is a unit step function.

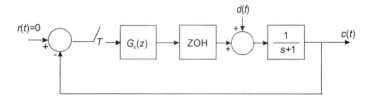

Figure P10-17

10-18 Determine the stability conditions of the digital systems that are represented by the following characteristic equations:

(a) $z^3 + 5z^2 + 3z + 2 = 0$

(b) $z^4 + 9z^3 + 3z^2 + 9z + 1 = 0$

(c) $z^3 - 1.5z^2 - 2z + 3 = 0$

(d) $z^4 - 1.55z^3 + 0.5z^2 - 0.5z + 1 = 0$

(e) $z^4 - 2z^3 + z^2 - 2z + 1 = 0$

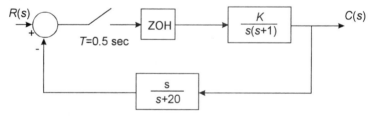

Figure P10-19

10-19 Plot the root loci in the z-plane for the system given in Fig. P10-19, and

(a) Determine the gain K for neutral stability.

(b) Find the gain K for $\zeta = 0.6$ and the undamped natural frequency of the dominant poles corresponding to this gain.

10-20 A closed loop sampled data system is given in Fig. P10-20 where G_0 is a zero-order-hold. It is desired to have a response such that the envelope of the output pulses is an exponentially decaying oscillation. What constraint does this impose on K? Assume that

$$\overline{G_0G_1}(z) = \frac{K(z + 0.71)}{(z - 1)(z - 0.37)}, \quad T = 1\,\text{sec}$$

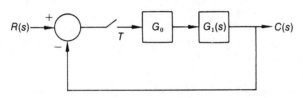

Figure P10-20

10-21 Obtain the output for the system shown in Fig. P10-21.

$$\overline{G_0G_1}(z) = \frac{Kz(z^2 - 0.736z + 0.276)}{(z - 1)(z - 0.368)(z - 0.146)}$$

Assume $r(t) = $ unit ramp.

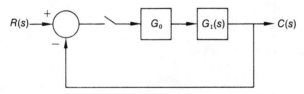

Figure P10-21

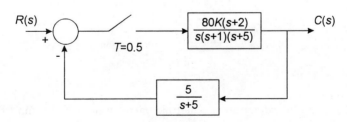

Figure P10-22

10-22 Plot the Nyquist diagram for the system given in Fig. P10-22, and

- **(a)** Determine the values of the gain K for which it is stable
- **(b)** Suppose it is desired that $M_m = 2$ dB. Determine the value of the gain K, the resonant frequency ω_m, and the velocity error constant K_v.

10-23 Obtain the Nyquist plots for the following pulse transfer function.

$$G(z) = \frac{z^2(1 - e^{-T})}{(z - 1)^2(z - e^{-T})}$$

Assume that $T = 1$ sec and then $T = 5$ sec. Does T affect the stability of the system?

10-24 Repeat the above problem for

$$\overline{G_0 G_1}\,(z) = \frac{K\,(z + 0.71)}{(z - 1)\,(z - 0.37)}$$

10-25 A control system is given in Fig. P10-25. What is the steady state response if $T = 1$ sec. Replace the ZOH by a low pass RC filter and let the forward loop transfer function be

$$G_1 = \frac{1}{s(s + 5)}$$

Assume a unit step input.

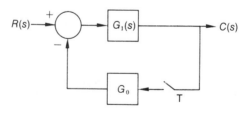

Figure P10-25

10-26 The block diagram of a digital control system is shown in Fig. P10-26. Find the range of the sampling period T for the system to be asymptotically stable.

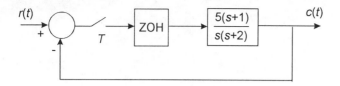

Figure P10-26

10-27 Design a digital compensator that can be substituted for the continuous compensator

$$G_c(s) = \frac{10(s+10)}{(s+2)(s^2+5s+2)}$$

10-28 Obtain the approximate digital transfer function for the continuous time lead network whose transfer function is

$$G_c(s) = \frac{s+1}{0.1s+1}$$

The following methods should be used: (i) forward difference integration, (ii) backward difference integration, (iii) trapezoidal rule with and without pre-warping, and (iv) zero-order-hold equivalence.

(a) For each case compute the phase lead given by the approximate transfer function at $z_1 = e^{j\omega_1 T}$ if $\omega_1 = 3$ rad/sec and $T = 0.25$ sec, and compare with the phase lead of the transfer function $G_c(j\omega)$ at the frequency $\omega = \omega_1$.

(b) Plot the Bode amplitude and phase plots of the approximate transfer functions over the frequency range 0.1 to 100 rad/sec.

Chapter 11

Non Linear Control Systems

11.1   Introduction

The analysis of linear control systems has been our primary goal so far. We assumed that the springs, motors, dampers, etc., all possessed linear characteristics in the range of operation of the control system as shown in Fig. 11-1. These systems were represented by linear differential equations with constant coefficients. The transient response, steady state response and stability were readily obtained by classical as well as state space techniques.

In practice all control systems have some degree of nonlinearity. This may be caused by hysteresis, backlash, saturation, etc. Sometimes nonlinear devices such as relays and deadband circuits may be intentionally introduced in order to obtain some desired feature. In cases where these nonlinearities affect the system performance in a significant way, they must be included in the system analysis.

When nonlinearities have to be included in the system analysis, then the principle of superposition is invalid and the methods developed so far may not be applicable. We must therefore develop different ways of investigating such systems. It must be stressed, however, that no general approach for analyzing nonlinear systems has been developed. Methods exist only for specific types of nonlinearities.

When the nonlinearities are large, generally numerical techniques are employed or the analog computer with nonlinear function generation is used. When the nonlinearities are small, methods are employed to linearize

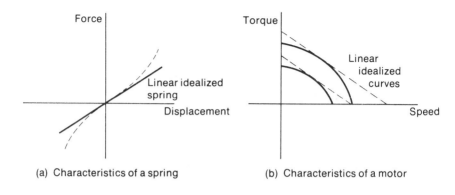

(a) Characteristics of a spring   (b) Characteristics of a motor

Figure 11.1: Examples of nonlinearities

the equations as a first step. These linearized equations are then solved to obtain the approximate behavior. The exact behavior of the system is considered as a deviation from this result.

The type of nonlinearities we shall consider in this chapter are mainly small nonlinearities. They will be analyzed by one of the following methods:

(1) Phase-plane and state space techniques

(2) Describing function techniques

The first method is mostly useful for second-order systems and involves the investigation of the output variable as a function of its derivative, i.e. the familiar state space variables. The second method is attractive since the classical techniques developed for linear control systems are used after the replacement of a non-linear element by its quasi-linear equivalent.

11.2   The Phase Plane-Method

The phase-plane method is useful for studying the transient behavior as well as the stability of second-order systems by applying the state space approach. In this method we take the system output displacement and velocity as its state variables so that the **state space** (normally referred to as the **phase plane**) is two-dimensional with the output along the abscissa and the output velocity along the ordinate. When the output displacement is plotted as a function of its velocity the resulting curve is called a **phase plane trajectory**. A family of trajectories depicting the behavior of the system, for different initial conditions, is called a **phase portrait**.

Consider the motion of a mass and spring

$$m\ddot{x} + kx = 0$$

which may be rewritten as

$$m\ddot{x}\dot{x} + kx\dot{x} = 0$$

or integrating with respect to time gives

$$\int_0^t m\dot{x}d\dot{x} + \int_0^t kxdx = 0$$

From this we obtain

$$\frac{m\dot{x}^2}{2} + \frac{kx^2}{2} = C \tag{11.1}$$

If we define $\dot{x} = y$, $a^2 = 2/m$, and $b^2 = 2/k$, then

$$\frac{y^2}{a^2} + \frac{x^2}{b^2} = C$$

where x, y are the state variables of the system. This equation simply states that the total energy, kinetic plus potential, of a system is conserved for an undamped system. For $k > 0$, Eq. (11-1) may be graphically interpreted as a family of ellipses as shown in Fig. 11-2(a). If $k < 0$, then Eq. (11-1) yields a family of hyperbolas as shown in Fig. 11-2(b). We notice that the point $(0,0)$ is either approached or circumscribed periodically. This point is called a **singular**, a **critical**, or an **equilibrium** point. If the motion is stable and purely periodic, the singular point is called a **vortex** or **center**. The singular point of Fig. 11-2(a) is a vortex. The critical point shown in the unstable motion of Fig. 11-2(b) is called a **saddle** point.

Linear Second Order Systems

A general second order system can be represented by the state equation

$$\begin{bmatrix} \dot{x}(t) \\ \dot{y}(t) \end{bmatrix} = \begin{bmatrix} a & b \\ c & d \end{bmatrix} \begin{bmatrix} x(t) \\ y(t) \end{bmatrix}, \quad \begin{bmatrix} x(0) \\ y(0) \end{bmatrix} = \begin{bmatrix} x_0 \\ y_0 \end{bmatrix} \tag{11.2}$$

or using matrix notation

$$\dot{\mathbf{x}}(t) = \mathbf{A}\mathbf{x}(t), \quad \mathbf{x}(0) = \mathbf{x}_0 \tag{11.3}$$

where

$$\mathbf{A} = \begin{bmatrix} a & b \\ c & d \end{bmatrix} \quad \text{and } \mathbf{x}(t) = \begin{bmatrix} x(t) \\ y(t) \end{bmatrix} \tag{11.4}$$

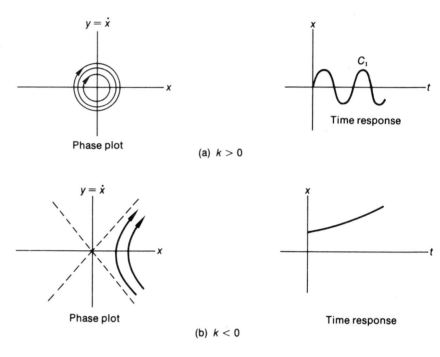

Figure 11.2: Phase plane plot for $m\ddot{x} + kx = 0$

From the discussion in Chapter 5, the general solution of this equation is $\mathbf{x}(t) = e^{\mathbf{A}t}\mathbf{x}_0$. To examine the solution in more detail we find the eigenvalues of $\mathbf{A}$ from the characteristic equation

$$\det(s\mathbf{I} - \mathbf{A}) = s^2 - \tau s + \delta = 0 \tag{11.5}$$

where trace $\mathbf{A} = \tau = a + d$ and $\det \mathbf{A} = \delta = ad - bc$. Thus the eigenvalues are

$$s_1, s_2 = \frac{\tau}{2} \pm \frac{1}{2}\sqrt{\tau^2 - 4\delta} \tag{11.6}$$

The eigenvalues may be classified into three distinct types.

Type 1 $|\tau| > 2\delta^{1/2}$. In this case the eigenvalues are real and distinct, and for convenience we will assume $s_1 > s_2$. Suppose we introduce the similarity transformation $\mathbf{T}$

$$\mathbf{T} = \begin{bmatrix} c & s_1 - a \\ c & s_2 - a \end{bmatrix}, \quad \text{with } \mathbf{T}^{-1} = \frac{1}{c(s_2 - s_1)} \begin{bmatrix} s_2 - a & -s_1 + a \\ -c & c \end{bmatrix} \tag{11.7}$$

where $\mathbf{z}(t) = \mathbf{T}\mathbf{x}(t)$. It is easy to see that

$$\dot{\mathbf{z}}(t) = \begin{bmatrix} s_1 & 0 \\ 0 & s_2 \end{bmatrix} \mathbf{z}(t) \qquad (11.8)$$

Solving Eq. (11-8) gives

$$\mathbf{z}(t) = \begin{bmatrix} e^{s_1 t} z_1(0) \\ e^{s_2 t} z_2(0) \end{bmatrix}$$

so that

$$\mathbf{x}(t) = \frac{1}{c(s_2 - s_1)} z_1(0) e^{s_1 t} \mathbf{v}_1 + \frac{1}{c(s_2 - s_1)} z_2(0) e^{s_2 t} \mathbf{v}_2 \qquad (11.9)$$

where $\mathbf{v}_1 = \begin{bmatrix} (s_2 - a) & -c \end{bmatrix}^T$ and $\mathbf{v}_2 = \begin{bmatrix} (-s_1 + a) & c \end{bmatrix}^T$ are the eigenvectors of $\mathbf{A}$.

If $s_1 > s_2 > 0$ then the phase-plane portrait is termed an **unstable node** and is as shown in Fig. 11-3(a), while if $s_2 < s_1 < 0$ then the resulting portrait will be a **stable node** as shown in Fig. 11-3(b). If however $s_2 < 0 < s_1$ then the resulting phase plane portrait is termed a **saddle point** as shown in Fig. 11-3(c).

Type 2 $|\tau| < 2\delta^{1/2}$. In this case the eigenvalues are complex conjugates which we will write as $s_1, s_2 = \mu \pm j\nu$. Taking the similarity transformation $\mathbf{T}$

$$\mathbf{T} = \begin{bmatrix} c & \mu - a \\ 0 & \nu \end{bmatrix}, \quad \text{with } \mathbf{T}^{-1} = \frac{1}{c\nu} \begin{bmatrix} \nu & a - \mu \\ 0 & c \end{bmatrix} \qquad (11.10)$$

where $\mathbf{z}(t) = \mathbf{T}\,\mathbf{x}(t)$. In this case

$$\dot{\mathbf{z}}(t) = \begin{bmatrix} \mu & -\nu \\ \nu & \mu \end{bmatrix} \mathbf{z}(t) \qquad (11.11)$$

and its solution is

$$\mathbf{z}(t) = R \begin{bmatrix} e^{\mu t} \cos(\nu t + \phi) \\ e^{\mu t} \sin(\nu t + \phi) \end{bmatrix}$$

Thus

$$\mathbf{x}(t) = \frac{R}{c\nu} e^{\mu t} \cos(\nu t + \phi) \mathbf{v}_1 + \frac{R}{c\nu} e^{\mu t} \sin(\nu t + \phi) \mathbf{v}_2 \qquad (11.12)$$

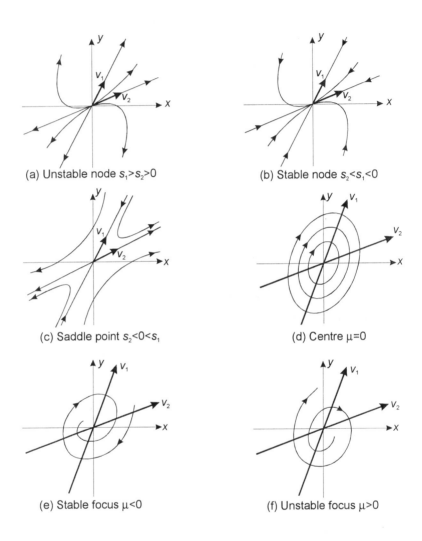

Figure 11.3: Phase plane trajectories for linear systems

where $\mathbf{v}_1 = [\begin{array}{cc} \nu & 0 \end{array}]^T$ and $\mathbf{v}_2 = [\begin{array}{cc} a - \mu & c \end{array}]^T$, and R and ϕ are determined by the initial condition $\mathbf{x}_0$.

If $\mu = 0$ then the phase plane portrait is termed a **center** and is shown in Fig. 11-3(d). However if $\mu < 0$ then the portrait is termed a **stable focus** while if $\mu > 0$ it is termed an **unstable focus**. Both these portraits are illustrated in Figs. 11-3(e) and 11-3(f).

Type 3 $|\tau| = 2\delta^{1/2}$. In this case there is only one eigenvalue $s = \tau/2$. If $b \neq 0$ and $c \neq 0$ we can take the transformation

$$\mathbf{T} = \left[\begin{array}{cc} a - d & 2b \\ 2c & 0 \end{array} \right] \tag{11.13}$$

and it can be shown that the solution gives rise to a phase plane portrait which is termed a **degenerate node**. Other possibilities are that $b = 0$ or $c = 0$, or both are zero. In these cases the portraits exhibit other forms of degenerate trajectories. However these cases are of little engineering interest and will not be considered further.

Having introduced the concept of a phase plot we shall relate the phase plane method to control system analysis. Consider the overall transfer function of a second-order control system

$$\frac{C(s)}{R(s)} = \frac{K}{As^2 + Bs + K} \tag{11.14}$$

This corresponds to the differential equation

$$A\frac{d^2c(t)}{dt^2} + B\frac{dc}{dt} + Kc(t) = Kr(t) \tag{11.15}$$

If we define $dx/dt = y$ and $c = x$ then Eq. (11-15) becomes

$$\dot{x} = y$$
$$\dot{y} = -\frac{K}{A}x - \frac{B}{A}y + \frac{K}{A}r(t) \tag{11.16}$$

We have seen above how these state equations and their phase plane plots may be constructed.

The characteristic equation for the control system can be seen to be

$$s^2 + \frac{B}{A}s + \frac{K}{A} = 0 \tag{11.17}$$

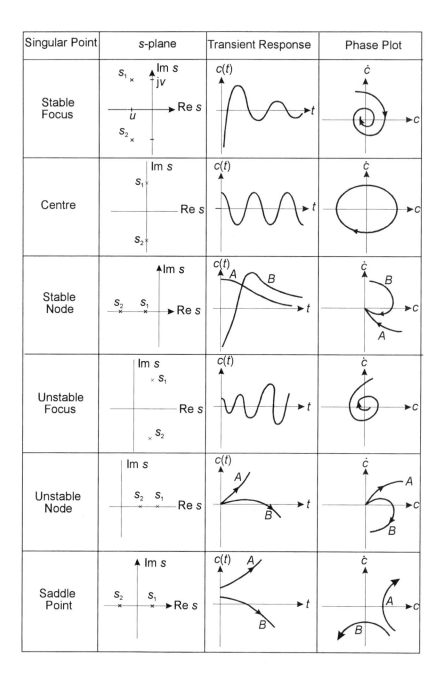

Figure 11.4: Phase trajectories for second-order linear system

The location of the system characteristic roots or eigenvalues on the s-plane can thus be related to the phase plane, as shown in Fig. 11-4. It can be observed from Eqs. (11-16) that

$$\frac{dy}{dx} = \frac{-\frac{K}{A}x - \frac{B}{A}y + \frac{K}{A}r}{y} \tag{11.18}$$

The phase plane plot can thus be obtained by directly integrating Eq. (11-18), which can be solved in terms of elementary functions.

Non Linear Second Order Systems

For the linear systems discussed above, if the point $(x, y) = (0, 0)$ on the phase plane is taken as the initial condition for Eq. (11-2) then the solution is $(x(t), y(t)) = (0, 0)$ for $t \geq 0$. Since the solution remains at the origin for all $t \geq 0$ it is referred to as an **equilibrium** point.

Consider now the non-linear second order state equations

$$\begin{aligned} \dot{x}(t) &= f_1(x(t), y(t)) \\ \dot{y}(t) &= f_2(x(t), y(t)) \end{aligned} \tag{11.19}$$

We say that a point (x_k, y_k) in the phase plane is an **equilibrium** point for this system if

$$f_1(x_k, y_k) = 0, \quad f_2(x_k, y_k) = 0 \tag{11.20}$$

Non-linear systems may have more than one isolated equilibrium point, whereas the linear system given in Eq. 11-3 has only one equilibrium point provided the matrix $\mathbf{A}$ is non-singular. As for linear systems, solutions of non-linear systems starting at (x_k, y_k) remain at this point for all $t > 0$.

Example 1 *Find the equilibrium points for the simple pendulum shown in Fig. 11-5, assuming the pivot has negligible friction.*

The equation of motion for this system can be shown to be given by

$$ml^2\ddot{\theta} + mgl\sin\theta = 0 \tag{11.21}$$

Introducing the state variables $x = \theta, y = \dot{x}$ we have

$$\begin{aligned} \dot{x} &= y = f_1(x, y) \\ \dot{y} &= -\frac{g}{l}\sin x = f_2(x, y) \end{aligned} \tag{11.22}$$

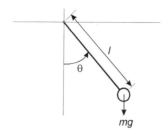

Figure 11.5: Simple pendulum

The equilibrium points are at the roots of $f_1(x, y) = f_2(x, y) = 0$ which are $(0, 0)$, $(\pm\pi, 0)$, $(\pm 2\pi, 0)$, ... Thus the pendulum has an infinite number of equilibrium points corresponding to the pendulum pointing either vertically upwards or downwards.

Example 2 *The differential equation*

$$\ddot{x} - \mu(1 - x^2)\dot{x} + x = 0 \qquad (11.23)$$

has been used by van der Pol to study the operation of electronic oscillators. Find the equilibrium points for this system.

Introducing the state variable $y = \dot{x}$ we have the state equations

$$\begin{aligned}
\dot{x} &= y = f_1(x, y) \\
\dot{y} &= \mu(1 - x^2)y - x = f_2(x, y)
\end{aligned} \qquad (11.24)$$

and from the definition above the equilibrium points are at the roots of $f_1(x, y) = f_2(x, y) = 0$. Solving these equations we find only one solution at the origin $(0, 0)$.

The phase plane portraits of non-linear systems can often be extremely complex compared with those for linear systems. In many instances these non linear systems may be linearized in small neighborhoods of their equilibrium points, and considerable information about their behavior can be determined by examining the phase plane portraits of the linearized systems.

To illustrate this we determine the phase plane portrait for the simple pendulum. The state equations for the pendulum are given in Example 1. As observed in Eq. (11-18) we can write a differential equation for the

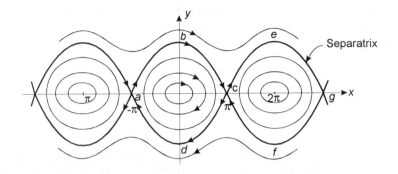

Figure 11.6: Phase plane portrait for simple pendulum

phase plane trajectories as

$$\frac{dy}{dx} = \frac{-\omega^2 \sin x}{y} \tag{11.25}$$

where $\omega^2 = g/l$. This equation can be directly integrated to give

$$\omega^2(\cos x - \cos x_0) = \frac{1}{2}(y^2 - y_0^2) \tag{11.26}$$

where (x_0, y_0) is the initial point for the trajectory. Typical phase plane trajectories are plotted in Fig. 11-6.

It will be observed from this figure that in the neighborhood of the equilibrium points at $(0,0)$ the phase portrait is a **center**, while at $(\pm\pi, 0)$ the portrait is a saddle point. It will also be observed that the phase plane is divided into separate regions by **separatrices**, which are the trajectories $abcdefg\cdots$ etc. In each of these regions defined by the separatrices the trajectories exhibit different qualitative oscillatory properties. It will also be noted that any trajectory starting in any one of these regions remains in it for all time $t > 0$. For this reason they are referred to as **invariant sets**. We will see below that by linearizing Eq. (11-22) about the equilibrium points the character of the phase plane portraits can be identified without the need for evaluating the complete family of phase plane trajectories.

Linearization About Equilibrium Points

Before considering the linearization of non-linear state equations about an equilibrium point we show it is sufficient to consider the equilibrium point to always be at the origin. Suppose (x_k, y_k) is an equilibrium away from the origin, and we define a new set of state variables $u = x - x_k$, $v = y - y_k$. Substituting into Eq. (11-19) gives

$$\dot{u}(t) = f_1(u(t) + x_k, v(t) + y_k) = F_1(u(t), v(t))$$
$$\dot{v}(t) = f_2(u(t) + x_k, v(t) + y_k) = F_2(u(t), v(t))$$

$$(11.27)$$

It will be observed that the equilibrium point at (x_k, y_k) is now translated to the origin in the new state space since $F_1(0,0) = F_2(0,0) = 0$. Consequently we will assume in our future discussions that the equilibrium point being considered has been translated to the origin.

Let us now consider the second order state variable system shown in Eq. (11-19) where it will be assumed f_1 and f_2 have derivatives of all orders in the variables x and y, and vanish at the origin. Developing f_1 and f_2 into power series at this point we have

$$\dot{x}(t) = ax(t) + by(t) + R_1(x, y)$$
$$\dot{y}(t) = cx(t) + dy(t) + R_2(x, y)$$

$$(11.28)$$

where the remainder terms R_1 and R_2 are series having only quadratic and terms of higher powers. Here we can identify the terms $a = f_{1x}(0,0)$, $b = f_{1y}(0,0)$, $c = f_{2x}(0,0)$, and $d = f_{2y}(0,0)$, where $f_{1x}(0,0)$, for example, denotes the partial derivative $\partial f_1 / \partial x$, evaluated at the origin. In a small neighborhood of the equilibrium point at the origin it will be observed that the terms R_1 and R_2 in Eq. (11-28) are small and may be neglected. Thus the behavior of the state equations in this neighborhood can be determined by investigating the system of **equations of first approximation**

$$\dot{x}(t) = ax(t) + by(t)$$
$$\dot{y}(t) = cx(t) + dy(t)$$

$$(11.29)$$

where we assume $ad - bc \neq 0$. Under the above conditions it can be shown that the behavior of the phase plane trajectories of the system described by Eq. (11-19) in the neighborhood of an equilibrium point is the same as the equation of first approximation except for the case of imaginary eigenvalues when they may become either a center or a focus.

Example 3 *Find the equations of first approximation for the simple pendulum about the equilibrium point $(\pi, 0)$ and determine the form of the phase plane portrait at this point.*

We first translate the equilibrium point $(\pi, 0)$ to the origin using the state variable transformation $u = x - \pi, v = y$. Thus we find

$$\dot{u} \;=\; \dot{x} = v$$

$$\dot{v} \;=\; \dot{y} = -\frac{g}{l}\sin(u + \pi) = \frac{g}{l}\sin u$$

Linearizing the above system equations about the origin we find the equation of first approximation

$$\dot{u} \;=\; v$$

$$\dot{v} \;=\; \frac{g}{l}u$$

or in matrix form

$$\left[\begin{array}{c} \dot{u}\,(t) \\ \dot{v}\,(t) \end{array}\right] = \left[\begin{array}{cc} 0 & 1 \\ \omega^2 & 0 \end{array}\right]\left[\begin{array}{c} u\,(t) \\ v\,(t) \end{array}\right]$$

where $\omega^2 = g/l$. The eigenvalues for this system are given by $s = \pm\omega$, and comparing this with the s-plane patterns shown in Fig. 11-4 shows the phase plane portrait in the neighborhood of the equilibrium point $(\pi, 0)$ is a saddle point. Observation of the phase portrait plotted in Fig. 11-6 shows that this is indeed the case.

Construction of Phase Plane Trajectories

If the non linear state equations for a system are time independent (autonomous) as is the case for the system shown in Eq. (11-19) we can obtain a differential equation directly describing the phase plane trajectories. From Eq. (11-19)

$$\frac{dy}{dx} = \frac{dy/dt}{dx/dt} = \frac{f_2\,(x, y)}{f_1\,(x, y)} \tag{11.30}$$

In some cases this equation can be integrated in terms of elementary functions so that the trajectory can be plotted on the phase plane. The pendulum equation shown in Eq. (11-25) gives an example of this possibility as it can be directly integrated. However, in many applications this cannot be done either due to the complexity of the equation or the form of the system nonlinearity. In such cases other methods have to be employed. For example, an analog or digital computer can be easily used for simulating the control system and constructing the phase portraits when the systems are linear. For systems having a nonlinear element a special function generator may be used which can be directly tied to the analog computer. Another straightforward and powerful approach to the construction of phase portrait trajectories is to obtain closed form or numerical solutions to the governing equations as functions of time and then to compute the output and its derivative at the same time t to obtain a state point on the trajectory.

Apart from numerical methods a graphical approach using the **isocline method** is useful for gaining a qualitative understanding of the phase plane portraits. To illustrate this method consider the control system with transfer function

$$\frac{C(s)}{R(s)} = \frac{K}{As^2 + K}$$

This is equivalent to the state equations

$$\dot{x} = y$$

$$\dot{y} = \frac{K(r - x)}{A}$$

where $x(t) = c(t)$. Thus from Eq. (11-30) we have

$$\frac{dy}{dx} = \frac{K(r - x)}{Ay} \tag{11.31}$$

If we plot y versus x to obtain the phase plane trajectories of a control system, then Eq. (11-31) is the slope of the trajectory on the phase plane at any point. For any *constant* slope m,

$$m = K(r - x)/Ay$$

or

$$y = (Kr/Am) - (K/Am)x \tag{11.32}$$

This then gives the equation of lines which all trajectories cross with the same slope. Such lines are called **isoclines**. Often phase-plane trajectories are obtained from isoclines. If r is a unit step, then Eq. (11-32) gives straight lines with slope $-K/Am$. All trajectories intersecting this line must have the same slope. The trajectories and isoclines for this example are shown in Fig. 11-7. We notice that the point $y = 0, x = 1$ is a singular point. The type of singular point is related to the roots of the characteristic equation. In this case the roots of the characteristic equation are pure imaginaries and the singular point is a vortex or center. Notice that this system does not have any damping.

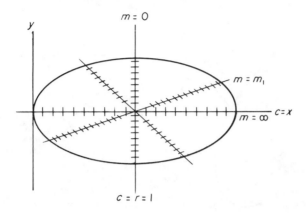

Figure 11.7: Isoclines and trajectories on the phase plane

Example 4 *Construct the phase plane trajectory for the system represented by the differential equation*

$$\ddot{x} + 2\zeta\omega\dot{x} + \omega^2 x = 0$$

where we take $y = \dot{x}$.

In this case

$$\frac{dy}{dx} = \frac{-2\zeta\omega y - \omega^2 x}{y}$$

Letting $dy/dx = m$ the equation for the isoclines is given by

$$y = \frac{-\omega^2}{2\zeta\omega + m}x$$

The isoclines for m in the range $(-\infty, \infty)$ are shown plotted on Fig. 11-
8(a). Suppose the system begins operation at point A for time $t = 0$. As
the trajectory must pass through point A with slope -1 we construct the
line Aa from point A with this slope as shown in Fig. 11-8(b). Similarly we
construct the line Ab with slope -1.2. The point B where the trajectory
crosses the $m/2\zeta\omega = -1.2$ isocline must lie within the interval ab. Provided
the isoclines are sufficiently close together a reasonable construction procedure is to assume B lies at the mid-point of the interval ab. This procedure
is repeated for points C, D, E, etc. until the trajectory is completed, as
shown in Fig. 11-8(a).

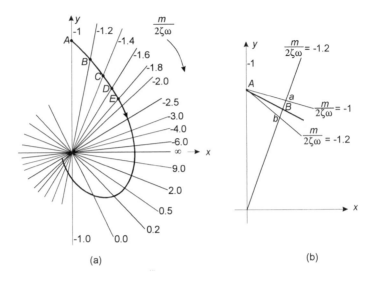

Figure 11.8: Phase plane trajectory construction using isocline method

Phase Plane Analysis of Piecewise Linear Systems

Often the non linearities present in a control system can be approximated by a piecewise linear function. In this case the phase plane plots are formed as a composite of the plots for several linear systems. In the example below analyzed by Ogata we examine the important case where the system has a crossover non-linearity. This commonly occurs in electric servo system amplifiers and in hydraulic servos when the reversing spool valve has some port overlap to prevent fluid leakage.

Example 5 *A control system having a crossover distortion non-linearity is shown in Fig. 11-9. Investigate the phase plane trajectory of this system for a unit step input assuming it is initially at rest and the slope $k < 1$.*

From Fig. 11-9(b) we see that the non-linearity can be represented by the equations

$$u = \begin{cases} ke & \text{for} \quad |e| < e_0 \\ e + e_0(k-1) & \text{for} \quad e > +e_0 \\ e - e_0(k-1) & \text{for} \quad e < -e_0 \end{cases} \qquad (11.33)$$

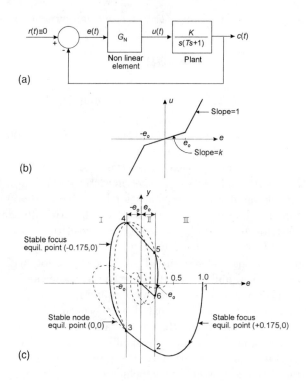

(a)

(b)

(c)

Figure 11.9: Control system with non-linear crossover distortion

Also, from Fig. 11-9(a) the equation of motion for this system can be seen to be

$$T\ddot{e}(t) + \dot{e}(t) + Ku(t) = T\ddot{r}(t) + \dot{r}(t) \qquad (11.34)$$

Since $r(t)$ is a step input, the initial conditions are $e(0) = 1$, and $\dot{e}(0) = 0$. As well, we have $\ddot{r}(t) = \dot{r}(t) = 0$ when $t > 0$ so that Eq. (11-34) becomes $T\ddot{e}(t) + \dot{e}(t) + Ku(t) = 0$, and we obtain the following three differential equations which are operative in the regions , I, II, and III shown on Fig. 11-9(c)

$$T\ddot{e}(t) + \dot{e}(t) + Ke(t) = -e_0 K(1-k) \quad \text{for } e < -e_0 \qquad (11.35)$$

$$T\ddot{e}(t) + \dot{e}(t) + Kke(t) = 0 \quad \text{for } |e| < e_0 \qquad (11.36)$$

and

$$T\ddot{e}(t) + \dot{e}(t) + Ke(t) = e_0 K(1-k) \quad \text{for } e > e_0 \qquad (11.37)$$

By choosing the state variable $y(t) = \dot{e}(t)$ we can write these equations in state variable form

$$\dot{e}(t) \;=\; y(t)$$

$$\dot{y}(t) \;=\; -\frac{1}{T}(y(t) + Ku(e))$$

and from Eqs. (11-35), (11-36) and (11-37) we see the equilibrium points are at $(-e_0(1 - k), 0)$, $(0, 0)$, and $(e_0(1 - k), 0)$ respectively.

Assuming the characteristic roots of Eqs. (11-35) and (11-37) are complex with negative real parts, and the characteristic roots of Eq. (11-36) are critically damped we can see that for large errors the system will be under damped, while for small errors it will be critically damped. The phase plane trajectory for $e(0) = 1$, $\dot{e}(0) = 0$ with the parameters $T = 1$, $K = 4$, $k = 0.0625$ and $e_0 = 0.2$ is shown in Fig. 11-9(c). The trajectories corresponding to Eqs. (11-35) and (11-37) are stable foci with equilibrium points at $(-0.175, 0)$ and $(0.175, 0)$ respectively, while the one corresponding to Eq. (11-36) is a stable node with equilibrium point at $(0, 0)$. Thus the portion of the trajectory indicated by the path 1-2 is part of a stable focus converging to the point $(0.175, 0)$. At point 2 the non linear gain changes to the value k and the corresponding trajectory having path 2-3 is part of a stable node converging to the point $(0, 0)$. At point 3 the trajectory again becomes a stable focus which is now converging to the point $(-0.175, 0)$. This process is repeated for path elements 3-4, 4-5, 5-6, etc. until the trajectory converges to the origin.

A special case occurs when $k = 0$ where the trajectory will converge to some point along the interval extending from $(-e_0, 0)$ to $(e_0, 0)$ and not necessarily to the origin $(0, 0)$ as is the situation when $k > 0$. Thus it can be seen that crossover distortion can lead to a deterioration of the operation of the system when it experiences small disturbances, and in the worst case if the crossover gain k is zero then it will fail to respond to disturbances of magnitude less than e_0. However so long as the magnitude of the disturbance is large the behavior of the system is as if the crossover distortion is absent.

Example 6 *Using the error and its derivative construct a phase-plane plot for the second-order system with saturation shown in Fig. 11-10(a).*

The transfer function is

$$\frac{C(s)}{E'(s)} = \frac{K_1}{s(s\tau + 1)}$$

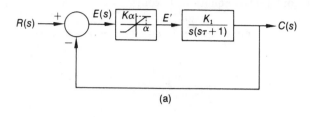

(a)

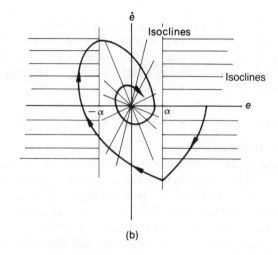

(b)

Figure 11.10: Phase plane diagram for second-order system with saturation

or

$$\frac{R(s) - E(s)}{E'(s)} = \frac{K_1}{s(s\tau + 1)}$$

The corresponding differential equation for $R(s) = 0$ is

$$\ddot{e} + \frac{1}{\tau}\dot{e} = -\frac{K_1}{\tau}e'$$

where

$$e' = \begin{cases} K_2 e & -\alpha < e < \alpha \\ K_2 \alpha & e \geq \alpha \\ -K_2 \alpha & e \leq -\alpha \end{cases}$$

The three differential equations become

$$\ddot{e} + \frac{1}{\tau}\dot{e} + \frac{K_1 K_2}{\tau}e = 0 \qquad -\alpha < e < \alpha$$

$$\ddot{e} + \frac{1}{\tau}\dot{e} = -\frac{K_1 K_2}{\tau}\alpha \qquad e \geq \alpha \tag{11.38}$$

$$\ddot{e} + \frac{1}{\tau}\dot{e} = \frac{K_1 K_2}{\tau}\alpha \qquad e \leq -\alpha$$

The isoclines for the last two equations may be shown to be straight horizontal lines. The isoclines for the first of the equations given in Eqs. (11-38) are also straight lines but are radial. Assuming that the first equation given above is underdamped, i.e. it has a stable focus, the phase trajectories may be constructed as shown in Fig. 11-10(b).

Phase Plane Analysis of Switching Servos

The non linear systems considered up to now have either had analytic non-linearities or have had piecewise continuous approximations to such non-linearities. The system considered in Example 5 having the piecewise linear distortion for small magnitudes of the signal $e(t)$, can be considered as an approximation of typical crossover non-linearities observed in operating plants. This idealization is valuable because of the fact that in each region the governing equations of motion are linear and thus amenable to analytic solution.

Another class of non-linear systems are those having discontinuities in the transfer characteristics of the non-linear elements. The most common example is when the servo-controller is a relay, although it is often valuable to examine linear control systems with saturation from this viewpoint when the loop gain is large so that the power amplifier saturates for small transient inputs. We now examine some examples of this class of control systems.

Example 7 *A control system with a relay is shown in Fig. 11-11. We wish to study this system via the phase-plane method.*

The transfer function $G_1(s)$ is given by

$$G_1(s) = \frac{C(s)}{E'(s)} = \frac{1}{s(Is + B)}$$

and the differential equation corresponding to this is

$$I\frac{d^2 c(t)}{dt^2} + B\frac{dc(t)}{dt} = e'(t) \tag{11.39}$$

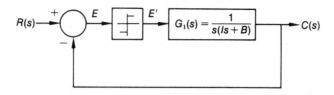

Figure 11.11: A second-order system with ideal relay

where

$$e' = \begin{cases} +A & e > 0 \\ -A & e < 0 \end{cases}$$

The differential equation given by Eq. (11-39) may be directly solved for each of the cases $e' = A$, and $e' = -A$, . For zero initial conditions we obtain the output position and velocity as

$$c(t) = \frac{A}{B}(t - \tau + \tau e^{-t/\tau})$$

$$v(t) = \dot{c}(t) = \frac{A}{B}(1 - e^{-t/\tau})$$

(11.40)

where $\tau = I/B$. Time may be eliminated by solving for t

$$t = -\tau \ln\left(1 - \frac{vB}{A}\right)$$

and then substituting back in Eq. (11-40), to obtain

$$c(t) = -\frac{A}{B}\tau \left[\ln\left(1 - \frac{v(t)B}{A}\right)\right] - \tau v(t)$$

(11.41)

If we take $x = cB/A\tau$, and $y = vB/A$, we have

$$x = -y - \ln(1 - y)$$

(11.42)

which is the equation for a phase trajectory as long as $e' = +A$. When $e < 0$, then $e' = -A$ and we may show that the phase trajectory is governed by

$$x = -y + \ln(1 + y)$$

(11.43)

If we had included initial conditions, then

$$x = -y - \ln(1-y) + (x_0 + y_0) + \ln(1-y_0) \quad \text{for} \quad e' = +A$$

$$x = -y + \ln(1+y) + (x_0 + y_0) - \ln(1+y_0) \quad \text{for} \quad e' = -A$$

(11.44)

These equations simply indicate that the trajectories shift horizontally by an amount corresponding to the contribution of the initial conditions. The time on the phase plot may be obtained by substituting for t into Eq.(11-44) where for $e' = +A$

$$t = \tau \left[(x - x_0) + (y - y_0) \right]$$

For $e' = -A$ we may show that

$$t = -\tau \left[(x - x_0) - (y - y_0) \right]$$

Assuming that the control system is subjected to a step input, we have

$$e_0 = r_0 - c_0 = 1$$

Since $e > 0$, we use the first of the equations in Eq. (11-44) to plot the trajectory from points 0 to 1 on the phase plot shown in Fig. 11-12. After point 1, since $e(t) = 1 - c(t)$, we see that $e(t)$ becomes negative. Since $e < 0$, we use the second of the equations in Eq. (11-44) to plot the trajectory from point 1 to 2 on the phase plot. At point 2 the system switches and the above steps are repeated. The time when $c = c_{max}$ is the time required for the system to overshoot the first time, and is also the time when $v = 0$. The amount of overshoot is a function of system damping.

We now examine two limiting cases which give insight into the operation of the system, these being when the input step is large and when it is small.

Example 8 *Consider the control system discussed in Example 7 and find its approximate phase plane trajectory for a step input $r(t)$, when $r(t) = R >> A\tau/B$ and when $r(t) = R << A\tau/B$ for $t > 0$.*

From Fig. 11-12 we see that $c(t) = r(t)$ when the phase plane trajectory reaches point 1, which is where the relay output sign changes from $+A$ to $-A$. Considering the first case where the step input $r(t) = R >> A\tau/B$ for $t > 0$, and recalling Eq. (11-40), it can be seen that at point 1

$$\frac{A\tau}{B} << r(t) = c(t) = \frac{A}{B}(t - \tau + \tau e^{-t/\tau})$$

(11.45)

This inequality will only be true when $t >> 2\tau$, and it can be seen that $v(t) \approx A/B$ under these circumstances. Thus we see that the velocity over

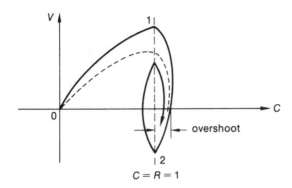

Figure 11.12: Phase plane plot for second-order system with ideal relay

a large part of the interval from points 0 to 1 is approximately constant and the phase plane trajectory asymptotically approaches the isocline at $v = A/B$. A typical phase plane trajectory satisfying these conditions is shown in Fig. 11-13(a).

In the second case we examine the phase plane response to a small step input when $r(t) = R << A\tau/B$ for $t > 0$. As above, $c(t) = r(t)$ when the phase plane trajectory reaches point 1, so that from Eq. (11-40) we have the inequality

$$\frac{A\tau}{B} >> r(t) = c(t) = \frac{A}{B}(t - \tau + \tau e^{-t/\tau}) \qquad (11.46)$$

This inequality is only satisfied when $t << 2\tau$, and under this condition it can be seen that $v(t) << A/B$. Thus we can simplify Eq. (11-41) by using a Taylor series expansion for the right hand side which gives

$$c(t) = \frac{A\tau}{B}\left[\frac{v(t)B}{A} + \frac{v^2(t)B^2}{2A^2}\right] - v(t)\tau$$

This simplifies to

$$c(t) = \frac{I}{2A}v^2(t)$$

It will be observed that the phase plane trajectory for this small disturbance is independent of the damping factor B, which implies that for small motions near the equilibrium point the system behaves as if there is no damping present. Compared with a linear control system having damping we see that this system will have a far more oscillatory response at low amplitudes and will take a considerably longer time for its oscillations to

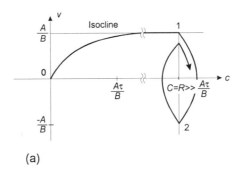

(a)

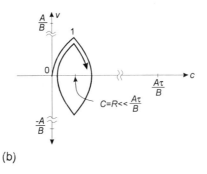

(b)

Figure 11.13: Phase plane trajectories when step input $r(t) \gg A\tau/B$ and when $r(t) \ll A\tau/B$

die away. A typical phase plane trajectory for this case is shown in Fig. 11-13(b).

Example 9 *If we include deadband in Example 7, then*

$$
e' = \begin{cases} +A & e > \alpha \\ 0 & -\alpha < e < \alpha \\ -A & e < -\alpha \end{cases}
$$

Construct a phase plane plot for this system.

For $e' > \alpha$ the previous equations apply. For $e' = 0$,

$$
I\frac{d^2c(t)}{dt^2} + B\frac{dc(t)}{dt} = 0
$$

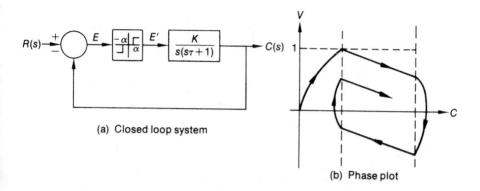

(a) Closed loop system

(b) Phase plot

Figure 11.14: Second-order system with relay having deadband

For c_0, v_0 initial conditions, we have

$$c(t) = c_0 + v_0(1 - e^{-t/\tau})\tau$$
$$v(t) = v_0 e^{-t/\tau}$$

(11.47)

Eliminating t, we obtain

$$c = c_0 + (v_0 - v)\tau$$

(11.48)

Again the previous definitions of x and y permit us to write

$$x = x_0 + (y_0 - y)$$

which is the equation of a straight line. The trajectory between $-\alpha < e < \alpha$ varies in a linear manner as shown in Fig. 11-14.

To date we have only considered examples of systems where the input to the non-linear element is the system error signal. We now consider an example of a system having proportional plus derivative control. This example will introduce the phenomenon of **chattering states** or **sliding regimes** which are peculiar to non-linear systems.

Example 10 *A control system having a relay and a proportional plus derivative controller is shown in Fig. 11-15. Using the error and its derivative construct a phase plane plot for this system.*

Since the transfer function of the system plant is

$$G_1(s) = \frac{1}{Is^2}$$

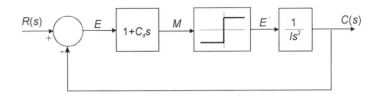

Figure 11.15: Second-order system with ideal relay and proportional plus derivative controller

the differential equation corresponding to it is

$$I\ddot{c}(t) = e^{'}(t) \tag{11.49}$$

From Fig. 11-15 $c(t) = r(t) - e(t)$ and assuming $r(t)$ is a unit step we find $\ddot{c}(t) = -\ddot{e}(t)$. Also defining the state variable $y(t) = \dot{e}(t)$ we obtain

$$\dot{e}(t) = y(t)$$

$$\dot{y}(t) = -\frac{1}{I}e^{'}(t) \tag{11.50}$$

where

$$e^{'}(t) = \begin{cases} +A & e + C_d y > 0 \\ \\ -A & e + C_d y < 0 \end{cases}$$

From the above, we obtain the equation describing the phase plane trajectories given by

$$\frac{dy}{de} = \begin{cases} -\dfrac{A}{Iy} & e + C_d y > 0 \\ \\ +\dfrac{A}{Iy} & e + C_d y < 0 \end{cases} \tag{11.51}$$

Solving this equation gives the equation for the phase plane trajectories

$$\frac{1}{2}\left(y^2(t) - y_0^2\right) = \begin{cases} -\dfrac{A}{I}(e(t) - e_0) & e + C_d y > 0 \\ \\ +\dfrac{A}{I}(e(t) - e_0) & e + C_d y < 0 \end{cases} \tag{11.52}$$

Such a trajectory is plotted in Fig. 11-16(a) from the initial conditions $e_0 = e_1, y_0 = 0$. It will be observed that the trajectory switches along the oblique switching line $e + C_d y = 0$, so that it follows the path $1 - 2 - 3$.

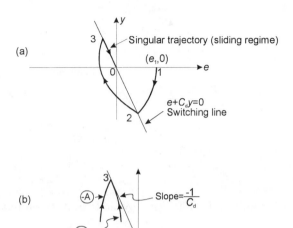

Figure 11.16: Phase plane trajectory for system with proportional plus derivative controller

When the phase plane trajectory reaches the switching line at point 3, a **chattering state** or **sliding regime** begins. An expanded view of the trajectory in the neighborhood of point 3 is shown in Fig. 11-16(b). At point 3, where the control system switches from a $+A$ to a $-A$ trajectory, say, the slope of the new trajectory is steeper than the slope of switching line. In this case the system will switch back and forth between the $-A$ and $+A$ trajectories at a rapid rate as it "slides" along the switching line $e + C_d y = 0$, toward the origin. This mode thus begins at the point on the switching line in the second or fourth quadrants where

$$\left. \frac{dy}{de} \right|_{\text{traj}} \geq \frac{1}{C_d} \tag{11.53}$$

From Eq. (11-51) we find the sliding regime occurs when

$$|y|_{\text{sliding}} \leq \frac{A C_d}{I} \tag{11.54}$$

For any point on the switching line where $|y| < |y|_{\text{sliding}}$ the phase plane trajectory is along the switching line

$$e + C_d y = 0$$

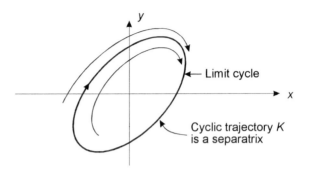

Figure 11.17: Limit cycle K forming cyclic trajectory on phase plane

so that it is governed by the differential equation

$$C_d \dot{e}(t) + e(t) = 0 \qquad (11.55)$$

Thus, from Eq. (11-55), it can be seen that the trajectory "slides" exponentially along the switching line as it converges to the origin.

11.3   Limit Cycles in Non-Linear Control Systems

Limit cycles occur quite commonly in practical control systems because of non linearities which are present. Often the magnitudes and frequencies of the oscillations are sufficiently small so that they can be neglected as they do not impair the operation of the system.

Suppose the operation of a control system can be described by the differential equations

$$\dot{x}(t) = f_1(x(t), y(t))$$
$$\dot{y}(t) = f_2(x(t), y(t)) \qquad (11.56)$$

A solution $(x(t), y(t))$ of these equations having a trajectory K in the phase plane is a limit cycle if it is **periodic** and also **isolated**. Thus a limit cycle must form a closed curve K in the phase plane separating the interior of the cycle from its exterior as shown in Fig. 11-17. For a periodic solution to be a limit cycle it is necessary for a small region to exist about K where no other periodic solution can exist. Thus it can be seen that the periodic solution given by Eq. (11-2), called a **center**, is not a limit cycle because irrespective

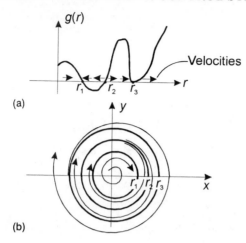

Figure 11.18: Non-linear system with limit cycles

of how small a region is taken about a candidate periodic solution there always exists other periodic solutions in this region, so that it is not isolated.

Because the trajectory K forms a closed curve in the phase plane, it is also a **separatrix**, so that the character of the trajectories inside curve K will differ from those on its outside. Limit cycles can be classified as stable, unstable, and semistable. If trajectories starting both inside and outside, in a small neighborhood of K, converge toward K as $t \to \infty$ then the limit cycle K is said to be **stable**. If on the other hand the trajectories converge toward K for $t \to -\infty$ then it is said to be **unstable**. Lastly if trajectories on one side of K converge toward it and diverge on the other side then it is said to be **semi-stable**.

To illustrate the above three classes of limit cycles consider the system described in polar form by the differential equations

$$\dot{\theta}(r) = -1$$

$$\dot{r}(t) = g(r)$$

where $g(r)$ is shown in Fig. 11-18(a). It will be noted from Fig. 11-18 that there are three limit cycles with radii of r_1, r_2, and r_3; these being the points where $g(r) = 0$. The radial velocities, which are also indicated on Fig. 11-18, show the limit cycles for r_1, r_2, and r_3, are stable, unstable and semi-stable respectively. The phase plane trajectories for this example are shown in Fig. 11-18(b).

The analytical tools available for the study of limit cycles are few. At best, some results concerning the presence or absence of limit cycles, proved

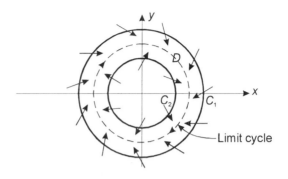

Figure 11.19: Poincare-Bendixson theorem

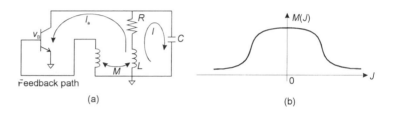

Figure 11.20: Transistor blocking oscillator

by Poincare and Bendixson, are useful.

Poincare and Bendixson Theorems

1. If outside a circle C_1 all the phase plane trajectories are converging, and inside a smaller circle C_2, having the same center as C_1, the paths are diverging, and as well there are no equilibrium points in the region $\mathcal{D}$ between these two circles, then there exists at least one limit cycle in region $\mathcal{D}$. This is illustrated in Fig. 11-19.

2. If for the system described by Eq. (11-56), the expression $\partial f_1/\partial x + \partial f_2/\partial y$ does not change sign in a region $\mathcal{D}$ of the phase-plane, then no closed trajectory can exist in $\mathcal{D}$.

Example 11 *The simplified circuit diagram for an electronic blocking oscillator is shown in Fig. 11-20. Obtain the equations of motion for this system and determine if it has a limit cycle assuming the mutual induc-*

tance $M(J)$ is a function as shown in Fig. 11-20(b), the tuned circuit resonant frequency $\omega_0 = 1/\sqrt{LC} = 1$, the transistor current $I_a = g_m v_B$, and $M(0) > RC/g_m$.

Applying Kirchoff's voltage law we obtain

$$Lp(I + I_a) + R(I + I_a) + \frac{1}{pC}I = 0$$

From the circuit diagram the inductor current $J = I + I_a$. Consequently we have

$$\ddot{J} + RC\dot{J} - I_a + J = 0$$

Also it can be seen that the transistor current I_a is given by

$$I_a = g_m v_B = g_m M(J)\dot{J}$$

Defining the state variables $x = J$ and $y = \dot{J}$ gives

$$
\begin{aligned}
\dot{x}(t) &= y(t) \\
\dot{y}(t) &= -\left[RC - g_m M\left(x(t)\right)\right] y(t) - x(t)
\end{aligned}
\tag{11.57}
$$

It can be observed that Eq. (11-57) has an equilibrium point at $(x, y) = (0, 0)$. Let us now consider a circle of small radius r_1 about the origin. It will be assumed that r_1 is sufficiently small so that $M(x) \approx M(0)$. Now since

$$r_1^2 = x^2 + y^2$$

we find that

$$r_1 \dot{r}_1 = x\dot{x} + y\dot{y}$$

where $x = r\cos\theta$ and $y = r\sin\theta$. From Eq. (11-57) we find that

$$\dot{r}_1 = (g_m M(0) - RC) r_1 \sin^2\theta \geq 0$$

Thus we find that at all points on this circle the trajectories are diverging.
Similarly we now consider a circle of large radius r_2 about the origin, so that $M(x) \approx 0$. In this case it can be shown that

$$\dot{r}_2 = (g_m M(x) - RC) r_2 \sin^2\theta \approx -RC r_2 \sin^2\theta$$

It can be seen that $\dot{r}_2 < 0$ for all non-zero θ. For trajectories passing through the points on the circle at $\theta = 0$ and $\theta = \pi$ it can be shown by other arguments that they also converge toward the origin. Since the only equilibrium point for this system is at the origin we see the conditions for

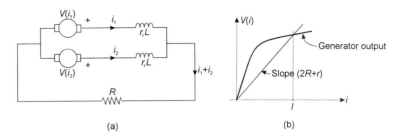

Figure 11.21: Parallel operation of series d.c. generators

Theorem 1 above are satisfied, so that we can conclude there exists at least one limit cycle in the region bounded by the concentric circles of radius r_1 and r_2, so that the circuit will oscillate.

To illustrate the application of the second theorem we examine an example attributed to Andronov.

Example 12 *Consider two series excited direct current generators connected in parallel across a load as shown in Fig. 11-21. Show there is an equilibrium point at the origin of the phase-plane and determine its stability assuming $\rho(0) = dV(0)/di > 2R+r$. Also show there exists no limit cycles in a neighborhood of the equilibrium point at the origin.*

Using Kirchoff's laws, the equations of motion for the system are

$$L\frac{di_1}{dt} = V(i_1) - i_1(R+r) - i_2R = f_1(i_1, i_2)$$

$$L\frac{di_2}{dt} = V(i_2) - i_2(R+r) - i_1R = f_2(i_1, i_2)$$

(11.58)

It can be easily seen that this system has an equilibrium point at $(i_1, i_2) = (0, 0)$. Linearizing the state equations about the origin gives

$$\begin{bmatrix} \dot{x}(t) \\ \dot{y}(t) \end{bmatrix} = \frac{1}{L} \begin{bmatrix} \rho(0) - (R+r) & -R \\ -R & \rho(0) - (R+r) \end{bmatrix} \begin{bmatrix} x(t) \\ y(t) \end{bmatrix}$$

It has eigenvalues at $s_1 = (\rho(0) - r)/L$ and $s_2 = (\rho(0) - 2R - r)/L$. Since it is assumed $\rho(0) > 2R + r$ we find both s_1 and s_2 lie in the right half plane so that the equilibrium point is an unstable node.

To apply Theorem 2 we evaluate the relation $\partial f_1/\partial i_1 + \partial f_2/\partial i_2$ in a region $\mathcal{D}$ about the origin. This is seen to be

$$\frac{\partial f_1}{\partial x_1} + \frac{\partial f_2}{\partial x_2} = \rho(i_1) - (R+r) + \rho(i_2) - (R+r)$$

(11.59)

where we define $\rho(i) = dV(i)/di$. From above we have $\rho(0) - (R+r) > R$, and since $V(i)$ is a smooth function of i there will always be a small region $\mathcal{D}$ about the origin where Eq. (11-59) is positive, so that by the second theorem no limit cycle can exist in this region. Thus we have shown that in this region about the origin the phase plane portrait will be an unstable node without limit cycles. Consequently the generators in the circuit of Fig. 11-21 will fail to operate in a stable manner where they equally share the load.

Lienard's Method

Lienard investigated the operation of second order systems in the phase-plane using a special transformation to define the state variables rather than taking $y(t) = \dot{x}(t)$ as shown in Example 11. To see his method consider the system

$$\ddot{x} + f(x)\dot{x} + g(x) = 0 \qquad (11.60)$$

We define the function $F(x)$ as

$$F(x) = \int_0^x f(x)dx$$

and then introduce the state variables $y = \dot{x} + F(x)$. Thus Eq. (11-60) is transformed to the state equations

$$\dot{x}(t) = y(t) - F(x(t))$$

$$\dot{y}(t) = -g(x(t))$$

We illustrate its application for finding the approximate period for the limit cycle of the van der Pol equation.

Example 13 *Van der Pol developed his equation*

$$\ddot{x}(t) - \epsilon(1 - x(t)^2)\dot{x}(t) + x(t) = 0$$

along similar lines to Example 11 to describe the operation of electronic oscillators. Using the Lienard method find the approximate phase-plane trajectory of the limit cycle and its period.

We have

$$F(x) = -\int_0^x \epsilon(1 - x^2)dx = -\epsilon\left(x - \frac{x^3}{3}\right)$$

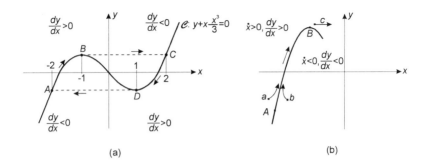

Figure 11.22: Limit cycle of van der Pol equation on Lienard phase plane

Defining the state variable

$$y = \frac{1}{\epsilon}\left[\dot{x} - \epsilon\left(x - \frac{x^3}{3}\right)\right]$$

we obtain

$$\dot{x}(t) = \epsilon\left[y(t) + x(t) - x(t)^3/3\right]$$

$$\dot{y}(t) = -x(t)/\epsilon$$

(11.61)

From Eq. (11-61) we find the equation describing the phase-plane trajectory

$$\frac{dy}{dx} = -\frac{1}{\epsilon^2}\frac{x}{y + x - \frac{x^3}{3}}$$

Supposing the parameter $\epsilon \gg 1$ it will be noted that $dy/dx \approx 0$ for points (x, y) in the phase-plane well away from the curve $\mathcal{C}$ defined by $y + x - x^3/3 = 0$. This curve is plotted on the phase-plane as shown in Fig. 11-22(a). The broken lines BC and DA are typical trajectories when the points (x, y) are distant from $\mathcal{C}$.

We now investigate the character of the phase-plane trajectories in the neighborhood of the curve $\mathcal{C}$ by considering the region AB. Suppose the trajectory is on curve $\mathcal{C}$ but moves slightly to the left to point a as shown in Fig. 11-22(b). Since $dy/dx > 0$ and $\dot{x}(t) > 0$ at this point the phase-plane motion will be as shown and will be forced back towards $\mathcal{C}$. A similar result occurs if the trajectory is disturbed to the right, to point b. In this region $dy/dx < 0$ and $\dot{x}(t) < 0$, so we have the trajectory moving from b towards $\mathcal{C}$, as shown. Consequently the trajectory must **chatter** along $\mathcal{C}$

from A to B. At point B, curve C has a turning point, and suppose the trajectory is disturbed to point c. Since $\dot{x} > 0$ and $dy/dx > 0$ at this point it can be seen that the trajectory is no longer driven towards C but is free to traverse along the path BC. The above process is repeated for region CD so that the limit cycle will approximately follow the path $ABCD$.

The period of the limit cycle can be estimated by evaluating the time to travel along the individual paths AB, BC, CD, and DA. The time taken to travel along the path AB is given by

$$t_B - t_A = \int_{t_A}^{t_B} dt = \int_{-2}^{-1} \frac{1}{\dot{x}} dx \qquad (11.62)$$

where $\dot{x}$ is evaluated along the path AB given by

$$y + x - \frac{x^3}{3} = 0$$

Differentiating this equation with respect to t gives

$$\frac{dy}{dt} + \frac{dx}{dt}(1 - x^2) = 0$$

Thus

$$\frac{dx}{dt} = \frac{x}{\epsilon}\frac{1}{1 - x^2} \qquad (11.63)$$

since $dy/dt = -x/\epsilon$. Substituting Eq. (11-63) into Eq. (11-62) and evaluating the integral gives

$$t_B - t_A = \epsilon \left(\frac{3}{2} + \ln\frac{1}{2} \right)$$

By symmetry $t_D - t_C = t_B - t_A$. On arcs BC and DA it can be seen $|\dot{x}(t)| >> 0$ so the time to travel along these paths is short compared with the interval $(t_B - t_A)$. Thus the period T of the limit cycle is approximately

$$T \approx 2(t_B - t_A) = \epsilon \left(3 + 2\ln\frac{1}{2} \right)$$

11.4 Describing Function Technique

If a linear system is excited with a sinusoidal function, the output is also a sinusoid with the same frequency but with a different amplitude and phase angle. However if a nonlinear element is excited with a sinusoidal function, then the output is nonsinusoidal, although periodic with the

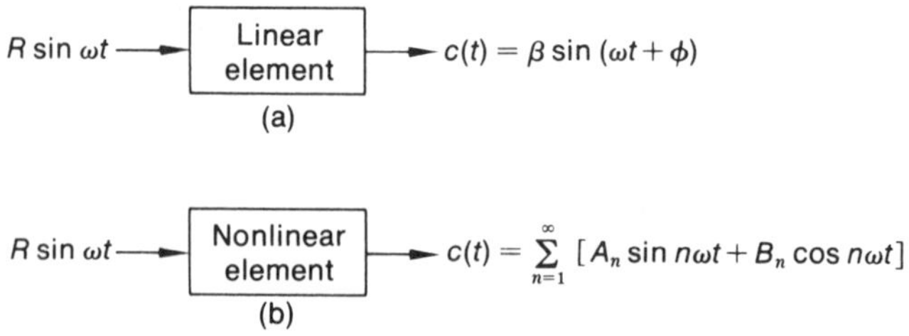

Figure 11.23: Linear and nonlinear response to a sinusoidal input

same period as the input. Let us assume that we represent the output in Fourier components. Then if the input $r(t) = r \sin \omega t$ and the output $c(t) \approx (1/T) \sum c_n e^{jn\omega t}$ the error ϵ becomes

$$\epsilon = c(t) - \frac{1}{T} \sum c_n e^{jn\omega t}$$

Assuming that $c(t)$ can be integrated over the range $-T/2$ to $T/2$, the mean square error becomes

$$\overline{\epsilon^2} = \frac{1}{T} \int_{-T/2}^{T/2} \epsilon^2 dt$$

which when minimized yields the coefficients c_n

$$c_n = \int_{-T/2}^{T/2} c(t) e^{-jn\omega t} dt$$

that are immediately recognized as the Fourier series coefficients[1]. We can conclude from this that the use of the sinusoidal input results in a particular linear equivalent element which minimizes the mean square difference between the actual element output and its approximation by the fundamental

---

[1] The complex Fourier series given in Section 4-4 can be transformed into a sum of real components by taking

$$c_n = \frac{T B_n}{2} - j \frac{T A_n}{2}$$

In this case

$$f(t) \quad = \quad \frac{1}{T} \sum_{n=-\infty}^{\infty} c_n e^{jn\omega t}$$

$$= \quad \frac{B_0}{2} + \sum_{n=1}^{\infty} (A_n \sin n\omega t + B_n \cos n\omega t)$$

where A_n and B_n are given in Eq. (11-65).

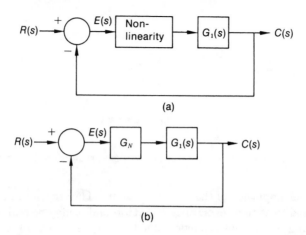

(a)

(b)

Figure 11.24: Replacing a nonlinear element by its describing function

component of the output wave. In the describing function technique it is assumed that the fundamental component of the output of the nonlinear element is the most significant and the higher harmonies may be neglected. This means that the behavior of the control system resembles that of a low pass filter. Indeed, many control systems tend to operate in this manner. The **transfer function** of the **nonlinear element** as shown in Fig. 11-23 is defined as

$$G_N = \frac{\text{Amplitude of the fundamental component of the output}}{\text{Amplitude of input sinusoid}}$$

Since the fundamental component is

$$A_1 \sin \omega t + B_1 \cos \omega t = \sqrt{A_1^2 + B_1^2} \sin(\omega t + \phi)$$

the transfer function G_N becomes

$$G_N = \frac{\sqrt{A_1^2 + B_1^2}}{R} e^{j\phi} \tag{11.64}$$

where

$$A_n = \frac{2}{T} \int_{-T/2}^{T/2} c(t) \sin n\omega t \ dt$$

$$\tag{11.65}$$

$$B_n = \frac{2}{T} \int_{-T/2}^{T/2} c(t) \cos n\omega t \ dt, \quad \phi = \arctan \frac{B_1}{A_1}$$

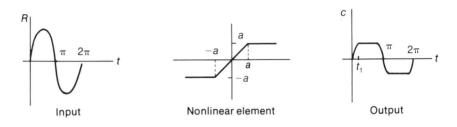

Figure 11.25: Characteristics of saturation or limiting

and R is the amplitude of the sinusoidal input. The transfer function G_N is also referred to as the **describing function** and it may be real or complex. If the nonlinear element is single valued, G_N is generally a function of the input amplitude. For more complex nonlinearities G_N may be a function of input amplitude and frequency.

Once G_N is obtained, it replaces the nonlinear element as shown in Figs. 11-24(a) and (b). The overall transfer function becomes

$$\frac{C(s)}{R(s)} = \frac{G_N G_1(s)}{1 + G_N G_1(s)} \qquad (11.66)$$

We may now obtain the system response by employing the classical methods developed for linear systems. However, before we consider closed loop systems let us derive G_N for some nonlinearities.

Example 14 *Obtain the describing function for the saturation (sometimes called limiting) function shown in Fig. 11-25.*

Since the output is an odd function about the origin, the cosine terms do not appear. Also since it is symmetrical about $\pi/2$, we may write

$$A_1 = \frac{4}{\pi} \int_0^{\pi/2} c(t) \sin \omega t \, d(\omega t)$$

From Fig. 11-25 the output is

$$c(t) = \begin{cases} R \sin \omega t & 0 < \omega t < \omega t_1 \\ a & \omega t_1 < \omega t < \dfrac{\pi}{2} \end{cases}$$

Substituting in A_1 and noting that $\sin \omega t_1 = a/R$, we have

$$A_1 = \frac{2R}{\pi} \left(\arcsin \frac{a}{R} + \frac{a}{R} \sqrt{1 - \left(\frac{a}{R} \right)^2} \right)$$

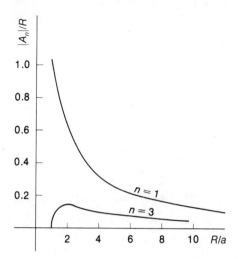

Figure 11.26: Comparison of the third harmonic with the fundamental for a limiter

and the describing function becomes

$$G_N = \frac{A_1}{R} = \frac{2}{\pi}\left(\arcsin\frac{a}{R} + \frac{a}{R}\sqrt{1 - \left(\frac{a}{R}\right)^2}\right)$$

which is a function of the input amplitude and the nonlinear element characteristics. The phase contribution is zero. Since only the fundamental component is used to define the transfer function, it has been assumed that the higher harmonic terms are small. That this indeed is the case is shown in Fig. 11-26. Here the relative magnitude of the fundamental is compared to the third harmonic which is seen to be considerably smaller.

Example 15 *The appearance of a relay or contactor is quite common in control systems. Obtain the describing function for (a) the ideal relay shown in Fig. 11-27, and (b) relay with deadband shown in Fig. 11-28.*

(a) Since the output is odd, we compute A_1

$$A_1 = \frac{4}{\pi}\int_0^{\pi/2} c(t)\sin\omega t\, d(\omega t)$$

where

$$c(t) = a, \qquad \text{for } 0 < \omega t < \frac{\pi}{2}$$

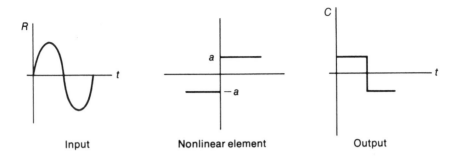

Figure 11.27: Characteristics of an ideal relay

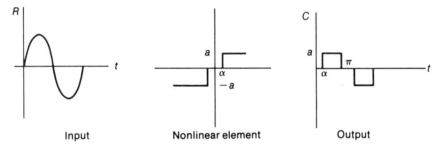

Figure 11.28: Characteristics of a relay with deadband

Substituting in A_1 we have

$$A_1 = \frac{4}{\pi} \int_0^{\pi/2} c(t) \sin \omega t \, d(\omega t) = \frac{4a}{\pi}$$

and the describing function becomes

$$G_N = \frac{A_1}{R} = \frac{4a}{\pi R}$$

Again this transfer function is a function of the input amplitude. There is however no phase shift.

(b) Now consider the relay with deadband shown in Fig. 11-28. The output is

$$c(t) = \begin{cases} 0 & \text{for } 0 < \omega t < \alpha \\ a & \text{for } \alpha < \omega t < \dfrac{\pi}{2} \end{cases}$$

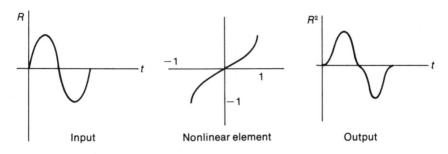

Figure 11.29: Characteristics of a quadratic function

The fundamental amplitude becomes

$$A_1 = \frac{4a}{\pi} \cos \alpha$$

From Fig. 11-28 we have

$$\sin \alpha = \frac{a}{R}, \qquad \cos \alpha = \sqrt{1 - \left(\frac{a}{R}\right)^2}$$

Substituting this, the transfer function becomes

$$G_N = \frac{A_1}{R} = \frac{4a}{\pi R}\sqrt{1 - \left(\frac{a}{R}\right)^2}$$

Example 16 *Obtain the describing function for the quadratic function (this could be the behavior of a spring) shown in Fig. 11-29.*

The output $c(t)$ is

$$c(t) = [R\sin \omega t]^2 \qquad 0 < \omega < \frac{\pi}{2}$$

The fundamental coefficient becomes

$$A_1 = \frac{4R^2}{\pi}\int_0^{\pi/2} \sin^3 \omega t \, d(\omega t) = \frac{8}{3}\frac{R^2}{\pi}$$

The describing function becomes

$$G_N = \frac{A_1}{R} = \frac{8}{3}\frac{R}{\pi}$$

Table 11-1 Characteristics and describing functions of some nonlinear elements.

Type	Input	Nonlinear Element	Output	Describing Function
Saturation with dead zone				$R < \alpha,\ G_N = 0$ $R > \dfrac{\beta}{k},\ G_N = \dfrac{2k}{\pi}\left[\beta_2 - \beta_1 - \dfrac{\sin 2\beta_2}{2} + \dfrac{\sin 2\beta_1}{2}\right]$ $A > R > \alpha,\ G_N = \dfrac{2k}{\pi}\left[\dfrac{\pi}{2} - \beta_1 - \dfrac{\sin 2\beta_1}{2}\right]$ $\beta_1 = \sin^{-1}\dfrac{\alpha}{R},\ \beta_2 = \sin^{-1}\dfrac{A}{R}$
Dead zone				$G_N = 0,\ R < \alpha$ $G_N = \dfrac{2k}{\pi}\left[\dfrac{\pi}{2} - \beta - \dfrac{\sin 2\beta}{2}\right]$ $R > \alpha$
Dead zone and hysteresis in a relay				$G_N = Me^{j\phi}$ $M = \dfrac{4A}{\pi R}\sin\left(\dfrac{\beta_2 - \beta_1}{2}\right)$ $\phi = \dfrac{\pi}{2} + \dfrac{\beta_1 + \beta_2}{2}$ $\beta_1 = \sin^{-1}\dfrac{b}{R}$ $\beta_2 = \sin^{-1}\dfrac{a}{R}$

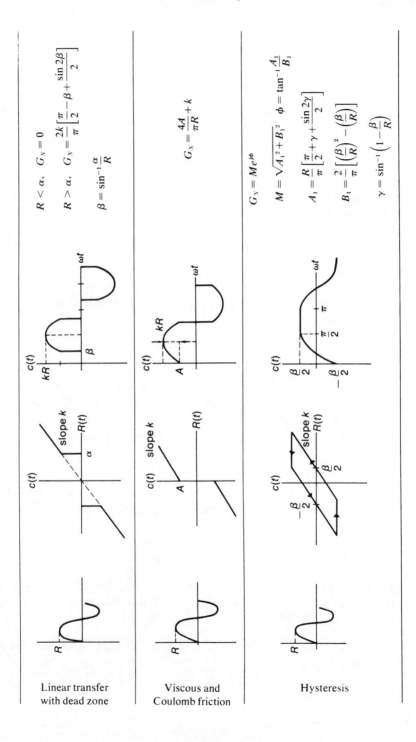

Linear transfer with dead zone

Viscous and Coulomb friction

Hysteresis

The describing function of several common nonlinear elements is shown in Table 11-1.

Since the transfer function of the nonlinear element is input amplitude dependent, the gain of G_N will constantly change with changing magnitude of the input signal. This will in turn produce a change in the gain of the closed loop system. For the closed loop system shown in Fig. 11-24(b), the overall transfer function is

$$\frac{C(s)}{R(s)} = \frac{G_N(E)G_1(s)}{1 + G_N(E)G_1(s)} \qquad (11.67)$$

The transfer function of the nonlinear element is a function of its input amplitude E. The characteristic equation for this system is

$$1 + G_N(E)G_1(s) = 0 \qquad (11.68)$$

and the roots of this equation determine the system stability. We shall investigate this stability using the Nyquist and Bode plots.

For the purposes of the Nyquist analysis, the characteristic equation may be rewritten as

$$G_1(s) = -\frac{1}{G_N(E)} \qquad (11.69)$$

The critical point $(-1, 0)$ used for linear system analysis has been replaced by $-1/G_N(E)$ which is a locus obtained as E varies. The stability on the Nyquist plot is investigated by observing the position of $G_1(s)$ relative to the locus of $-1/G_N(E)$.

For the Bode plot we write the characteristic equation

$$G_1(s)G_N(E) = -1 \qquad (11.70)$$

and observe $G_N(E)$ may contribute both magnitude and phase. Therefore as E varies, the 0 dB line on the Bode plot moves up or down. The amount this line moves is equal to $G_N(E)$ measured in decibels. In addition the phase plot may move up or down as E varies. This phase shift is independent of frequency and only depends on the magnitude of E.

Example 17 *Consider the third-order control system in Fig. 11-30 where the nonlinear element is a simple relay. Study system stability using the Nyquist and Bode plots.*

Note that the amplitude of the input to the nonlinear element is E, therefore

$$G_N = \frac{4a}{\pi E}$$

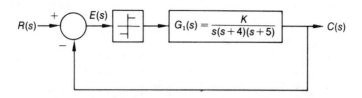

Figure 11.30: Closed loop system with an ideal relay

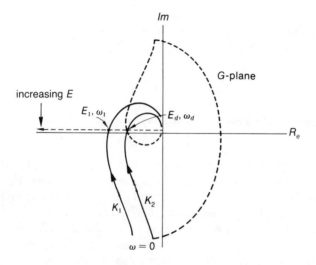

Figure 11.31: Nyquist plot of $G_1(s) = K/s(s+4)(s+5)$ and the locus of $-\pi E/4a$ for $0 \le E \le \infty$

The Nyquist plot of $G_1(s)$ for three values of gain is shown in Fig. 11-31. Superimposed on this plot also is the locus of $-1/G_N(E)$ where

$$-\frac{1}{G_N(E)} = -\frac{\pi E}{4a}$$

As the error E gets large, $-1/G_N(E)$ moves away from the origin and as the error E gets small, $-1/G_N(E)$ approaches the origin.

Let us assume that the error of the system at some point is E_1 and that the gain is K_1. The frequency corresponding to the point where the locus and Nyquist plot intersect is ω_1 as indicated on Fig. 11-31. This point is where the system oscillates. Now let us suppose that the error

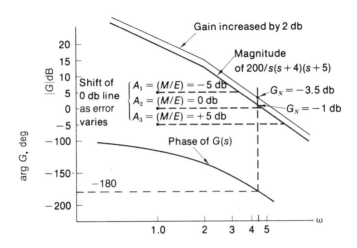

Figure 11.32: Effect of ideal relay on third-order system

signal decreases, possibly due to a sudden change in input, to E_d, then the system becomes unstable since $-\pi E_d/4a$ is now enclosed. Since the system gets unstable the error begins to increase and the locus of $-1/G_N(E)$ begins to move to the left. This continues until the $-1/G_N(E)$ contour moves back to the Nyquist plot for gain K_1 and the system stabilizes. Clearly then the system has a stable oscillation about the point E_1 at a frequency of ω_1. If now the system gain is decreased to K_2, we can use the previous argument to show that the system shall still exhibit stable oscillations but with a smaller amplitude.

The Bode plot is shown in Fig. 11-32. The plot of $G_1(s)$ is constructed for $K = 200$. The line A_2 corresponds to $4a/\pi E = 1$. If the error E is decreased, then $4a/\pi E > 1$ and $|G_N|$ will have the effect of lowering the 0 dB line. Notice that $G_N(E)$ makes no contribution to the system phase. Let us begin with the system operating at $G_N = 5$ dB. Here the system is unstable, therefore the amplitude of the oscillations shall increase thereby increasing E. This in turn will decrease $|G_N|$ whereby the 0 dB line will move up on the phase plot. This continues until the 0 dB line intersects the $|G_1(s)|$ plot at a phase shift of -180 deg. As a matter of fact, the system exhibits stable oscillations about the point of intersection of $G_N = -1$ dB and the gain plot for $K = 200$. Let us now increase the system gain by 2 dB. The point of intersection of $|G_N| = -1$ dB and the $G_1(s)$ plot for the new gain is unstable. Therefore the amplitude of oscillations shall increase so that E becomes larger. This decreases $|G_N|$ to a new stable value of

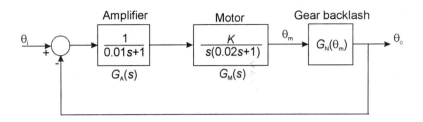

Figure 11.33: Nonlinear servo system having an output gear-box with back-lash

−3.5 dB. The system oscillations therefore stabilize but at an increased amplitude. The magnitude may be obtained from

$$\frac{4a}{\pi E} = -3.5 \text{ dB} = 0.67$$

or

$$E = 1.91a$$

The exact value of E depends upon a which is dependent upon the characteristics of the nonlinear element.

It can thus be seen that the describing function is a useful tool for predicting whether the non-linear system will exhibit limit cycle oscillations. It is possible to determine whether these are either stable or unstable. In the above example it was shown that the limit cycle oscillation has a stable amplitude $E = 1.91a$ and its frequency is 4.6 rad/sec.

If, using the describing function method, the system shown in Fig. 11-24(b) is stable then the closed loop sinusoidal response can be determined using Eq. (11-67). It will be noted, in this case, that just as the gain and phase of G_N are dependent upon the amplitude of E, the control ratio $C(j\omega)/R(j\omega)$ gain and phase responses will also depend upon E.

The next example illustrates the application of the describing function method for a servo system with an output gear train having backlash. When the backlash is friction limited the output motion ceases the instant the input begins to reverse so that it can be represented by a hysteresis non-linearity as shown in Table 11-1. When the gear train drives an inertial load the backlash is inertially limited and the describing function differs from the one given for hysteresis in Table 11-1.

Example 18 *Consider the servo system shown in Fig. 11-33 where the load is driven by the motor through a gear train having a backlash of* 0.5

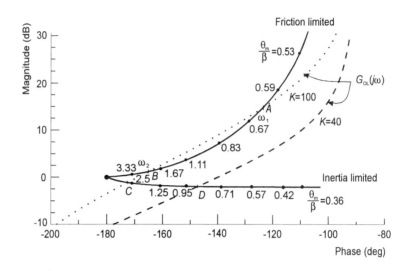

Figure 11.34: Nichol plot of $G_{OL}(j\omega)$ for $K = 100$ and $K = 40$, together with locii of $-1/G_N(\theta_m)$ for friction limited and inertia limited backlash

degree. Study the stability of the system using the Nichol chart for the two cases where the motor is driving a frictional load, and when it is driving an inertial load. Find the amplitude of any oscillations which are present.

The Nichol plots of the transfer function $G_{OL}(s) = G_M(s)G_A(s)$ for $K = 100$ and $K = 40$ are shown in Fig. 11-34. Superimposed upon this plot are the loci of $-1/G_N(\theta_m)$ for both friction limited and inertia limited backlash. In practice these two loci are extreme cases as physical systems will generally have both inertia and friction present so that the actual describing function locus will lie somewhere between these two limits. We consider both of these cases in turn to determine the possible range of system behavior. It will be noted that for both cases the loci of $-1/G_N(\theta_m)$ move away from the critical point at 0 dB and -180 degree, although the phase and amplitude variation differ widely for each case.

Examining the case where the backlash is friction limited we see that the $-1/G_N(\theta_m)$ locus intersects $G_{OL}(j\omega)$ contour at points A and B. Thus the system can oscillate at each of these points with frequencies $\omega_1 = 20$ rad/sec and $\omega_2 = 52.6$ rad/sec. Closer examination shows that only one of these points gives a stable limit cycle. To see this suppose the system is oscillating at point A and the amplitude decreases slightly to $\theta_m/\beta = 0.59$. Since the critical point is no longer encircled by $G_{OL}(j\omega)$ the oscillations will die away and the system will be stable. If however the amplitude was to increase slightly to $\theta_m/\beta = 0.67$, say, then the system becomes unstable and the oscillation amplitude would continue to increase until point B was reached. This point will give rise to oscillations of stable amplitude since if the amplitude increases slightly the critical point will no longer be encircled and the amplitude will die away so that the system oscillates at point B. Thus point B corresponds to the operating conditions for a stable limit cycle having an amplitude of $\theta_m/\beta = 2.34$ or the peak oscillation amplitude of the motor output $\theta_m = 1.17$ degrees.

If we were to decrease the gain to $K = 40$, it will be observed that the plot $G_{OL}(j\omega)$ no longer intersects with the friction limited describing function. In this case the describing function method suggests that the servo-system is stable for all amplitudes of the motor output θ_m.

The other extreme case is where the backlash non-linearity is inertially limited. This case is also shown on Fig. 11-34, and it can be seen that the frequency plots $G_{OL}(j\omega)$ for $K = 100$ and $K = 40$ intersect the inertia limited locus $-1/G_N(\theta_m)$ at points C and D respectively. Thus we see the system will oscillate with an amplitude of $\theta_m/\beta = 3.5$ at point C when $K = 100$, and at point D with $\theta_m/\beta = 0.87$ when $K = 40$. Consequently a sustained output chatter will occur in both cases, even if the amplitude of oscillation is small and not objectionable. These oscillations are characteristic of servo-systems with inertially controlled backlash. For this reason small servo systems having gear trains are often constructed using anti-backlash gears.

It was mentioned earlier that the describing function is an attractive method since the techniques developed for linear systems may be used with slight modification as we have seen. Also this method allows us to analyze higher-order control systems. The chief drawback however is that it is limited to sinusoidal inputs.

11.5 Stability Criteria

In recent years, considerable interest has been centered upon the stability analysis of nonlinear control systems without solving the actual governing equations. Much of the work has been based upon the work of A. M. Lyapunov who developed his theory at the turn of the century. He developed a very fundamental theory to ascertain the stability of a system via energy considerations. In recent times methods developed by V. M. Popov have also been applied to control systems.

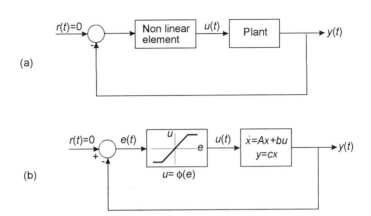

Figure 11.35: Non-linear control system

A typical example of a system to which these methods can be applied is shown in Fig. 11-35(a). It is usually assumed in the analysis that the **nonlinear element** has no energy storage elements, while all the dynamics of the system are concentrated in the **plant**. An example of a system which fits this model is shown in Fig. 11-35(b). Here the plant is assumed to be linear and to be represented by its state equations

$$\dot{\mathbf{x}}(t) = \mathbf{A}\mathbf{x}(t) + \mathbf{b}u(t)$$
$$\tag{11.71}$$
$$y(t) = \mathbf{c}\mathbf{x}(t)$$

The non-linearity is characterized by the input-output function $\varphi(e)$, such as the saturation function shown in the above figure.

In the ensuing discussion let us assume the system is described by the state equation

$$\dot{\mathbf{x}}(t) = \mathbf{f}(\mathbf{x}(t)) \qquad \mathbf{x}(t_0) = \mathbf{x}_0 \tag{11.72}$$

and that it has an **equilibrium** or **singular point** at $\mathbf{x} = 0$. In the Lyapunov method the behavior of the system, after it is perturbed from this equilibrium point, is investigated.

Concept of Stability

Before we get involved any further, we need to clearly understand the meaning of **stability**. In general, the stability of a nonlinear system is dependent upon the range of the state vector as well as the perturbation. If the system is subjected to small perturbations and it stays within an infinitesimal region about the singular point, the system has **local stability**. If a system is perturbed and, thereby displaced to any point within a finite region, after which it returns to the singular point, then this system has **finite stability**. Finally, if the finite region encompasses the entire state space, then the system is said to have **global stability**. If a system, initially beginning from any point in the region of stability, approaches the singular point as time goes to infinity, it is said to be **asymptotically stable**. Here when we speak of stability, we understand it in the sense given by Lyapunov.

The equilibrium point $\mathbf{x} \equiv \mathbf{0}$ of Eq. (11-72) is **stable in the sense of Lyapunov** (or just **stable**) if when the initial condition is subjected to a small disturbance so that $\|\mathbf{x}(t_0)\| < \delta$ then the system transient response is also very small for all time $t > t_0$, and satisfies $\|\mathbf{x}(t)\| < \epsilon$, where $\epsilon > 0$ is any number and δ is a function ϵ. Here $\|\mathbf{x}(t)\|$ denotes the length of the vector $\mathbf{x}$, given by

$$\|\mathbf{x}(t)\| = \sqrt{x_1^2(t) + \cdots + x_n^2(t)}$$

Further a system is **asymptotically stable in the sense of Lyapunov** if it is stable and if $\lim_{t \to \infty} \|\mathbf{x}(t)\| = 0$.

It can be seen that the Lyapunov definition of stability is closely related to the concept of a continuous function at a point. For we say a function $\mathbf{f}(\mathbf{x})$ is continuous at a point $\mathbf{x}_0$ if arbitrary small changes $\delta\mathbf{x}$ about $\mathbf{x}_0$, always lead to small changes $\delta\mathbf{f}$ in the function about the value $\mathbf{f}(\mathbf{x}_0)$.

The usefulness of the Lyapunov stability method from an engineering standpoint is that often we can find a finite region (as opposed to infinitesimal) in the state space where the solutions are stable in the sense of Lyapunov. In this case if the system is disturbed to any point in this region it will always remain in this region and may converge to the equilibrium point from whence it started.

The Lyapunov Function

Consider the second-order system of Fig. 11-36 governed by

$$m\ddot{x}(t) + B\dot{x}(t) + kx(t) = 0$$

Assuming that the system is underdamped, the phase plot of this system is shown in Fig. 11-37. Now we form the sum of the potential and kinetic energy of the system

$$V = k_1 x_1^2 + k_2 x_2^2$$

where $x_1 = x$, $x_2 = \dot{x}$ and $k_1 = k/2$, $k_2 = m/2$ are positive constants. We note that if the system whose total energy is given by V is perturbed

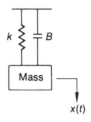

Figure 11.36: Second-order linear system

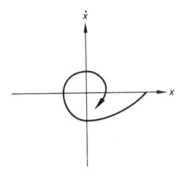

Figure 11.37: Phase plane plot for under damped second-order linear system

from its rest position, it will oscillate and eventually return to its rest position. Actually the energy increase of the system, due to a perturbation, is dissipated so that

$$\dot{V} = -Bx_2^2$$

Such a system is said to be asymptotically stable.

Fundamental to Lyapunov's stability criterion is the **Lyapunov function** V. Such a function has three important properties. The *first* is that V is positive for all nonzero values of the state variables that are used to define V in some region about the origin. The *second* is that $V = 0$ only if all the state variables are zero. When a scalar function, such as V, has these two properties it is said to be **positive definite**. If V is given by

$$V = x_1^2 + x_2^2$$

then $V(x_1, x_2)$ is shown in Fig. 11-38 or alternatively in 3-dimensions as

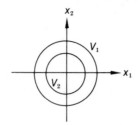

Figure 11.38: Plot of constant $V(x_1, x_2)$ loci in two dimensions

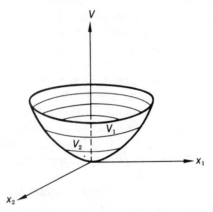

Figure 11.39: Plot of constant $V(x_1, x_2)$ loci

shown in Fig. 11-39. If V is given by

$$V = \sum_{i=1}^{n} x_i^2 \qquad (11.73)$$

then we have an nth-order system but V cannot be geometrically illustrated for $n > 3$.

We have seen examples above of functions which are positive definite for all points in the state space. This is not always the case, for the function $V = x_1^2 + x_2^5 + x_2^2$ will be negative for $x_1 = 0$ and $x_2 < -1$. This function is thus only positive definite for a small region about the origin, in the state space. Other functions may be positive, but not positive definite, as shown by the function $V = (x_1 + x_2)^2$. this function is not positive definite because $V = 0$ for all non-zero points where $x_1 = -x_2$.

The *third* important property is that the time derivative of $V(x)$ along the solution of Eq. (11-72) which is given by

$$\dot{V}(\mathbf{x}) = \frac{d}{dt}V(\mathbf{x}) = \frac{\partial V}{\partial \mathbf{x}}\mathbf{f}(\mathbf{x}) \qquad (11.74)$$

must satisfy $\dot{V}(\mathbf{x}) \le 0$ for all times $t > t_0$, and all points in the region of the state space where $V(\mathbf{x})$ is positive definite.

Lyapunov Stability Theorem

The stability criterion of Lyapunov is stated in terms of the Lyapunov function $V(\mathbf{x})$. Let us assume that a Lyapunov function for Eq. (11-72) can be found for a small region about the origin. Then the criterion states that the origin is **stable** in the sense of Lyapunov.

In addition if there exists a Lyapunov function $V(\mathbf{x})$ which also satisfies $\dot{V}(\mathbf{x}) < 0$ in a region about the origin then the criterion states that the origin is **asymptotically stable** in the sense of Lyapunov. Often the function V turns out to be quite different from that shown in Eqs.(11-73). Generally, it is quite difficult to obtain these V functions. Much research has been directed at ways of constructing or obtaining the Lyapunov functions. It should be noted that if a Lyapunov function can be found then the above results indicate the system is (asymptotic) stable. The converse statement however, that stability implies there exists a Lyapunov function is not always true, so that the inability to find a Lyapunov function does not imply the system is unstable.

To see how the above general results can be obtained consider the case of a second order system whose state space is the phase plane and suppose there exists a Lyapunov function $V(x,y)$ as defined above. From the definition of Lyapunov stability given above we take a circle of radius ϵ, as shown in Fig. 11-40, with its center at the origin which is assumed to be an equilibrium point. From the definition $V(x,y)$ is positive definite so a set of constant $V(x,y)$ contours will exist as shown in the above figure. It can be seen that the contour $V(x,y) = V_2$ remains inside the circle of radius ϵ and just grazes it at points A and B. It can also be seen that another circle can be inscribed within the contour $V(x,y) = V_2$, where its radius η is a function of ϵ. Now supposing all initial conditions $(x(t_0), y(t_0))$ are taken inside this latter circle, then $V(x_0, y_0) < V_2$. Also since $V(x,y)$ is a Lyapunov function

$$\dot{V}(x(t), y(t)) \le 0$$

so we can see that for all time $t > t_0$

$$V(x(t), y(t)) \le V(x_0, y_0) < V_2$$

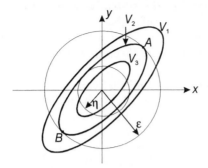

Figure 11.40: Outline of proof of Lyapunov stability theorem

Thus the trajectories $(x(t), y(t))$ remain inside the circle of radius ϵ for all time.

To establish the asymptotic stability condition we note $\dot{V}(x(t), y(t)) < 0$ so it is clear that $V(x(t), y(t)) \to 0$ as $t \to \infty$. Since the only point (x, y) where $V(x, y) = 0$ is the origin we can conclude that $x(t) \to 0$ and $y(t) \to 0$ as $t \to \infty$.

Example 19 *Using the sum of the potential and kinetic energies for V, show that the following mechanical system with a nonlinear spring is stable*

$$m\ddot{x} + a\dot{x} + k\left(x - \frac{x^3}{6}\right) = 0$$

The kinetic energy is $\frac{1}{2}m\dot{x}^2$ and the potential energy is

$$\text{Potential Energy} = \int_0^x k\left(y - \frac{y^3}{6}\right) dy = \frac{kx^2}{2}\left(1 - \frac{x^2}{12}\right)$$

The Lyapunov function becomes

$$V = \frac{1}{2}m\dot{x}^2 + \frac{kx^2}{2}\left(1 - \frac{x^2}{12}\right)$$

which is positive definite if $|x| \leq 12$. A singular point for the system is the origin. We now form $\dot{V}$

$$
\begin{aligned}
\dot{V} &= m\dot{x}\ddot{x} + kx\dot{x}\left(1 - \frac{x^2}{6}\right) \\
&= -a\dot{x}^2 - kx\dot{x}\left(1 - \frac{x^2}{6}\right) + kx\dot{x}\left(1 - \frac{x^2}{6}\right) = -a\dot{x}^2
\end{aligned}
$$

so that $\dot{V} \leq 0$. Therefore the system is stable in the sense of Lyapunov. In this example we cannot infer that the system is asymptotically stable since $\dot{V}$ is not a function of x and $\dot{x}$. Actually a different choice of V or the invoking of a different theorem shows that the system indeed is asymptotically stable.

A useful technique for obtaining a Lyapunov function is illustrated in the next example. Here $V(x, y)$ is only partially defined in terms of a general function, and then we look to obtain this function so that $\dot{V}(x, y) \leq 0$. If this function assures $V(x, y)$ is also positive definite then it is of the Lyapunov type.

Example 20 *Consider the non-linear control system discussed in Example 7, and show it is stable in the sense of Lyapunov by showing that there exists a Lyapunov function.*

The equation of motion for this system is

$$I\ddot{c}(t) + B\dot{c}(t) = e'(t) = -A\,\mathrm{sgn}\,c(t)$$

where

$$\mathrm{sgn}\,c(t) = \begin{cases} +1 & c(t) > 0 \\ -1 & c(t) < 0 \end{cases}$$

Defining the state variables $x(t) = c(t)$, $y(t) = \dot{c}(t)$ we obtain the state equations

$$\begin{aligned} \dot{x}(t) &= y(t) \\ \dot{y}(t) &= -\frac{B}{I}y(t) - \frac{A}{I}\,\mathrm{sgn}\,x(t) \end{aligned} \tag{11.75}$$

and find they have an equilibrium point at $(x, y) = (0, 0)$. As mentioned above let us consider

$$V(x, y) = f(x) + \frac{1}{2}y^2$$

where $f(x)$ will be chosen to ensure $\dot{V} \leq 0$. From Eq. (11-75)

$$\begin{aligned} \dot{V}(x, y) &= \begin{bmatrix} f'(x) & y \end{bmatrix} \begin{bmatrix} y \\ -\frac{B}{I}y - \frac{A}{I}\,\mathrm{sgn}\,x \end{bmatrix} \\ &= \left(f'(x) - \frac{A}{I}\,\mathrm{sgn}\,x \right) y - \frac{B}{I}y^2 \end{aligned}$$

We see that if $f'(x) - A\,\mathrm{sgn}\,x = 0$ then $\dot{V}(x,y) \leq 0$. Solving for $f(x)$ we find $f(x) = (A/I)\,x\,\mathrm{sgn}\,x$, so that

$$V(x,y) = \frac{A}{I}x\,\mathrm{sgn}\,x + \frac{1}{2}y^2$$

Thus we find $V(x,y)$ is positive definite and $\dot{V}(x,y) \leq 0$, so that it is a Lyapunov function. This means that the control system is stable in the sense of Lyapunov.

Example 21 *The equations of a rotating spacecraft are given by*

$$I_x\dot{\omega}_x + (I_z - I_y)\omega_y\omega_z \;=\; -B\omega_x$$

$$I_z\dot{\omega}_z + (I_y - I_x)\omega_x\omega_z \;=\; -B\omega_z$$

$$I_y\dot{\omega}_y + (I_x - I_z)\omega_x\omega_z \;=\; -B\omega_y$$

where I_x, I_y, I_z are principal inertias, B is a positive damping constant, and ω_x, ω_y, ω_z are angular rates. Show that the point $(0,0,0)$ is asymptotically stable.

We consider the scalar function V as

$$V = \frac{1}{2}I_x\omega_x^2 + \frac{1}{2}I_y\omega_y^2 + \frac{1}{2}I_z\omega_z^2$$

which is positive definite since V is positive for all values of ω_x, ω_y, ω_z except $\begin{bmatrix} \omega_x & \omega_y & \omega_z \end{bmatrix}^T = \mathbf{0}$. Forming $\dot{V}$

$$
\begin{aligned}
\dot{V} &= I_x\dot{\omega}_x\omega_x + I_y\dot{\omega}_y\omega_y + I_z\dot{\omega}_z\omega_z \\[4pt]
&= -(I_z - I_y)\omega_x\omega_y\omega_z - (I_x - I_z)\omega_x\omega_y\omega_z - (I_y - I_x)\omega_x\omega_y\omega_z \\[4pt]
&\quad -B\left(\omega_x^2 + \omega_y^2 + \omega_z^2\right) \\[4pt]
&= -B\left(\omega_x^2 + \omega_y^2 + \omega_z^2\right)
\end{aligned}
$$

so that $\dot{V} < 0$, and is zero only at $\begin{bmatrix} \omega_x & \omega_y & \omega_z \end{bmatrix}^T = \mathbf{0}$. Therefore the system is asymptotically stable.

Stability of Linearized Systems

The results we now present gives a strong justification for the intense study of linear control systems which was presented in earlier chapters.

Earlier in this chapter we asserted that the stability behavior of a second order control system can be determined from its linearized state equations. We now show that this is true for higher order systems. Suppose the state equations for a non-linear control system are given by

$$\dot{\mathbf{x}}(t) = \mathbf{f}(\mathbf{x}(t)) = \mathbf{A}\mathbf{x}(t) + \mathbf{g}(\mathbf{x}(t)) \tag{11.76}$$

where $\mathbf{x}(t)$ is an n-dimensional state vector, and $\mathbf{g}(\mathbf{x})$ is an n-dimensional function of $\mathbf{x}$, with $\mathbf{g}(\mathbf{0}) = \mathbf{0}$. We show that if all the eigenvalues of $\mathbf{A}$ have negative real parts and $\|\mathbf{g}(\mathbf{x})\| / \|\mathbf{x}\| \to 0$ as $\|\mathbf{x}\| \to 0$ then the equilibrium point at $\mathbf{x} = \mathbf{0}$ is asymptotically stable in the sense of Lyapunov.

It can been shown that if the eigenvalues of matrix $\mathbf{A}$ all have negative real parts then there exists a positive definite matrix $\mathbf{B}$ for which

$$\mathbf{A}^T\mathbf{B} + \mathbf{B}^T\mathbf{A} = \mathbf{C} \tag{11.77}$$

where $\mathbf{C}$ is negative definite[2]. Let us take the candidate Lyapunov function to be the positive definite function

$$V(\mathbf{x}) = \mathbf{x}^T\mathbf{B}\mathbf{x}$$

Differentiating this function with respect to time t gives

$$\begin{aligned}\dot{V}(\mathbf{x}) &= \dot{\mathbf{x}}^T\mathbf{B}\mathbf{x} + \mathbf{x}^T\mathbf{B}\dot{\mathbf{x}}\\ &= (\mathbf{A}\mathbf{x} + \mathbf{g}(\mathbf{x}))^T\mathbf{B}\mathbf{x} + \mathbf{x}^T\mathbf{B}(\mathbf{A}\mathbf{x} + \mathbf{g}(\mathbf{x}))\\ &= \mathbf{x}^T\mathbf{C}\mathbf{x} + \mathbf{g}(\mathbf{x})^T\mathbf{B}\mathbf{x} + \mathbf{x}^T\mathbf{B}\mathbf{g}(\mathbf{x})\end{aligned}$$

---

[2] An $n \times n$ matrix $\mathbf{A}$ is said to be positive (negative) definite if $\mathbf{x}^T\mathbf{A}\mathbf{x} > 0$ (< 0) for all non-zero vectors $\mathbf{x}$. Sylvester has shown a matrix $\mathbf{A}$

$$\mathbf{A} = \begin{bmatrix} a_{11} & a_{12} & \cdots & a_{1n} \\ a_{21} & a_{22} & \cdots & a_{2n} \\ \vdots & \vdots & & \vdots \\ a_{n1} & a_{n2} & \cdots & a_{nn} \end{bmatrix}$$

is positive (negative) definite if and only if the determinants

$$a_{11},\ \begin{vmatrix} a_{11} & a_{12} \\ a_{21} & a_{22} \end{vmatrix},\ \begin{vmatrix} a_{11} & a_{12} & a_{13} \\ a_{21} & a_{22} & a_{23} \\ a_{31} & a_{32} & a_{33} \end{vmatrix},\ \cdots,\ \begin{vmatrix} a_{11} & a_{12} & \cdots & a_{1n} \\ a_{21} & a_{22} & \cdots & a_{2n} \\ \vdots & \vdots & & \vdots \\ a_{n1} & a_{n2} & \cdots & a_{nn} \end{vmatrix}$$

are all positive (negative).

Now for every $\mathbf{x}$ such that $\|\mathbf{x}\|$ is sufficiently small we can see that the terms $g(\mathbf{x})^T \mathbf{Bx} + \mathbf{x}^T \mathbf{B}g(\mathbf{x})$ satisfy

$$\left| g(\mathbf{x})^T \mathbf{Bx} + \mathbf{x}^T \mathbf{B}g(\mathbf{x}) \right| < \alpha \mathbf{x}^T \mathbf{x}$$

where $\alpha > 0$ is dependent upon $\|\mathbf{x}\|$, and $\alpha \to 0$ as $\|\mathbf{x}\| \to 0$. Thus

$$\dot{V} \le \mathbf{x}^T [\mathbf{C} + \alpha(\|\mathbf{x}\|)\mathbf{I}]\mathbf{x}$$

and for $\|\mathbf{x}\|$ sufficiently small it can be seen that $\dot{V}(\mathbf{x}) < 0$ since $\mathbf{C}$ is negative definite. Thus the system described by Eq. (11-76) is asymptotically stable.

This result is useful as it shows that if a system can be linearized about an equilibrium point then its stability in the sense of Lyapunov can be determined from the stability of the linearized equation. Since virtually all practical control systems exhibit non-linearities to a greater or lesser extent this result gives confidence that the techniques developed in earlier chapters for totally linear systems can be applied to their practical counterparts. Thus in attacking a control problem we first see if it can be linearized, and if so, we use the above result and any of the classical methods such as root locus, Nyquist, or pole-placement techniques for its design.

Frequency Domain Methods

Frequency domain methods have been shown to be extremely useful for designing linear control systems. Popov pioneered the extension of these techniques for non-linear systems, where he has shown that provided a frequency domain criterion is satisfied then the system is asymptotically stable. These results have been extensively generalized to a large class of non-linearities, and have led to the development of a further class of so called **circle stability criteria**.

To see Popov's criterion let us consider the non-linear control system shown in Fig. 11-41. Here it is assumed the plant $G(s)$ is linear, while any non-linearities are represented by the feedback elements and are denoted by the function $\varphi(\cdot)$. It will also be assumed that the input $r(t) \equiv 0$. The transfer characteristic of the non-linearity is assumed to lie in the first and third quadrants as shown in Fig. 11-42, and to be bounded by the abscissa and a line of slope k passing through the origin. It will also be assumed that it is single-valued. Under these assumptions we can state the criterion.

Popov Criterion

If the following conditions are satisfied:

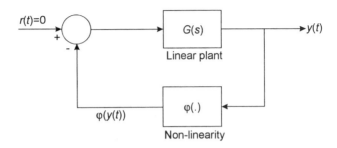

Figure 11.41: Non-linear control system for Popov criterion

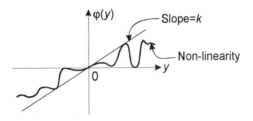

Figure 11.42: Restricted region for non-linearity

(a) $G(s)$ is stable, i.e. all poles lie in the left half of the s-plane

(b) the non-linearity satisfies

$$0 \leq \frac{\varphi(y)}{y} \leq k$$

for some $k > 0$, i.e. $\varphi(\cdot)$ lies in the sector $[0, k]$

(c) there exists a real number $q \geq 0$ such that

$$\text{Re}[(1 + jq\omega)G(j\omega)] + \frac{1}{k} > 0 \qquad \text{for all } \omega \leq 0 \qquad (11.78)$$

Then the system is stable and all outputs $y(t)$ approach zero asymptotically as $t \to \infty$.

We will not attempt to prove this result but will illustrate how it can be applied.

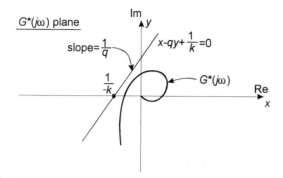

Figure 11.43: Permitted region for $G^*(j\omega)$ to satisfy Popov criterion

Example 22 *Suppose the control system shown in Fig. 11-41 has a linear plant with transfer function*

$$G(s) = \frac{s+3}{s^2 + 7s + 10}$$

Find the limit on the slope of the non-linearity such that the system is stable.

Evaluating $\mathrm{Re}[(1 + jq\omega)G(j\omega)] + 1/k$ we find

$$\mathrm{Re}[(1 + jq\omega)G(j\omega)] + \frac{1}{k} = 4\omega^2 + 30 + q\omega^2(\omega^2 + 11) + \frac{1}{k}(\omega^4 + 29\omega^2 + 100)$$

This expression can be seen to be positive for all $\omega > 0$, and for all $q > 0$, irrespective of the value of k, so long as $0 < k < \infty$. Thus so long as the non-linearity is single valued, passes through the origin, and lies anywhere in the first and third quadrants as shown in Fig. 11-42, the system is stable and $|y(t)| \to 0$ as $t \to \infty$.

Graphical Interpretation of Popov Criterion

To establish a graphical interpretation of the Popov criterion let us expand the first term in the inequality relation (11-78) given above

$$\mathrm{Re}[(1 + jq\omega)G(j\omega)] = G_R(\omega) - q\omega G_I(\omega)$$

where $G_R(\omega) = \mathrm{Re}\, G(j\omega)$ and $G_I(\omega) = \mathrm{Im}\, G(\omega)$. Defining the modified frequency response function as

$$G^*(j\omega) = G_R(\omega) + j\omega G_I(j\omega)$$

the relation (11-78) can be written as

$$G_R^*(\omega) - qG_I^*(\omega) + \frac{1}{k} > 0 \qquad (11.79)$$

where $G_R^*(\omega) = G_R(\omega)$ and $G_I^*(\omega) = \omega G_I(\omega)$. In the $G^*(j\omega)$ plane we plot the Popov line defined by

$$G_R^*(\omega) = -\frac{1}{k} + qG_I^*(\omega) \qquad (11.80)$$

as shown in Fig. 11-43. The relation (11-79) indicates that the plot $G^*(j\omega)$ must lies to the right of the line defined by Eq. (11-80) for some $q \geq 0$.

To illustrate the graphical application of the Popov criterion suppose the **modified frequency responses** $G^*(j\omega)$ for two different systems are as shown in Figs. 11-44(a) and (b) and the non-linearity lies inside the sector $[0, k]$. In Fig. 11-44(a) the intersect point $-1/k_1$ lies to the left of $G^*(j\omega)$ and a Popov line can be constructed for $q \geq 0$. Thus the system is asymptotically stable. The largest non-linear sector $[0, k_2]$ can also be obtained by constructing a Popov line tangential to the $G^*(j\omega)$ locus as shown on this same figure.

In Fig. 11-44(b) a second example is illustrated. In this case if the non-linear sector is $[0, k_3]$ then no conclusion can be drawn from the Popov criterion because no Popov line can be drawn passing through the point at $-1/k_3$. The largest non-linear sector for this example which satisfies the Popov criterion is obtained by constructing the Popov line which passes through the intersect point at $-1/k_4$.

11.6   Stability Region for Non-Linear Systems

For second order control systems the phase plane method is valuable because it not only gives information about its stability but also gives information about the global behavior of the system for large disturbances. With higher order non-linear control systems the Lyapunov method, not only can be used for investigating their stability, but can also be used for obtaining information about their behavior in finite regions about an equilibrium point. In this section we state a basic result without proof which can be used to determine such regions.

The following result which is applicable to the system described by Eq. (11-72) has been shown by Hale: Suppose $V(\mathbf{x})$ is a scalar function of the state vector $\mathbf{x}$, and has a continuous derivative. Let Ω_ℓ be a region in the

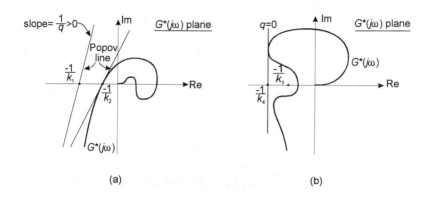

Figure 11.44: Graphical application of Popov criterion

state space for which $V(\mathbf{x}) < \ell$ and suppose Ω_ℓ is bounded (i.e. lies inside a sphere centered at the origin having a finite radius). We also assume that $\dot{V}(\mathbf{x}) = [\partial V(\mathbf{x})/\partial \mathbf{x}]\mathbf{f}(\mathbf{x}) \leq 0$ for all points in the region Ω_ℓ. Under these assumptions every solution of Eq. (11-72) starting with an initial condition $\mathbf{x}_0$ in Ω_ℓ remains in Ω_ℓ for all time $t > t_0$. If, in addition, we find that the only point in Ω_ℓ, where $\dot{V}(\mathbf{x}) = 0$ and for which a trajectory passing through the point has $\dot{V}(\mathbf{x}(t)) = 0$ for all $t > t_0$, is the origin then every solution starting in Ω_ℓ must converge asymptotically to the origin.

Example 23 *A non-linear magnetic bearing may be described by the equation*

$$\ddot{x} + b\dot{x} + x - x^3 = 0$$

Investigate the stability of this system for the equilibrium point at $x = 0$ using the Lyapunov method and find a region of stability for the system.

Defining the state variable $y = \dot{x}$ we obtain the state equations

$$\dot{x} = y$$
$$\dot{y} = -by - x + x^3$$

(11.81)

This system has equilibrium points at $(x, y) = (0,0)$, $(-1,0)$, $(1,0)$. Since this system can be considered as a mass connected to a non-linear spring

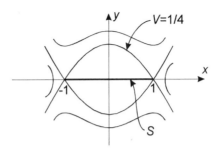

Figure 11.45: Region of stability for magnetic bearing

we will take $V(x, y)$ to be the total energy

$$
\begin{aligned}
V(x, y) &= \frac{1}{2}y^2 + \int_0^x \left(x - x^3\right) dx \\
&= \frac{1}{2}y^2 + \frac{1}{2}x^2 - \frac{1}{4}x^4
\end{aligned}
\tag{11.82}
$$

Now $V(x, y) > 0$ for a small neighborhood of the origin, and also $\dot{V}(x, y) = -by^2 \leq 0$. Hence from the Lyapunov theorem given in the previous section the system is stable in the sense of Lyapunov.

We now apply the result of Hale given above using Eq. (11-82) and taking $\ell = 1/4$. Plotting the curve for $V(x, y) = 1/4$ as shown in Fig. 11-45 shows the region $\Omega_{1/4}$ is bounded. In addition we see that $\dot{V}(x, y) \leq 0$ for all points inside $\Omega_{1/4}$. Consequently if the system is disturbed so that it starts operation at some point inside $\Omega_{1/4}$ then the state space trajectory will remain in the region for all time $t > t_0$.

Also we find that the points in $\Omega_{1/4}$ where $\dot{V}(x, y) = 0$ lie on the x-axis in the range $(-1, +1)$, and are denoted by S in Fig. 11-45. From Eq. (11-81) we see that trajectories passing through these points in S no longer give $\dot{V}(x, y) = 0$, except for the one passing through the origin. Consequently we find that all trajectories starting in $\Omega_{1/4}$ converge asymptotically to the origin.

It can be shown that other choices of the function $V(\mathbf{x})$ can be used to construct different and possibly larger regions of stability.

11.7 Summary

Since systems in real life are not linear, we have briefly discussed nonlinearities in this chapter. It has been assumed that the nonlinearities are small. Systems with small linearities were analyzed by two methods, via the phase-plane technique, and the describing function technique.

The phase-plane method involves the plotting of the variable versus its derivative with time as a parameter. This method is useful for studying the transient behavior as well as the stability of a system. Its chief drawback is its applicability to primarily second-order systems. It should be remembered however, that the number of second-order systems describing physical phenomena is very substantial.

The describing function technique involved the representation of the nonlinear element by its quasi-linearized form. The input to the nonlinear element was assumed to be a sinusoid. The output was represented by Fourier coefficients. It was assumed that the fundamental component is the most important part of the output. The transfer function, or the describing function, of the nonlinear element was defined as G_N,

$$G_N = \frac{\text{Fundamental amplitude of output}}{\text{Amplitude of sinusoidal input}}$$

Whereas this method was deemed attractive by virtue of the application of classical techniques already developed, its chief drawback was the constraint that the input must be a sinusoid.

The stability criterion of Lyapunov and Popov were introduced. The Lyapunov criterion is based on energy considerations. If a suitable scalar function V can be obtained, the stability of a system can be ascertained from V and $\dot{V}$. The difficulty lies in the fact that V is not readily available.

Following this the Popov stability criterion was considered. This approach utilizes a modified frequency response function $G^*(j\omega)$, derived from the plant frequency response function $G(j\omega)$, which must lie to the right of the Popov line if the system is to be stable. Because of its close similarities to the frequency response methods for linear systems many of the techniques used for their design can be adapted to non-linear control system design. It can be shown that the Popov and Lyapunov criterions are closely related because if there exists a Lyapunov function for a system then the Popov criterion is satisfied and vice versa.

Finally we examined the use of the Lyapunov method for constructing finite regions of stability and asymptotic stability. The ability to characterize these regions of stability is extremely important as control systems are often subject to disturbances. In these cases it is important to know the maximum range of these disturbances for which the system will still return to its operating conditions. Unfortunately since the approach is based upon the Lyapunov method the results are conservative. However by using a number of scalar functions $V(x)$ it is often possible to extend the stability region. Nevertheless the results obtained are extremely useful as they give considerable insight into the robustness of a control system subject to disturbances.

11.8   References

1. J. J. D'Azzo, C. A. Houpis, *Linear Control System Analysis and Design — Conventional and Modern*, Mc Graw Hill Book Co., New York, 1988.

2. S. Shinners, *Modern Control System Theory and Applications*, Addison Wesley Publ. Co., Reading, Mass, 1978.

3. L. Timothy, B. Bona, *State Space Analysis: An Introduction*, Mc Graw Hill Book Co., New York, 1980.

4. D. Luenberger, *Introduction to Dynamic Systems*, John Wiley and Sons, Inc., New York, 1979.

5. L. S. Pontryagin, *Ordinary Differential Equations*, Addison Wesley Publ. Co., Reading, Mass, 1962.

6. N. Minorsky, *Non Linear Oscillations*, Van Nostrand, Princeton N. J., 1962. (Reprinted by R. G. Krieger Publ. Co., Inc., Malabu, Florida, 1987).

7. D. F. Lawden, *Mathematics of Engineering Systems,* Methuen & Co. Ltd., London, 1961.

8. W. Hurewicz, *Lectures on Ordinary Differential Equations*, The M.I.T. Press, Cambridge, Mass, 1964.

9. J. Gibson, *Non Linear Automatic Control*, Mc Graw Hill Book Co., New York, 1963.

10. A. L. Greensite, *Elements of Modern Control Theory — Volume 1*, Spartan Books, Washington, DC, 1970.

11. K. Ogata, *Modern Control Engineering*, Prentice-Hall Inc., Englewood Cliffs, N. J., 1970.

12. V. V. Nemytskii, V. V. Stepanov, *Qualitative Theory of Differential Equations*, Princeton Univ. Press, Princeton, N. J., 1960.

13. S. Lefschetz, *Differential Equations: Geometric Theory*, John Wiley and Sons, New York, 1962.

14. J. K. Hale, *Ordinary Differential Equations*, John Wiley and Sons, Inc., New York, 1969.

15. W. Hahn, *Theory and Application of Liapunov's Direct Method*, Prentice Hall, Inc., Englewood Cliffs, N. J., 1963.

16. M. A. Aizerman, F. R. Gantmacher, *Absolute Stability of Regulator Systems*, Holden-Day, Inc., San Francisco, 1964.

17. L. Cesari, *Asymptotic Behavior and Stability Problems*, Springer-Verlag, Berlin, 1959.

18. B. van der Pol, "Force Oscillations in a Circuit with Non-Linear Resistance", Phil. Mag, Vol. 3 (1927), pp. 65-80.

19. R. J. Kochenburger, "A Frequency Response Method for Analyzing and Synthesizing Contactor Servomechanisms", Trans. AIEE, Vol. 69 (1950), pp. 270-284.

20. V. M. Popov, "Absolute Stability of Non-Linear Systems of Automatic Control", Automat. Remote Control (USSR), Vol. 22, (1962) pp. 857-875.

11.9   Problems

11-1 Draw a phase plane portrait of the system defined by

$$\dot{x}_1 \;=\; x_1 + 3x_2$$

$$\dot{x}_2 \;=\; x_1 + x_2$$

(a) Using the analytical solution

(b) Using the method of isoclines

11-2 For the system whose equation of motion is given below, find the state equations when the state variable y is defined as $y = \dot{x}$, and find the form of the phase plane portrait. Draw the trajectory on the phase plane for the initial conditions $x(0) = 0$, $y(0) = 5$.

$$\ddot{x}(t) + 8\dot{x}(t) + 100x(t) = 100$$

11-3 For the following linear systems classify the equilibrium point and discuss their stability
    (a) $\dot{x} = x + 3y,$ $\dot{y} = -6x + 7y$
    (b) $\dot{x} = x + 3y + 5,$ $\dot{y} = -6x + 7y$
    (c) $\dot{x} = 3x + y + 5,$ $\dot{y} = -x + y - 6$

11-4 Derive the equation of a pendulum. Linearize the equation and solve it. Under what conditions are the assumptions of linearization violated?

11-5 For the control system shown in Fig. P11-5, obtain c versus $\dot{c}$. Assume the input to be a step.

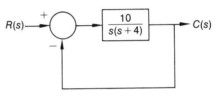

Figure P11-5

11-6 Could you analyze Problem 11-24 using phase-plane techniques? Explain.

11-7 Verify Eq. (11-44) and thence solve for the time t for the state to move from (x_0, y_0) to (x, y).

11-8 For each of the following systems find their isolated equilibrium points, and classify them:
    (a) $\dot{x} = -6y + 3xy - 8$ $\dot{y} = 10y^2 + 2x^2$
    (b) $\ddot{x} - \epsilon(1 - x^2)\dot{x} + x = 0$
    (c) $\ddot{x} + b\dot{x} + x - x^3 = 0$
    (d) $\ddot{x} - \dot{x}^3 + 5x + 10 = 0$

11-9 The populations of a host $H(t)$ and a parasite $P(t)$ are described by the equations

$$\frac{dH}{dt} = aH - bPH$$

$$\frac{dP}{dt} = cP - d\frac{P^2}{H}$$

where $H > 0$, and a, b, c, d are positive constants. Find and classify the equilibrium points, and use the method of isoclines to find the phase plane portrait.

11-10 Derive the equation of the isoclines for the system described by Eq. (11-38).

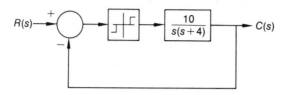

Figure P11-11

11-11 Problem 11-5 is modified with the nonlinear element as shown in Fig. P11-11. How has the system changed so far as overshoot is concerned? Assume a step input.

11-12 Could Problem 11-11 be solved if the input is a ramp?

11-13 Obtain the phase-plane plot for the system shown in Fig. P11-13.

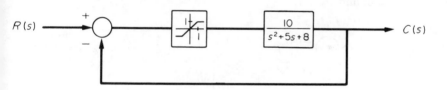

Figure P11-13

11-14 Can we say something about the steady state error for the last three problems? [Remember Problem 11-11 is a Type 1 system and Problem 11-13 is a type 0 system.]

11-15 The block diagram of a control system with non-linear damping is shown in Fig. P11-15(a) and the input-output relationship of the non-linear element G_N is shown in Fig. P11-15(b). Suppose the system is initially at rest and that a unit step input is applied. Construct the phase plane trajectories for the system.

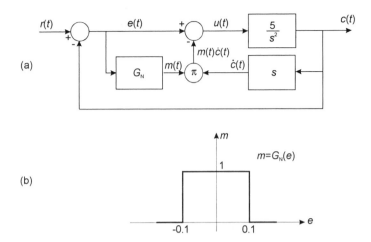

Figure P11-15

11-16 A position servo-system having coulomb friction is shown in Fig. P11-16(a). Coulomb friction is approximately independent of the magnitude of the velocity, but only depends upon its sign. Construct the phase plane trajectories in the $e(t)$, $y(t) = \dot{e}(t)$ plane when the initial conditions are $e(t) = 4$, $y(t) = 0$, assuming $r(t) \equiv 0$, $K = 5$, $J = 3$, and $F_c = 1$.

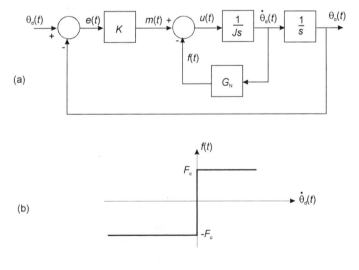

Figure P11-16

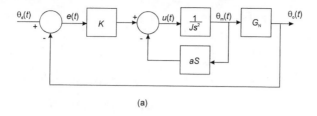

(a)

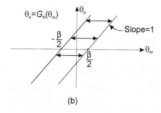

(b)

Figure P11-17

11-17 A position servo system having an output gear box with backlash is shown in Fig. P11-17. Construct the phase plane trajectories in the $e(t)$, $y(t) = \dot{e}(t)$ plane when the initial conditions are $e(t) = 4$, $y(t) = 0$, $r(t) \equiv 0$, $K = 5$, $J = 1$, $\beta = 1$, and $a = 0.2$, and determine if a limit cycle exists.

11-18 Construct a phase plane portrait for the system shown in Fig. P11-18 when $\Delta = 0.1$ assuming $\theta_d(t)$ is a unit step input, and the coordinates are $e(t)$ and $y(t) = \dot{e}(t)$.

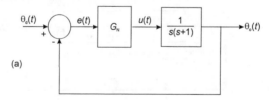

(a)

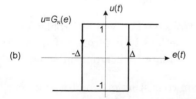

(b)

Figure P11-18

11-19 Using the Lienard transformation construct the phase plane portrait for the systems

(a)   described in Problem 11-16
(b) $\ddot{x} + x\dot{x} + 5x = 0$
(c) $\ddot{x} + (1 - x^2)\dot{x} + x = 0$

11-20 Find the limit cycles and determine their periods for the system

$$\dot{x} = y + (1 - x^2 - y^2)$$

$$\dot{y} = -x + y(1 - x^2 - y^2)$$

11-21 Verify the describing function for hysteresis in Table 11-1.

11-22 For the quadratic spring shown in Fig. 11-29 derive the magnitude of the third and fifth harmonics and compare to the first harmonic. Are you justified in neglecting the third and fifth harmonics? Show that the nth harmonic has an amplitude given by

$$B_n = \frac{-8}{(n+2)n(n-2)\cdots 1} \frac{R^2}{\pi}$$

11-23 For Example 17, how does the Nyquist and Bode plots change if the relay has deadband?

11-24 Using the describing function technique study the behavior of the control system shown in Fig. P11-24.

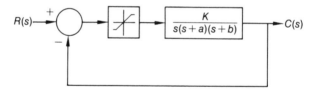

Figure P11-24

11-25 Consider the control system given in Problem 11-17 and apply the describing function method to determine if there are any limit cycle oscillations? If so, estimate their amplitude and frequency.

11-26 A servo-mechanism consisting of a motor and gear train at the output is shown in Fig. P11-26, where the non-linearity $G_N(\theta_m)$ is an inertia limited backlash due to gear tolerances. Using the describing function method determine the amplitude and frequency of any limit cycle assuming the backlash $\beta = 1$ degree.

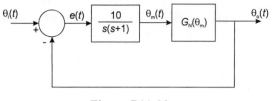

Figure P11-26

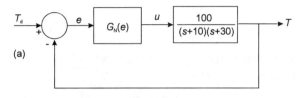

(a)

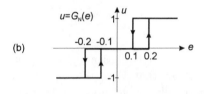

(b)

Figure P11-27

11-27 A heating control system where the relay has hysteresis and dead-band is shown in Fig. P11-27(a) and the transfer characteristic of the relay is given in Fig. P11-27(b). Investigate whether any limit cycles are present and if so determine their amplitude and frequency.

11-28 Use the sum of the kinetic and potential energies for the function V. For a simple pendulum with damping obtain the conditions for asymptotic stability.

11-29 A control system with a non-linear feedback gain is described by $\ddot{y} + b\dot{y} + y - y^2 = 0$. Taking $x_1 = y$ and $x_2 = \dot{y}$ the state equations are

$$\dot{x}_1 = x_2$$

$$\dot{x}_2 = -bx_2 - (x_1 - x_1{}^2)$$

Suppose the function V is defined as the total energy of the system

$$V(x_1, x_2) = \frac{1}{2}x_2^2 + \int_0^{x_1}(x - x^2)dx$$

Show that it is a Lyapunov function at the origin. Also find the largest region of stability for the system using this function.

11-30 *Quadratic Form Method.* Suppose a control system has the state equation with n state variables given by

$$\dot{\mathbf{x}} = \mathbf{f}(\mathbf{x})$$

where $\mathbf{f}(0) = 0$, and suppose $\mathbf{f}(\mathbf{x})$ can be written as

$$f_1(\mathbf{x}) = f_{11}(x_1) + \cdots + f_{1n}(x_n)$$
$$\vdots$$
$$f_n(\mathbf{x}) = f_{n1}(x_1) + \cdots + f_{nn}(x_n)$$

Assume that $\lim_{x_j \to 0} [f_{ij}(x_j)/x_j]$ exists. The state equations can be written as

$$\dot{\mathbf{x}} = \mathbf{F}(\mathbf{x})\,\mathbf{x} = \begin{bmatrix} f_{11}(x_1)/x_1 & \cdots & f_{1n}(x_n)/x_n \\ \vdots & & \vdots \\ f_{n1}(x_1)/x_1 & \cdots & f_{nn}(x_n)/x_n \end{bmatrix} \begin{bmatrix} x_1 \\ \vdots \\ x_n \end{bmatrix}$$

The trial Lyapunov function is chosen to be $V(\mathbf{x}) = \mathbf{x}^T \mathbf{B} \mathbf{x}$ with $\mathbf{B}$ positive definite. Show that $V(\mathbf{x})$ is a Lyapunov function, and the system is asymptotically stable, provided there exists a negative definite matrix $\mathbf{C}(\mathbf{x})$ for all points $\mathbf{x}$ in a small neighborhood of the origin for which $\mathbf{F}^T(\mathbf{x})\mathbf{B} + \mathbf{B}^T \mathbf{F}(\mathbf{x}) = \mathbf{C}(\mathbf{x})$.

11-31 *Krasovskii's Method.* Suppose a control system has the state equation with n state variables given by $\dot{\mathbf{x}} = \mathbf{f}(\mathbf{x})$. Suppose the function $\mathbf{f}(\mathbf{x})$ is non-zero for all $\mathbf{x}$ in a neighborhood of the origin excepting for $\mathbf{f}(0) = 0$. Take as a trial Lyapunov function

$$V(\mathbf{x}) = \dot{\mathbf{x}}^T \dot{\mathbf{x}} = \mathbf{f}^T(\mathbf{x})\mathbf{f}(\mathbf{x})$$

which is positive definite.

(a) Show that the system is asymptotically stable if

$$\frac{\partial \mathbf{f}(\mathbf{x})}{\partial \mathbf{x}}^T + \frac{\partial \mathbf{f}(\mathbf{x})}{\partial \mathbf{x}}$$

is negative definite in a neighborhood of the origin.

(b) Consider the non-linear control system shown in Fig. P11-31. Applying the Krasovskii method, show the system is asymptotically stable if

$$a > 0, \quad \frac{\partial g_1(x_1)}{\partial x_1} > 0, \quad \text{and} \quad 4a\frac{\partial g_1(x_1)}{\partial x_1} - \left[1 - \frac{\partial g_2(x_2)}{\partial x_2}\right]^2 > 0$$

in some neighborhood of the origin $(0,0)$.

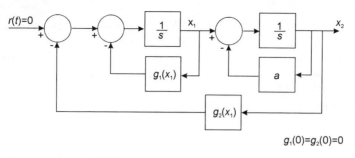

Figure P11-31

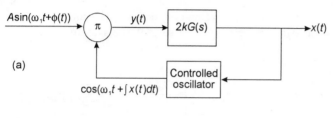

(a)

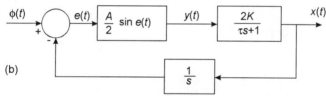

(b)

Figure P11-32

11-32 A phase locked loop is a control system for synchronizing a controlled oscillator to a periodically varying input signal $A \sin(\omega t + \phi(t))$ as shown in Fig. P11-32(a). The output of the controlled oscillator is given by

$$b(t) = \cos\left(\omega_1 t + \int x(t)dt\right)$$

This signal is instantaneously multiplied with the input signal to give

$$y(t) = A \sin\left(\omega_1 t + \phi(t)\right) \cos\left(\omega_1 t + \int x(t)dt\right)$$

This signal is fed to the transfer function $G(s)$ which acts as a low pass filter with a cutoff frequency ω_c. Show that if $2\omega_c \ll \omega_1$ then

$$y(t) \approx \frac{A}{2} \sin\left(\phi(t) - \int x(t)dt\right)$$

and show the equivalent block diagram representing this system is as given in Fig. P11-32(b). Taking $G(s) = 1/(\tau s + 1)$ where $\omega_c = 1/\tau$, obtain a set of state equations for the system and apply Krasovskii's method to demonstrate that the system is stable under appropriate conditions. Using the obtained Lyapunov function find the region of stability.

11-33 A unity feedback control system is shown in Fig. P11-33 where it contains a linear plant and a non linear controller. Using Popov's method determine the value of k defining the maximum slope of the non-linearity sector for $G_N(e)$ so that the system is stable.

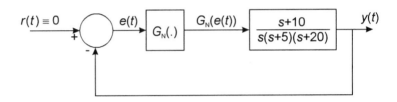

Figure P11-33

11-34 A non-linear control system is shown in Fig. P11-34(a).

(a) Using Popov's method find an upper limit for gain k of the non-linearity which assures stability.

(b) Demonstrate the system shown in Fig. P11-34(b) is equivalent to the one shown in Fig. P11-34(a). Apply Popov's method to the equivalent system shown in Fig. P11-34(b) for $a = 5$ and show the non-linearity gain k is increased over the value obtained in (a) above.

[This method is often referred to as a pole-shifting technique, and it can be used to obtain increased values for the maximum slope k, of the non-linearity, which assures stability.]

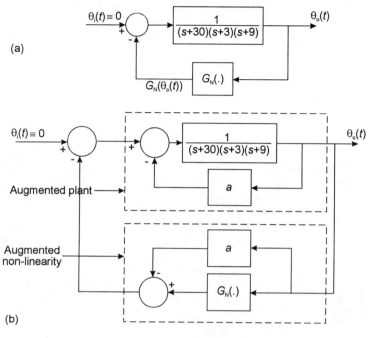

Figure P11-34

Chapter 12

Systems with Stochastic Inputs

12.1   Introduction

We have been studying systems that are subjected to deterministic inputs that are also Laplace transformable. In many real systems, however, the input is either in a form that is not easily Laplace transformable or more importantly has a structure that varies in a random manner. A signal which has a random component is generally referred to as a stochastic signal. Sometimes the random behavior can be characterized by a particular type of probability distribution such as Gaussian or Weibull distribution, but quite often this is not possible.

In this chapter we are interested in developing an approach to describe and analyze linear systems that are subjected to inputs that are stochastic in nature. Since the approach is based on an understanding of elementary probability theory we will briefly review fundamental concepts and terminology which are used to describe information that has some probabilistic content.

A random variable x is said to have a **probability density function** $p(x)$ such that

$$\int_{x_1}^{x_2} p(x)dx = P(x_1 < x \le x_2)$$

where $P(x_1 < x \le x_2)$ is the probability that x lies between x_1 and x_2. A property of the probability density function $p(x)$ is that

$$\int_{-\infty}^{\infty} p(x)dx = 1$$

The **probability distribution function** $F(x)$ is defined by

$$F(x_1) = P(x \leq x_1)$$

where $F(x_1)$ is the probability that x is equal to or less than x_1. It should be noted that $F(\infty) = 1$.

The **mean, expected,** or **average** value of x is defined as

$$E(x) = \bar{x} = \int_{-\infty}^{\infty} x\, p(x)dx$$

If necessary the mean could also be described over a more limited range. A general expression useful for expressing several important indices describing probability distribution functions is given by the nth moment about the mean,

$$E\left\{(x - \bar{x})^n\right\} = \int_{-\infty}^{\infty} (x - \bar{x})^n p(x)dx \qquad (12.1)$$

We note that when $n = 0$ the above integral goes to 1 and when $n = 1$ the integral goes to zero. When $n = 2$ we have

$$\sigma^2 = E\left\{(x - \bar{x})^2\right\} = \int_{-\infty}^{\infty} (x - \bar{x})^2 p(x)dx$$

where σ is the standard deviation and σ^2 is called the **variance**.

A useful and interesting probability density function is the **Gaussian** or **normal** distribution described by

$$p(x) = \frac{1}{\sigma\sqrt{2\pi}} e^{-(x-\bar{x})^2/2\sigma^2} \qquad (12.2)$$

For the Gaussian distribution the third moment about the mean is zero which simply means that the distribution is symmetrical about the mean or average value.

When there is only one random variable the probability distribution function is often prefaced by the word *first*. In many engineering applications, however, we are often concerned with two variables that have random content. In this case, the **joint probability density function** is described such that

$$\int_{-\infty}^{\infty} \int_{-\infty}^{\infty} p(x, y)\, dx\, dy = 1$$

and the **mean** or **expected** value is

$$E(xy) = \overline{xy} = \int_{-\infty}^{\infty} \int_{-\infty}^{\infty} xy\, p(x,y) dx dy$$

A random time function $x(t)$ can be viewed from two perspectives. Firstly if we observe it over time it will vary in a random manner from one instant to the next. The alternative way of viewing it is to imagine that at each time t, $x(t)$ is a random variable. If we compute an average, in the former case it is referred to as a **time average**, and in the latter it is referred to as an **ensemble average**. In many engineering applications we assume that the statistical properties of $x(t)$ do not change by a shift in time origin or we say it is a **time stationary variable**. Further, we assume that the time average and ensemble average are equal. This assumption is referred to as the **ergodic hypothesis** and allows us to write

$$E\{x(t)\} = \int_{-\infty}^{\infty} x\, p(x) dx = \lim_{T \to \infty} \frac{1}{2T} \int_{-T}^{T} x(t) dt \qquad (12.3)$$

Since signals of interest in control systems are generally functions of time it is convenient to use time averages. In the remainder of this chapter we will assume that the ergodicity of variables hold and compute time averages rather than ensemble averages.

12.2 Signal Properties

The type of signals that have been discussed in the previous section have important and useful properties both in the time domain and frequency domain. In this section we review these properties before discussing the response of linear systems that are subjected to such signals.

Time Domain Properties

For a signal which is **ergodic**, the expected value of the signal is given by

$$\overline{x(t)} = \lim_{T \to \infty} \frac{1}{2T} \int_{-T}^{T} x(t) dt$$

The expected value of $x(t)x(t + \tau)$ is given by

$$E\{x(t)x(t+\tau)\} = \overline{x(t)x(t+\tau)} = \phi_{xx}(\tau) = \lim_{T \to \infty} \frac{1}{2T} \int_{-T}^{T} x(t)x(t+\tau) dt$$

$$(12.4)$$

This expected value is called the **autocorrelation function** and is denoted by $\phi_{xx}(\tau)$. It gives information about the relationship or correlation of two samples of the same signal shifted by time τ. We will see that if $x(t)$ and $x(t+\tau)$ are completely independent of each other then $\phi_{xx}(\tau) = 0$ for all $\tau \neq 0$. We note that if $\tau = 0$ then Eq. (12-4) gives the variance and

$$\phi_{xx}(0) = \sigma_x^2$$

Since the correlation of a signal is strongest to itself, i.e. $\tau = 0$, $\phi_{xx}(0)$ is always greater than $\phi_{xx}(\tau)$. We also note that the autocorrelation function is an even function and $\phi_{xx}(\tau) = \phi_{xx}(-\tau)$.

The expected value of two signals $x(t)$ and $y(t+\tau)$ is given by

$$E\left\{x(t)y(t+\tau)\right\} = \overline{x(t)y(t+\tau)} = \phi_{xy}(\tau) = \lim_{T \to \infty} \frac{1}{2T} \int_{-T}^{T} x(t)y(t+\tau)dt$$

$$(12.5)$$

where $\phi_{xy}(\tau)$ is the **crosscorrelation function** and yields information as to the dependence or correlation of $x(t)$ to $y(t)$. Again if $x(t)$ and $y(t)$ are completely independent then $\phi_{xy}(\tau) = 0$. If the two signals have a finite and nonzero value for the crosscorrelation function, the signals are said to be **linearly correlated**. More importantly, the crosscorrelation function will have frequencies that appear in *both* $x(t)$ and $y(t)$ with the amplitude being dependent upon the amplitude, phase and frequencies in the two signals. The crosscorrelation function is not an even function, and does not relate to the variance as was the case with the autocorrelation function, although we note that $\phi_{xy}(\tau) = \phi_{yx}(-\tau)$.

Example 1 *Obtain the autocorrelation function if $x(t) = A\cos\omega t$.*

Since the autocorrelation function is even, we have

$$\phi_{xx}(\tau) = \lim_{T \to \infty} \frac{1}{T} \int_0^T A^2 \cos\omega t \cos\omega(t+\tau)\, dt = \frac{A^2}{2}\cos\omega\tau$$

Example 2 *Obtain the crosscorrelation function for*

$$x(t) = A\cos\omega_1 t \quad and \quad y(t) = B\cos\omega_2 t,$$

for the two cases, $\omega_1 = \omega_2 = \omega$, and $\omega_1 \neq \omega_2$.

The crosscorrelation function becomes

$$\phi_{xy}(\tau) = \lim_{T \to \infty} \frac{1}{2T} \int_{-T}^{T} AB\cos\omega_1 t \cos\omega_2(t+\tau)\, dt$$

If $\omega_1 = \omega_2 = \omega$ this reduces to the autocorrelation function and

$$\phi_{xy}(\tau) = \frac{AB}{2} \cos \omega \tau$$

If $\omega_1 \neq \omega_2$ we find that the integral yields

$$\phi_{xy}(\tau) = 0$$

Example 3 *Obtain the autocorrelation function if $x(t) = x_1(t) + x_2(t)$, where $x_1(t) = A \cos \omega_1 t$, $x_2(t) = B \cos \omega_2 t$ and $\omega_1 \neq \omega_2$.*

The autocorrelation function becomes

$$\phi_{xx}(\tau) = \phi_{x_1 x_1}(\tau) + \phi_{x_1 x_2}(\tau) + \phi_{x_2 x_1}(\tau) + \phi_{x_2 x_2}(\tau)$$

The results from the last example show that $\phi_{x_1 x_2}(\tau) = \phi_{x_2 x_1}(\tau) = 0$ and from the first example we obtain the remaining functions so that

$$\phi_{xx}(\tau) = \frac{A^2}{2} \cos \omega_1 \tau + \frac{B^2}{2} \cos \omega_2 \tau$$

We note the autocorrelation function picks up both the frequencies that appear in the signal $x(t)$.

Frequency Domain Properties

Assuming that the time functions given by Eq. (12-4) and Eq. (12-5) are Fourier transformable we can define the Fourier transform of the autocorrelation and crosscorrelation functions as

$$\Phi_{xx}(j\omega) = \int_{-\infty}^{\infty} \phi_{xx}(\tau) e^{-j\omega\tau} \, d\tau \qquad (12.6)$$

$$\Phi_{xy}(j\omega) = \int_{-\infty}^{\infty} \phi_{yx}(\tau) e^{-j\omega\tau} \, d\tau \qquad (12.7)$$

where $\Phi_{xx}(j\omega)$ is called the **power-spectral density** (or power-density) function and $\Phi_{xy}(j\omega)$ is called the **cross-spectral density** (or cross-density) function. The inverse Fourier transforms are likewise defined as

$$\phi_{xx}(\tau) = \frac{1}{2\pi} \int_{-\infty}^{\infty} \Phi_{xx}(j\omega) e^{j\omega\tau} \, d\omega \qquad (12.8)$$

$$\phi_{xy}(\tau) = \frac{1}{2\pi} \int_{-\infty}^{\infty} \Phi_{xy}(j\omega) e^{j\omega\tau} \, d\omega \qquad (12.9)$$

An interesting and useful interpretation of Eq. (12-8) is obtained for the case when $\tau = 0$. We noted earlier that when $\tau = 0$, $\phi_{xx}(0) = \sigma_x^2$ so that Eq. (12-8) reduces to

$$\sigma_x^2 = \frac{1}{2\pi} \int_{-\infty}^{\infty} \Phi_{xx}(j\omega)d\omega \qquad (12.10)$$

which states that the variance times 2π is equal to the area under the power-spectral density curve.

Example 4 *Consider the time function $x(t) = A\cos\omega_1 t$ whose autocorrelation function is*

$$\phi_{xx}(\tau) = \frac{A^2\cos\omega_1\tau}{2}$$

Find the power-spectral density function for this signal.

The power-spectral density becomes

$$\begin{aligned}
\Phi_{xx}(j\omega) &= \int_{-\infty}^{\infty} \frac{A^2\cos\omega_1\tau}{2}e^{-j\omega\tau}d\tau \\
&= \int_{-\infty}^{\infty} \frac{A^2}{4}\left(e^{-j(\omega-\omega_1)\tau} + e^{-j(\omega+\omega_1)\tau}\right)d\tau
\end{aligned}$$

From the properties of Fourier transforms we have

$$\Phi_{xx}(j\omega) = \frac{A^2\pi}{2}\left[\delta\left(\omega - \omega_1\right) + \delta\left(\omega + \omega_1\right)\right]$$

Since the power-spectral density is a real function of ω it is often written as

$$\Phi_{xx}(j\omega) = \Phi_{xx}(\omega) = \frac{A^2\pi}{2}\left[\delta\left(\omega - \omega_1\right) + \delta\left(\omega + \omega_1\right)\right]$$

The autocorrelation functions, and the Fourier transforms of some standard signals are listed in Table 12-1. In practice the signals under analysis may not look like those shown in Table 12-1. Under such situations the best approach is to calculate the autocorrelation function for various values of τ and then fit a mathematical function through the data. The power-spectral density function is then obtained by taking the Fourier transform of the fitted mathematical function.

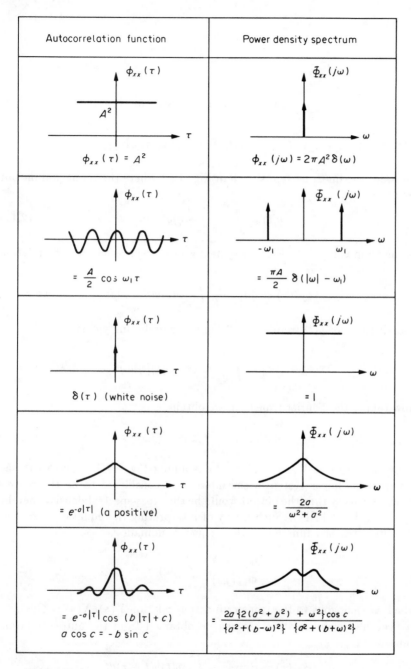

Autocorrelation function	Power density spectrum
$\phi_{xx}(\tau)$ A^2 $\phi_{xx}(\tau) = A^2$	$\Phi_{xx}(j\omega)$ $\phi_{xx}(j\omega) = 2\pi A^2 \delta(\omega)$
$\phi_{xx}(\tau)$ $= \dfrac{A}{2}\cos \omega_1 \tau$	$\Phi_{xx}(j\omega)$ $-\omega_1 \quad \omega_1$ $= \dfrac{\pi A}{2}\delta(|\omega| - \omega_1)$
$\phi_{xx}(\tau)$ $\delta(\tau)$ (white noise)	$\Phi_{xx}(j\omega)$ $= 1$
$\phi_{xx}(\tau)$ $= e^{-a|\tau|}$ (a positive)	$\Phi_{xx}(j\omega)$ $= \dfrac{2a}{\omega^2 + a^2}$
$\phi_{xx}(\tau)$ $= e^{-a|\tau|}\cos(b|\tau| + c)$ $a\cos c = -b\sin c$	$\Phi_{xx}(j\omega)$ $= \dfrac{2a\{2(a^2 + b^2) + \omega^2\}\cos c}{\{a^2 + (b-\omega)^2\}\ \{a^2 + (b+\omega)^2\}}$

Table 12-1 Autocorrelation functions and their power density spectra

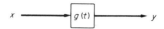

Figure 12.1: Linear system

12.3   Input-Output Relationships

Consider the linear system shown in Fig. 12-1 where the input and output are related by

$$y(t) = \int_{-\infty}^{\infty} x(t - \eta)g(\eta)d\eta \qquad (12.11)$$

as shown in Chapter 3. For this system we can express the crosscorrelation function as

$$\phi_{xy}(\tau) = \lim_{T \to \infty} \frac{1}{2T} \int_{-T}^{T} x(t)y(t + \tau)dt$$

Substituting for $y(t)$ from Eq. (12-11) and simplifying yields

$$\phi_{xy}(\tau) = \int_{-\infty}^{\infty} \phi_{xx}(\tau - \eta)g(\eta)d\eta$$

Now taking the Fourier transform we obtain

$$\Phi_{xy}(j\omega) = G(j\omega)\Phi_{xx}(j\omega) \qquad (12.12)$$

Comparing this to Eq. (12-11) and examining Fig. 12-1 allows us to say that for this linear system if the input were considered to be the power-spectral density then the output would be the cross-spectral density thereby satisfying Eq. (12-12). Such a way of interpreting the equation allows us to write a transfer function in the frequency domain as

$$\frac{\Phi_{xy}(j\omega)}{\Phi_{xx}(j\omega)} = G(j\omega)$$

which is analogous to the transfer function relation $Y(s)/X(s) = G(s)$.

Another useful relationship can be obtained from the autocorrelation function

$$\phi_{yy}(\tau) = \lim_{T \to \infty} \frac{1}{2T} \int_{-T}^{T} y(t)y(t + \tau)dt$$

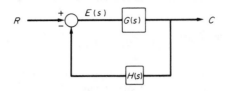

Figure 12.2: Closed loop control system

Substituting for $y(t)$ from Eq. (12-11) and simplifying yields

$$\phi_{yy}(\tau) = \int_{-\infty}^{\infty} \int_{-\infty}^{\infty} \phi_{xx}(\tau + \eta - \beta)g(\eta)g(\beta)d\eta \, d\beta$$

Now taking the Fourier transform we obtain

$$\Phi_{yy}(j\omega) = G(j\omega)G(-j\omega)\Phi_{xx}(j\omega)$$

or

$$\Phi_{yy}(j\omega) = |G(j\omega)|^2 \, \Phi_{xx}(j\omega) \tag{12.13}$$

This equation relates the power-spectral densities of the input and output. It can also be used for obtaining the variance of the output by taking the inverse Fourier transform of Eq. (12-13) and setting $\tau = 0$ to yield

$$\sigma_y^2 = \Phi_{yy}(0) = \frac{1}{2\pi} \int_{-\infty}^{\infty} |G(j\omega)|^2 \, \Phi_{xx}(j\omega)d\omega \tag{12.14}$$

We note that whereas this gives the variance of the output, Eq. (12-10) is used for obtaining the variance of the input.

The previous approach allows us to write useful relationships involving various signals in a feedback control system. Consider the control system shown in Fig. 12-2. The output and input if represented in the Laplace domain are related by $C(s)/R(s) = G(s)/[1 + G(s)H(s)]$. This immediately suggests, and can be rigorously derived using the previous approach, that

$$\Phi_{cr}(j\omega) = \frac{G(j\omega)}{1 + G(j\omega)H(j\omega)}\Phi_{rr}(j\omega) \tag{12.15}$$

This equation relates the cross-spectral density $\Phi_{cr}(j\omega)$ of the input and output signal to the power-spectral density of the input signal using the overall transfer function expressed in the frequency domain. Similarly, the error-input cross-spectral density is

$$\Phi_{er}(j\omega) = \frac{1}{1 + G(j\omega)H(j\omega)}\Phi_{rr}(j\omega) \tag{12.16}$$

The power-spectral density of the input and output are related by

$$\Phi_{cc}(j\omega) = \left| \frac{G(j\omega)}{1 + G(j\omega)H(j\omega)} \right|^2 \Phi_{rr}(j\omega) \qquad (12.17)$$

A particular type of a random signal, useful for system analysis and closely approximated by many practical situations, is referred to as **white noise**. The autocorrelation function of white noise can be mathematically derived and is found to be

$$\phi_{xx}(\tau) = N\delta(\tau)$$

where $\delta(\tau)$ is a delta function. White noise is only correlated with itself at $\tau = 0$. The power-spectral density of white noise is thus a constant,

$$\Phi_{xx}(j\omega) = N$$

Theoretically, this means that it would require infinite energy to produce white noise. Clearly, such a signal does not exist physically although approximations do exist. For a particular transfer function of the type shown in Fig. 12-1 we have

$$G(j\omega) = \frac{\Phi_{xy}(j\omega)}{\Phi_{xx}(j\omega)}$$

and

$$|G(j\omega)| = \sqrt{\frac{\Phi_{yy}(j\omega)}{\Phi_{xx}(j\omega)}}$$

So if the input is considered to be white noise we have

$$G(j\omega) = \frac{\Phi_{xy}(j\omega)}{N}, \quad |G(j\omega)| = \sqrt{\frac{\Phi_{yy}(j\omega)}{N}}$$

which means that given the cross-spectral density of the input-output from experimental data, we can determine the transfer function of the system under consideration if its input signal is white noise. Also, given the power-spectral density of the output we can obtain the magnitude of the transfer function. It is interesting to note that if $N = 1$, the cross-spectral density function $\Phi_{xy}(j\omega)$ equals the system transfer function, or alternatively the system impulse response $g(\tau) = \phi_{xy}(\tau)$.

With the advent of advanced digital signal processors and the fast Fourier transform algorithm, instruments have been developed for laboratory use for measuring the transfer function of linear systems, by injecting white noise into their inputs and measuring the cross-correlation function $\phi_{xy}(\tau)$. The dynamic signal analyzer manufactured by Tektronix for use

Figure 12.3: Dynamic signal analyser

with a desk top computer is shown in Fig. 12-3. It can be used for carrying out the measurements discussed above and then by using post-measurement software supplied with the machine a transfer function $G(j\omega)$ can be fitted to the experimental data.

Example 5 *The input signal to a unity feedback control system has an autocorrelation function that has been determined to be $\phi_{rr}(\tau) = e^{-0.5|\tau|}$. If the forward loop transfer function is*

$$G(s) = \frac{1}{s(s+2)}$$

what is the power-spectral density of the output?

The power-spectral density of the input is obtained by taking the Fourier transform of $\phi_{rr}(\tau)$ (given in Table 12-1) which is

$$\Phi_{rr}(j\omega) = \frac{1}{\omega^2 + 0.25}$$

The power-spectral density of the output, from Eq. (12-17), becomes

$$\Phi_{cc}(j\omega) = \frac{4}{(1+\omega^2)^2(1+4\omega^2)}$$

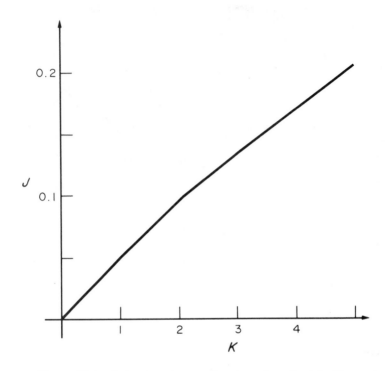

Figure 12.4: Output variance of system described in Example 6

Example 6 *A unity feedback control system has a forward loop transfer function of $G(s) = K/s(s + A)$ and is subjected to input whose autocorrelation is given by $\phi_{rr}(\tau) = e^{-0.5|\tau|}$. Obtain an expression for the output variance and solve for $A = 10$ and $K = 5$.*

The input power-spectral density is given by

$$\phi_{rr}(\tau) = \frac{4}{\omega^2 + 4}$$

and the magnitude of the overall transfer function is

$$\left| \frac{G(j\omega)}{1 + G(j\omega)} \right| = \frac{K}{\sqrt{(K - \omega^2)^2 + \omega^2 A^2}}$$

The variance of the output becomes

$$J = \overline{c(t)}^2 = \frac{1}{2\pi} \int_{-\infty}^{\infty} \frac{4}{\omega^2 + 4} \left| \frac{K^2}{(K - \omega^2)^2 + \omega^2 A^2} \right| d\omega$$

This integral is difficult to compute in closed form and therefore a numerical approach is used. The results are shown in Fig. 12-4. For $A = 10$ and $K = 5$ we obtain $\overline{c\left(t\right)^2} = 0.20688$.

12.4   Linear Correlation

In many systems there is not only an input of interest but also noise inputs that may enter the system at different locations. Generally, this noise or interference is unwelcome and it is desirable to relate the power emanating from this noise source to the signal. Consider the unity feedback control system shown in Fig. 12-5 where $r(t)$ is the input and $n(t)$ some noise interference or an input that accounts for some inherent system nonlinearity. Assuming that the system as shown behaves in a linear way and $n(t)$ and $r(t)$ are uncorrelated, then the power-spectral density of the output due to $r(t)$ and $n(t)$ is given by

$$\Phi_{cc}\left(j\omega\right) = \left|\frac{G_1\left(j\omega\right)G_2\left(j\omega\right)}{1 + G_1\left(j\omega\right)G_2\left(j\omega\right)}\right|^2 \Phi_{rr}\left(j\omega\right)$$

$$+ \left|\frac{G_2\left(j\omega\right)}{1 + G_1\left(j\omega\right)G_2\left(j\omega\right)}\right|^2 \Phi_{nn}\left(j\omega\right)$$

The **linear correlation** ρ is defined as the square root of the fraction of the signal power output to the total power output and can be obtained from the previous equation

$$\rho^2 = \left|\frac{G_1\left(j\omega\right)G_2\left(j\omega\right)}{1 + G_1\left(j\omega\right)G_2\left(j\omega\right)}\right|^2 \frac{\Phi_{rr}\left(j\omega\right)}{\Phi_{cc}\left(j\omega\right)}$$

$$= 1 - \left|\frac{G_2\left(j\omega\right)}{1 + G_1\left(j\omega\right)G_2\left(j\omega\right)}\right|^2 \frac{\Phi_{nn}\left(j\omega\right)}{\Phi_{cc}\left(j\omega\right)}$$

We note that if $\phi_{nn}\left(j\omega\right) = 0$, i.e. there is no noise, the linear correlation is unity. Often the **signal to noise ratio** (SNR) is also defined using linear correlation as

$$\text{SNR} = \frac{|G_1(j\omega)|^2\Phi_{rr}(j\omega)}{\Phi_{nn}(j\omega)} = \frac{\rho^2}{1 - \rho^2}$$

Again, if there is no noise, the SNR is infinite.

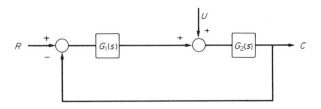

Figure 12.5: System with disturbance

12.5   Summary

Some fundamental concepts of probability theory have been presented in order to introduce the idea of **correlation** or **coherence**. The **auto-correlation** function is seen to give information about dependency of a function with itself at a later time whereas the **crosscorrelation** gives the dependency between two functions. The Fourier transforms of the auto-correlation and crosscorrelation functions yield the **power-spectral** and **cross-spectral** densities.

For a linear system it is seen that the power-spectral and cross-spectral densities of the input and output of a control system are related through the system transfer function expressed in the frequency domain. Finally, a figure of merit relating the signal to noise power is introduced. We note that unlike systems subjected to deterministic inputs, a one-to-one correspondence between input and output is not available. Instead the *average* input and output are related. This treatment of the operation of control systems with stochastic (noise) inputs is necessarily brief, as there is a vast body of knowledge on this subject. Two topics of great importance in control systems, which have not been discussed due to lack of space, are Kalman filtering and the Linear Quadratic (LQ) control problem. Useful references on these and other related topics are given below.

12.6   References

1. W. B. Davenport, W. L. Root, *An Introduction to the Theory of Random Signals and Noise*, Mc Graw Hill Book Co. Inc., New York, 1958.

2. A. Papoulis, *Probability, Random Variables and Stochastic Processes*, Mc Graw Hill Book Co. Inc., New York, 1965.

3. E. Wong, *Stochastic Processes in Information and Dynamical Systems*, Mc Graw Hill Book Co. Inc., New York, 1971.

4. B. D. O. Anderson, J. B. Moore, *Optimal Control, Linear Quadratic Methods*, Prentice-Hall International, Inc., Englewood Cliffs, N. J., 1989.

5. K. J. Astrom, B. Wittenmark, *Computer-Controlled Systems*, Prentice-Hall International, Inc., Englewood Cliffs, N. J., 1990.

6. K. J. Astrom, *Introduction to Stochastic Control Theory*, Academic Press, New York, 1970.

7. A. H. Jazwinski, *Stochastic Processes and Filtering Theory*, Academic Press, New York, 1970.

12.7   Problems

12-1 Obtain the autocorrelation for a signal described by (a) $x(t) = 10$, (b) $x(t) = 5 \sin 2t$ for $\tau = 0$ and 1.

12-2 Obtain the crosscorrelation for two signals described by $x_1(t) = \sin 2t$ and $x_2(t) = 4 \cos 2t$ for $\tau = 0$ and 1.

12-3 Obtain the cross-spectral density for the crosscorrelation function described in Problem 12-2.

12-4 The input to a unity feedback control system is described by $r(t) = 5 \sin 10t$ and the forward loop transfer function is $G(s) = 1/(s+2)$. What is the variance of the output?

12-5 Verify the results of Example 6.

12-6 For the system described by Fig. 12-5, assume $G_2 = 10$, $G_1 = 1/(s+2)$, $n(t) = 0.1 \sin t$ and $r(t) = 5 \sin 10t$. What is the SNR for this system.

12-7 Verify the derivation of Eq. (12-12).

12-8 Verify the derivation of Eq. (12-13).

12-9 A unity feedback control system has a forward loop transfer function of $G(s) = K/s(s+A)$ and is subjected to an input whose autocorrelation is given by $e^{-2\tau}$. Plot the output variance as a function of K.

Chapter 13

Adaptive Control Systems

13.1   Introduction

In earlier chapters, we discussed two advanced concepts for control systems, namely pole placement design using state space methods and stochastic control systems. In this chapter, we shall introduce another advanced control system design technique, that is, the **adaptive control** approach.

Adaptive control has been developed for several decades since it was proposed in 1950s. But unfortunately, there is no widely accepted definition of adaptive control. Usually, an adaptive control system is regarded as a system that can automatically adjust controller settings to accommodate process changes so that the controlled system displays the desired performance.

Adaptive control was originally proposed to cope with the difficulties encountered in controlling systems having varying process dynamics. Many practical examples can be cited to illustrate such difficulties. One particular example is the variation of the short-period dynamics of an aircraft as its flight speed varies. This variation causes difficulties for the design of a conventional controller. The controller settings, which are designed for one particular operating condition, will work well for this condition, but may not work well for other operating conditions, and may even lead to system instability. However, these difficulties can often be handled by use of adaptive control.

Adaptive control has been extensively researched over the last decade, as is indicated by the vast reference literature and by several recently published

659

books. The interested reader is referred to the books listed at the end of the chapter for a detailed analysis of adaptive systems.

In this chapter we briefly present the basics of adaptive control. Section 13-2 will review various concepts associated with this topic. One important concept is the **certainty equivalence principle**. Using this principle various controller design methods can be combined with a number of parameter estimation techniques to generate a wide spectrum of adaptive control algorithms. In Section 13-3 we will review several controller design methods for known parameter cases. Selected parameter estimation methods and adaptive control algorithms will be presented in the following two sections. Closed loop systems that incorporate adaptive control are highly nonlinear and time-varying, and thus are difficult to analyze. A reasonably complete theoretical treatment is well beyond the scope of this book, and therefore only a few statements regarding stability and convergence for the selected adaptive control algorithms are given in Section 13-6, which is followed in the next section by an application example.

As mentioned above, adaptive control systems are usually highly nonlinear and time varying. For this reason adaptive control algorithms are normally implemented using computers, so that it is convenient to use discrete time models of the type discussed in Chapter 10, for the process to be controlled.

13.2   Adaptive Control Methods

There are various types of adaptive control systems resulting from different adaptation strategies. Broadly speaking, adaptive control systems could include **gain scheduling, auto-tuning, model reference adaptive systems, self-tuning regulators**, etc. However, in practice, adaptive control more often refers to the model reference adaptive systems, the self-tuning regulators and their various modified versions. In this section a few issues regarding the adaptive control systems will be discussed.

Gain Scheduling

In many practical systems, there indeed exists a particular variable (or variables) upon which the process dynamics depend. If a measurement of this variable is available, then it can be used to determine the corresponding characteristics of the process dynamics and to change the controller parameters accordingly. Originally, this scheme was used to tune the controller gain so as to accommodate changes in the characteristics of the process dynamics. Therefore this approach is called **gain scheduling**.

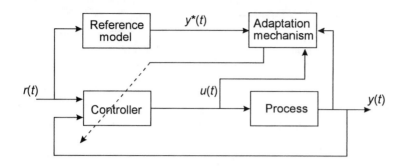

Figure 13.1: Block diagram of model reference adaptive system

Gain scheduling is basically an open loop compensation scheme for the process dynamics variations in the sense that there is no feedback from a performance measure of the closed loop system to compensate for an incorrect schedule. Because the process dynamics variations are often not known with high precision, perfect compensation for varying conditions can hardly be expected. Nevertheless, gain scheduling is a very useful technique for reducing the effects of process dynamics variations. In fact, it is the foremost method used for handling parameter variations in aircraft flight control systems. A very important issue in the design of gain scheduling systems is to find suitable scheduling variables. After the scheduling variables have been determined, the regulator parameters are selected for a number of operating conditions. The regulator is thus tuned for each operating condition.

Model Reference Adaptive Systems

The model reference adaptive control systems is one of the main approaches to adaptive control. It was originally proposed to tune the controller parameters in such a way that the process output follows the output of a pre-specified reference model. The basic principle is illustrated in Fig. 13-1.

As shown in the above figure, the model reference adaptive system consists of two loops. One loop, the inner loop which is composed of an unknown process and an ordinary controller, provides ordinary feedback control, while another loop, the outer loop which is composed of a reference model and adaptation mechanism, adjusts the parameters of the controller in the inner loop. The desired performance of the closed loop system is expressed in terms of the reference model, which gives the desired response to a command signal. The margin of error between the outputs of the process

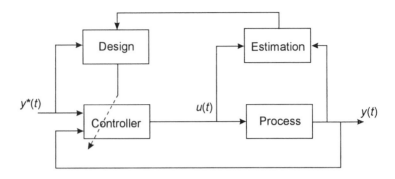

Figure 13.2: Block diagram of self-tuning regulator

and the reference model is used to regulate the parameters of the controller.

Self-Tuning Regulators

The self-tuning regulator is another of the main approaches to adaptive control. It is based on the idea of separating the estimation of unknown parameters from the design of the controller. The system automatically tunes the controller to the desired performance. The basic principle of the self-tuning regulators is illustrated in Fig. 13-2.

Similar to the model reference adaptive control system, the self-tuning regulator system shown above can also be regarded as having two loops. The inner loop is an ordinary linear feedback loop which includes the unknown process and an ordinary regulator. The outer loop consists of a recursive parameter estimator and an on-line controller design mechanism. The parameters of the regulator are adjusted on the basis of the underlying controller design calculations. One important feature of the self-tuning regulator scheme is its flexibility with regard to the choice of the underlying design and estimation methods. Various combinations of controller design and estimation methods will lead to different types of self-tuning schemes.

It can be seen in our discussion of the model reference adaptive control system and the self-tuning regulator, the structure of the process of the system has been assumed to be known and only the parameters of the structure are unknown. Therefore in the discussion of these systems they are more correctly described as parameter adaptive control systems. Structure adaptive control systems are far beyond the scope of our present discussions.

Direct and Indirect Adaptive Controllers

The original model reference adaptive control system is generally referred to as a **direct** scheme, since in this approach, the process model is re-parameterized so that it can be expressed in terms of the controller parameters. In this scheme, the controller parameters, rather than process parameters, are directly estimated. The original self-tuning regulator is referred to as an **indirect** scheme, since in this scheme the regulator parameters are obtained indirectly via a design procedure. In this scheme, the parameters of the process model are first estimated and the controller parameters are then calculated from the estimated process parameters. However, it should be noted that the terms "direct" or "indirect" only refer to whether the controller parameters are directly or indirectly estimated. It is possible for the model reference adaptive system to be implemented with an indirect scheme, while the self-tuning regulator can be implemented with a direct scheme for estimating the controller parameters.

The Certainty Equivalence Principle

In our discussion of the model reference adaptive control system and the self-tuning regulator, the controller parameters or the process parameters are estimated on-line. The estimated parameters are then used as if they are the true ones. In other words, the uncertainty attaching to each of these estimated parameters is not considered. The principle involved, which is called the **certainty equivalence principle**, has played a key role in the development of adaptive control systems. According to this principle, many existing controller design methods for known processes can be suitably used in the development of adaptive control systems. Therefore, in the next section, we shall review a few controller design methods, such as model reference controllers and pole placement controllers.

13.3   Controller Design Methods

There exist many controller design methods for known linear time-invariant systems both in the continuous and in the discrete time domain. Here we will focus our attention on discrete time domain design methods since this is in line with the application requirement for using modern computer technology, especially microcomputers. Two types of controller design methods will be reviewed.

In the following discussions we consider the process to be controlled as being modeled by the discrete form of the differential equations discussed in Section 3-4. To assist in these discussions we introduce the backward shift operator, q^{-1}, which is defined as

$$q^{-1}x(t) = x(t-1)$$

Using this operator we assume that a linear time invariant single-input single-output system is described by the equation

$$y(t) + a_1 q^{-1} y(t) + \cdots + a_n q^{-n} y(t) = q^{-d}\left[b_0 u(t) + b_1 q^{-1} u(t)\right.$$

$$\left. + \cdots + b_m q^{-m} u(t)\right]$$

where $y(t)$ and $u(t)$ are the process output and input respectively, and it is assumed that $m + d \le n$. This equation can be represented more concisely as

$$A\left(q^{-1}\right) y(t) = B\left(q^{-1}\right) u(t) \tag{13.1}$$

where

$$A\left(q^{-1}\right) = 1 + a_1 q^{-1} + \cdots + a_n q^{-n}$$

$$B\left(q^{-1}\right) = q^{-d}\left(b_0 + b_1 q^{-1} + \cdots + b_m q^{-m}\right) = q^{-d} B'\left(q^{-1}\right)$$

The above representation of a system or process is often referred to as an ARMA (auto regressive moving average) model. It will be assumed in the remainder of this chapter that the polynomials $A\left(q^{-1}\right)$ and $B\left(q^{-1}\right)$ are relatively prime, or equivalently, they have no common zeros.

Model Reference Controllers

Suppose the process to be controlled is linear and time-invariant and has a single-input and a single-output as described by Eq. (13-1).

Our objective is to design a feedback controller such that the output $\{y(t)\}$ of the process, when driven by a command signal $\{r(t)\}$, tracks a desired reference model output $\{y^*(t)\}$ driven by the same command input. The reference model is assumed to be described by the following stable equation,

$$E\left(q^{-1}\right) y^*(t) = q^{-d} H(q^{-1}) r(t) \tag{13.2}$$

where

$$H\left(q^{-1}\right) = h_0 + h_1 q^{-1} + \cdots + h_l q^{-l}$$

$$E\left(q^{-1}\right) = 1 + e_1 q^{-1} + \cdots + e_l q^{-l}$$

It can be easily seen that the reference model has the transfer function

$$C\left(z^{-1}\right) = \frac{z^{-d} H\left(z^{-1}\right)}{E\left(z^{-1}\right)}$$

where the denominator polynomial $E\left(z^{-1}\right)$ is assumed to be stable. In other words, all its zeros lie inside the unit circle of the z-plane.

In order to achieve our control objective, we firstly express the process model Eq. (13-1) in a **predictor** form with $E\left(q^{-1}\right)y(t+d)$ as its output, and then choose a suitable control law so that $E\left(q^{-1}\right)y(t+d)$ is equal to $H\left(q^{-1}\right)r(t)$. Consequently the process output will track the reference model output. It can be easily verified that the process model Eq. (13-1) can be expressed in a predictor form as

$$E\left(q^{-1}\right)y(t+d) = \alpha\left(q^{-1}\right)y(t) + \beta\left(q^{-1}\right)u(t) \tag{13.3}$$

where

$$\alpha\left(q^{-1}\right) \;=\; G\left(q^{-1}\right)$$

$$\beta\left(q^{-1}\right) \;=\; F\left(q^{-1}\right)B'\left(q^{-1}\right)$$

For the polynomial $E\left(q^{-1}\right)$ and relatively prime polynomials $A\left(q^{-1}\right)$ and q^{-d} it is shown in elementary algebra textbooks that the polynomials $F\left(q^{-1}\right)$ and $G\left(q^{-1}\right)$ are unique and satisfy

$$E\left(q^{-1}\right) = F\left(q^{-1}\right)A\left(q^{-1}\right) + q^{-d}G\left(q^{-1}\right) \tag{13.4}$$

where $F\left(q^{-1}\right)$ and $G\left(q^{-1}\right)$ are of order $(\ell+d-1)$ and $(\ell+n-1)$ respectively.

To derive Eq. (13-3) we multiply Eq. (13-1) by $F\left(q^{-1}\right)$

$$F\left(q^{-1}\right)A\left(q^{-1}\right)y(t) = q^{-d}F\left(q^{-1}\right)B'\left(q^{-1}\right)u(t) \tag{13.5}$$

Substituting Eq.(13-4) into Eq. (13-5), we have

$$E\left(q^{-1}\right)y(t) = q^{-d}G\left(q^{-1}\right)y(t) + q^{-d}F\left(q^{-1}\right)B'\left(q^{-1}\right)u(t)$$

that is,

$$E\left(q^{-1}\right)y(t+d) \;=\; G\left(q^{-1}\right)y(t) + F\left(q^{-1}\right)B'\left(q^{-1}\right)u(t)$$

$$\;=\; \alpha\left(q^{-1}\right)y(t) + \beta\left(q^{-1}\right)u(t)$$

Based on this predictor model, we choose the control law such that

$$\alpha\left(q^{-1}\right)y(t) + \beta\left(q^{-1}\right)u(t) = H\left(q^{-1}\right)r(t) \tag{13.6}$$

In this case the output of the closed loop system is given by

$$E\left(q^{-1}\right)y(t+d) \;=\; H\left(q^{-1}\right)r(t)$$

$$\;=\; E\left(q^{-1}\right)y^*(t+d) \tag{13.7}$$

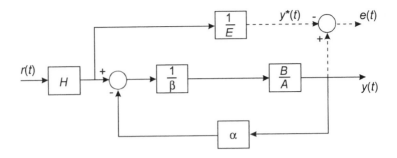

Figure 13.3: Block diagram of model reference controller

which demonstrates that $y(t)$ tracks $y^*(t)$. The block diagram for the resultant closed loop control system is illustrated in Fig. 13-3.

It should be noted that in order to guarantee stability of the closed loop system, the zeros of the polynomial $B(z^{-1})$ are required to be inside the unit circle of the z-plane. This is quite restrictive in many applications. The following pole placement controller will overcome this difficulty.

Pole Placement Controllers

Consider the process model as in Eq. (13-1)

$$A(q^{-1}) y(t) = B(q^{-1}) u(t) \tag{13.8}$$

Our control objective is to design a feedback controller such that the closed loop system has the desired poles, which are specified by the zeros of a polynomial $A^*(z^{-1})$.

Goodwin and Sin have shown that provided the polynomials $A(q^{-1})$ and $B(q^{-1})$ are relatively prime, then there exist unique polynomials $L(q^{-1})$ and $P(q^{-1})$, both of order $(n-1)$ such that

$$A(q^{-1}) L(q^{-1}) + B(q^{-1}) P(q^{-1}) = A^*(q^{-1}) \tag{13.9}$$

where $A^*(q^{-1})$ is an arbitrarily chosen polynomial of order $(2n-1)$ representing the desired closed loop pole locations. Eq. (13-9) is usually called the **Diophantine** equation.

If we choose the control law as

$$L(q^{-1}) u(t) = P(q^{-1}) \{y^*(t) - y(t)\} \tag{13.10}$$

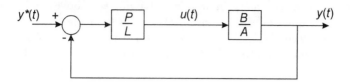

Figure 13.4: Block diagram of pole placement controller

where $y^*(t)$ is the desired reference output, then substituting Eq. (13-10) into Eq. (13-8) leads to

$$A\left(q^{-1}\right) L\left(q^{-1}\right) y(t) = B\left(q^{-1}\right) L\left(q^{-1}\right) u(t)$$

$$= B\left(q^{-1}\right) P\left(q^{-1}\right) \{y^*(t) - y(t)\}$$

that is,

$$\left[A\left(q^{-1}\right) L\left(q^{-1}\right) + B\left(q^{-1}\right) P\left(q^{-1}\right)\right] y(t) = B\left(q^{-1}\right) P\left(q^{-1}\right) y^*(t)$$

Using Eq. (13-9) yields

$$A^*\left(q^{-1}\right) y(t) = B\left(q^{-1}\right) P\left(q^{-1}\right) y^*(t) \qquad (13.11)$$

This equation describes the operation of the closed loop system which it will be noted has its closed loop poles defined by $A^*\left(q^{-1}\right)$. The block diagram of the closed loop system can be shown using Eq. (13-10) to be given by Fig. 13-4.

For the sake of closed loop stability it should be noted the closed loop poles must be chosen to lie within the unit circle on the z-plane.

In many applications, the process is often subject to external disturbances. We will show here that the pole placement controller can be designed in such a way that external deterministic disturbances can be rejected. Moreover, we will also show how perfect tracking can be achieved for certain classes of desired reference outputs.

Pole Placement Controller with Disturbance Rejection

Consider a process model with external disturbances described by

$$A\left(q^{-1}\right) \bar{y}(t) = B\left(q^{-1}\right) u(t), \quad y(t) = \bar{y}(t) + d(t) \qquad (13.12)$$

where the disturbance $d(t)$ satisfies

$$D\left(q^{-1}\right) d(t) = 0 \qquad (13.13)$$

There are many disturbances in practice satisfying the above condition. For instance, the sinusoidal disturbance $d(t) = \sin \omega_0 t$ satisfies Eq. (13-13) for $D\left(q^{-1}\right) = 1 - 2\cos \omega_0 T q^{-1} + q^{-2}$, where T is the sampling period.

We also define the class of desired reference outputs $\{y^*(t)\}$ as those satisfying

$$R\left(q^{-1}\right) y^*(t) = 0 \qquad (13.14)$$

For example, the constant reference output indeed belongs to this class with $R\left(q^{-1}\right) = 1 - q^{-1}$.

Multiplying Eq. (13-12) by $D\left(q^{-1}\right)$, and combining with Eq. (13-13) it can be rewritten as

$$A\left(q^{-1}\right) D\left(q^{-1}\right) y(t) = B\left(q^{-1}\right) D\left(q^{-1}\right) u(t) \qquad (13.15)$$

If we choose the control law as

$$L\left(q^{-1}\right) R\left(q^{-1}\right) D\left(q^{-1}\right) u(t) = P\left(q^{-1}\right) \{y^*(t) - y(t)\} \qquad (13.16)$$

where $L\left(q^{-1}\right)$ and $P\left(q^{-1}\right)$ satisfy the following Diophantine equation

$$A\left(q^{-1}\right) D\left(q^{-1}\right) R\left(q^{-1}\right) L\left(q^{-1}\right) + B\left(q^{-1}\right) P\left(q^{-1}\right) = A^*\left(q^{-1}\right) \quad (13.17)$$

then we obtain the equation describing the operation of the closed loop system as

$$A^*\left(q^{-1}\right) \{y(t) - y^*(t)\} = 0 \qquad (13.18)$$

This can be easily shown as follows. Substituting the control law Eq. (13-16) into the process model Eq. (13-15) leads to

$$A\left(q^{-1}\right) D\left(q^{-1}\right) R\left(q^{-1}\right) L\left(q^{-1}\right) y(t) = B\left(q^{-1}\right) P\left(q^{-1}\right) \{y^*(t) - y(t)\}$$

Using Eq. (13-17) yields

$$A^*\left(q^{-1}\right) y(t) = B\left(q^{-1}\right) P\left(q^{-1}\right) y^*(t) \qquad (13.19)$$

Noting that $R\left(q^{-1}\right) y^*(t) = 0$, Eq. (13-19) can be rewritten as

$$A^*\left(q^{-1}\right) y(t) \;=\; B\left(q^{-1}\right) P\left(q^{-1}\right) y^*(t)$$

$$+ A\left(q^{-1}\right) D\left(q^{-1}\right) L\left(q^{-1}\right) R\left(q^{-1}\right) y^*(t)$$

which leads to

$$A^*\left(q^{-1}\right) \{y(t) - y^*(t)\} = 0$$

as desired.

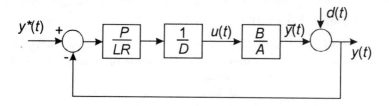

Figure 13.5: Block diagram of pole placement controller with disturbance rejection

It should be noted that the solution of Eq. (13-17) further requires that $R\left(q^{-1}\right) D\left(q^{-1}\right)$ and $B\left(q^{-1}\right)$ are also relatively prime. The block diagram for this system is shown in Fig. 13-5.

It can be seen that the corresponding modes of the disturbance model $D\left(q^{-1}\right)$ are placed in the denominator polynomial of the controller to cancel the disturbance. This is usually called the **internal model principle**.

13.4   System Parameter Estimation

In the discussion of controller design, it is assumed that the process model, including the model structure and parameters, is known a priori. However in many cases, it is not feasible to assume a knowledge of the process parameters, and moreover, these parameters are often subject to variation based on changing operating conditions. Therefore it is desirable for the values of these parameters to be obtained from input-output measurements on the process. Usually this procedure is called **parameter estimation**. Broadly speaking, there exist two approaches to parameter estimation — **off-line** and **on-line**. Off-line methods use all the data available prior to analysis. Consequently, they usually give estimates with high precision and set no time limit for the process of analysis. In contrast to the off-line case, the on-line methods give estimates recursively as the measurements are obtained within the time limit imposed by the sampling period. Consequently, the on-line methods are more often used in adaptive control, especially in the case of time-varying processes. In this section we will present one off-line estimation algorithm and two on-line estimation algorithms. The off-line estimation algorithm is the well-known least squares method.

Least Squares Estimation

The **principle of least squares** was formulated by Gauss at the end of the eighteenth century for determining the orbits of planets. According to this principle, the unknown parameters of a mathematical model should be chosen in such a way that the sum of the squares of the differences between the parameter values actually observed and their computed values, multiplied by numbers that measure the degree of precision, is minimized.

Consider the process model in ARMA form

$$A\left(q^{-1}\right) y(t) = B\left(q^{-1}\right) u(t) \tag{13.20}$$

where

$$A\left(q^{-1}\right) = 1 + a_1 q^{-1} + \cdots + a_n q^{-n}$$

$$B\left(q^{-1}\right) = b_1 q^{-1} + \cdots + b_n q^{-n}$$

Note that the time delay q^{-d} has been included in the expression, and some of the coefficients $b_i, i = 1, \cdots, n$ could be zero.

It is easy to see that Eq. (13-20) can be rewritten in **regression form** as

$$y(t) = \phi(t-1)^T \theta_0 \tag{13.21}$$

where

$$\phi(t-1)^T = \begin{bmatrix} y(t-1) & \cdots & y(t-n) & u(t-1) & \cdots & u(t-n) \end{bmatrix}$$

is called the **regressor** and

$$\theta_0 = \begin{bmatrix} -a_1 & \cdots & -a_n & b_1 & \cdots & b_n \end{bmatrix}^T$$

is the unknown **parameter vector**.

If we define the **parameter estimate** as $\hat{\theta}$, the output estimate is

$$\hat{y}(t) = \phi(t-1)^T \hat{\theta} \tag{13.22}$$

Also the **estimation error**

$$\varepsilon(t) = y(t) - \hat{y}(t) \tag{13.23}$$

Using the above notation the least squares problem can be stated concisely. It is to find $\hat{\theta}$ such that the cost function

$$J\left(\hat{\theta}\right) = \frac{1}{2}\sum_{i=1}^{t} \varepsilon(i)^2 \tag{13.24}$$

is minimized.

Assume that a sequence of inputs $\{u(1), \cdots, u(t)\}$ has been applied to the process, and that the corresponding sequence of outputs $\{y(1), \cdots, y(t)\}$ has been observed. Let

$$Y(t) \;=\; [\; y(1) \quad \cdots \quad y(t) \;]^T$$

$$E(t) \;=\; [\; \varepsilon(1) \quad \cdots \quad \varepsilon(t) \;]^T$$

$$\Phi(t) \;=\; [\; \phi(0) \quad \cdots \quad \phi(t-1) \;]^T$$

Then minimization of the cost function shows the solution of the least squares problem occurs when

$$\Phi(t)^T \Phi(t)\hat{\theta} = \Phi(t)^T Y(t) \tag{13.25}$$

If the $\Phi(t)^T \Phi(t)$ is nonsingular, then the minimum is unique and is given by

$$\hat{\theta} = \left[\Phi(t)^T \Phi(t)\right]^{-1} \Phi(t)^T Y(t) \tag{13.26}$$

The simple proof of this minimization problem is left as an exercise for the reader.

In adaptive control, however, the observations are obtained sequentially. It is therefore desirable to also compute the estimates sequentially by using a **recursive** relationship. In the following, two recursive estimation schemes are presented without derivation for the process model given in Eq. (13-21).

Recursive Projection Algorithm

$$\hat{\theta}(t) = \hat{\theta}(t-1) + \frac{\phi(t-1)}{c + \phi(t-1)^T \phi(t-1)} \left[y(t) - \phi(t-1)^T \hat{\theta}(t-1)\right]$$

$$\tag{13.27}$$

where $\hat{\theta}(0)$ is given and c is an arbitrarily chosen constant with $c > 0$.

The elementary properties of the projection algorithm can be summarized as follows:

(i) $\left\|\hat{\theta}(t) - \theta_0\right\| \le \left\|\hat{\theta}(t-1) - \theta_0\right\| \le \left\|\hat{\theta}(0) - \theta_0\right\|,\ t \ge 1$

(ii) $\displaystyle \lim_{t \to \infty} \frac{e(t)}{\left[c + \phi(t-1)^T \phi(t-1)\right]^{1/2}} = 0$

   where $e(t) = y(t) - \phi(t-1)^T \hat{\theta}(t-1)$

(iii) $\lim\limits_{t\to\infty} \left\| \hat{\theta}(t) - \hat{\theta}(t-k) \right\| = 0$, for any finite k.

As in Chapter 11, $\|\mathbf{x}(t)\|$ denotes the length (or **norm**) of a vector $\mathbf{x}(t)$ given by

$$\|\mathbf{x}(t)\| = \sqrt{x_1(t)^2 + \cdots + x_n(t)^2}$$

Recursive Least Squares Algorithm

$$\hat{\theta}(t) = \hat{\theta}(t-1) + \frac{P(t-1)\phi(t-1)}{1+\phi(t-1)^T P(t-1)\phi(t-1)} \left[y(t) - \phi(t-1)^T \hat{\theta}(t-1) \right]$$
(13.28)

$$P(t) = P(t-1) - \frac{P(t-1)\phi(t-1)\phi(t-1)^T P(t-1)}{1+\phi(t-1)^T P(t-1)\phi(t-1)}$$
(13.29)

where $\hat{\theta}(0)$ is given and $P(0)$ is any positive definite matrix.

The elementary properties of the recursive least squares algorithm can also be summarized as follows:

(i) $\left\| \hat{\theta}(t) - \theta_0 \right\|^2 \le k_1 \left\| \hat{\theta}(0) - \theta_0 \right\|^2, t \ge 1$ where k_1 is a suitable constant determined by the initial covariance matrix $P(0)$

(ii) $\lim\limits_{t\to\infty} \dfrac{e(t)}{\left[1+\phi(t-1)^T P(t-1)\phi(t-1) \right]^{1/2}} = 0$

(iii) $\lim\limits_{t\to\infty} \left\| \hat{\theta}(t) - \hat{\theta}(t-k) \right\| = 0$, for any finite k.

It can be seen that the properties of both recursive parameter estimation algorithms are very similar. These properties are usually referred to as the **convergence properties** of the algorithms. The recursive least squares algorithm generally converges more rapidly than the recursive projection algorithm. Also, the recursive least squares algorithm can be used essentially unaltered with noisy signals. Therefore in practice, the recursive least squares algorithm is preferred for applications. However, it should be noted that the above recursive least squares algorithm cannot be used for time-varying systems because the covariance matrix $P(t)$ will approach zero as time t goes towards infinity. A number of modified versions of the least squares algorithm have been proposed to handle this difficulty. They include the least squares algorithm with selective data weighting, the least squares algorithm with covariance resetting, the least squares algorithm with covariance modification, etc. All of these modified algorithms have similar properties to the original least squares algorithm. Readers interested in further details about these algorithms are referred to the book by Goodwin and Sin listed at the end of the chapter.

Parameter Convergence

In our discussion of the properties of the recursive parameter estimation algorithms given above, we have not considered the question of convergence of the estimated parameter values to their true values. In general, such convergence cannot be guaranteed unless certain **persistent excitation** conditions are satisfied. The persistent excitation conditions usually require that the input signals contain a sufficiently broad spectrum of frequencies so as to excite the process dynamics of the system at these frequencies. In many cases, the reference input signal does not have such a sufficiently rich spectrum of frequencies, so that an auxiliary probing signal is needed to achieve parameter convergence. The detailed discussion of the persistent excitation concept is beyond the scope of this book. The reader is referred to the references at the end of the chapter for further discussion on this matter.

13.5   Adaptive Control Algorithms

The design of an adaptive control system based on the certainty equivalence principle is conceptually simple. A natural approach is to select one of a number of parameter estimation algorithms and to combine it with one of the many control design methods. Therefore a wide spectrum of algorithms can be generated for adaptive control systems, depending on the methods chosen for parameter estimation and for control design. In this section, we shall mainly present two schemes. One is the model reference adaptive control scheme, which is implemented in direct form. Another is the pole placement adaptive control scheme, which is implemented in indirect form. In addition, a modified version of the pole placement adaptive control scheme which rejects deterministic disturbances will also be considered.

Model Reference Adaptive Control (Direct Approach)

Essentially, the model reference adaptive control approach discussed here is the adaptive version of the model reference control scheme described in Section 13-3. Consider the same process model as in Section 13-3

$$A\left(q^{-1}\right)y(t) = q^{-d}B'\left(q^{-1}\right)u(t) \tag{13.30}$$

We recall the reference model

$$E\left(q^{-1}\right)y^*(t) = q^{-d}H\left(q^{-1}\right)r(t) \tag{13.31}$$

We also recall that the process model Eq. (13-30) can be expressed in predictor form as

$$E\left(q^{-1}\right)y(t + d) = \alpha\left(q^{-1}\right)y(t) + \beta\left(q^{-1}\right)u(t) \tag{13.32}$$

where

$$\alpha\left(q^{-1}\right) = G\left(q^{-1}\right), \quad \beta\left(q^{-1}\right) = F\left(q^{-1}\right)B'\left(q^{-1}\right)$$

and

$$E\left(q^{-1}\right) = F\left(q^{-1}\right)A\left(q^{-1}\right) + q^{-d}G\left(q^{-1}\right) \tag{13.33}$$

The corresponding model reference optimal control law is

$$\alpha\left(q^{-1}\right)y(t) + \beta\left(q^{-1}\right)u^*(t) = H\left(q^{-1}\right)r(t) \tag{13.34}$$

It can be seen that the above control law depends on the knowledge of the process parameters. However in the context of adaptive control, those parameters are supposed to be unknown. Therefore the certainty equivalence form of the Eq.(13-34) should be used with $\alpha\left(q^{-1}\right)$ and $\beta\left(q^{-1}\right)$ replaced by their estimates, and one of the parameter estimation methods discussed in the last section can be used to generate these estimates. This is shown below.

Dividing Eq.(13-32) by β_0 gives

$$\begin{aligned}\frac{1}{\beta_0}E\left(q^{-1}\right)y(t + d) &= \frac{1}{\beta_0}\alpha\left(q^{-1}\right)y(t) + \frac{1}{\beta_0}\beta\left(q^{-1}\right)u(t) \\ &= \alpha'\left(q^{-1}\right)y(t) + \left[\beta'\left(q^{-1}\right) + 1\right]u(t)\end{aligned} \tag{13.35}$$

where

$$\begin{aligned}\alpha'\left(q^{-1}\right) &= \frac{\alpha_0}{\beta_0} + \frac{\alpha_1}{\beta_0}q^{-1} + \cdots + \frac{\alpha_{\ell+n-1}}{\beta_0}q^{-(\ell+n-1)} \\ &= \alpha'_0 + \alpha'_1 q^{-1} + \cdots + \alpha'_{\ell+n-1}q^{-(\ell+n-1)}\end{aligned}$$

Similarly

$$\beta'\left(q^{-1}\right) = \beta'_1 q^{-1} + \cdots + \beta'_{\ell+m+d-1}q^{-(\ell+m+d-1)}$$

Therefore, the Eq.(13-35) can be rewritten as

$$u(t) = \bar{\phi}(t)^T \theta_0 \tag{13.36}$$

where

$$\theta_0 = \begin{bmatrix} -\alpha'_0 & \cdots & -\alpha'_{\ell+n-1} & -\beta'_1 & \cdots & -\beta'_{\ell+m+d-1} & 1/\beta_0 \end{bmatrix}^T \quad (13.37)$$

$$\bar{\phi}(t)^T = \begin{bmatrix} y(t) & \cdots & y(t-\ell-n+1) & u(t-1) & \cdots \\ & u(t-\ell-m-d+1) & y_a(t+d) \end{bmatrix} \quad (13.38)$$

with $y_a(t+d) = E(q^{-1}) y(t+d)$.

Similarly, the optimal control law Eq. (13-34) can be rewritten as

$$u^*(t) = \phi'(t)^T \theta_0 \quad (13.39)$$

where

$$\phi'(t)^T = \begin{bmatrix} y(t) & \cdots & y(t-\ell-n+1) & u(t-1) & \cdots \\ & u(t-\ell-m-d+1) & r_a(t) \end{bmatrix} \quad (13.40)$$

with $r_a(t) = H(q^{-1}) r(t)$.

Therefore, the following adaptive control algorithm can be implemented.

Model Reference Adaptive Control Algorithm

Step 1: Estimate the parameter vector $\hat{\theta}$ based on Eq.(13-36) using the following recursive least squares method

$$\hat{\theta}(t) = \hat{\theta}(t-1) + \frac{P(t-d)\,\bar{\phi}(t-d)}{1 + \bar{\phi}(t-d)^T P(t-d)\,\bar{\phi}(t-d)} \times$$
$$\left[u(t-d) - \bar{\phi}(t-d)^T \hat{\theta}(t-d) \right] \quad (13.41)$$

$$P(t-d+1) = P(t-d) - \frac{P(t-d)\,\bar{\phi}(t-d)\,\bar{\phi}(t-d)^T P(t-d)}{1 + \bar{\phi}(t-d)^T P(t-d)\,\bar{\phi}(t-d)} \quad (13.42)$$

where $\hat{\theta}(0)$ is given, and $P(0)$ is any positive definite matrix. It should be noted that the regressor $\bar{\phi}(t)$ is defined in Eq. (13-38).

Step 2: Calculate the control law according to

$$u(t) = \phi'(t)^T \hat{\theta}(t) \quad (13.43)$$

where the regressor $\phi'(t)$ is defined in Eq. (13-40).

Step 3: Then for the next sampling period, go back to Step 1 and repeat.

Example 1 *Apply the model reference adaptive control algorithm described above to the process having the transfer function*

$$G\left(z^{-1}\right) = \frac{0.5242z^{-1} - 0.429z^{-2}}{1 - 1.9048z^{-1} + 0.9048z^{-2}}$$

with the sampling period $T = 0.1$ sec. After $t = 20$ seconds the process changes and is described by the transfer function

$$G\left(z^{-1}\right) = \frac{0.5242z^{-1} - 0.429z^{-2}}{1 - 2.1052z^{-1} + 1.1052z^{-2}}$$

which has one pole ($z = 1.1052$) outside the unit circle on the z-plane.

The operation of the adaptive control algorithm has been simulated with the reference model chosen as

$$C\left(z^{-1}\right) = \frac{0.0952z^{-1}}{1 - 0.9048z^{-1}}$$

The command signal $r(t)$ has been chosen as a square wave of unit amplitude and a period of 20 seconds. The initial control parameter vector has been chosen as zero.

The performance of the model reference adaptive control algorithm is shown in Fig. 13-6. It can be seen that the algorithm can automatically adapt to the unknown process and also tune its controller settings to account for the variations of the process. In addition it can be observed that, asymptotically perfect tracking is achieved.

Pole Placement Adaptive Control (Indirect Approach)

As discussed in Section 13-3, the model reference control approach places a restrictive assumption on the location of the zeros of the polynomial $B\left(z^{-1}\right)$ in the process model. That is, all the zeros of the process model must lie inside the unit circle on the z-plane. This restriction carries over to the adaptive case. Therefore, as in Section 13-3, we introduce pole placement adaptive control to overcome this difficulty. Consider the following linear time-invariant process

$$A\left(q^{-1}\right) y\left(t\right) = B\left(q^{-1}\right) u\left(t\right) \qquad (13.44)$$

where

$$A\left(q^{-1}\right) = 1 + a_1 q^{-1} + \cdots + a_n q^{-n}$$

$$B\left(q^{-1}\right) = b_1 q^{-1} + \cdots + b_n q^{-n}$$

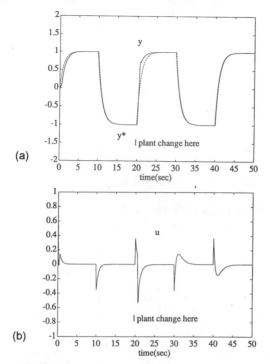

Figure 13.6: Model reference adaptive control (a) Plant output and reference model output (b) Plant control signal

In general some of the coefficients b_i, $i = 1, \ldots, n$ of the polynomial $B\left(q^{-1}\right)$ can be zero.

Recall the control law

$$L\left(q^{-1}\right) u(t) = P\left(q^{-1}\right) \{y^*(t) - y(t)\} \qquad (13.45)$$

where $y^*(t)$ is a desired reference output, and $L\left(q^{-1}\right)$ and $P\left(q^{-1}\right)$ satisfy the Diophantine equation

$$A\left(q^{-1}\right) L\left(q^{-1}\right) + B\left(q^{-1}\right) P\left(q^{-1}\right) = A^*\left(q^{-1}\right) \qquad (13.46)$$

where $A^*\left(q^{-1}\right)$ is the desired closed loop polynomial.

The solution to the above equation requires the assumption that the process model polynomials $A\left(q^{-1}\right)$ and $B\left(q^{-1}\right)$ are relatively prime, although they are unknown. The process model Eq. (13-44) can be rewritten in regression form as

$$y(t) = \phi(t-1)^T \theta_0 \qquad (13.47)$$

where

$$\phi(t-1)^T = [\ y(t-1) \quad \cdots \quad y(t-n) \quad u(t-1) \quad \cdots \quad u(t-n)\ ]$$

$$\theta_0 = [\ -a_1 \quad \cdots \quad -a_n \quad b_1 \quad \cdots \quad b_n\ ]^T$$

Then combining the recursive least squares parameter estimation algorithm with the pole placement controller design method, we obtain the following indirect pole placement adaptive control algorithm.

Pole Placement Adaptive Control Algorithm

Step 1: Estimate the parameter vector $\hat{\theta}$ based on Eq. (13-47) using the following recursive least squares method

$$\hat{\theta}(t) = \hat{\theta}(t-1) + \frac{P(t-1)\phi(t-1)}{1+\phi(t-1)^T P(t-1)\phi(t-1)} \times$$
$$\left[y(t) - \phi(t-1)^T \hat{\theta}(t-1)\right] \tag{13.48}$$

$$P(t) = P(t-1) - \frac{P(t-1)\phi(t-1)\phi(t-1)^T P(t-1)}{1+\phi(t-1)^T P(t-1)\phi(t-1)} \tag{13.49}$$

where $\hat{\theta}(0)$ is given, and $P(0)$ is any given positive definite matrix.

Step 2: Obtain $\hat{A}(q^{-1})$ and $\hat{B}(q^{-1})$ from $\hat{\theta}$, where

$$\hat{A}(q^{-1}) = 1 + \hat{a}_1 q^{-1} + \cdots + \hat{a}_n q^{-n}$$

$$\hat{B}(q^{-1}) = \hat{b}_1 q^{-1} + \cdots + \hat{b}_n q^{-n}$$

Step 3: Solve the following Diophantine equation

$$\hat{A}(q^{-1})\hat{L}(q^{-1}) + \hat{B}(q^{-1})\hat{P}(q^{-1}) = A^*(q^{-1}) \tag{13.50}$$

to update the controller polynomials $\hat{L}(q^{-1})$ and $\hat{P}(q^{-1})$.

Step 4: Implement the following adaptive control law

$$\hat{L}(q^{-1})u(t) = \hat{P}(q^{-1})\{y^*(t) - y(t)\} \tag{13.51}$$

Step 5: Then for the next sampling period, go back to Step 1 and repeat.

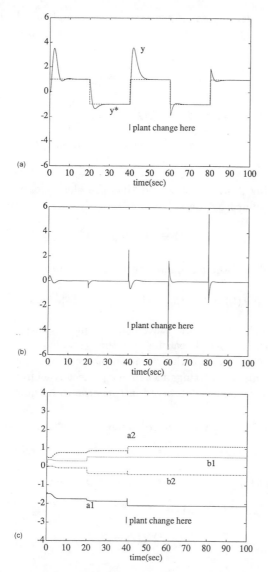

Figure 13.7: Pole placement adaptive control (a) Plant output and reference input (b) Plant control input (c) Estimated plant parameters

Example 2 *Consider the same process as in Example 1. In this case it will be assumed the process change occurs at $t = 40$ seconds. That is, the process parameter vector is*

$$\theta_0 = \begin{cases} \begin{bmatrix} 1.9048 & -0.9048 & 0.5242 & -0.429 \end{bmatrix}^T & \text{for } t < 40 \text{ sec.} \\ \\ \begin{bmatrix} 2.1052 & -1.1052 & 0.5242 & -0.429 \end{bmatrix}^T & \text{for } t \geq 40 \text{ sec.} \end{cases}$$

Investigate the performance of the pole placement adaptive control algorithm.

In the following it will be assumed that the closed loop poles are co-located and are defined by the polynomial

$$A^* \left(z^{-1} \right) = \left(1 - 0.9z^{-1} \right)^3$$

and the initial value of the parameter vector $\hat{\theta}$ is

$$\hat{\theta}(0) = \begin{bmatrix} -1.5 & -1.5 & 0.1 & 0.1 \end{bmatrix}^T$$

The performance of the pole placement adaptive control algorithm is then as shown in Fig. 13-7. It can be seen that the algorithm can identify the unknown process and also the variation of the process, and automatically adjust its controller settings to achieve perfect tracking. In addition, the parameter convergence is observed in this case to be quite rapid.

Pole Placement Adaptive Control with Disturbance Rejection

Consider a process model with disturbance as in Section 13-3

$$A \left(q^{-1} \right) \bar{y}(t) = B \left(q^{-1} \right) u(t), \quad y(t) = \bar{y}(t) + d(t) \tag{13.52}$$

where the disturbance $d(t)$ satisfies

$$D \left(q^{-1} \right) d(t) = 0 \tag{13.53}$$

with $\deg \left\{ D \left(q^{-1} \right) \right\} = \ell$.

It has been shown in Section 13-3 that the Eq. (13-52) can be rewritten as

$$A \left(q^{-1} \right) D \left(q^{-1} \right) y(t) = B \left(q^{-1} \right) D \left(q^{-1} \right) u(t) \tag{13.54}$$

The class of desired reference outputs $y^* (t)$ are assumed to satisfy

$$R \left(q^{-1} \right) y^* (t) = 0 \tag{13.55}$$

Recall the control law is

$$L\left(q^{-1}\right) R\left(q^{-1}\right) D\left(q^{-1}\right) u(t) = P\left(q^{-1}\right) \{y^*(t) - y(t)\} \qquad (13.56)$$

where $L\left(q^{-1}\right)$) and $P\left(q^{-1}\right)$ satisfy the following Diophantine equation

$$A\left(q^{-1}\right) D\left(q^{-1}\right) R\left(q^{-1}\right) L\left(q^{-1}\right) + B\left(q^{-1}\right) P\left(q^{-1}\right) = A^*\left(q^{-1}\right) \quad (13.57)$$

In this equation $A^*\left(q^{-1}\right)$ is the polynomial defining the position of the closed loop poles.

Since the process and the disturbance parameters are unknown, we will combine the recursive least squares estimation method with the pole placement control design to obtain the corresponding adaptive control algorithm. Depending on how the parameter estimation algorithm is implemented, four algorithms can be developed for this system. Here we just present one of them. The interested reader is referred to the work of Feng and Palaniswami for other algorithms. Let

$$A'\left(q^{-1}\right) = A\left(q^{-1}\right) D\left(q^{-1}\right)$$

$$B'\left(q^{-1}\right) = B\left(q^{-1}\right) D\left(q^{-1}\right)$$

then Eq.(13-54) can be written as

$$A'\left(q^{-1}\right) y(t) = B'\left(q^{-1}\right) u(t) \qquad (13.58)$$

This equation can also be rewritten in regression form as

$$y(t) = \phi\left(t - 1\right)^T \theta_0 \qquad (13.59)$$

where

$$\phi(t-1)^T = [\; y(t-1) \quad \cdots \quad y(t-n-\ell) \quad u(t-1)$$

$$\cdots \quad u(t-n-\ell) \;]$$

$$\theta_0 = [\; -a_1' \quad \cdots \quad -a_{n+\ell}' \quad b_1' \quad \cdots \quad b_{n+\ell}' \;]^T$$

We then have the following algorithm:

Pole Placement Adaptive Control Algorithm with Disturbance Rejection

Step 1: Estimate the parameter vector $\hat{\theta}$ based on Eq.(13-59) using the following recursive least squares method

$$\hat{\theta}(t) \;=\; \hat{\theta}(t-1) + \frac{P(t-1)\,\phi(t-1)}{1 + \phi(t-1)^T\,P(t-1)\,\phi(t-1)} \times$$
$$\left[y(t) - \phi(t-1)^T\,\hat{\theta}(t-1)\right] \tag{13.60}$$

$$P(t) = P(t-1) - \frac{P(t-1)\,\phi(t-1)\,\phi(t-1)^T\,P(t-1)}{1 + \phi(t-1)^T\,P(t-1)\,\phi(t-1)} \tag{13.61}$$

where $\hat{\theta}(0)$ is given, and $P(0)$ is any given positive definite matrix.

Step 2: Obtain $\hat{A}'\left(q^{-1}\right)$ and $\hat{B}'\left(q^{-1}\right)$ from $\hat{\theta}$.

Step 3: Generate $\hat{A}\left(q^{-1}\right)$, $\hat{B}\left(q^{-1}\right)$ and $\hat{D}\left(q^{-1}\right)$ from $\hat{A}'\left(q^{-1}\right)$ and $\hat{B}'\left(q^{-1}\right)$.

Step 4: Solve the following Diophantine equation

$$\hat{A}\left(q^{-1}\right)\hat{D}\left(q^{-1}\right)\hat{L}\left(q^{-1}\right)R\left(q^{-1}\right) + \hat{B}\left(q^{-1}\right)\hat{P}\left(q^{-1}\right) = A^*\left(q^{-1}\right) \tag{13.62}$$

to update the controller polynomials $\hat{L}\left(q^{-1}\right)$ and $\hat{P}\left(q^{-1}\right)$.

Step 5: Implement the following adaptive control law

$$\hat{L}\left(q^{-1}\right)\hat{D}\left(q^{-1}\right)R\left(q^{-1}\right)u(t) = \hat{P}\left(q^{-1}\right)\{y^*(t) - y(t)\} \tag{13.63}$$

Step 6: Then for the next sampling period, go back to Step 1 and repeat.

A simulated example will show the performance of the above algorithm.

Example 3 *The process is a first order model whose transfer function is*

$$G\left(q^{-1}\right) = \frac{0.1903252q^{-1}}{1 - 0.9048374q^{-1}}$$

with the sampling period $T = 0.1$ sec. The output disturbance is a sine wave signal

$$d(t) = 0.3\sin 6.435t$$

Investigate the operation of the pole placement adaptive control algorithm for this system.

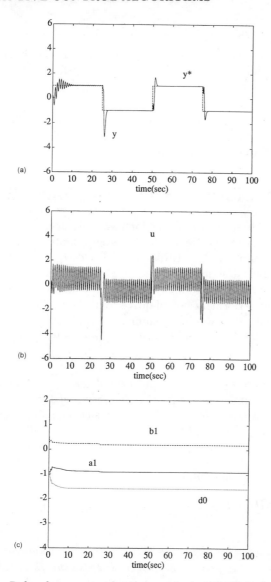

Figure 13.8: Pole placement adaptive control with disturbance rejection (a) Plant output and reference input (b) Plant control signal (c) Estimated plant and disturbance parameters

It can be seen that $D\left(q^{-1}\right) = 1 + d_0 q^{-1} + q^{-2}$ with $d_0 = -1.6$. In the following the initial parameters have been taken as $\hat{a}_1 = -0.8$, $\hat{b}_1 = 0.5$, and $\hat{d}_1 = -1.0$. In addition the closed loop roots are defined by the polynomial $A^*\left(q^{-1}\right) = \left(1 - 0.5q^{-1}\right)^7$. The simulation results for the above system are as shown in Fig. 13-8.

It can be seen that the adaptive control algorithm can effectively cancel the disturbance and achieve perfect tracking. In addition, the parameters can be observed to converge rapidly.

13.6 Stability of Adaptive Controllers

Adaptive control has been developed for several decades. However, rigorous proofs of stability and convergence for the basic adaptive control algorithms did not appear until the late 1970s. Indeed, closed loop adaptive control systems of the type described above are highly nonlinear and time-varying, and thus are very difficult to analyze. The detailed discussion of stability, convergence and performance of adaptive control systems is far beyond the scope of this book. Here we simply summarize the stability and convergence results for the adaptive control algorithms presented in the last section.

Theorem 1 *(Model Reference Adaptive Control)*
   Suppose the process order, n, and time delay, d, are known, and that the process is minimum phase (that is, the zeros of the process are inside the unit circle on the z-plane). Then the model reference adaptive control algorithm defined by Eqs. (13-41) to (13-43), when applied to the process given by Eq. (13-30) yields

(i)   *bounded $\{y(t)\}$ and $\{u(t)\}$*

(ii) $\lim_{t\to\infty} \{y(t) - y^*(t)\} = 0.$

Theorem 2 *(Pole Placement Adaptive Control)*
   Suppose the process order, n, is known, and the polynomials $A\left(q^{-1}\right)$ and $B\left(q^{-1}\right)$ are relatively prime. Moreover, suppose the estimated polynomials $\hat{A}\left(q^{-1}\right)$ and $\hat{B}\left(q^{-1}\right)$ are also relatively prime, then the pole placement adaptive control algorithm defined by Eqs. (13-48) to (13-51), when applied to the process given by Eq. (13-44), gives

(i)   *bounded $\{y(t)\}$ and $\{u(t)\}$*

(ii) $\lim_{t\to\infty} \left\{ A^*\left(q^{-1}\right) y(t) - \hat{B}\left(q^{-1}\right) \hat{P}\left(q^{-1}\right) y^*(t) \right\} = 0.$

It should be noted in the pole placement adaptive control algorithm that the solution to the Diophantine equation given in Eq. (13-50) requires that $\hat{A}\left(q^{-1}\right)$ and $\hat{B}\left(q^{-1}\right)$ are also relatively prime. Therefore in general a modified parameter estimation algorithm is needed to guarantee this property. The same problem exists in the pole placement adaptive control with disturbance rejection.

Theorem 3 *(Pole Placement Adaptive Control with Disturbance Rejection)*

Suppose the process order, n, and the disturbance model order, ℓ, are known, and the polynomials $A\left(q^{-1}\right) D\left(q^{-1}\right) R\left(q^{-1}\right)$ and $B\left(q^{-1}\right)$ are relatively prime. Moreover, suppose the estimated polynomials $\hat{B}\left(q^{-1}\right)$ and $\hat{A}\left(q^{-1}\right) \hat{D}\left(q^{-1}\right) R\left(q^{-1}\right)$ are also relatively prime, then the pole placement adaptive control algorithm defined by Eqs. (13-60) to (13-63), when applied to the process given by Eq. (13-52), leads to

(i) *bounded $\{y\left(t\right)\}$ and $\{u\left(t\right)\}$*

(ii) $\lim_{t\to\infty}\{y\left(t\right) - y^*\left(t\right)\} = 0.$

13.7   An Application Example

In this section, an industrial application of adaptive control is illustrated. The experimental hardware of the system is composed of a Personal Computer with 12-bit A/D and D/A facilities, power amplifier and a d.c. motor with suitable sensing mechanisms. The velocity of the d.c. motor is chosen as the controlled variable $y\left(t\right)$. Therefore if we neglect the dynamics, of the power amplifier, the sensing mechanism and the A/D and D/A converters, then the controlled process can be modeled as a first order plant with the d.c. motor velocity as its output and the computer output voltage as its input. Its transfer function can be expressed as

$$G(s) = \frac{b}{s + a}$$

With the sampling period $T = 0.04$ sec., the corresponding discrete time transfer function can be denoted by

$$G\left(z^{-1}\right) = \frac{b_1 z^{-1}}{1 - a_1 z^{-1}}$$

The following polynomials are chosen

$$R(z) = 1 - z^{-1}$$

$$A^*\left(z\right) = \left(1 - 1.9155 z^{-1} + 0.9231 z^{-2}\right)\left(1 - 0.7261 z^{-1}\right)$$

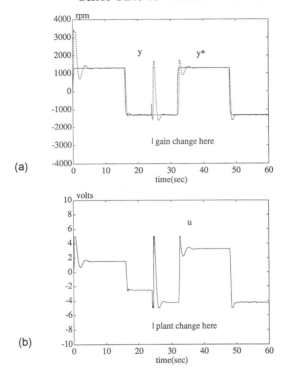

Figure 13.9: Pole placement adaptive control of d.c. motor (a) Plant output and reference input (b) Plant control signal

and the reference output $y^*(t)$ is chosen to be the output of the low pass filter

$$G_R(z) = \frac{0.3297z^{-1}}{1 - 0.6703z^{-1}}$$

driven by a square wave command signal of amplitude 1300 rpm and period 32 seconds.

The initial estimated parameter vector is chosen as zero. The process gain is decreased by a factor of two after the elapsed time $t = 25$ seconds. The performance of the pole placement adaptive control algorithm is then as shown in Fig. 13-9.

It will be observed that the algorithm automatically identifies the unknown process and its variations, and adaptively tunes its controller to achieve good tracking.

13.8   Summary

In this chapter, we have presented various adaptive control algorithms. It will be noted from the discussion that the certainty equivalence principle plays a key role in the development of adaptive control systems. Based on this principle, a wide spectrum of adaptive control algorithms can be generated by combining different parameter estimation methods such as the recursive projection algorithm or the recursive least squares algorithm with various control design approaches such as the model reference control or the pole placement control, as well as PID control, linear quadratic design, amongst others.

13.9   References

1. H. P. Whitaker, J. Yamrom, A. Kezer, "Design of Model Reference Adaptive Control Systems for Aircraft", Report R-164, Instrumentation Lab., MIT, Cambridge, Mass., 1958.

2. K. J. Astrom, B. Wittenmark, *Adaptive Control*, Addison Wesley Publ. Co., Reading, Mass., 1989.

3. G. C. Goodwin, K. S. Sin, *Adaptive Filtering, Prediction and Control*, Prentice Hall Inc., Englewood Cliffs, New Jersey, 1984.

4. K. S. Narendra, A. M. Annaswamy, *Stable Adaptive Systems*, Prentice Hall Inc., Englewood Cliffs, New Jersey, 1989.

5. G. Feng, M. Palaniswami, "Unified Treatment of Internal Model Principle Based Adaptive Control Algorithms", Int. J. Control, Vol. 54, No. 4 (1991) pp. 883-901.

13.10   Problems

13-1 Design a model reference controller for the following process

$$\left(1 - 2q^{-1}\right) y\left(t\right) = 2q^{-1}u\left(t\right)$$

where the reference model is

$$\left(1 - 0.5q^{-1}\right) y^*\left(t\right) = 0.5q^{-1}r\left(t\right)$$

13-2 Design a feedback control system to place the closed loop poles of the following process at the origin

$$\left(1 - 4q^{-1} + 3q^{-2}\right) y\left(t\right) = \left(q^{-1} - 2q^{-2}\right) u\left(t\right)$$

13-3 Show that Eq. (13-25) is the solution of the least squares problem. That is, it is the solution which minimizes the cost function defined by Eq. (13-24).

13-4 Consider the process with the model

$$\left(1 - 1.5q^{-1}\right) y\left(t\right) = 0.5q^{-1}u\left(t\right)$$

Of the model reference adaptive control and the pole placement adaptive control algorithms, which would you propose to use? Why?

13-5 Consider the process with the model

$$\left(1 - 1.7q^{-1} + 0.72q^{-2}\right) y\left(t\right) = \left(q^{-1} - 0.8q^{-2}\right) u\left(t\right)$$

Of the model reference adaptive control and the pole placement adaptive control algorithms, which would you propose to use? Why?

13-6 Develop a pole-placement adaptive control algorithm which can achieve zero tracking error, that is, $y\left(t\right) - y^*\left(t\right) = 0$ as t approaches infinity.

13-7 Develop a model reference adaptive control algorithm using the recursive projection estimation method.

Appendix A

Laplace and $\mathcal{Z}$-Transforms

A.1   Laplace Transforms

The use of Laplace transforms for solving ordinary linear differential equations of the type found in many engineering applications leads to a simple and mostly algebraic methodology which provides all the transient and steady state solutions of the equations. More importantly, this approach allows one to use the same approach in handling the input and initial conditions. The Laplace transform is basically the transformation of the linear equations and functions from the time domain to a complex plane where the resulting equations are algebraic. Once the necessary manipulations are performed on the complex plane using these algebraic equations, then the equations undergo an inverse Laplace transformation to yield the required solutions in the time domain. In this section we will provide sufficient information to apply the Laplace transform technique but leave the esoteric and more detailed discussions to more advanced texts.

We begin our consideration by noting that the functions of interest are piece-wise continuous, and defined for $t \geq 0$. For function $f(t)$ which is zero for $t < 0$ and defined for $t \geq 0$, the Laplace transform (sometimes called the one-sided Laplace transform) which is assumed to exist is defined as

$$F(s) = \mathcal{L}\left[f(t)\right] = \int_0^\infty f(t)e^{-st}\,dt \tag{A.1}$$

where $s = \sigma + j\omega$ is an introduced complex variable. We note that s has the dimensions of inverse time or frequency and is often called the **complex frequency**. It can be shown that the Laplace transform $F(s)$ is **uniquely related** to the time function $f(t)$, so that if $F(s)$ is given there is only one time function $f(t)$ to which it corresponds. This can be written as

$$f(t) = \mathcal{L}^{-1}[F(s)] = \frac{1}{2\pi j} \int_{c-j\infty}^{c+j\infty} F(s)e^{st}ds \tag{A.2}$$

where c is a real number such that when Re $s > c$ the integral given in Eq. (A-1) converges so that $F(s)$ exists.

The Laplace transform has a number of useful properties which are used in this book.

Linearity

Examination of Eq. (A-1) allows us to recognize the Laplace transform satisfies the property of linearity,

$$\mathcal{L}[f_1(t) + f_2(t)] = F_1(s) + F_2(s) \tag{A.3}$$

If $f(t)$ is multiplied by a constant then

$$\mathcal{L}[af(t)] = aF(s) \tag{A.4}$$

Differentiation

The Laplace transform of a derivative can be obtained by using integration by parts. Let us take the transform of $df(t)/dt$ and write

$$\mathcal{L}\left[\frac{df(t)}{dt}\right] = \int_0^\infty \frac{df(t)}{dt} e^{-st} dt$$

$$= \int_0^\infty d[f(t)] e^{-st} \tag{A.5}$$

If we let

$$u = f(t), \quad dv = e^{-st} dt$$

so that

$$du = d[f(t)], \quad v = -\frac{1}{s} e^{-st}$$

then substitution in Eq. (A-1) yields

$$F(s) = \left[-\frac{f(t)}{s} e^{-st}\right]_0^\infty + \frac{1}{s} \int_0^\infty d[f(t)] e^{-st}$$

$$= \frac{f(0)}{s} + \frac{1}{s} \int_0^\infty d[f(t)] e^{-st}$$

Substituting Eq. (A-5) above yields

$$\mathcal{L}\left[\frac{df(t)}{dt}\right] = sF(s) - f(0) \qquad (A.6)$$

which gives the Laplace transform of the first derivative. For the nth derivative the Laplace transform is given by

$$\mathcal{L}[f^{(n)}(t)] = s^n F(s) - s^{n-1} f(0) - s^{n-2} f^{(1)}(0) - \cdots - f^{(n-1)}(0) \qquad (A.7)$$

where $f^{(k)}(0)$ is the kth derivative of $f(t)$ evaluated at $t = 0$.

Integration

A similar procedure to that used for differentiation allows us to write for the first integral

$$\mathcal{L}\left[f^{(-1)}(t)\right] = \mathcal{L}\left[\int f(t)dt\right] = \frac{F(s)}{s} + \frac{f^{(-1)}(0)}{s} \qquad (A.8)$$

For the nth integral of $f(t)$ denoted by $f^{(-n)}(t)$, the Laplace transform is given by

$$\mathcal{L}\left[f^{(-n)}(t)\right] = \frac{F(s)}{s^n} + \frac{f^{(-1)}(0)}{s^n} + \frac{f^{(-2)}(0)}{s^{n-1}} + \cdots + \frac{f^{(-n)}(0)}{s} \qquad (A.9)$$

Time-Displacement

The unit step function is denoted by $H(t)$ and suppose the time displacement is $t_0 > 0$. Then the Laplace transform of the delayed time function $f(t - t_0)H(t - t_0)$ is given by

$$\mathcal{L}\left[f(t - t_0)H(t - t_0)\right] = e^{-t_0 s} F(s) \qquad (A.10)$$

s-Plane Displacement

Suppose the Laplace transform of $f(t)$ is displaced by a distance a in the s-plane. It can be shown that

$$\mathcal{L}\left[e^{at} f(t)\right] = F(s - a) \qquad (A.11)$$

Convolution

Let the time functions $f(t)$ and $g(t)$ have Laplace transforms $F(s)$ and $G(s)$ respectively. Then the convolution integral has the Laplace transform given by

$$\mathcal{L}\left[\int_0^t f(t)g(t - \tau)d\tau\right] = F(s)G(s) \qquad (A.12)$$

Complex Convolution

If time functions $f(t)$ and $g(t)$ have Laplace transforms $F(s)$ and $G(s)$ respectively then it can be shown

$$\mathcal{L}\left[f(t)g(t)\right] = \frac{1}{2\pi j} \int_{\sigma_1 - j\infty}^{\sigma_1 + j\infty} F(\lambda)G(s - \lambda)d\lambda$$

where the integration is carried out along the vertical line Re $\lambda = \sigma_1$. The variable s and real number σ_1 must satisfy the conditions $\sigma_1 > c_f$ and Re $s > \sigma_1 + c_g$, where c_f and c_g are the minimum numbers as defined in Eq. (A-2) for which the functions $f(t)$ and $g(t)$ have Laplace transforms.

Final and Initial Value Theorems

Suppose $f(t)$ has the Laplace transform $F(s)$ and the limits

$$\lim_{t \to 0} f(t) \text{ and } \lim_{t \to \infty} f(t)$$

exist. Then the **final value theorem** is

$$f(\infty) = \lim_{t \to \infty} f(t) = \lim_{s \to 0}[sF(s)] \tag{A.13}$$

and the **initial value theorem** is

$$f(0) = \lim_{t \to 0} f(t) = \lim_{t \to \infty}[sF(s)] \tag{A.14}$$

A list of the more useful Laplace transforms is given in Table A-1.

A.2 $\mathcal{Z}$-Transforms

Suppose $f^*(t)$ is the sampled version of $f(t)$ then its Laplace transform leads to an infinite series of exponential terms. It is then advantageous to use the $\mathcal{Z}$-transform substitution defined as

$$z = e^{sT} \tag{A.15}$$

which is a simple change of variable from s to z. The value of T is given by the time between successive pulses and is referred to as the **sampling time**. The $\mathcal{Z}$-transform substitution basically transforms from the s-domain to the z-domain which is also a complex plane. The transformation given by Eq. (A-15) allows us to write

$$F(z) = \mathcal{Z}\left[f^*(t)\right] = F^*(s)\big|_{s = \ln z / T} \tag{A.16}$$

Table A-1

Time function	Laplace transform	$\mathcal{Z}$-Transform
$\delta(t)$	1	1
$\delta(t - nT)$	e^{-snT}	z^{-n}
$AH(t)$	$\dfrac{A}{s}$	$\dfrac{Az}{z-1}$
t	$\dfrac{1}{s^2}$	$\dfrac{Tz}{(z-1)^2}$
t^{n-1}	$\dfrac{(n-1)!}{s^n}$	$\lim\limits_{a\to 0}(-1)^{n-1}\dfrac{d^{n-1}}{da^{n-1}}\left(\dfrac{z}{z-e^{-aT}}\right)$
e^{-at}	$\dfrac{1}{s+1}$	$\dfrac{z}{z-e^{-aT}}$
$\dfrac{1}{b-a}\left(e^{-at}-e^{-bt}\right)$	$\dfrac{1}{(s+a)(s+b)}$	$\dfrac{1}{b-a}\left(\dfrac{z}{z-e^{-aT}}-\dfrac{z}{z-e^{-bT}}\right)$
$\dfrac{1}{a}(1-e^{-at})$	$\dfrac{1}{s(s+a)}$	$\dfrac{1}{a}\dfrac{(1-e^{-aT})z}{(z-1)(z-e^{-aT})}$
te^{-at}	$\dfrac{1}{(s+a)^2}$	$\dfrac{Tze^{-aT}}{(z-e^{-aT})^2}$
$\sin\omega t$	$\dfrac{\omega}{s^2+\omega^2}$	$\dfrac{z\sin\omega T}{z^2-2z\cos\omega T+1}$
$\cos\omega t$	$\dfrac{s}{s^2+\omega^2}$	$\dfrac{z(z-\cos\omega T)}{z^2-2z\cos\omega T+1}$
$e^{-at}\sin\omega t$	$\dfrac{\omega}{(s+a)^2+\omega^2}$	$\dfrac{ze^{-aT}\sin\omega T}{z^2-2ze^{-aT}\cos\omega T+e^{-2aT}}$
$e^{-at}\cos\omega t$	$\dfrac{s+a}{(s+a)^2+\omega^2}$	$\dfrac{z(z-e^{-aT}\cos\omega T)}{z^2-2ze^{-aT}\cos\omega T+e^{-2aT}}$
$\dfrac{e^{-at}-e^{-bt}}{b-a}$	$\dfrac{1}{(s+a)(s+b)}$	$\dfrac{z(e^{-aT}-e^{-bT})}{(b-a)(z-e^{-aT})(z-e^{-bT})}$
$\sinh at$	$\dfrac{a}{s^2-a^2}$	$\dfrac{z\sinh aT}{(z-e^{aT})(z-e^{-aT})}$
$\cosh at$	$\dfrac{s}{s^2-a^2}$	$\dfrac{z(z-\cosh aT)}{(z-e^{aT})(z-e^{-aT})}$

Thus if the Laplace transform $F^*(s)$ of a discrete or sampled data function $f(nT)$ is given by

$$F^*(s) = \sum_{n=0}^{\infty} f(nT)e^{-nsT} \tag{A.17}$$

then the $\mathcal{Z}$-transform is given by

$$F(z) = \sum_{n=0}^{\infty} f(nT)z^{-n} \tag{A.18}$$

For many forms of $f(t)$ this expression converges.

Since $F(z)$ is analytic and therefore converges for $|z| \geq R$ it is known from the theory of complex variables that it can be expanded into a Laurent series

$$F(z) = \sum_{n=0}^{\infty} a_{-n}z^{-n} \tag{A.19}$$

where

$$a_{-n} = \frac{1}{2\pi j} \int_C z^{n-1} F(z)dz$$

and C is any closed contour outside $|z| = R$. Since the expansion is unique we can identify a_{-n} with $f(nT)$ in Eq. (A-18), and as a consequence it can be concluded

$$f(nT) = \frac{1}{2\pi j} \int_C z^{n-1} F(z)dz \tag{A.20}$$

Eq. (A-20) can be easily evaluated by the **method of residues,** namely

$$f(nT) = \sum_{\substack{\text{Poles of} \\ z^{n-1} F(z)}} \text{Res}\left[z^{n-1} F(z)\right]$$

The use of the $\mathcal{Z}$-transforms is facilitated by taking advantage of several properties which are:

Linearity

$$\mathcal{Z}[f_1^*(t) + f_2^*(t)] = F_1(z) + F_2(z) \tag{A.21}$$

Multiplication by constant

$$\mathcal{Z}[af^*(t)] = aF(z) \tag{A.22}$$

Time translation

$$\mathcal{Z}[f^*(t - mT)] = z^{-m}F(z))$$ (A.23)

z-Plane translation

Let $f_a(t) = e^{at}f(t)$ then

$$\mathcal{Z}[f_a^*(t)] = F(ze^{aT})$$ (A.24)

In addition to the above properties there are two theorems that are useful for obtaining the initial and final values. These are

Initial Value Theorem

$$f^*(0) = \lim_{z \to \infty} F(z)$$ (A.25)

Final Value Theorem

$$f^*(\infty) = \lim_{z \to 1} \left(\frac{z-1}{z} \right) F(z)$$ (A.26)

This theorem, useful for obtaining the steady state value of $f^*(t)$, is valid only if a final value exists. A list of the more useful $\mathcal{Z}$-transforms is given in Table A-1.

A.3   References

1. H. S. Carslaw, J. B. Jaeger, *Operational Methods in Applied Mathematics*, Oxford University Press, Oxford, 2nd Ed. 1948.

2. J. C. Jaeger, *An Introduction to the Laplace Transform with Engineering Applications*, Methuen and Co. Ltd., London, 1961.

3. W. Kaplan, *Operational Methods for Linear Systems*, Addison Wesley Publ. Co., Reading, Mass., 1962.

4. E. J. Jury, *Theory and Application of the Z-Transform Method*, John Wiley and Sons Inc., New York, 1964.

5. B. C. Kuo, *Digital Control Systems*, Holt, Rinehart and Winston, Inc., New York 1980.

Appendix B

Symbols, Units and Analogous Systems

B.1   Systems of Units

By international agreement the International System of Units (abbreviated SI units) has been adopted for expressing the values of physical quantities. This system consists of seven **basic units** shown in Table B-1 and two **supplementary units** given in Table B-2.

All other physical quantities are derived from these base units and their units are called **derived units**. To see this we observe that velocity is distance travelled per unit time, and has the SI units m/s, while acceleration has the SI units m/s^2. Other examples of derived units in the SI system are spring stiffness, viscous damping, specific heat, thermal resistance, electric potential, electric resistance and inductance.

While the last five units in Table B-1 are now universally accepted, there are a number of other systems of units which are widely used for the first two. The **British absolute units** are based upon the foot (ft), pound (lb) and second (s) as the base units of length, mass, and time respectively, while the **British gravitational** (or engineering) system, is based upon the foot (ft), pound force (lbf), and second (s) as base units. In this system mass is a derived unit. The relationship between the SI units and British units can be derived from the equivalencies:

$$1 \text{ ft} = 0.3048 \text{ m (exact)}$$

$$1 \text{ lb} = 0.4536 \text{ kg (approx)}$$

Table B-1 SI Base Units

Physical Quantity	SI Units	
	Name	Symbol
Length	meter	m
Mass	kilogram	kg
Time	second	s (sec)
Electric current	ampere	A
Thermodynamic temperature	kelvin	K
Amount of substance	mole	mol (kg-mol)
Luminous intensity	candela	cd

Table B-2 SI Supplementary Units

Physical Quantity	SI Units	
	Name	Symbol
Phase angle	radian	rad
Solid angle	steradian	sr

The unit of mass is the slug where 1 slug = 32.174 lb in the British gravitational units.

B.2   Symbols and Units

The symbols and names used for various derived units are given in Table B-3. This is not an exhaustive list however.

To convert from one set of derived units to another it is useful to use the method shown by the following example:

$$1 \text{ kPa} = 1000 \left[\frac{N}{m^2} \right] \left[\frac{1 \text{ kgf}}{9.807 \text{ N}} \right] \left[\frac{1 \text{ lbf}}{0.4536 \text{ kgf}} \right] \left[\frac{0.3048^2 \text{ m}^2}{1 \text{ ft}^2} \right] \left[\frac{1 \text{ ft}^2}{12^2 \text{ in}^2} \right]$$

$$= \frac{1000 \times 0.3048^2}{9.807 \times 0.4536 \times 12^2} \frac{\text{lbf}}{\text{in}^2} = 0.1450 \text{ lbf/in}^2$$

Table B-3 Symbols and Names for Derived Units

Physical Quantity	Symbol	SI Units	British Gravitational Units
Mechanical Translation			
Force	F, f	newton (N)	pound (lbf)
Displacement	x	metre (m)	foot (ft)
Velocity	v	m/s	ft/s
Acceleration	a	m/s^2	ft/s^2
Mass	M	kilogram (kg)	slug
	M	N/m s^{-2}	lbf/ft s^{-2}
Stiffness	K	N/m	lbf/ft
Damping	B	N/m s^{-1}	lbf/ft s^{-1}
Mechanical Rotation			
Torque	T	N m	lbf ft
Angle	θ	rad	rad
Angular speed	ω	rad/s	rad/s
Angular acceleration	α	rad/s^2	rad/s^2
Moment of inertia	I, J	kg m^2	slug ft^2
Stiffness	K	N m/rad	lbf ft/rad
Damping	B	N m/rad s^{-1}	lbf ft/rad s^{-1}
Hydraulic			
Force	F	N	lbf
Displacement	x	m	ft
Density	ρ	kg/m^3	slug/ft^3
Pressure	P	N/m^2	lbf/ft^2
Mass flow rate	q	kg/s	slug/s
Thermal			
Heat energy	Q	joule(J)	BTU
Temperature	T	kelvin (K or C)	F or R
Specific heat	c	J/(kg K)	BTU/(lb F)
Heat flow rate	q	watt(W)	BTU/s
Thermal capacitance	C	J/K	BTU/F
Thermal resistance	R	K/W	F/(BTU s^{-1})
Electrical			
Voltage	e, v	volt (V)	Same as SI units
Current	i	ampere (A)	
Inductance	L	henry (H)	
Capacitance	C	farad (F)	
Resistance	R	ohm (Ω)	
Conductance	G	siemens (S)	

B.3 Comparison of Variables in Analogous Systems

Table B-4 Variables in Analogous Systems

System	Potential	Flow	Dissipation	Capacitive	Inductive
Electrical	Voltage, v	Current, i	Resistance, R	Capacitance, C	Inductance, L
Mechanical (translational)	Velocity, v	Force, F	Viscous friction, B	Mass, M	Translational stiffness (spring), k
Mechanical (rotational)	Angular velocity, ω	Torque, T	Viscous friction, B	Moment of inertia, J	Rotational stiffness (spring), k
Fluid	Pressure, P	Volumetric flow, q	Fluid resistance, R	Fluid capacitance, C	Fluid inertance, I
Thermal	Temperature, T	Heat flow, q	Thermal resistance, R	Thermal capacitance, C	

B.4 References

1. _____, "*SI Units and Recommendations for the Use of Their Multiples and of Certain Other Units*", International Standards Organization, ISO 1000-1981.

Appendix C

Fundamentals of Matrix Theory

C.1   Introduction

The word **matrix** is used to denote an orderly array of objects called elements. These elements are mathematical symbols, constants, complex numbers, functions, etc. A matrix **A** is denoted by

$$
\mathbf{A} = \begin{bmatrix}
a_{11} & a_{12} & \cdots & a_{1m} \\
a_{21} & a_{22} & \cdots & a_{2m} \\
\vdots & \vdots & & \vdots \\
a_{n1} & a_{n2} & \cdots & a_{nm}
\end{bmatrix}
\tag{C.1}
$$

where a_{ij} are the elements of **A**. The entire array of elements is sometimes referred to as $[a_{ij}]$. The subscripts follow a numerical sequence such that the first subscript i refers to the effect, whereas the second refers to the cause. The term $a_{ij}x_j$ causes the effect y_i. If there are n effects and m causes, then the matrix is an $n \times m$ matrix. The number of columns is equal to m and the number of rows to n.

A matrix having one column or row can be interpreted as a **vector**. For example the 3×1 column matrix,

$$
\mathbf{a} = \begin{bmatrix}
a_1 \\
a_2 \\
a_3
\end{bmatrix}
$$

can be thought of as a vector with components a_1, a_2, a_3 in the orthogonal system, or three-dimensional space. If the number of elements is larger than three, then we must think of a vector in a higher dimensional space, at least conceptually. A column matrix with n elements is a vector in **n-dimensional space.**

A rectangular matrix may be thought of as consisting of m vectors. The matrix

$$\mathbf{A} = \begin{bmatrix} a_{11} & \cdots & a_{1m} \\ \vdots & & \vdots \\ a_{n1} & \cdots & a_{nm} \end{bmatrix}$$

can be interpreted to consist of m vectors,

$$\mathbf{A} = \begin{bmatrix} \mathbf{b}_1 & \mathbf{b}_2 & \cdots & \mathbf{b}_m \end{bmatrix}$$

where each vector has n elements,

$$\mathbf{b}_i = \begin{bmatrix} a_{1i} \\ a_{2i} \\ \vdots \\ a_{ni} \end{bmatrix}$$

Vectors, like matrices, combine linearly. A vector may be formed by adding several vectors each having the same dimension. Let $\mathbf{a}_1, \mathbf{a}_2, \ldots$ be vectors, then a vector $\mathbf{c}$ can be formed such that

$$\mathbf{c} = \alpha_1 \mathbf{a}_1 + \alpha_2 \mathbf{a}_2 + \cdots$$

where $\alpha_1, \alpha_2, \ldots$ are constants. Such an equation is called a linear combination of vectors.

When the number of rows and columns of a matrix are unequal, it is a **rectangular matrix**, and when they are equal, it is a **square matrix**. Every square matrix has associated with it a scalar quantity called the **determinant** of the matrix. It is designated as det $\mathbf{A}$ or simply as $|\mathbf{A}|$. For a matrix $\mathbf{A}$ the determinant is written as

$$\det \mathbf{A} = |\mathbf{A}| = \begin{vmatrix} a_{11} & a_{12} & \cdots & a_{1n} \\ a_{21} & a_{22} & \cdots & a_{2n} \\ \vdots & \vdots & & \vdots \\ a_{n1} & a_{n2} & \cdots & a_{nn} \end{vmatrix}$$

If the square matrix is $n \times n$, then the determinant is called an nth-order determinant. The value of the determinant can be defined in terms of its **minors** and **cofactors**. The determinant obtained by deleting the ith row and jth column is called the ijth minor of det $\mathbf{A}$ and is denoted by

$|\mathbf{M}_{ij}|$. This minor is associated with, or belongs to, the element a_{ij}. As an example, consider a third-order determinant,

$$\det \mathbf{A} = a_{11}\,|\mathbf{M}_{11}| - a_{12}\,|\mathbf{M}_{12}| + a_{13}\,|\mathbf{M}_{13}|$$

where

$$\mathbf{M}_{11} = \begin{bmatrix} a_{22} & a_{23} \\ a_{32} & a_{33} \end{bmatrix}, \quad \mathbf{M}_{12} = \begin{bmatrix} a_{21} & a_{23} \\ a_{31} & a_{33} \end{bmatrix}, \quad \mathbf{M}_{13} = \begin{bmatrix} a_{21} & a_{22} \\ a_{31} & a_{32} \end{bmatrix}$$

If the diagonal elements of the minor are the diagonal elements of $|\mathbf{A}|$, then the minor is a **principal minor**. In the above expansion, only $|\mathbf{M}_{11}|$ is a principal minor.

Associated with each minor $|\mathbf{M}_{ij}|$ is a cofactor C_{ij} such that

$$C_{ij} = (-1)^{i+j}\,|\mathbf{M}_{ij}|$$

Using this, the determinant of $\mathbf{A}$ becomes

$$\det \mathbf{A} = |\mathbf{A}| = \sum_{i=1}^{n} a_{ij} C_{ij} \quad j = 1, 2, \ldots, n$$

$$= \sum_{j=1}^{n} a_{ij} C_{ij} \quad i = 1, 2, \ldots, n$$

(C.2)

C.2   Matrix Algebra

In combining matrices, the following rules must be observed:

Equality. The equality of matrices requires that all the elements in the same relative position of each matrix be equal. Obviously for matrices to be equal, the dimensions of one matrix must be equal to the dimensions of the other matrix. If an $n \times m$ matrix $\mathbf{A}$ is equal to another $n \times m$ matrix $\mathbf{B}$,

$$\mathbf{A} = \mathbf{B}$$

then

$$a_{ij} = b_{ij} \text{ for all } i \text{ and } j$$

Addition and subtraction. Two matrices must have the same dimensions in order to be added or subtracted. If two matrices $\mathbf{A}$ and $\mathbf{B}$ are combined,

$$\mathbf{A} \pm \mathbf{B} = \mathbf{C}$$

then

$$a_{ij} \pm b_{ij} = c_{ij} \text{ for all } i \text{ and } j$$

In addition and subtraction the associative as well as commutative laws apply.

Multiplication of matrices. Multiplication of a matrix by a scalar is achieved by multiplying each element by the scalar. If

$$\mathbf{C} = k\mathbf{A}$$

where k is a constant, then we have

$$c_{ij} = ka_{ij} \text{ for all } i \text{ and } j$$

For two matrices to be multiplied, the number of columns of the first matrix must be equal to the number of rows of the second matrix. If $\mathbf{A}$ is $n \times m$, then $\mathbf{B}$ must be $m \times p$ for multiplying them together. The resulting product is an $n \times p$ matrix. Let

$$\mathbf{A} = \begin{bmatrix} a_{11} & \cdots & a_{1m} \\ \vdots & & \vdots \\ a_{n1} & \cdots & a_{nm} \end{bmatrix}, \quad \mathbf{B} = \begin{bmatrix} b_{11} & \cdots & b_{1p} \\ \vdots & & \vdots \\ b_{m1} & \cdots & b_{mp} \end{bmatrix}$$

Then if

$$\mathbf{C} = \mathbf{AB}$$

the elements of $\mathbf{C}$ become

$$\mathbf{C} = \begin{bmatrix} (a_{11}b_{11} + a_{12}b_{21} + \cdots + a_{1m}b_{m1}) & \cdots & (a_{11}b_{1p} + \cdots + a_{1m}b_{mp}) \\ \vdots & & \vdots \\ (a_{n1}b_{11} + a_{n2}b_{21} + \cdots + a_{nm}b_{m1}) & \cdots & (a_{n1}b_{1p} + \cdots + a_{nm}b_{mp}) \end{bmatrix}$$

The operation of obtaining the ijth term of $\mathbf{C}$ involves taking the sum of all the terms obtained by multiplying the ith row of the first matrix with the jth column of the second matrix. Note that there are no restrictions on the number of rows of the first matrix or the number of columns of the second matrix. This simply determines the dimension of the new matrix.

Matrix multiplication does not obey the law of commutation, i.e. in general,

$$\mathbf{AB} \neq \mathbf{BA}$$

Therefore when a matrix $\mathbf{A}$ is multiplied such that

$$\mathbf{C} = \mathbf{BA}$$

we say that $\mathbf{A}$ is **premultiplied** by $\mathbf{B}$. If instead

$$\mathbf{C} = \mathbf{AB}$$

then $\mathbf{A}$ is said to be **postmultiplied** by $\mathbf{B}$.

Division.   The division by a matrix is perhaps the most difficult operation. If

$$\mathbf{C} = \mathbf{AB}$$

and we are given $\mathbf{C}$ and $\mathbf{A}$, how do we obtain $\mathbf{B}$? We can answer this by premultiplying by $\mathbf{A}^{-1}$

$$\mathbf{A}^{-1}\mathbf{C} = \mathbf{A}^{-1}\mathbf{AB}$$

where $\mathbf{A}^{-1}$ is the **inverse** of $\mathbf{A}$ and is analogous to division. Note that this is valid only if $\mathbf{A}$ is a square matrix and $\det \mathbf{A} \neq 0$. The term $\mathbf{A}^{-1}\mathbf{A}$ is unity, i.e.

$$\mathbf{A}^{-1}\mathbf{A} = \mathbf{I} = \begin{bmatrix} 1 & 0 & \cdots & 0 \\ 0 & 1 & \cdots & 0 \\ \vdots & \vdots & & \vdots \\ 0 & 0 & \cdots & 1 \end{bmatrix} \tag{C.3}$$

where $\mathbf{I}$ is called the **unit matrix**. Substituting

$$\mathbf{IB} = \mathbf{B} = \mathbf{A}^{-1}\mathbf{C}$$

The value of $\mathbf{A}^{-1}$ is obtained by satisfying Eq. (C-3).

C.3   Types of Matrices

There are many square matrices having special properties that make them unique. Some of the ones useful to us are listed below.

Diagonal matrix.   The principal diagonal of this matrix has nonzero elements. All the other elements are zero.

$$\text{Diagonal matrix } \mathbf{A} = \begin{bmatrix} a_{11} & 0 & \cdots & 0 \\ 0 & a_{22} & \cdots & 0 \\ \vdots & \vdots & & \vdots \\ 0 & 0 & \cdots & a_{nn} \end{bmatrix}$$

This matrix will also be denoted by $\mathbf{A} = \text{diag}\,[a_{11}, a_{22}, \ldots, a_{nn}]$.

Null matrix. This matrix has all its elements equal to zero. This is sometimes called a zero matrix.

Unit matrix. Sometimes called an identity matrix, this matrix is a diagonal matrix except that all the diagonal elements are unity. The identity matrix when multiplied by another matrix leaves the matrix unchanged.

$$\mathbf{IA} = \mathbf{AI} = \mathbf{A}$$

Transpose matrix. The transpose of a matrix is obtained by interchanging the rows and columns of a matrix. The transpose matrix is denoted by $\mathbf{A}^T$. If

$$\mathbf{A} = \begin{bmatrix} a_{11} & a_{12} & \cdots & a_{1n} \\ a_{21} & a_{22} & \cdots & a_{2n} \\ \vdots & \vdots & & \vdots \\ a_{n1} & a_{n2} & \cdots & a_{nn} \end{bmatrix}$$

Then

$$\mathbf{A}^T = \begin{bmatrix} a_{11} & a_{21} & \cdots & a_{n1} \\ a_{12} & a_{22} & \cdots & a_{n2} \\ \vdots & \vdots & & \vdots \\ a_{1n} & a_{2n} & \cdots & a_{nn} \end{bmatrix}$$

Symmetric matrix. A matrix is symmetric if it is equal to its transpose, i.e.

$$a_{ij} = a_{ji} \text{ for all } i, j$$

If the matrix coefficients a_{ij} are complex and

$$a_{ij} = \bar{a}_{ji} \text{ for all } i, j$$

then it is said to be **hermitian**.

Skew-symmetric matrix. If the transpose of the matrix is equal to the negative of the matrix, it is skew symmetric,

$$a_{ij} = -a_{ji} \text{ for all } i, j$$

Orthogonal matrix. An orthogonal matrix is one which when multiplied by its transpose yields the identity matrix,

$$\mathbf{A}\mathbf{A}^T = \mathbf{I}$$

which implies

$$\mathbf{A}^T = \mathbf{A}^{-1}$$

Matrices defining coordinate transformations are generally orthogonal.

Triangular matrix. When all the elements above or below the main diagonal are zero, then the matrix is a triangular matrix. If the elements above the main diagonal are zero, the matrix is a **lower triangular** matrix. If the elements below the main diagonal are zero, the matrix is an **upper triangular** matrix. An example of a lower triangular matrix is

$$\mathbf{A} = \begin{bmatrix} 1 & 0 & 0 & 0 \\ 2 & 2 & 0 & 0 \\ 3 & 0 & 4 & 0 \\ 1 & 5 & 2 & 3 \end{bmatrix}$$

Adjoint matrix. This matrix is the transpose of a matrix formed by replacing each element by its cofactor. If C_{ij} is the cofactor of the element a_{ij}, then the adjoint matrix is

$$\operatorname{adj} \mathbf{A} = [C_{ji}] \tag{C.4}$$

It is denoted by adj $\mathbf{A}$. Consider a matrix $\mathbf{A}$,

$$\mathbf{A} = \begin{bmatrix} 0 & 0 & 1 \\ 2 & 1 & 2 \\ 3 & 1 & 1 \end{bmatrix}$$

The matrix formed by the cofactors is

$$\mathbf{C} = \begin{bmatrix} -1 & 4 & -1 \\ 1 & -3 & 0 \\ -1 & 2 & 0 \end{bmatrix}$$

The adjoint becomes

$$\operatorname{adj} \mathbf{A} = \mathbf{C}^T = \begin{bmatrix} -1 & 1 & -1 \\ 4 & -3 & 2 \\ -1 & 0 & 0 \end{bmatrix}$$

Nonsingular matrix. If the determinant of a matrix is nonzero, then the matrix is nonsingular. If

$$\det \mathbf{A} = 0$$

then it is a singular matrix.

Inverse matrix. The inverse $\mathbf{A}^{-1}$ of a matrix $\mathbf{A}$ must satisfy the property that

$$\mathbf{A}^{-1}\mathbf{A} = \mathbf{I}$$

Not all matrices have an inverse. It can be shown that all nonsingular matrices possess an inverse. The inverse of a matrix can be constructed by satisfying Eq. (C-3). Here we shall investigate an alternative way.

Consider a matrix $\mathbf{A}$ whose determinant $|\mathbf{A}|$ is

$$|\mathbf{A}| = \sum_{j=1}^{n} a_{ij} C_{ij}, \quad i = 1, 2, \ldots, n$$

The cofactor does not contain any elements of the ith row of $\mathbf{A}$. If instead of a_{ij} we had a_{kj}, then from the properties of determinants it can be shown

$$\sum_{j=1}^{n} a_{kj} C_{ij} = 0, \quad k \neq i$$

The two previous equations can be combined to yield

$$\sum_{j=1}^{n} a_{kj} C_{ij} = \delta_{ik} |\mathbf{A}|$$

where

$$\delta_{ik} = \begin{cases} 0 & i \neq k \\ 1 & i = k \end{cases}$$

This may be rewritten as

$$[a_{ij}]\,[C_{ji}] = |\mathbf{A}|\,\mathbf{I}$$

or

$$\mathbf{A}\ \text{adj}\ \mathbf{A} = |\mathbf{A}|\,\mathbf{I}$$

or

$$\mathbf{I} = \frac{\mathbf{A}\ \text{adj}\ \mathbf{A}}{|\mathbf{A}|}$$

Since $\mathbf{AA}^{-1} = \mathbf{I}$, we have

$$\mathbf{A}^{-1} = \frac{\text{adj } \mathbf{A}}{|\mathbf{A}|} \tag{C.5}$$

and the inverse exists if and only if $|\mathbf{A}| \neq 0$.

Complex matrices. When all or some of the elements of a matrix are complex numbers, then the matrix is a complex matrix. The conjugate of a complex matrix is obtained by taking the conjugate of each element.

Partitioned matrices. The elements of a matrix can not only be numbers and functions by may be other matrices. Consider the matrix

$$\mathbf{A} = \begin{bmatrix} a_{11} & a_{12} & a_{13} & a_{14} \\ a_{21} & a_{22} & a_{23} & a_{24} \\ a_{31} & a_{32} & a_{33} & a_{34} \\ a_{41} & a_{42} & a_{43} & a_{44} \end{bmatrix}$$

This may be rewritten as

$$\mathbf{A} = \begin{bmatrix} \mathbf{B}_{11} & \mathbf{B}_{12} \\ \mathbf{B}_{21} & \mathbf{B}_{22} \end{bmatrix}$$

where

$$\mathbf{B}_{11} = \begin{bmatrix} a_{11} & a_{12} \\ a_{21} & a_{22} \end{bmatrix} \quad \mathbf{B}_{12} = \begin{bmatrix} a_{13} & a_{14} \\ a_{23} & a_{24} \end{bmatrix}$$

$$\mathbf{B}_{21} = \begin{bmatrix} a_{31} & a_{32} \\ a_{41} & a_{42} \end{bmatrix} \quad \mathbf{B}_{11} = \begin{bmatrix} a_{33} & a_{34} \\ a_{43} & a_{44} \end{bmatrix}$$

The rules of matrix algebra apply to partitioned matrices without modifications. Extra care should be exercised to insure that the various matrices to be combined are conformable.

C.4   Matrix Calculus

Having defined the algebraic operations of matrices we now consider operations involving calculus. In this section we shall be concerned with differentiation, integration, and transformation of a matrix. The transformation will be from the time domain to the complex s- and z-domains and vice versa.

Differentiation and integration. Consider a matrix $\mathbf{A}(t)$ whose elements are functions of time. Then the derivative of $\mathbf{A}(t)$ is obtained by taking the derivative of each element of $\mathbf{A}(t)$,

$$\frac{d\mathbf{A}(t)}{dt} = \left[\frac{da_{ij}}{dt}\right] \qquad (C.6)$$

The derivative of a product of two matrices is

$$\frac{d}{dt}\left[\mathbf{A}(t)\mathbf{B}(t)\right] = \mathbf{A}(t)\frac{d\mathbf{B}(t)}{dt} + \frac{d\mathbf{A}(t)}{dt}\mathbf{B}(t)$$

The derivative of the sum or difference is

$$\frac{d}{dt}\left[\mathbf{A}(t) \pm \mathbf{B}(t)\right] = \frac{d\mathbf{A}(t)}{dt} \pm \frac{d\mathbf{B}(t)}{dt}$$

The integral of a matrix is obtained by taking the integral of each element of the matrix.

$$\int \mathbf{A}(t)dt = \left[\int a_{ij}dt\right] \qquad (C.7)$$

The integral of the sum or difference is

$$\int \left[\mathbf{A}(t) \pm \mathbf{B}(t)\right]dt = \int \mathbf{A}(t)dt \pm \int \mathbf{B}(t)dt$$

The integral of a product is taken after the two matrices are multiplied.

Laplace and $\mathcal{Z}$-transform. We shall assume that the matrix $\mathbf{A}$ is such that its Laplace and $\mathcal{Z}$-transforms exist. Then the Laplace transform of $\mathbf{A}$ is simply obtained by taking the Laplace transform of each element of $\mathbf{A}$,

$$\mathcal{L}\left[\mathbf{A}(t)\right] = \left[\mathcal{L}a_{ij}\right] \qquad (C.8)$$

The Laplace transform of the sum of two matrices is

$$\mathcal{L}\left[\mathbf{A}(t) + \mathbf{B}(t)\right] = \mathcal{L}\left[\mathbf{A}(t)\right] + \mathcal{L}\left[\mathbf{B}(t)\right]$$

The Laplace transform of the derivative of a matrix may be written as

$$\mathcal{L}\left[\frac{d\mathbf{A}(t)}{dt}\right] = \left[\mathcal{L}\frac{da_{ij}}{dt}\right]$$

The inverse Laplace transform of a matrix is obtained by taking the inverse of each element,

$$\mathbf{A}(t) = \mathcal{L}^{-1}\left[\mathbf{A}(s)\right] = \left[\mathcal{L}^{-1}a_{ij}\right] \qquad (C.9)$$

The $\mathcal{Z}$-transform of matrix requires that a matrix be defined at discrete time steps,

$$\mathbf{A}(nT) = [a_{ij}(nT)] \qquad (C.10)$$

The $\mathcal{Z}$-transform of $\mathbf{A}(nT)$ can be obtained by taking the $\mathcal{Z}$-transform of each element,

$$\mathbf{A}(z) = \mathcal{Z}\left[\mathbf{A}(nT)\right] = [\mathcal{Z}\left[a_{ij}(nT)\right]] \qquad (C.11)$$

The $\mathcal{Z}$-transform of the sum or difference of two matrices is equal to the sum or difference of the $\mathcal{Z}$-transforms of the matrices. If the matrix $\mathbf{A}(nT)$ is defined at $\mathbf{A}[(n+1)T]$, then it can be shown that

$$\mathcal{Z}\left[\mathbf{A}\left((n+1)T\right)\right] = z\mathbf{A}(z) - z\mathbf{A}(0)$$

The inverse $\mathcal{Z}$-transform is obtained by taking the inverse $\mathcal{Z}$-transform of each element,

$$\mathbf{A}(nT) = \mathcal{Z}^{-1}\mathbf{A}(z) = \left[\mathcal{Z}^{-1}(a_{ij}(z))\right]$$

The rules developed for the $\mathcal{Z}$-transform and the inverse $\mathcal{Z}$-transform apply to each element of the matrix without modification.

C.5   Linear Algebraic Equations

Consider a set of linear algebraic equations

$$\begin{aligned}
a_{11}x_1 + a_{12}x_2 + \cdots + a_{1n}x_n &= y_1 \\
&\ \ \vdots \\
a_{n1}x_1 + a_{n2}x_2 + \cdots + a_{nn}x_n &= y_n
\end{aligned} \qquad (C.12)$$

where all the a's and y's are known. We wish to determine the x's that satisfy these equations. It is worth noting that the a's and y's can be constants, functions of s or functions of z. Now Eq. (C-12) can be represented in matrix form

$$\mathbf{Ax} = \mathbf{y} \qquad (C.13)$$

where $\mathbf{A}$ is an $n \times n$ matrix, $\mathbf{x}$ is an $n \times 1$ column matrix or vector, and $\mathbf{y}$ is an $n \times 1$ column matrix or vector. The solution to Eq. (C-13) can be obtained using Cramer's rule

$$x_k = \sum_{i=1}^{n} \frac{C_{ik}y_i}{|\mathbf{A}|} \qquad (C.14)$$

where C_{ik} is the cofactor of a_{ik}. The numerator of Eq. (C-14) is the determinant of $\mathbf{A}$ with the kth column replaced by the column formed by the right side of Eq. (C-12). Note that a solution exists only if $\mathbf{A}$ is a nonsingular matrix. A slightly different formulation of Eq. (C-14) results if we premultiply Eq. (C-13) by $\mathbf{A}^{-1}$,

$$\mathbf{A}^{-1}\mathbf{A}\mathbf{x} = \mathbf{A}^{-1}\mathbf{y}$$

and since

$$\mathbf{A}^{-1} = \frac{\text{adj } \mathbf{A}}{|\mathbf{A}|}$$

we have

$$\mathbf{x} = \frac{\text{adj } \mathbf{A}}{|\mathbf{A}|}\mathbf{y} \qquad (C.15)$$

It is left for you as an exercise to show that the use of Eq. (C-15) in the previous example gives the same result.

If the right-hand side of Eq. (C-13) is zero,

$$\mathbf{A}\mathbf{x} = 0 \qquad (C.16)$$

then the equations are said to be homogeneous. When this happens the numerator of Eq. (C-15) vanishes and if $|\mathbf{A}|$ does not vanish, then the set of equations has a trivial solution. If the determinant does vanish, then two or more vectors, whose elements are the elements of the columns of $\mathbf{A}$, are linearly related.

C.6   Characteristic Equations and Eigenvectors

Since the behavior of a linear system is dependent upon the characteristic values (or roots) of a system, we will develop some fundamental concepts pertaining to the characteristic equation. Consider the matrix equation,

$$\mathbf{y} = \mathbf{A}\mathbf{x}$$

This equation may be thought of as one that transforms the vector $\mathbf{x}$ into the vector $\mathbf{y}$. In linear systems we wish to know if a vector $\mathbf{x}$ exists such that the vector $\mathbf{y}$ is linearly related to $\mathbf{x}$, i.e.

$$\mathbf{y} = \lambda\mathbf{x}$$

Substituting this in the matrix equation we have

$$\mathbf{y} = \mathbf{A}\mathbf{x} = \lambda\mathbf{x} \qquad (C.17)$$

This becomes a **characteristic value** problem. The values of λ that satisfy Eq. (C-17) for $x_i \neq 0$, are the characteristic values (or eigenvalues) of $\mathbf{A}$. The vector solutions $x_i \neq 0$ are called the eigenvectors of $\mathbf{A}$. Eq. (C-17) can be expressed as

$$[\lambda \mathbf{I} - \mathbf{A}]\,\mathbf{x} = \mathbf{0} \tag{C.18}$$

where

$$[\lambda \mathbf{I} - \mathbf{A}] = \begin{bmatrix} \lambda - a_{11} & -a_{12} & \cdots & -a_{1n} \\ -a_{21} & \lambda - a_{22} & \cdots & -a_{2n} \\ \vdots & \vdots & & \vdots \\ -a_{n1} & -a_{n2} & \cdots & \lambda - a_{nn} \end{bmatrix}$$

The system of homogeneous equations has a nontrivial solution if

$$\det(\lambda \mathbf{I} - \mathbf{A}) = 0 \tag{C.19}$$

This is called the **characteristic equation** of matrix $\mathbf{A}$, and the polynomial on the left is of nth degree in λ. These values of λ satisfy Eq. (C-17). Consider

$$\mathbf{A} = \begin{bmatrix} 0 & 1 \\ -2 & -3 \end{bmatrix}$$

Then

$$|\lambda \mathbf{I} - \mathbf{A}| = \begin{vmatrix} \lambda & -1 \\ 2 & \lambda + 3 \end{vmatrix} = \lambda^2 + 3\lambda + 2$$

The characteristic equation, denoted by $\Delta(\lambda) = 0$, is

$$\Delta(\lambda) = \lambda^2 + 3\lambda + 2 = 0$$

The roots of $\Delta(\lambda) = 0$ are $\lambda_1 = -2$, $\lambda_2 = -1$ and are called the eigenvalues. Depending upon $\Delta(\lambda)$, these eigenvalues may be real, imaginary, distinct, or repeated.

Since each root of the characteristic equation satisfies Eq. (C-18), we have

$$[\lambda_i \mathbf{I} - \mathbf{A}]\,\mathbf{x}_i = \mathbf{0} \tag{C.20}$$

where $\mathbf{x}_i$ refers to $\mathbf{x}$ evaluated for $\lambda = \lambda_i$. The vectors that are the solutions to this equation are the **eigenvectors** of $\mathbf{A}$.

C.7 Functions of a Matrix

We shall now consider some aspects of matrix polynomials and infinite series after which we shall introduce functions of a matrix and the Cayley-Hamilton theorem. This theorem is very useful in solving matrix differential equations.

When a square matrix is raised to a power it obeys the following rules

$$\mathbf{A}^n \mathbf{A}^m = \mathbf{A}^{n+m}$$

$$(\mathbf{A}^n)^m = \mathbf{A}^{nm}$$

$$\mathbf{A}^0 = \mathbf{I}$$

where n and m are positive integers. They may be negative only if the matrix $\mathbf{A}$ is nonsingular.

A matrix polynomial $P(\mathbf{A})$ is defined as

$$P(\mathbf{A}) = \alpha_n \mathbf{A}^n + \alpha_{n-1} \mathbf{A}^{n-1} + \cdots + \alpha_1 \mathbf{A} + \alpha_0 \mathbf{I} \qquad \text{(C.21)}$$

where the α's are scalar quantities. The factorization of a matrix polynomial follows the rule of scalar polynomials. If

$$P(\mathbf{A}) = \mathbf{A}^2 + 3\mathbf{A} + 2\mathbf{I}$$

then factorizing yields

$$P(\mathbf{A}) = (\mathbf{A} + 2\mathbf{I})(\mathbf{A} + \mathbf{I})$$

An infinite series of matrices may be written as

$$S(\mathbf{A}) = \alpha_0 \mathbf{I} + \alpha_1 \mathbf{A} + \alpha_2 \mathbf{A}^2 + \cdots \qquad \text{(C.22)}$$

If the eigenvalues of $\mathbf{A}$ are given by $\lambda_1, \lambda_2, \ldots, \lambda_n$ and the series defined by

$$S(\lambda_i) = \alpha_0 \lambda_i + \alpha_1 \lambda_i + \alpha_2 \lambda_i^2 + \cdots \qquad \text{(C.23)}$$

converges, then the series defined by Eq. (C-22) is also convergent. A convergent series that appears quite often in the solution of matrix differential equations is the exponential function series

$$e^{\pm A} = \mathbf{I} \pm \mathbf{A} + \frac{\mathbf{A}^2}{2!} \pm \frac{\mathbf{A}^3}{3!} + \cdots \qquad \text{(C.24)}$$

Cayley-Hamilton Theorem

We can now state the Cayley-Hamilton theorem which is useful in the definition of any matrix function. The Cayley-Hamilton theorem states: *that every square matrix satisfies its own characteristic equation.* Consider $\mathbf{A}$ to be a 2×2 matrix, then we know that the characteristic equation of $\mathbf{A}$ is

$$\Delta(\lambda) = \lambda^2 + \beta_1 \lambda + \beta_2 = 0$$

The Cayley-Hamilton theorem states that

$$\Delta(\mathbf{A}) = \mathbf{A}^2 + \beta_1 \mathbf{A} + \beta_2 \mathbf{I} = 0$$

We can demonstrate the theorem in a straightforward manner for the following example. Let

$$\mathbf{A} = \begin{bmatrix} 0 & 1 \\ -3 & -4 \end{bmatrix}$$

then

$$|\lambda \mathbf{I} - \mathbf{A}| = \begin{vmatrix} \lambda & -1 \\ 3 & \lambda+4 \end{vmatrix} = \lambda^2 + 4\lambda + 3$$

and the characteristic equation is

$$\Delta(\lambda) = \lambda^2 + 4\lambda + 3 = 0$$

The Cayley-Hamilton theorem states that

$$\Delta(\mathbf{A}) = \mathbf{A}^2 + 4\mathbf{A} + 3\mathbf{I} = 0$$

Is this true? We can find out by direct substitution

$$\Delta(\mathbf{A}) = \begin{bmatrix} 0 & 1 \\ -3 & -4 \end{bmatrix}\begin{bmatrix} 0 & 1 \\ -3 & -4 \end{bmatrix} + 4\begin{bmatrix} 0 & 1 \\ -3 & -4 \end{bmatrix} + 3\begin{bmatrix} 1 & 0 \\ 0 & 1 \end{bmatrix}$$

$$= \begin{bmatrix} -3 & -4 \\ 12 & 13 \end{bmatrix} + \begin{bmatrix} 0 & 4 \\ -12 & -16 \end{bmatrix} + \begin{bmatrix} 3 & 0 \\ 0 & 4 \end{bmatrix} = 0$$

which illustrates the contention of the theorem.

For a given matrix polynomial, the Cayley-Hamilton theorem may be used to reduce the order of the polynomial. Consider a polynomial

$$N(\mathbf{A}) = \mathbf{A}^3 + 2\mathbf{A}^2 + \mathbf{A} + 5\mathbf{I} \tag{C.25}$$

where

$$\mathbf{A} = \begin{bmatrix} 0 & 1 \\ -1 & -2 \end{bmatrix}$$

How can be find a simpler form for $N(\mathbf{A})$? First we obtain the characteristic polynomial,

$$\Delta(\lambda) = \lambda^2 + 2\lambda + 1$$

From the Cayley-Hamilton theorem we know that

$$\Delta(\mathbf{A}) = \mathbf{A}^2 + 2\mathbf{A} + \mathbf{I} = 0$$

Solving for $\mathbf{A}^2$, we have

$$\mathbf{A}^2 = -(2\mathbf{A} + \mathbf{I})$$

Forming $\mathbf{A}^3$,

$$\mathbf{A}^3 = \mathbf{A} \cdot \mathbf{A}^2 = -\mathbf{A}(2\mathbf{A} + \mathbf{I}) = -2\mathbf{A}^2 - \mathbf{AI}$$

Now we substitute for $\mathbf{A}^2$ once again

$$\mathbf{A}^3 = -2[-2\mathbf{A} - \mathbf{I}] - \mathbf{AI}$$

$$= 3\mathbf{A} + 2\mathbf{I}$$

We can substitute these into Eq. (C-25) and

$$N(\mathbf{A}) = (3\mathbf{A} + 2\mathbf{I}) + 2(-2\mathbf{A} - \mathbf{I}) + \mathbf{A} + 5\mathbf{I}$$

or

$$N(\mathbf{A}) = 5\mathbf{I} \qquad (C.26)$$

This equation is equivalent to Eq. (C-25) and is certainly much simpler. This result can be generalized in that any polynomial of an $n \times n$ matrix $\mathbf{A}$ can be reduced to a polynomial of order $(n - 1)$. Consider a polynomial $N(\mathbf{A})$,

$$N(\mathbf{A}) = \mathbf{A}^m + \alpha_{m-1}\mathbf{A}^{m-1} + \cdots + \mathbf{I} \qquad (C.27)$$

We would like to obtain a simpler expression for $N(\mathbf{A})$. This can be done by forming

$$\frac{N(\lambda)}{\Delta(\lambda)} = Q(\lambda) + \frac{R(\lambda)}{\Delta(\lambda)} \qquad (C.28)$$

where $\Delta(\lambda)$ is the characteristic polynomial of $\mathbf{A}$. $Q(\lambda)$ is a unique polynomial, and $R(\lambda)$ is the remainder. Rewriting this

$$N(\lambda) = \Delta(\lambda)Q(\lambda) + R(\lambda) \qquad (C.29)$$

Now the corresponding matrix expression is

$$N(\mathbf{A}) = \Delta(\mathbf{A})Q(\mathbf{A}) + R(\mathbf{A})$$

and since $\Delta(\mathbf{A}) = 0$ from the Cayley-Hamilton's theorem, we have

$$N(\mathbf{A}) = R(\mathbf{A}) \qquad (C.30)$$

which is the required simpler form of $N(\mathbf{A})$.

As an example let

$$N(\mathbf{A}) = \mathbf{A}^3 + 5\mathbf{A}^2 + 18\mathbf{A} + 7\mathbf{I}$$

and suppose the characteristic polynomial of a 2×2 matrix is

$$\Delta(\lambda) = \lambda^2 + 3\lambda + 2$$

Then

$$\frac{N(\lambda)}{\Delta(\lambda)} = \lambda + 2 + \frac{10\lambda + 3}{\lambda^2 + 3\lambda + 2}$$

The remainder is $R(\lambda) = 10\lambda + 3$. Therefore the simpler form of $N(\mathbf{A})$ is

$$N(\mathbf{A}) = 10\mathbf{A} + 3\mathbf{I}$$

We can go a step further. If $\mathbf{A}$ is an $n \times n$ matrix having discrete eigenvalues $\lambda_1, \lambda_2, \ldots, \lambda_n$ and we have a function $F(\mathbf{A})$, then we can reduce this function to $R(\mathbf{A})$,

$$F(\mathbf{A}) = R(\mathbf{A}) \tag{C.31}$$

where $R(\mathbf{A})$ is defined as

$$R(\mathbf{A}) = h_0\mathbf{I} + h_1\mathbf{A} + h_2\mathbf{A}^2 + \cdots + h_{n-1}\mathbf{A}^{n-1} \tag{C.32}$$

The coefficients $h_0, h_1, \ldots, h_{n-1}$ are obtained by satisfying the equations

$$F(\lambda_i) = R(\lambda_i), \quad i = 1, 2, \ldots, n \tag{C.33}$$

The proof of this closely follows our previous arguments and shall be omitted here.

If the eigenvalues of $\mathbf{A}$ are not distinct, then the previous formulation is slightly changed. Let us assume that $\mathbf{A}$ is a 3×3 matrix and the roots are

$$\lambda_1, \quad \lambda_2, \quad \text{and } \lambda_3 = \lambda_2$$

Then

$$R(\mathbf{A}) = h_0\mathbf{I} + h_1\mathbf{A} + h_2\mathbf{A}^2 \tag{C.34}$$

where h_0, h_1, h_2 are determined by satisfying

$$F(\lambda_1) = R(\lambda_1)$$

$$F(\lambda_2) = R(\lambda_2)$$

$$\left.\frac{dF}{d\lambda}\right|_{\lambda=\lambda_2} = \left.\frac{dR(\lambda)}{d\lambda}\right|_{\lambda=\lambda_2} \tag{C.35}$$

The consequence of the previous discussion becomes very important when we select $F(\mathbf{A})$ such that

$$F(\mathbf{A}) = e^{\mathbf{A}t} \tag{C.36}$$

where $e^{\mathbf{A}t}$ is the convergent exponential function series defined previously. It happens to be the solution to linear matrix differential equations. As an example, consider

$$\mathbf{A} = \begin{bmatrix} 0 & 1 \\ -3 & -4 \end{bmatrix}$$

where the function $F(\mathbf{A})$ is given by

$$F(\mathbf{A}) = e^{\mathbf{A}t}$$

We would like to reduce $F(\mathbf{A})$ to a polynomial

$$F(\mathbf{A}) = R(\mathbf{A}) = h_0\mathbf{I} + h_1\mathbf{A}$$

which is a first-order polynomial. If this is done, we have managed to replace an infinite series by a first-order polynomial for the solution of a linear matrix differential equation. The eigenvalues of $\mathbf{A}$ are

$$\lambda_1 = -3, \quad \lambda_2 = -1$$

Since these are distinct, we form

$$F(-3) = e^{-3t} = h_0 - 3h_1$$

$$F(-1) = e^{-t} = h_0 - h_1$$

Solving these

$$h_0 = \frac{1}{2}(3e^{-t} - e^{-3t}), \quad h_1 = \frac{1}{2}(e^{-t} - e^{-3t})$$

Therefore

$$F(\mathbf{A}) = R(\mathbf{A}) = \frac{1}{2}(3e^{-t} - e^{-3t})\mathbf{I} + \frac{1}{2}(e^{-t} - e^{-3t})\mathbf{A}$$

Substitution yields

$$F(\mathbf{A}) = e^{\mathbf{A}t} = \begin{bmatrix} \frac{1}{2}(3e^{-t} - e^{-3t}) & \frac{1}{2}(e^{-t} - e^{-3t}) \\ -\frac{3}{2}(e^{-t} - e^{-3t}) & \frac{1}{2}(-e^{-t} + 3e^{-3t}) \end{bmatrix}$$

As a further example, if the matrix is given by

$$A = \begin{bmatrix} -1 & 1 \\ 0 & -1 \end{bmatrix}$$

then the characteristics roots are

$$\lambda_1 = -1, \quad \lambda_2 = -1$$

which are not distinct. We therefore must satisfy Eqs. (C-34) and (C-35)

$$\left.\frac{dF(\lambda)}{d\lambda}\right|_{\lambda_1} = \left.\frac{dR(\lambda)}{d\lambda}\right|_{\lambda_1}$$

$$F(\lambda_1) = R(\lambda_1)$$

The derivatives are

$$\frac{dF(\lambda)}{d\lambda} = te^{\lambda t}, \quad \frac{dR(\lambda)}{d\lambda} = h_1$$

We form two equations

$$e^{-t} = h_0 - h_1$$

$$te^{-t} = h_1$$

Solving them gives

$$h_0 = e^{-t}(1+t), \quad h_1 = te^{-t}$$

Therefore

$$F(\mathbf{A}) = e^{\mathbf{A}t} = (1+t)e^{-t}\mathbf{I} + te^{-t}\mathbf{A}$$

Substitution yields

$$F(\mathbf{A}) = \begin{bmatrix} e^{-t} & te^{-t} \\ 0 & e^{-t} \end{bmatrix}$$

C.8   References

1. G. Strang, *Linear Algebra and Its Applications*, Harcourt Brace Jo-vanovich, Publishers, San Diego, Calif., 1988.

2. F. R. Gantmacher, *The Theory of Matrices, Vol. 1 and 2*, Chelsea Publishing Co.,New York, 1959.

3. R. Bellman, *Introduction to Matrix Analysis*, McGraw Hill Book Co., New York, 1960.

Appendix D

Computer Software for Control

Table D-1 Computer Software

Program	Vendor	Comments
ACET	Information and Control Systems, 28 Research Drive, Hampton, VA 23666	Simulation and design package. Incorporates linear and some non-linear design techniques.
ACSL	Mitchell and Gauthier Assoc., 73 Junction Square Drive, Concord, MA 01742	PC to supercomputer platforms. Non-linear simulations using state variable models and block diagrams. Linear system analysis and design methods.
CC	System Technology Inc., 13766 S. Hawthorne Blvd., Hawthorne, CA 90250	PC platform, Command driven linear system analysis and design. Multiloop and state variable, s- and z-domain.
Ctrl-C	System Control Technology Inc, 2300 Geng Road, Palo Alto, CA 94303	Workstation platform, matrix and state variable analysis. Linear system design.

Table D-1 Computer Software (contd.)

Program	Vendor	Comments
CSMP	IBM Corp., New York, NY	Mainframe platform. Linear and non-linear simulation.
CSSL	Simulation Services, Chatsworth, CA	Mainframe platform. Linear and non-linear simulation.
Easy-5	Boeing Computer Services, PO Box 24346, Seattle, WA	Mainframe and work-station platforms. Block diagram structure analysis. Linear and non-linear continuous and discrete simulation.
MATLAB	The Math Works Inc, 24 Prime Parkway, Natick, MA 01760	PC and workstation platforms. Classical and state space control tools for linear systems. (Used with toolboxes: Control system, signal processing, robust control, system identification.)
MATRIX-X	Integrated Systems, Inc., 2500 Mission College Blvd., Santa Clara, CA 95054	Mainframe and workstation platforms. Classical and state space control tools. Both discrete and continuous systems.
SABER	Analogy, Inc., PO Box 1669, Beaverton, OR 97075	Workstation platforms, graphical display on PC. Simulates non-linear systems in discrete and continuous domains. Can include electrical circuit models.
SIMNON	SSPA Systems, PO Box 24001, S-40022 Goteberg, Sweden	PC platform. Simulation of linear and non-linear discrete and continuous systems.
SIMULAB	The Math Works, Inc., 24 Prime Parkway, Natick, MA 01760	PC platform. Non-linear system simulation.

D.1 References

D. Frederick, et al., "*The Extended List of Control Software*", Electronics, Vol. 61, No. 6 (1988) p. 77.

Index

723

695457